INTRODUCTION TO THEORETICAL MECHANICS

INTERNATIONAL SERIES IN PURE AND APPLIED PHYSICS

LEONARD I. SCHIFF, *Consulting Editor*

Allis and Herlin Thermodynamics and Statistical Mechanics
Becker Introduction to Theoretical Mechanics
Clark Applied X-Rays
Collin Field Theory of Guided Waves
Evans The Atomic Nucleus
Finkelnburg Atomic Physics
Ginzton Microwave Measurements
Green Nuclear Physics
Gurney Introduction to Statistical Mechanics
Hall Introduction to Electron Microscopy
Hardy and Perrin The Principles of Optics
Harnwell Electricity and Electromagnetism
Harnwell and Livingood Experimental Atomic Physics
Harnwell and Stephens Atomic Physics
Henley and Thirring Elementary Quantum Field Theory
Houston Principles of Mathematical Physics
Hund High-frequency Measurements
Kennard Kinetic Theory of Gases
Lane Superfluid Physics
Leighton Principles of Modern Physics
Lindsay Mechanical Radiation
Livingston and Blewett Particle Accelerators
Middleton An Introduction to Statistical Communication Theory
Morse Vibration and Sound
Morse and Feshbach Methods of Theoretical Physics
Muskat Physical Principles of Oil Production
Present Kinetic Theory of Gases
Read Dislocations in Crystals
Richtmyer, Kennard, and Lauritsen Introduction to Modern Physics
Schiff Quantum Mechanics
Seitz The Modern Theory of Solids
Slater Introduction to Chemical Physics
Slater Quantum Theory of Atomic Structure, Vol. I
Slater Quantum Theory of Atomic Structure, Vol. II
Slater Quantum Theory of Matter
Slater and Frank Electromagnetism
Slater and Frank Introduction to Theoretical Physics
Slater and Frank Mechanics
Smythe Static and Dynamic Electricity
Stratton Electromagnetic Theory
Thorndike Mesons: A Summary of Experimental Facts
Townes and Schawlow Microwave Spectroscopy
White Introduction to Atomic Spectra

The late F. K. Richtmyer was Consulting Editor of the series from its inception in 1929 to his death in 1939. Lee A. DuBridge was Consulting Editor from 1939 to 1946; and G. P. Harnwell from 1947 to 1954.

Introduction to Theoretical Mechanics

ROBERT A. BECKER, Ph.D.

Associate Professor of Physics
University of Illinois

McGRAW-HILL BOOK COMPANY, INC.

New York Toronto London

1954

INTRODUCTION TO THEORETICAL MECHANICS

Library of Congress Catalog Card Number 54-6718

VIII

04231

PREFACE

This book has evolved from a course on the subject which I have given at the University of Illinois for the last five or six years. The course is a two-semester one meeting three times a week and is required, in their junior or senior year, of all undergraduates in the curriculum of engineering physics. In addition, a good proportion of the incoming graduate students in physics customarily enroll in at least the second semester's work. This is especially true if the prior background of the student in this particular field of classical physics is considered to be inadequate. Since classical mechanics is a basis for most other advanced courses in physics, the student should acquire a good deal of facility with this fundamental subject before attempting to undertake more advanced work. In this book the only preparation expected of the student is that obtained from thorough courses in elementary physics and calculus.

The methods of vectors are employed rather extensively throughout the text. However, no previous preparation on the part of the student in this regard is assumed, and an introduction to the subject of vector analysis, adequate for the present text, is presented in Chap. 1. Similarly no prior knowledge of elementary differential equations is necessary, although it must be admitted that a subject such as mechanics necessarily involves a certain dependence on this important branch of mathematics. However, the organization of the book is such that those portions in which a slight knowledge of differential equations is of advantage occur in later chapters of the book. Even here the mathematical tools are developed where needed and frequent reference is made to Appendix 2 in which a very brief introduction to the subject of ordinary differential equations is presented.

The emphasis of the book is quite definitely toward the solution of problems and, although an effort has been made to include a few very easy ones in each chapter, some of the exercises are rather difficult and are calculated to require a good deal of ingenuity on the part of the student. Indeed, it is not to be expected that all students will be able to solve all of the 400-odd problems in the book without assistance. However, the serious student who rises to the challenge presented by some of the problems is certain to find himself amply rewarded. The problem emphasis is easily justified, since the surest way to cultivate an ability to do physical reasoning is to apply it. Very little indeed can be learned in a course in mechanics, or physics in general, by the majority of students if the course is purely of a lecture type in which the solving of problems plays but a minor role.

There are upwards of eighty rather carefully selected examples which are worked out in the text material of the chapters. In addition to amplifying the mathematical steps in these solutions, a serious attempt has been made to present extensive details of the physical reasoning involved in the problem. In studying these examples the student is strongly advised against simply reading through the solution given. Rather should he first read only the statement of the example and, following this, attempt to set up the problem himself. In this way the student's difficulty with certain aspects of the case will become much more apparent to him, details which might have passed unnoticed had he contented himself with merely reading through the solution given.

The book is arranged so that, with the possible exception of Chap. 4, topics occur in the order of increasing difficulty as to both mathematical maturity and physical insight required. For this reason such subjects as central field motion, accelerated coordinate systems, general rigid body motion, Lagrangian methods, vibrating systems having several degrees of freedom, and wave motion are relegated to the latter half of the book. Although the treatment is primarily intended for a two-semester course on mechanics, the arrangement and order of the topics presented is such that the first nine chapters suffice to meet the demands of most one-semester courses on the subject.

Certain features of the book reflect the trend of modern physics. In connection with oscillatory motion in one dimension brief mention is made of nonlinear systems, a topic of ever-increasing importance in modern technology. In the chapter dealing with theorems concerning systems of particles, the case of a body in which the mass is varying (witness the rocket) is considered, and the procedure for setting up the equation of motion for such a situation is described. More space than is usually customary is devoted to the subject of general rigid body rotations in space. This is in keeping with the present wide interest in the fields of magnetic resonance and microwave spectroscopy, with their obvious applications of this class of motions.

Generalized coordinate methods are not introduced until after rigid body motion is considered. I am strongly of the opinion that the student should be taken through the latter material once without the use of the more sophisticated procedures so as to acquire more of a feeling than might otherwise have been gained for the way in which the forces are acting and for the selection of suitable coordinate systems. Not to be overlooked, also, is the fact that fairly complicated nonholonomic problems frequently can be successfully attacked step by step with the less sophisticated methods (witness Prob. 12-19). Such cases often present difficulty when Lagrangian procedures involving the use of Lagrangian multipliers (not discussed in this text) are employed.

Vibrating systems of several degrees of freedom are considered in the light of normal coordinates. One system, the vibrating string, having a large number of degrees of freedom is treated both from the normal coordinate and traveling wave points of view.

In conclusion I wish to mention my great indebtedness to the Cambridge University Press, and to Ginn and Company for graciously granting me permission for the use of certain of the problems in the text. Those marked C in the text are taken from the Cambridge publications: "Statics," "Dynamics," and "Higher Mechanics," all by Lamb; "Dynamics of a Particle," and "Elementary Rigid Dynamics," both by Routh; Ramsey's "Dynamics"; and "Mechanics," by Love. Many of these problems are reprinted by these authors from former Cambridge examinations. A few problems, marked J, are taken from Jeans' "Theoretical Mechanics," published by Ginn and Company. The remainder of the problems are either of my own composition or are taken from former examination lists that have been used at the University of Illinois. Some also have been suggested by certain interested individuals.

Finally I wish to thank my colleagues for many helpful suggestions. I especially wish to thank Professor Ronald Geballe of the University of Washington, and Professors A. T. Nordsieck and C. P. Slichter of the University of Illinois for their valuable criticisms, suggestions, and comments.

Robert A. Becker

CONTENTS

CHAPTER 1

FUNDAMENTAL PRINCIPLES

1-1. Introductory Remarks. In the present text many basic concepts will be assumed to be possessed intuitively by the student. Such geometrical terms as *position* and *length* have familiar connotations to all from everyday experience. To some extent these notions are rendered more precise by secondary school mathematics. The first lends meaning to the location of a point in space. The second provides a common basis for describing the distance along a prescribed path between two such points. If the element of *time* is added and if one inquires into the rate at which the distance is traversed, the discussion becomes *kinematical*. *Kinematics*, it may be said, is the geometry of motion. Typical kinematical quantities are *velocity* and *acceleration*. The addition of the concepts of *mass* and *force*, which are physical quantities, brings the considerations under the heading of *mechanics*. The concepts of mass and force are employed in any elementary text on physics and will be quantitatively defined later in the present volume in terms of Newton's laws.

In order to complete the list of elementary concepts, it is necessary to mention two terms which are frequently employed in the discussion of mechanical problems. These are the *particle* and the *body*. The first of these, an idealized construct which is convenient in many problems, is a mass which has no size associated with it. In brief it is a geometrical point which possesses mass. The body, on the other hand, in general possesses both mass and extent.

1-2. Coordinate Systems. A typical mechanical problem, as applied to a given system, is to determine the configuration of that system as a function of time. If it consists of a number of particles, the general problem will be to specify the positions and velocities of all the particles in terms of time as the independent variable. In practice, however, it may be sufficient to determine a much smaller amount of information. For example, it may be desired to know the way in which the velocity of one of the particles will vary as its position in space is varied.

In order to attack any problem of this nature, it is necessary first to select an appropriate coordinate system. We limit ourselves at this early stage to the familiar *rectangular* system of the type shown in Fig. 1-1.

The system $Oxyz$ has its yz plane in the plane of the paper, with y positive to the right and z positive upward. The x axis points out from the paper toward the reader and is positive in that direction.

In selecting suitable coordinates for a problem it is convenient to retain only the minimum number of distinct coordinates necessary to describe the motion completely. For example, if a particle is free to move in one plane, such as a table top, clearly only two coordinates will be necessary. We may choose the plane of the table to be the xy plane, as in Fig. 1-2. Suppose the particle is at point P at a given instant.

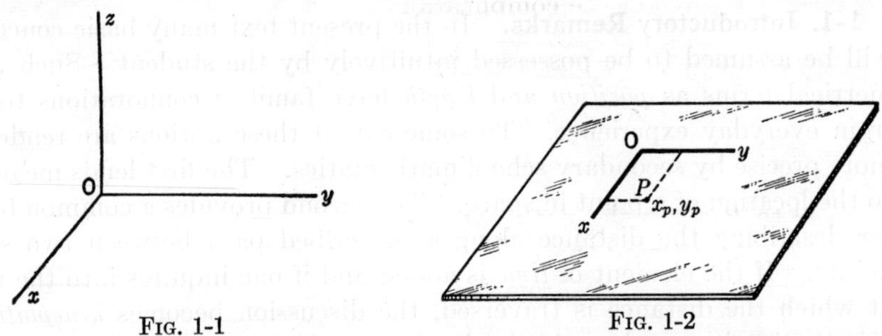

FIG. 1-1

FIG. 1-2

Its coordinates are the particular values, x_p and y_p, of x and y. In this case, in which the path is restricted to one plane and the system consists of but one particle, the system is completely specified by the knowledge, as functions of time, of the two coordinates, x and y, of the particle. If two particles were present, both confined to the plane, the specification of the system would require four coordinates, the x and y coordinates of each particle. The removal of the restriction confining the path of the two particles to the plane would require the addition of the z coordinate for each particle.

When the motion is not permitted to extend freely in three dimensions, the system is said to be subject to constraints. In the instance of the particles on the table top there exists one constraint the equation of which is $z = 0$ for all time. This is a particular example of the general condition expressed by

$$f(x,y,z) = 0 \tag{1-1}$$

A constraint described by Eq. (1-1) is called an *integrable constraint*. The term *integrable* is employed here since the differential relation expressing the fact that z is not allowed to vary is $dz = 0$, an expression which is readily integrated to $z = \text{const.}$ The constant of integration is zero in the present example since the table top is in the plane $z = 0$. Relations such as Eq. (1-1) enable one, at the outset of a problem, to reduce the number of distinct coordinates which are required in order to describe the system involved in the problem. The number of coordinates eliminated is just equal to the number of the relations of the type of Eq. (1-1) which may be present.

Constraints also exist which are of the *nonintegrable* type, that is, the equations of these limitations involve differential coefficients in a manner such that they cannot be integrated. Consequently no coordinates may be eliminated by means of these relationships. Attention will be called to these again later in the text (cf. Sec. 13-1).

Simultaneously with the choice of a coordinate system careful attention must be paid to its state of motion. In Sec. 1-20 some of the complications attending an injudicious selection of coordinates will be considered.

1-3. Linear Velocity and Acceleration. Consider a particle which is experiencing a rectilinear displacement from O to P along the path shown in Fig. 1-3. At a time t the particle is at a distance s measured from point O along this path. During the subsequent increment Δt of time the particle moves through a distance Δs. The quantity $\Delta s/\Delta t$ is called the *average velocity* (time average) during the interval Δt. The *instantaneous linear velocity* at point s is then defined as

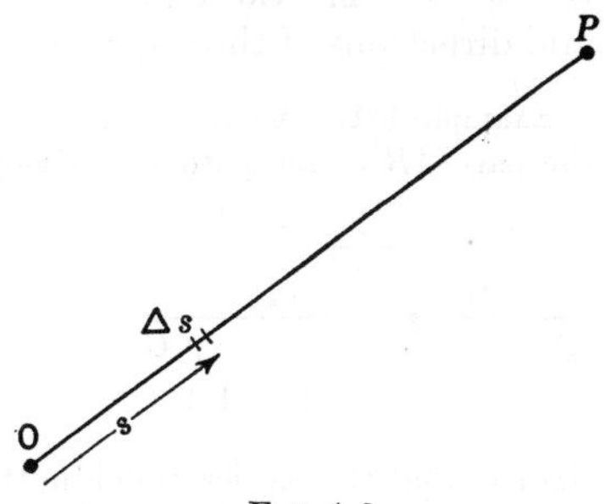

Fig. 1-3

$$v_s = \lim_{t \to 0} \frac{\Delta s}{\Delta t} = \frac{ds}{dt} = \dot{s} \tag{1-2}$$

In Eq. (1-2) the symbol $\dot{s}$ ($= ds/dt$) has been introduced. It is read "s dot." Extensive use will be made of this notation. The term *speed* is often employed to denote the magnitude of the velocity.

A second kinematical quantity which requires definition is the *acceleration.* Suppose that at point s the particle has a velocity v_s. During the time Δt thereafter, the velocity changes by an amount Δv_s. The *average acceleration* during this interval Δt is thus $\Delta v_s/\Delta t$, from which we are immediately able to define the *instantaneous linear acceleration* at point s to be

$$a_s = \lim_{\Delta t \to 0} \frac{\Delta v_s}{\Delta t} = \frac{dv_s}{dt} = \frac{d^2 s}{dt^2} = \ddot{s}$$

Here several equivalent symbols for the acceleration have been stated. The quantity $\ddot{s}$, for example, is read "s double dot."

It is sometimes convenient to employ the terms *average velocity* and *average acceleration* in the larger sense of being the time average of these quantities during the entire time T of the translation from O to P. These

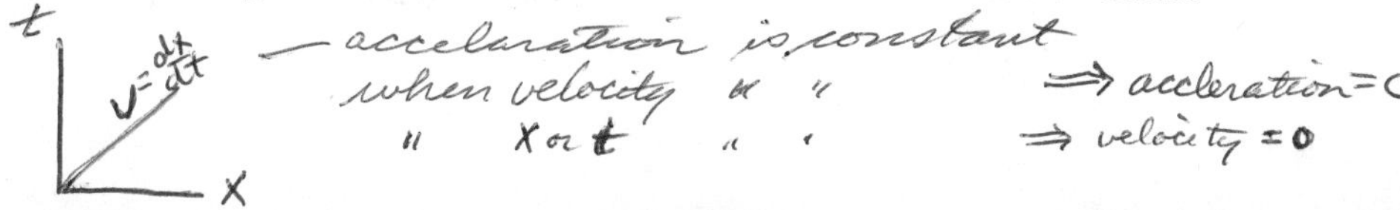

averages are

$$\bar{v}_s = \frac{OP}{T} \qquad \bar{a}_s = \frac{v_P - v_O}{T} \tag{1-3}$$

where in each case the bar signifies that the time average is meant. In the second of Eqs. (1-3) v_O is the velocity in the s direction at point O (the initial velocity), and v_P is the velocity in the s direction at point P.

The terms *uniform velocity* and *uniform acceleration* will also be encountered. By *uniform* velocity and acceleration is meant that the magnitudes and directions of these quantities remain constant throughout the motion.

Example 1-1. A particle starts toward D, from rest at point A (Fig. 1-4). During the part AB of the path (a distance x_1) the particle has a uniform acceleration a_1, during the time when the particle is between B and C (a distance x_2) there is no acceleration, and during the third interval, between C and D, there is a uniform acceleration $-a_3$, where a_3 is a positive quantity. The negative sign signifies that the acceleration is in the direction of decreasing x, that is, a deceleration. The magnitude of a_3 is such that the particle will just be brought to rest at D. At what times will the particle arrive at points B, C, and D?

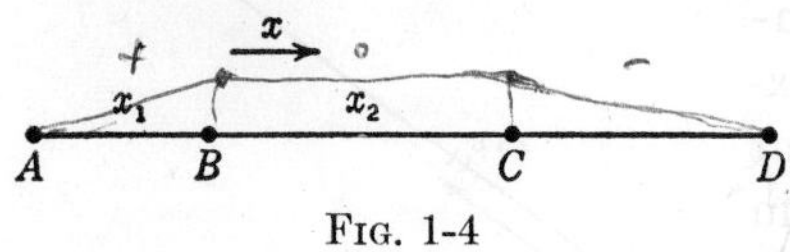

Fig. 1-4

We note first that the known quantities in the problem are a_1, a_3, x_1, and x_2, and the end results must be expressed in terms of these. For the first step of the motion the origin of x is chosen to be at A, where the particle is located at $t = 0$. The equation of motion is

$$\ddot{x} = a_1 \tag{1-4}$$

Integrating once with respect to time, we have

$$\dot{x} = a_1 t + c_1 \tag{1-5}$$

where c_1 is a constant of integration. The constant c_1 can be determined from the boundary condition that at $t = 0$ the velocity $\dot{x} = 0$ also. Thus $c_1 = 0$. Integrating a second time, we obtain

$$x = \frac{a_1 t^2}{2} \tag{1-6}$$

the constant of integration again being zero since $x = 0$ at $t = 0$. Thus t_1, the time required by the particle to traverse the distance x_1, becomes

$$t_1 = \sqrt{\frac{2x_1}{a_1}} \tag{1-7}$$

Similarly the time t_2 required to pass from B to C can easily be found since the velocity is unchanged during the interval $BC(= x_2)$. From Eqs. (1-5) and (1-7)

$$\dot{x}_B = a_1 t_1 = \sqrt{2a_1 x_1} \tag{1-8}$$

Thus the total time required by the particle to go from A to C is

$$t_{AC} = t_1 + t_2 = \sqrt{\frac{2x_1}{a_1}} + \frac{x_2}{\sqrt{2a_1 x_1}} \tag{1-9}$$

In the third interval the motion is governed by the equation

$$\ddot{x} = -a_3 \tag{1-10}$$

from which, selecting new origins of x and t,

$$\dot{x} = -a_3 t + \dot{x}_B = -a_3 t + \sqrt{2a_1x_1}$$

and

$$t_3 = \frac{\sqrt{2a_1x_1}}{a_3}$$

and where the second equality follows since at $t = t_3$ the velocity is zero. Thus the total time t_{AD} required by the particle to pass from A to D becomes

$$t_{AD} = t_1 + t_2 + t_3 = \sqrt{\frac{2x_1}{a_1}} + \frac{x_2}{\sqrt{2a_1x_1}} + \frac{\sqrt{2a_1x_1}}{a_3} \tag{1-11}$$

1-4. Angular Velocity and Acceleration. In a manner very similar to that for the corresponding linear quantities we are able to define *angular* velocity and acceleration. Consider a particle which suffers a translation along a segment AB of a circle with center at O in Fig. 1-5. During this translation it undergoes a displacement through $\angle AOB$. We may define in the same way, as for the linear case, time rates at which the angular displacement is carried out. Thus at any angle θ the *angular velocity* (or simply *angular speed,* if the magnitude alone is being referred to) in radians per second and the *angular acceleration* in radians per second per second are defined, respectively, as

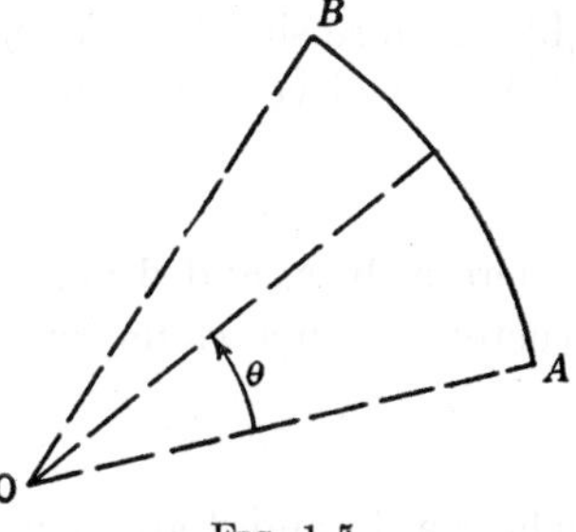

FIG. 1-5

$$\omega = \lim_{\Delta t \to 0} \frac{\Delta\theta}{\Delta t} = \frac{d\theta}{dt} = \dot{\theta} \qquad \alpha = \lim_{\Delta t \to 0} \frac{\Delta\omega}{\Delta t} = \frac{d\omega}{dt} = \frac{d^2\theta}{dt^2} = \ddot{\theta} \tag{1-12}$$

ELEMENTS OF VECTOR ANALYSIS

1-5. Vectors and Scalars.[1] Two[2] classes of quantities are of great importance in elementary mechanics. These are *vectors* and *scalars.* A

[1] It was perceived very early (cf. "The Collected Works of J. Willard Gibbs", Vol. II, Longmans, Green & Co., Inc., New York, 1928) that certain physical quantities could be represented by directed segments having definite components in a given coordinate system. The relations among these directed segments themselves, rather than their components, in many cases furnished expressions of physical laws which did not depend upon any one coordinate system, a noteworthy advance indeed. A notation was developed, and the rules of manipulation of these quantities were worked out. The resulting framework is what is now known as *vector analysis.* Mathematicians have since put these procedures on a more rigorous basis.

[2] In certain more advanced physical problems the two notions of vectors and scalars

scalar quantity is one that has magnitude but has no direction associated with it. Typical examples are *temperature*, *mass*, *density*, and *energy*. The mathematical manipulation of scalar quantities follows the rules of ordinary algebra. Vector[1] quantities, on the other hand, are characterized by both magnitude and direction, examples of which are *displacement*, *velocity*, and *acceleration*. It will be seen later that *angular velocity* and *angular acceleration* also are vectors. Vector quantities obey the so-called *parallelogram law* of addition. The general manipulation of vectors follows a mathematical procedure called *vector analysis*.

In dealing with vectors and vector equations it is necessary to introduce a new notation. If a quantity b is a vector quantity, it is denoted by the symbol $\mathbf{b}$, that is, it is written in boldface type. In writing a vector symbol by hand this procedure is obviously not convenient. The handwriting procedure is to place a small arrow over the symbol, as $\vec{b}$. The absolute magnitude of the vector $\mathbf{b}$ is represented by the symbol $|\mathbf{b}|$ or, more simply, b in ordinary type.

A typical vector equation may be written

$$\mathbf{a} + \mathbf{b} = \mathbf{c} + \mathbf{d} \tag{1-13}$$

where $\mathbf{a}$, $\mathbf{b}$, $\mathbf{c}$, and $\mathbf{d}$ are vectors. It is to be noted that each term is a vector. It makes no sense to write an equation of the form

$$\mathbf{a} + \mathbf{b} = m + \mathbf{d}$$

where $\mathbf{a}$, $\mathbf{b}$, and $\mathbf{d}$ are vectors and m is a scalar quantity. It will be seen shortly that vectors and scalars can be combined by processes involving multiplication, addition, and subtraction, yielding new vectors (occasionally such a combination will result in a scalar). Consequently in Eq. (1-13) $\mathbf{a}$, $\mathbf{b}$, $\mathbf{c}$, and $\mathbf{d}$ can each be any such combination of vectors and scalars, and the equation will still be correct in form, provided only that these combinations are themselves vectors.

Finally it should be pointed out that when two vectors are equal they are equal in both magnitude and direction. Thus in an equation of the form of (1-13), not only does the sum $\mathbf{a} + \mathbf{b}$ have the same magnitude as that of $\mathbf{c} + \mathbf{d}$, but both sums have the same direction as well.

1-6. Composition of Displacements. Parallelogram Law of Addition. The rules for the manipulation of vector quantities display some interesting resemblances to the laws of ordinary algebra. Stated briefly, the laws of algebraic addition and multiplication are:

---

are insufficient. For example, in the consideration of the strains produced by stresses in anisotropic media the methods of *tensor analysis* must be employed.

[1] A better definition of a vector is provided by its transformation properties (cf. Sec. 1-11).

I. *Commutative law for addition:* $a + b = b + a$.
II. *Associative law for addition:* $a + (b + c) = (a + b) + c$.
III. *Commutative law for multiplication:* $ab = ba$.
IV. *Associative law for multiplication:* $(ab)c = a(bc)$.
V. *Distributive law for multiplication:* $a(b + c) = ab + ac$.

The simplest illustration of the addition of two or more vectors is given by the compounding of displacements. In Fig. 1-6 we take the initial point to be O. The vector $\mathbf{a}$ represents a displacement from O to A. In the same manner the vector $\mathbf{b}$ represents a displacement from A to B, and we see that the end result is a translation from O to B, which we may represent by the vector $\mathbf{c}$, drawn from the initial point of $\mathbf{a}$ to the terminal point of $\mathbf{b}$. In this sense we say that the displacement $\mathbf{c}$ is the sum of the displacements $\mathbf{a}$ and $\mathbf{b}$. This is written as

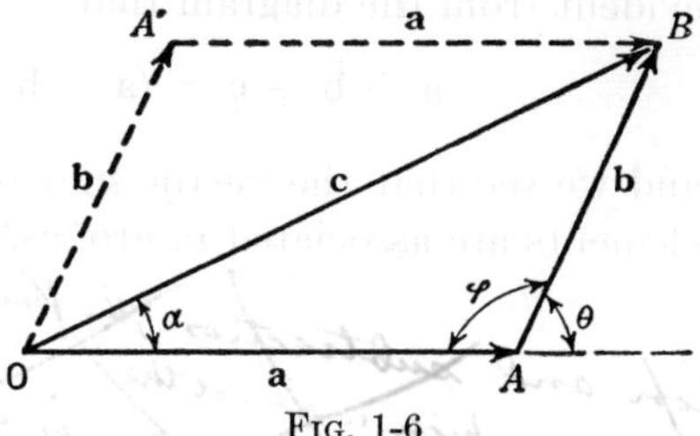

FIG. 1-6

$$\mathbf{c} = \mathbf{a} + \mathbf{b} \tag{1-14}$$

The magnitude c, of $\mathbf{c}$, can be obtained from a and b by means of the cosine law of trigonometry, as

$$c^2 = a^2 + b^2 - 2ab \cos \varphi = a^2 + b^2 + 2ab \cos \theta \tag{1-15}$$

where the second form, employing the angle θ between the two vectors $\mathbf{a}$ and $\mathbf{b}$, is that conventionally used in the addition of vectors. The vector $\mathbf{c}$ is said to be the resultant of $\mathbf{a}$ and $\mathbf{b}$.

It is apparent from Fig. 1-6 that the same final result would have been obtained if first the displacement $\mathbf{b}$ from O to A' had been executed, followed by the displacement $\mathbf{a}$ from A' to B. Thus

$$\mathbf{c} = \mathbf{b} + \mathbf{a}$$

and we have the analog, in vector analysis, of the commutative law of algebraic addition. This is summed up in the equation

$$\mathbf{c} = \mathbf{b} + \mathbf{a} = \mathbf{a} + \mathbf{b} \tag{1-16}$$

The geometrical figure in Fig. 1-6 is a parallelogram and is the reason for the assertion that the addition of vectors follows the parallelogram law. Equation (1-14) is a shorthand way of expressing the fact that the magnitude of the sum of $\mathbf{a}$ and $\mathbf{b}$ can be obtained by the parallelogram law of addition, that is, by Eq. (1-15). The direction of $\mathbf{c}$ is, of course, obtainable from those of $\mathbf{a}$ and $\mathbf{b}$ by means of elementary trigonometry.

It is not difficult to show that vector addition is associative, again in analogy to algebraic addition. To do this, we take the vectors $\mathbf{a}$, $\mathbf{b}$, and $\mathbf{c}$ and construct a figure (not necessarily plane) having these vectors as consecutive sides. Their sum, as shown in Fig. 1-7, is the vector drawn from the initial point of the first to the terminal point of the last. The property of associativity is demonstrated with the help of the diagonal drawn from A to C $(= \mathbf{a} + \mathbf{b})$ and that from B to D $(= \mathbf{b} + \mathbf{c})$. It is evident from the diagram that

$$\mathbf{a} + \mathbf{b} + \mathbf{c} = (\mathbf{a} + \mathbf{b}) + \mathbf{c} = \mathbf{a} + (\mathbf{b} + \mathbf{c}) = \mathbf{d} \tag{1-17}$$

and we see that the vector sum is independent of the way in which its elements are associated in groups.

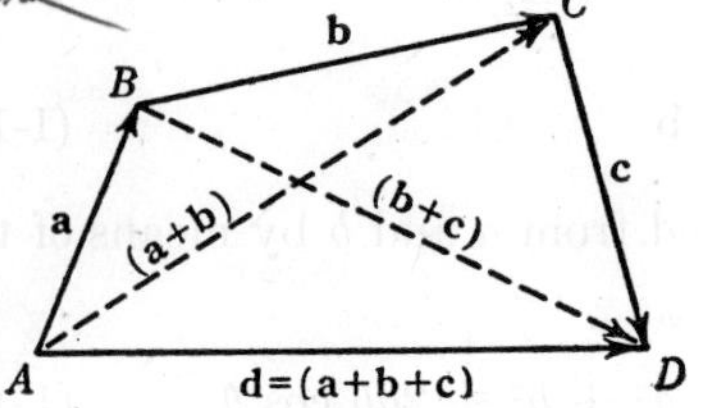

FIG. 1-7

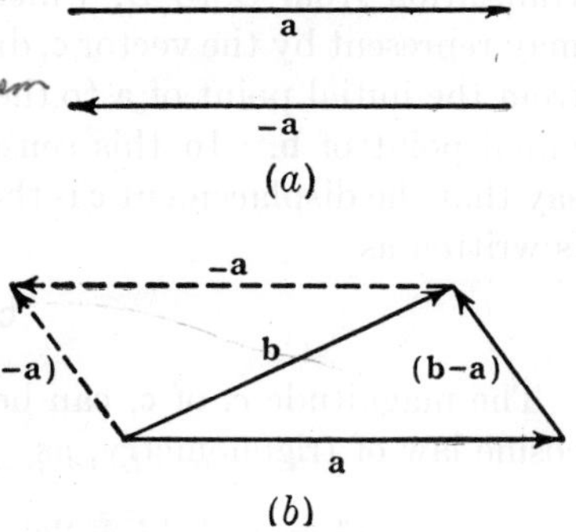

FIG. 1-8

If $\mathbf{a}$ is a given vector (Fig. 1-8a), then $-\mathbf{a}$ is a vector of the same length but opposite in direction. Inspecting Fig. 1-8b, we are able to define the difference between two vectors $\mathbf{b}$ and $\mathbf{a}$, drawn from the same origin, as being that vector which is drawn from the end of $\mathbf{a}$ to the end of $\mathbf{b}$. That this is true is easily shown, since $\mathbf{b} - \mathbf{a}$ is the vector which, when added to $\mathbf{a}$, yields $\mathbf{b}$.

A little reflection will enable the student to realize that the sum or difference of two vectors is another vector which is always coplanar with the original vectors, since all three must be joined end to end.

1-7. Multiplication of a Vector by a Scalar. If m is a scalar and $\mathbf{a}$ is a vector, then

$$m\mathbf{a} = \mathbf{a}m$$

represents a vector m times as long as the original vector $\mathbf{a}$ and having the same direction if m is a positive number. If m is negative, then the vector $m\mathbf{a}$ has a direction opposite to $\mathbf{a}$. We may carry this reasoning further, and it is not difficult to see that if m and n both are scalars we may write

$$(m + n)\mathbf{a} = m\mathbf{a} + n\mathbf{a} \tag{1-18}$$

Equation (1-18) expresses the fact that the operation of multiplying a vector by scalars is distributive over addition of the scalars.

A survey of Fig. 1-9 reveals a similar situation in the case of multiplication of the sum of two vectors by a single scalar. We are given two vectors **a** and **b** originating at a common point O, the sum of which is represented by the diagonal **c** of the parallelogram $OACB$. The side OA' represents a vector $m\mathbf{a}$, which is m times the length of **a**, and the side OB' represents $m\mathbf{b}$, which is a vector m times the length of **b**. It is evident that the sum of $m\mathbf{a}$ and $m\mathbf{b}$ will be represented by the diagonal OC' of the parallelogram $OA'C'B'$. Furthermore, the sides OA' and OB' are proportional to OA and OB, with a proportionality factor m, and the angle between OA' and OB' is the same as that between OA and OB. Thus $OA'C'B'$ is similar to $OABC$, and the diagonal OC' lies along OC, having a length equal to m times that of the latter. These results may be expressed as

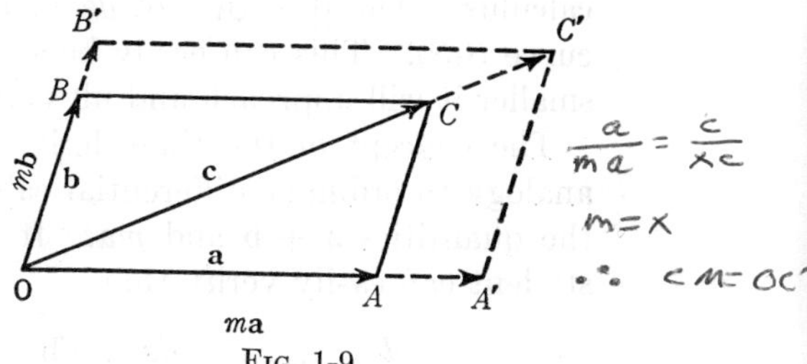

FIG. 1-9

$$m\mathbf{a} + m\mathbf{b} = m(\mathbf{a} + \mathbf{b}) = m\mathbf{c} \tag{1-19}$$

which states that the operation of multiplying the sum of two vectors by a single scalar is distributive over the addition of the vectors. This result can be extended to any number of vectors, since, from (1-14), vectors may be combined in pairs to form a single vector.

The operation of dividing a vector **a** by a scalar m can be accomplished by considering it to be equivalent to multiplying by the reciprocal of m.

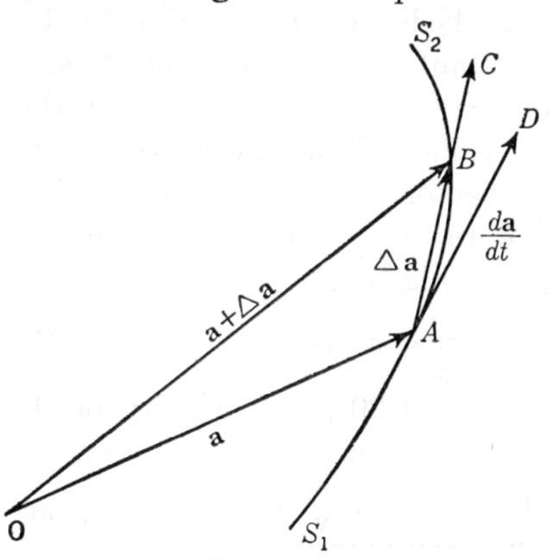

FIG. 1-10

1-8. Derivative of a Vector with Respect to a Single Scalar Variable. In Fig. 1-10 **a** is an arbitrary vector which is a function of a single scalar independent variable, for example, time, such that its tip moves along a curve S_1S_2. In a time Δt the vector changes in length and direction from OA to OB. In analogy to ordinary calculus we now call the new vector $\mathbf{a} + \Delta\mathbf{a}$, in which $\Delta\mathbf{a}$ is represented in Fig. 1-10 by the chord AB. It is clear that, in general, $\Delta\mathbf{a}$ does not have the direction of **a**. On dividing $\Delta\mathbf{a}$ by Δt, the result is a vector with the direction of $\Delta\mathbf{a}$ but which has the magnitude of $\Delta\mathbf{a}$ divided by the scalar Δt and which may be represented by the line AC. The derivative is defined by proceeding to the limit in which Δt becomes very small. We have

$$\frac{d\mathbf{a}}{dt} = \lim_{\Delta t \to 0} \frac{\Delta \mathbf{a}}{\Delta t}$$

which is entirely similar to the definition of the derivative in elementary calculus. The direction of $d\mathbf{a}/dt$ is that of the tangent, at A, to the curve S_1S_2. This can easily be seen, since as AB becomes smaller and smaller it will approach and ultimately coincide in direction with AD.

The concept of the time derivative of a vector can be extended, in analogy to ordinary differential calculus, to cover a like operation upon the quantities $\mathbf{a} + \mathbf{b}$ and $m\mathbf{a}$. If all quantities are functions of t, the student can easily verify that

$$\frac{d}{dt}(\mathbf{a} + \mathbf{b}) = \frac{d\mathbf{a}}{dt} + \frac{d\mathbf{b}}{dt} \qquad \frac{d}{dt}(m\mathbf{a}) = \mathbf{a}\frac{dm}{dt} + m\frac{d\mathbf{a}}{dt} \tag{1-20}$$

It should be apparent that the foregoing results apply equally well to the derivative of a vector with respect to any single scalar independent variable and need not be restricted to the time derivative alone.

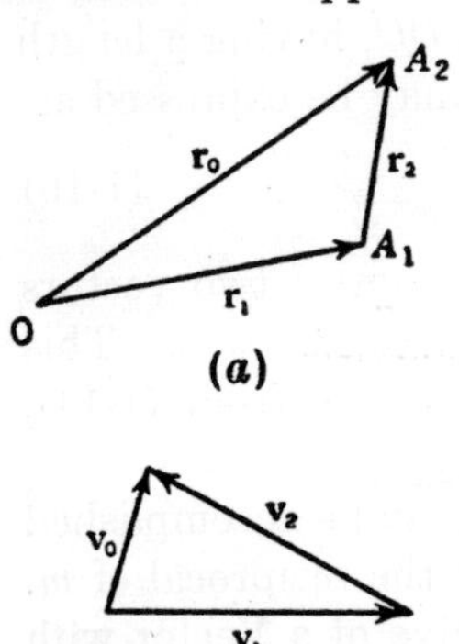

(a)

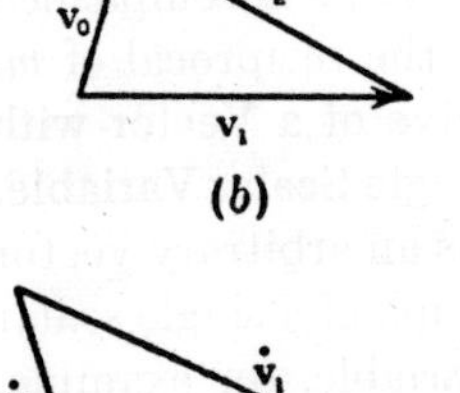

(b)

(c)

Fig. 1-11

1-9. Moving Reference Systems. The compounding of velocities and accelerations is important for the class of problems in which one point is moving relative to a second point, while at the same time both are moving with respect to a third point. In Fig. 1-11a, points A_1 and A_2 are moving relative to O, while at the same time A_2 is moving relative to A_1. At a given instant the positions of A_1 and A_2 with reference to O are represented by the instantaneous values of the radius vectors $\mathbf{r}_1$ and $\mathbf{r}_0$. The vector $\mathbf{r}_2$ specifies the position of A_2 relative to A_1. The vector equation of the triangle in Fig. 1-11a is

$$\mathbf{r}_0 = \mathbf{r}_1 + \mathbf{r}_2 \tag{1-21}$$

Employing Eq. (1-20), the velocity of A_2 relative to O can be obtained by differentiating Eq. (1-21). We have

$$\dot{\mathbf{r}}_0 = \dot{\mathbf{r}}_1 + \dot{\mathbf{r}}_2 \qquad \text{or} \qquad \mathbf{v}_0 = \mathbf{v}_1 + \mathbf{v}_2 \tag{1-22}$$

in which the equivalent notation $\dot{\mathbf{r}}_0 = \mathbf{v}_0$, and so on, is employed. Equation (1-22) shows that the velocity vectors form a closed triangle (see Fig. 1-11b). The instantaneous directions of the velocity vectors are the directions in which the tips of the corresponding radius vectors are moving at that instant.

The content of Eq. (1-22) is of sufficient importance to warrant a verbal statement of its meaning:

VI. *If a point A_1 moves with a velocity* $\mathbf{v}_1$ *relative to a reference point O, and if a second point A_2 moves with a velocity* $\mathbf{v}_2$ *relative to A_1, then the velocity of A_2 relative to O is obtained by taking the vector sum of* $\mathbf{v}_1$ *and* $\mathbf{v}_2$.

The accelerations can be found in the same way by differentiating Eq. (1-22). We obtain

$$\dot{\mathbf{v}}_0 = \dot{\mathbf{v}}_1 + \dot{\mathbf{v}}_2 \qquad (1\text{-}23)$$

demonstrating that the accelerations also form a closed vector triangle (Fig. 1-11*c*).

Example 1-2. A motorboat capable of a speed of 20 mi/hr starts to go across a river from a point *A* on one bank (see Fig. 1-12). The speed of the water is 10 mi/hr. It is further required that the prow of the boat will always be pointed in a direction perpendicular to the velocity of the stream. Where will it land on the opposite bank?

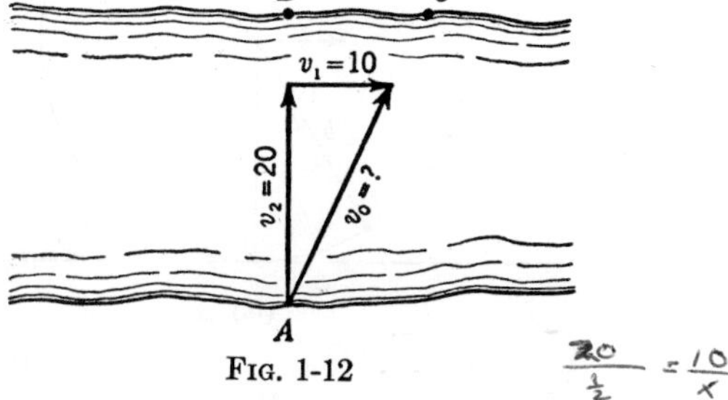

Fig. 1-12

The actual velocity $\mathbf{v}_0$ of the boat, relative to an observer standing on a bank, is the resultant of the velocity of the stream and that of the boat relative to the stream. The first of these has the magnitude 10 mi/hr and is pointed downstream at all times, while the second has the magnitude 20 mi/hr and is pointed directly across at all times. The resultant $\mathbf{v}_0$ is the hypotenuse of the velocity triangle shown in Fig. 1-12. Moreover, since the velocities of the stream and of the boat relative to the stream are both uniform, their resultant will also be uniform. This means that the boat will actually land at point *C*, along the vector $\mathbf{v}_0$ produced. The triangle *ABC* is similar to the triangle of the velocity vectors so the distance *BC* can easily be found, provided the width *AB* of the stream is known. If the river is $\frac{1}{2}$ mi wide, the distance *BC* that the boat drifts downstream is $\frac{1}{4}$ mi.

Example 1-3. A wheel of radius *b* is rolling without sliding along a road with a velocity $\mathbf{v}_1$ (not necessarily constant). It is desired to calculate the velocity, relative to an observer standing on the road, of a point on the rim.

In Fig. 1-13*a*, the center of the wheel is A_1, the arbitrary point on the rim is A_2, and the point on the wheel which is in contact with the road is *C*. The velocity of A_1 relative to the observer is $\mathbf{v}_1$ and is directed to the left. If $\mathbf{v}_2$ is the velocity of A_2 relative to A_1, then $\mathbf{v}_0$, the resultant of $\mathbf{v}_1$ and $\mathbf{v}_2$, is, by virtue of Eq. (1-22), the velocity of A_2 relative to the observer. The quantity $\mathbf{v}_1$, the velocity of the center of the wheel along the road, is given. Also $\mathbf{v}_2$, the velocity of A_2 on the rim relative to the center A_1, is always tangent to the wheel at A_1. It remains only to find the magnitude of $\mathbf{v}_2$.

Since all points on the rim have the same speed relative to A_1, we are free to choose any point on the rim which is convenient in order to compute this magnitude. Suppose we select the point *C* on the rim, which is momentarily in contact with the road. A time Δt previously, this point was at *D*, and during this time it shifted from *D* to *C*, a distance on the wheel (relative to A_1) of $b\,\Delta\theta$. Now during the same time the wheel moved along the road a distance Δs, where $\Delta s = b\,\Delta\theta$ in magnitude (since no slipping is permitted). Thus, dividing by Δt, and proceeding to the limit as Δt becomes small, we have

$$\frac{ds}{dt} = b\,\frac{d\theta}{dt} \tag{1-24}$$

Now, by definition, ds/dt is the scalar magnitude of $\mathbf{v}_1$, the velocity of the center of the wheel along the road. Also $b\,d\theta/dt$ is the magnitude of $\mathbf{v}_2$, the velocity of a point on the rim relative to the center. Thus we have that $\mathbf{v}_2$ is a vector having the same length as $\mathbf{v}_1$, and consequently $\mathbf{v}_0$ for point A_2 is the diagonal of a parallelogram having equal sides, as shown in Fig. 1-13*a*.

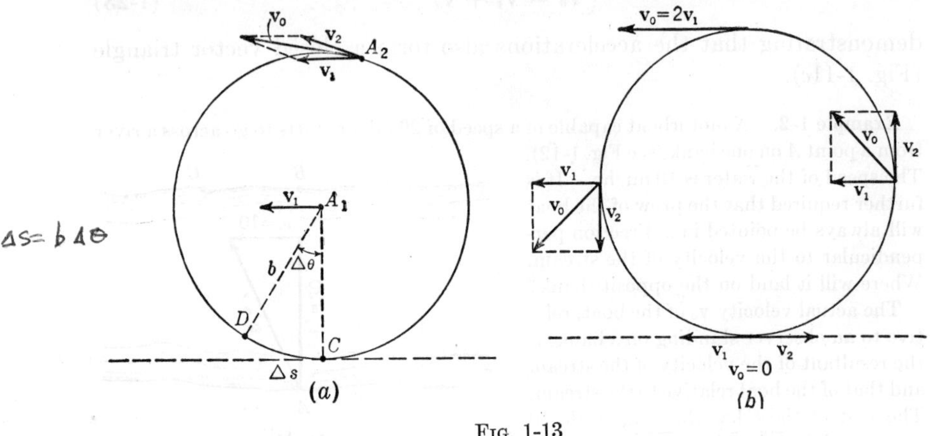

FIG. 1-13

The velocities of certain other points on the rim relative to the road can be seen by consulting Fig. 1-13*b*. It is apparent, for example, that $\mathbf{v}_0$ for the top of the wheel is just $2\mathbf{v}_1$ and that the point on the wheel instantaneously in contact with the road is momentarily at rest with respect to the road.

Example 1-4. With the wheel of Example 1-3, it is now desired to find the acceleration of a point on the rim as perceived by the observer on the road.

The acceleration of the center of the wheel relative to the road is $\dot{\mathbf{v}}_1$, which is a vector pointing either in the same direction as $\mathbf{v}_1$ or in the opposite direction, depending, respectively, on whether $\mathbf{v}_1$ is increasing or decreasing in magnitude with time. The situation is not so simple in the case of $\dot{\mathbf{v}}_2$, the acceleration of A_2 with reference to A_1, since $\mathbf{v}_2$ is changing in direction and will also be changing in magnitude if $\mathbf{v}_1$ is changing.

In Fig. 1-14*a* we see that during a time Δt the wheel has rotated through an angle $\Delta\theta$, and the velocity relative to the center A_1 of the point A_2 on the rim has changed from $\mathbf{v}_2$ to $\mathbf{v}_2 + \Delta\mathbf{v}_2$. In Fig. 1-14*b* is shown an enlarged view of the vector triangle consisting of the sides $\mathbf{v}_2$, $\Delta\mathbf{v}_2$, and $\mathbf{v}_2 + \Delta\mathbf{v}_2$. Clearly, since $\mathbf{v}_2$ is always tangent to the wheel, the angle between $\mathbf{v}_2$ and $\mathbf{v}_2 + \Delta\mathbf{v}_2$ is also $\Delta\theta$. Moreover it is evident that $\Delta\mathbf{v}_2$ may be written

$$\Delta\mathbf{v}_2 = \Delta\mathbf{v}_2' + \Delta\mathbf{v}_2'' \tag{1-25}$$

This is done by marking off along $\mathbf{v}_2 + \Delta\mathbf{v}_2$ a length A_2E equal in magnitude to A_2B. Evidently $\Delta\mathbf{v}_2'$ ($= EF$) is a vector the magnitude of which is the change in the magnitude of $\mathbf{v}_2$. This latter increment will be present only if the center of the wheel is being accelerated, that is, only if $\dot{\mathbf{v}}_1$ is not zero. Dividing Eq. (1-25) by Δt and passing to the limit as Δt becomes small, we have

$$\dot{\mathbf{v}}_2 = \dot{\mathbf{v}}_2' + \dot{\mathbf{v}}_2'' \tag{1-26}$$

The direction of $\dot{\mathbf{v}}_2'$ is along the tangent, and is counterclockwise if the wheel is being accelerated to the left. If the wheel is being accelerated to the right, the direction of $\dot{\mathbf{v}}_2'$ is tangent to the wheel in a clockwise direction. The vector $\dot{\mathbf{v}}_2'$ is different from zero only if the velocity of the center of the wheel along the road is changing. The direction of $\dot{\mathbf{v}}_2''$ is along the inward normal to the rim and is the familiar centripetal acceleration obtained when a point is traveling along a circular path.

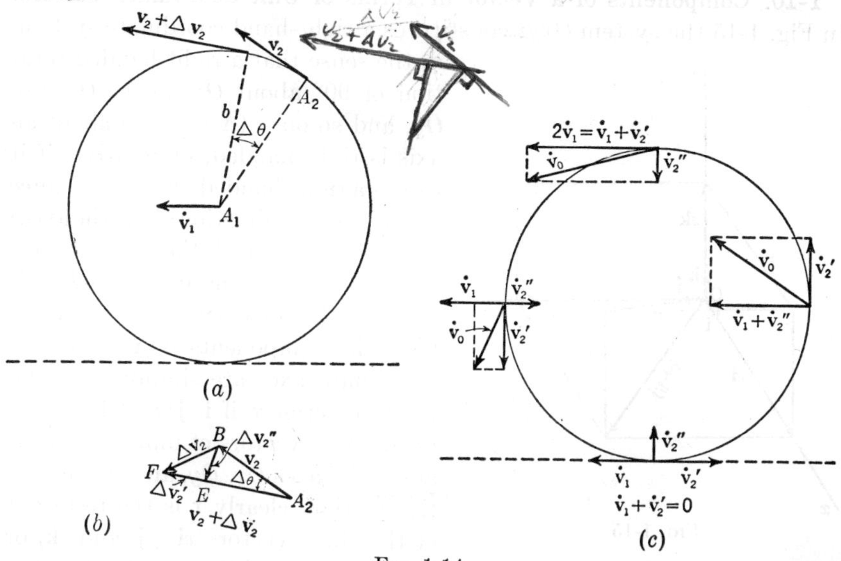

Fig. 1-14

In Fig. 1-14*b*, if $\Delta\theta$ is small, we may write the scalar equation

$$\Delta v_2'' = v_2\,\Delta\theta \tag{1-27}$$

Dividing by Δt and passing to the limit as Δt approaches zero, we obtain

$$\dot{v}_2'' = v_2\dot{\theta} = \frac{v_2^{\,2}}{b} \tag{1-28}$$

where the second equality follows from Eq. (1-24) and from the fact that $\dot{s} = v_1 = v_2$ in magnitude.

The process of finding $\dot{\mathbf{v}}_2'$ is not difficult. The change $\Delta v_2'$ during the time Δt is just the change in the magnitude of $\mathbf{v}_2$. But $\mathbf{v}_2$ and $\mathbf{v}_1$ are equal in magnitude at all times if there is no slipping. Consequently $\dot{\mathbf{v}}_2'$ is equal in magnitude to $\dot{\mathbf{v}}_1$.

It is apparent that the acceleration of any point on the rim relative to the road will be the sum of three vectors, as

$$\dot{\mathbf{v}}_0 = \dot{\mathbf{v}}_1 + \dot{\mathbf{v}}_2' + \dot{\mathbf{v}}_2'' \tag{1-29}$$

Here $\dot{\mathbf{v}}_1$ is the acceleration of the center of the wheel, $\dot{\mathbf{v}}_2'$ is the tangential acceleration of the point on the rim relative to the center (which is equal in magnitude to $\dot{\mathbf{v}}_1$), and finally $\dot{\mathbf{v}}_2''$ is the centripetal accleration of the point on the rim. The centripetal acceleration has the magnitude v_2^2/b, or, what is equivalent, v_1^2/b, and arises because $\mathbf{v}_2$ is changing in direction.

Accelerations of certain points on the rim are depicted in Fig. 1-14c. It is of interest to note that, although the instantaneous velocity of the point on the wheel in contact with the road is zero, its acceleration is not zero. It has the magnitude v_1^2/b, and is directed toward the center. These notions will be important later when the motions of rigid bodies are considered.

1-10. Components of a Vector in Terms of Unit Coordinate Vectors. In Fig. 1-15 the system $Oxyz$ constitutes a right-hand coordinate system, in the sense that a right-handed rotation of 90° about Oz carries Ox into Oy, and so on. (A rotation about an axis is right-handed, or positive, if it causes a right-handed screw to advance in the positive direction along the axis.) Selecting a point P, the radius vector $\mathbf{r}$ is drawn from the origin to P. If the coordinates of P are x, y, and z, then the components of $\mathbf{r}$ along the coordinate axes are simply x, y, and z. Furthermore if $\mathbf{i}$, $\mathbf{j}$, and $\mathbf{k}$ are defined to be vectors of unit magnitude along the axes Ox, Oy, and Oz, respectively, then clearly $\mathbf{r}$ is the resultant of the three vectors $x\mathbf{i}$, $y\mathbf{j}$, and $z\mathbf{k}$, or simply $\mathbf{r}$ has these component vectors in terms of the coordinate axes. Thus

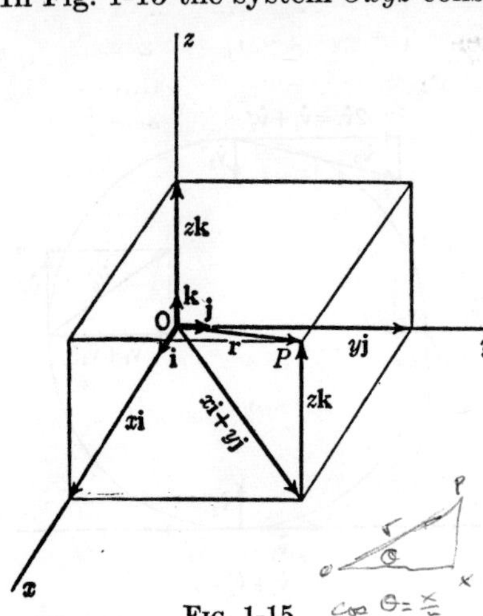

FIG. 1-15

$$\mathbf{r} = x\mathbf{i} + y\mathbf{j} + z\mathbf{k} \tag{1-30}$$

The results are entirely the same for an arbitrary vector $\mathbf{a}$. In terms of its components and the unit coordinate vectors it can be expressed as

$$\mathbf{a} = a_x\mathbf{i} + a_y\mathbf{j} + a_z\mathbf{k} \tag{1-31}$$

It is evident that the unit vectors must be dimensionless. This is easy to see in Eq. (1-30) since $\mathbf{r}$ has the dimensions of length and x, y, and z have also (cf. Sec. 1-21).

The time derivative of any vector $\mathbf{a}$, when expressed in terms of its components, can be written

$$\dot{\mathbf{a}} = \dot{a}_x\mathbf{i} + \dot{a}_y\mathbf{j} + \dot{a}_z\mathbf{k} \tag{1-32}$$

In the same way the velocity and acceleration of a particle whose radius vector is $\mathbf{r}$ can be expressed, respectively, as

$$\dot{\mathbf{r}} = \dot{x}\mathbf{i} + \dot{y}\mathbf{j} + \dot{z}\mathbf{k} \qquad \ddot{\mathbf{r}} = \ddot{x}\mathbf{i} + \ddot{y}\mathbf{j} + \ddot{z}\mathbf{k} \tag{1-33}$$

It was pointed out in Sec. 1-5 that, if two vectors are equal, they have the same magnitude and direction. Evidently, therefore, in view of the present discussion they have the same components. This has an important application in connection with a vector equation in which the left side can be replaced by a single vector quantity and the right side also by a single vector quantity, for then, if the equality is to hold, it is necessary that the x component of the left side be equal to the x component of the right side, the y component of the left to the y component of the right, and the z component of the left to the z component of the right. Thus, such a vector equation is equivalent to three scalar equations.

The rectangular components of a vector **a** can also be expressed in terms of the direction cosines of the vector. The direction cosines are the cosines of the angles included between the vector itself and its components along the coordinate axes. As an example we may again consider Fig. 1-15. The direction cosines of **r** in terms of the magnitudes of **r** and its components are x/r, y/r, and z/r (since $x = r \cos POx$, etc.). Clearly these results will apply to any vector, and it need not pass through the origin. Thus, multiplying and dividing Eq. (1-31) by a, we are able to write

$$\mathbf{a} = a\left(\frac{a_x}{a}\mathbf{i} + \frac{a_y}{a}\mathbf{j} + \frac{a_z}{a}\mathbf{k}\right) = a(\alpha\mathbf{i} + \beta\mathbf{j} + \gamma\mathbf{k}) \qquad (1\text{-}34)$$

where the symbols α, β, and γ stand for the direction cosines of **a** with respect to the coordinate axes ($\alpha = a_x/a$, etc.).

Occasionally it is necessary to know the component of a vector **a** along a given direction **s**. From the material in the preceding sections it is evident that, if the angle (a,s) between **a** and **s** and the magnitude of **a** are both known, this projection is simply $a \cos (a,s)$. This can also be written in terms of the rectangular components of **a**. Since the projection of **a** in any direction is equal to the sum of the projections of the rectangular components of **a** in that same direction, it follows that

$$a_s = a_x \cos (s,x) + a_y \cos (s,y) + a_z \cos (s,z) \qquad (1\text{-}35)$$

The quantity a_s is the magnitude of the projection of **a** along **s**, and (s,x) is the angle included between the direction of **s** and the Ox axis, and so on. In terms of the direction cosines α, β, and γ of **a**, Eq. (1-35) becomes

$$a_s = a[\alpha \cos (s,x) + \beta \cos (s,y) + \gamma \cos (s,z)] \qquad (1\text{-}36)$$

where $\alpha = \cos (a,a_x)$, $\beta = \cos (a,a_y)$, $\gamma = \cos (a,a_z)$.

1-11.[1] **Transformation of Vector Components from One Set of Cartesian Axes to Another.** As the example of a vector we choose the dis-

[1] May be omitted in a first reading.

tance between points P_1 and P_2 (Fig. 1-16). These points, respectively, possess the coordinates x_1, y_1, z_1 and x_2, y_2, z_2 in a coordinate system $Oxyz$ and the coordinates x_1', y_1', z_1' and x_2', y_2', z_2' in a system $O'x'y'z'$. The distance between P_1 and P_2 may be represented by a displacement **a**

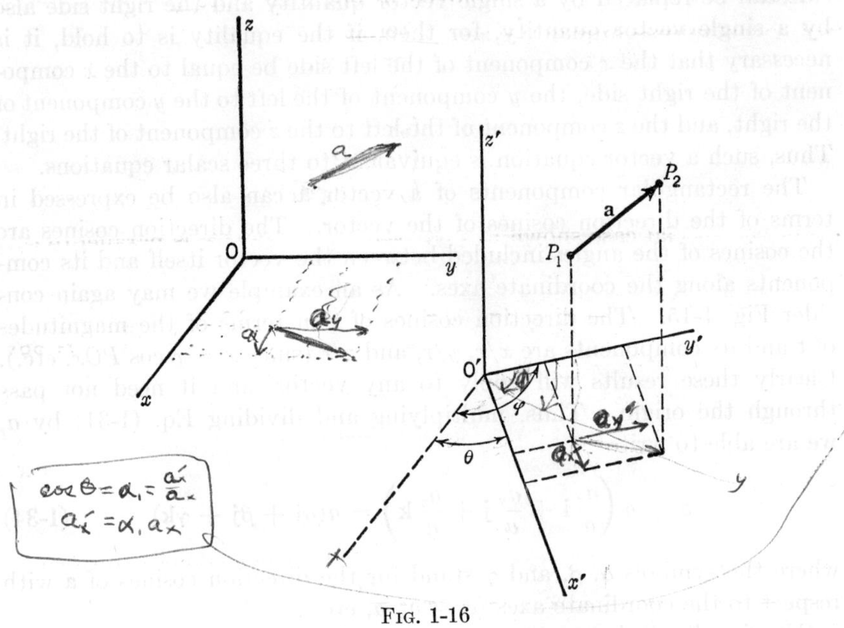

Fig. 1-16

having a definite direction in space, and a magnitude a. The direction cosines of the various primed axes are presented in Table 1-1. The

Table 1-1. Direction Cosines Involved in Transformation of the Components of a Vector from One Rectangular Coordinate System to Another

		Ox	Oy	Oz
		$a_x \equiv x_2 - x_1$	$a_y \equiv y_2 - y_1$	$a_z \equiv z_2 - z_1$
$O'x'$	$a_{x'} \equiv x_2' - x_1'$	α_1	β_1	$\gamma_1\ (= 0)$
$O'y'$	$a_{y'} \equiv y_2' - y_1'$	α_2	β_2	$\gamma_2\ (= 0)$
$O'z'$	$a_{z'} \equiv z_2' - z_1'$	$\alpha_3\ (= 0)$	$\beta_3\ (= 0)$	$\gamma_3\ (= 1)$

symbols are interpreted such that α_1, for example, is the cosine of the angle between the directions of Ox and $O'x'$, α_2 the cosine of the angle between Ox and $O'y'$, and so on. Furthermore it is evident that α_1 is also the cosine of the angle between a_x and $a_{x'}$, the components, respectively, of **a** along the Ox and the $O'x'$ axes, and so on. (The numbers in parentheses are the values of the direction cosines which obtain for the particular case of $O'z'$ parallel to Oz, discussed below.) It is apparent, also, that a_x is identical to $x_2 - x_1$, $a_{z'}$ to $z_2' - z_1'$, and so on.

Equation (1-35) may be employed in order to write the components of **a**, in the primed system $O'x'y'z'$, in terms of the components of **a** in the unprimed system $Oxyz$ and the direction cosines listed in Table 1-1. We have

$$\begin{aligned} x_2' - x_1' &= (x_2 - x_1)\alpha_1 + (y_2 - y_1)\beta_1 + (z_2 - z_1)\gamma_1 \\ y_2' - y_1' &= (x_2 - x_1)\alpha_2 + (y_2 - y_1)\beta_2 + (z_2 - z_1)\gamma_2 \\ z_2' - z_1' &= (x_2 - x_1)\alpha_3 + (y_2 - y_1)\beta_3 + (z_2 - z_1)\gamma_3 \end{aligned} \tag{1-37}$$

For the special case shown in Fig. 1-16, in which $O'z'$ is parallel to Oz, Eqs. (1-37) become

$$\begin{aligned} x_2' - x_1' &= (x_2 - x_1)\alpha_1 + (y_2 - y_1)\beta_1 \\ y_2' - y_1' &= (x_2 - x_1)\alpha_2 + (y_2 - y_1)\beta_2 \\ z_2' - z_1' &= z_2 - z_1 \end{aligned} \tag{1-38}$$

Assuming that, in the $Oxyz$ system of axes,

$$(x_2 - x_1)^2 + (y_2 - y_1)^2 + (z_2 - z_1)^2 \equiv a_x^2 + a_y^2 + a_z^2 = a^2 \tag{1-39}$$

we desire to show that the transformation equations, (1-38), ensure that in the $O'x'y'z'$ system the square of the magnitude of **a** can be expressed as

$$(x_2' - x_1')^2 + (y_2' - y_1')^2 + (z_2' - z_1')^2 \equiv a_{x'}^2 + a_{y'}^2 + a_{z'}^2 = a^2 \tag{1-40}$$

In order to do this we first substitute Eqs. (1-38) in (1-40). The left side of (1-40) becomes

$$(x_2 - x_1)^2\alpha_1^2 + (y_2 - y_1)^2\beta_1^2 + 2(x_2 - x_1)(y_2 - y_1)\alpha_1\beta_1 + (x_2 - x_1)^2\alpha_2^2 \\ + (y_2 - y_1)^2\beta_2^2 + 2(x_2 - x_1)(y_2 - y_1)\alpha_2\beta_2 + (z_2 - z_1)^2$$

or

$$(x_2 - x_1)^2(\alpha_1^2 + \alpha_2^2) + (y_2 - y_1)^2(\beta_1^2 + \beta_2^2) \\ + 2(x_2 - x_1)(y_2 - y_1)(\alpha_1\beta_1 + \alpha_2\beta_2) + (z_2 - z_1)^2 \tag{1-41}$$

It is evident from the figure that $\alpha_1 = \cos\theta$ and $\alpha_2 = \cos\varphi = -\sin\theta$. Consequently

$$\alpha_1^2 + \alpha_2^2 = 1 \tag{1-42}$$

In a similar fashion

$$\beta_1^2 + \beta_2^2 = 1 \tag{1-43}$$

That

$$\alpha_1\beta_1 + \alpha_2\beta_2 = 0 \tag{1-44}$$

follows from the fact (cf. Prob. 1-20) that the left-hand side of Eq. (1-44) is just the cosine of the angle between $O'x'$ and $O'y'$ which is zero. Accordingly, simplifying (1-41) by means of (1-42) to (1-44) and combining with Eq. (1-40), we obtain, finally,

$$(x_2' - x_1')^2 + (y_2' - y_1')^2 + (z_2' - z_1')^2 = (x_2 - x_1)^2 + (y_2 - y_1)^2 + (z_2 - z_1)^2 = a^2 \quad (1\text{-}45)$$

and the relationship which it was desired to verify.

It is interesting, in passing, to write the transformation equations which are the inverse of Eqs. (1-37). These are

$$\begin{aligned} x_2 - x_1 &= (x_2' - x_1')\alpha_1 + (y_2' - y_1')\alpha_2 + (z_2' - z_1')\alpha_3 \\ y_2 - y_1 &= (x_2' - x_1')\beta_1 + (y_2' - y_1')\beta_2 + (z_2' - z_1')\beta_3 \\ z_2 - z_1 &= (x_2' - x_1')\gamma_1 + (y_2' - y_1')\gamma_2 + (z_2' - z_1')\gamma_3 \end{aligned} \quad (1\text{-}46)$$

The present material furnishes a better definition of a vector than the simple statements made in Sec. 1-5. A quantity which may be represented by means of the components of a directed segment, such as a_x, a_y, and a_z above for the distance between two points, which components transform by means of equations of the type of (1-37) to components such as $a_{x'}$, $a_{y'}$, and $a_{z'}$, is defined to be a vector. (For more general transformation properties in connection with arbitrary coordinate systems, other references may profitably be consulted.)

1-12. Scalar, or Dot, Product of Two Vectors. An important vector operation is based upon the idea of projecting a vector **a** along the direction of some other given vector **c**. This has frequent application, as will be seen later, in finding the work done by a given force acting through a specified displacement. The operation consists in first projecting the vector **a** along **c** and then multiplying the magnitude of this result by the magnitude of **c**. The procedure is called that of taking the *dot*, or *scalar, product* of the vectors **a** and **c**. The result is a scalar quantity. Thus

$$\mathbf{a} \cdot \mathbf{c} = ac \cos (a,c) \quad (1\text{-}47)$$

where the symbol (a,c) denotes the angle between the directions of **a** and **c**. It is clear, since

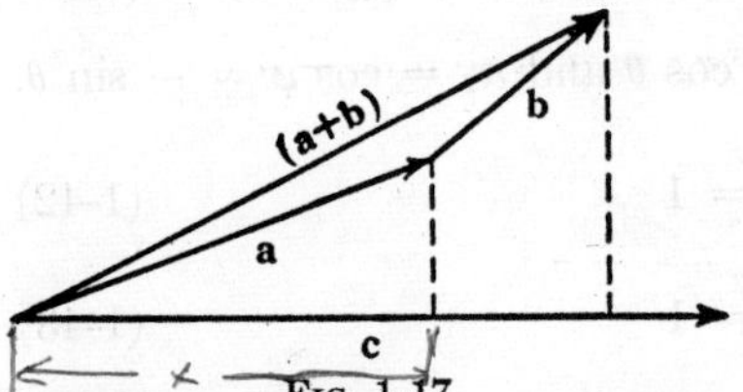

$$\mathbf{c} \cdot \mathbf{a} = ca \cos (a,c)$$

that

$$\mathbf{a} \cdot \mathbf{c} = \mathbf{c} \cdot \mathbf{a} \quad (1\text{-}48)$$

Thus the operation of taking the dot product is commutative.

Fig. 1-17

That the dot product is distributive over vector addition can be seen by a study of Fig. 1-17. The projection of **a** on **c** is simply $\mathbf{a} \cdot \mathbf{c}/c$, and that of **b** on **c** is $\mathbf{b} \cdot \mathbf{c}/c$. Furthermore, from Fig. 1-17, since the projection of **a** + **b** on **c** is the same as the sum of the individual projections, we may write

$$\frac{(\mathbf{a}+\mathbf{b})\cdot\mathbf{c}}{c} = \frac{\mathbf{a}\cdot\mathbf{c}}{c} + \frac{\mathbf{b}\cdot\mathbf{c}}{c} \qquad \text{or} \qquad (\mathbf{a}+\mathbf{b})\cdot\mathbf{c} = \mathbf{a}\cdot\mathbf{c} + \mathbf{b}\cdot\mathbf{c} \tag{1-49}$$

and the property that the operation of taking the dot product is distributive over vector addition is demonstrated.

It is worthwhile to notice that the operation of the dot product can be performed only once, for example, $\mathbf{a}\cdot\mathbf{b}\cdot\mathbf{c}$ has no meaning. This is so because $\mathbf{a}\cdot\mathbf{b}$ is a scalar quantity, and its dot product with $\mathbf{c}$ is not defined. Therefore the question of associativity does not arise. The scalar product differs from the ordinary algebraic multiplication of two numbers because of the multiplication by the cosine of an angle. If $\mathbf{a}$ and $\mathbf{c}$ are perpendicular, the dot product is zero regardless of the magnitudes of $\mathbf{a}$ and $\mathbf{c}$. Conversely, if the dot product is zero, either we must have one of the vectors equal to zero or else the two are perpendicular.

The derivative of the dot product of two vectors $\mathbf{a}$ and $\mathbf{b}$, each of which is a function of a single scalar variable t, can easily be shown to be

$$\frac{d}{dt}(\mathbf{a}\cdot\mathbf{b}) = \frac{d\mathbf{a}}{dt}\cdot\mathbf{b} + \mathbf{a}\cdot\frac{d\mathbf{b}}{dt} \equiv \mathbf{b}\cdot\frac{d\mathbf{a}}{dt} + \mathbf{a}\cdot\frac{d\mathbf{b}}{dt} \tag{1-50}$$

The proof of this is left to the student.

It is useful to consider the various dot products which can be made up from the unit coordinate vectors. Evidently

$$\angle(i,j) = \angle(i,k) = \angle(j,k) = \frac{\pi}{2} \qquad \angle(i,i) = \angle(j,j) = \angle(k,k) = 0 \tag{1-51}$$

whence

$$\mathbf{i}\cdot\mathbf{i} = \mathbf{j}\cdot\mathbf{j} = \mathbf{k}\cdot\mathbf{k} = 1 \qquad \mathbf{i}\cdot\mathbf{j} = \mathbf{i}\cdot\mathbf{k} = \mathbf{j}\cdot\mathbf{k} = 0 \tag{1-52}$$

In view of Eq. (1-52) the dot product of two vectors $\mathbf{a}$ and $\mathbf{b}$ referred to their components will be

$$\mathbf{a}\cdot\mathbf{b} = (a_x\mathbf{i} + a_y\mathbf{j} + a_z\mathbf{k})\cdot(b_x\mathbf{i} + b_y\mathbf{j} + b_z\mathbf{k}) = a_xb_x + a_yb_y + a_zb_z \tag{1-53}$$

all other terms being zero. In particular

$$\mathbf{a}\cdot\mathbf{a} = a^2 = a_x^2 + a_y^2 + a_z^2 \tag{1-54}$$

1-13. Vector, or Cross, Product of Vectors. The *vector*, or *cross, product* of two vectors, in contrast to the case of the dot product, is itself another vector. If $\mathbf{c}$ is the vector which is the cross product of two other vectors $\mathbf{a}$ and $\mathbf{b}$ (in this order), it is written

$$\mathbf{c} = \mathbf{a}\times\mathbf{b} \tag{1-55}$$

and the scalar magnitude of **c** is

$$|\mathbf{c}| = c = ab \sin (a,b) \tag{1-56}$$

The direction of the vector **c** is perpendicular to the plane defined by **a** and **b** and is such that the rotation through the angle (a,b) of a right-hand screw advancing along **c** carries **a** into **b**. [In this rotation, angle (a,b) is not more than 180°.] In Fig. 1-18, h is a line drawn perpendicular to **a** from the tip of **b** and is therefore the altitude of the parallelogram $OABC$. But

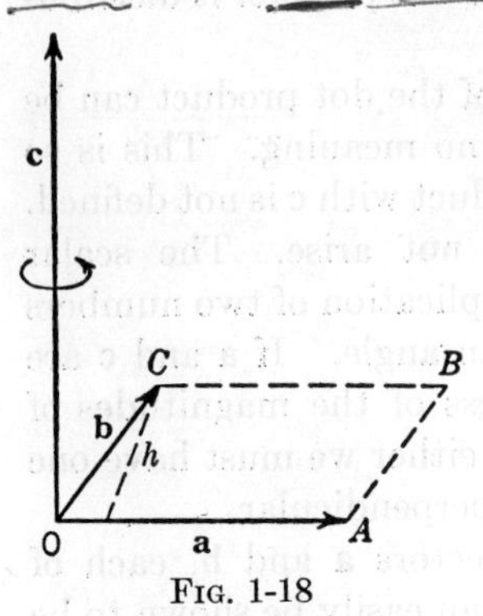

Fig. 1-18

$$h = b \sin (a,b)$$

and therefore, from Eq. (1-56), the magnitude of the cross product is $ab \sin (a,b)$, which is the area of the parallelogram $OABC$.

From the definition of the cross product it is evident that the product $\mathbf{b} \times \mathbf{a}$ has the same magnitude as $\mathbf{a} \times \mathbf{b}$ but the rotation which carries **b** into **a** is opposite to that which carried **a** into **b**, and therefore

$$\mathbf{b} \times \mathbf{a} = -\mathbf{a} \times \mathbf{b} = -\mathbf{c} \tag{1-57}$$

Consequently the operation of taking the vector product of two vectors is not commutative. However, as in the case of the dot product, the cross product is distributive over vector addition. Thus if **a**, **b**, and **c** are three vectors, it is possible to show that

$$\mathbf{a} \times (\mathbf{b} + \mathbf{c}) = \mathbf{a} \times \mathbf{b} + \mathbf{a} \times \mathbf{c} \tag{1-58}$$

The derivative of the cross product of two vectors **a** and **b**, each of which is a function of a single scalar variable t, can be shown to be

$$\frac{d}{dt}(\mathbf{a} \times \mathbf{b}) = \frac{d\mathbf{a}}{dt} \times \mathbf{b} + \mathbf{a} \times \frac{d\mathbf{b}}{dt} \tag{1-59}$$

This is stated without proof and may easily be verified by the student.

The cross products among the unit coordinate vectors **i**, **j**, **k** are often useful. The definition gives immediately

$$\begin{aligned} &\mathbf{i} \times \mathbf{i} = \mathbf{j} \times \mathbf{j} = \mathbf{k} \times \mathbf{k} = 0 \qquad && \mathbf{i} \times \mathbf{j} = \mathbf{k} = -\mathbf{j} \times \mathbf{i} \\ &\mathbf{k} \times \mathbf{i} = \mathbf{j} = -\mathbf{i} \times \mathbf{k} && \mathbf{j} \times \mathbf{k} = \mathbf{i} = -\mathbf{k} \times \mathbf{j} \end{aligned} \tag{1-60}$$

By employing the results of Eqs. (1-60), it is possible to express the cross product of two vectors in terms of the components of the vectors. If

$$\mathbf{a} = a_x\mathbf{i} + a_y\mathbf{j} + a_z\mathbf{k} \qquad \text{and} \qquad \mathbf{b} = b_x\mathbf{i} + b_y\mathbf{j} + b_z\mathbf{k}$$

then

$$\mathbf{a} \times \mathbf{b} = (a_yb_z - a_zb_y)\mathbf{i} + (a_zb_x - a_xb_z)\mathbf{j} + (a_xb_y - a_yb_x)\mathbf{k} \tag{1-61}$$

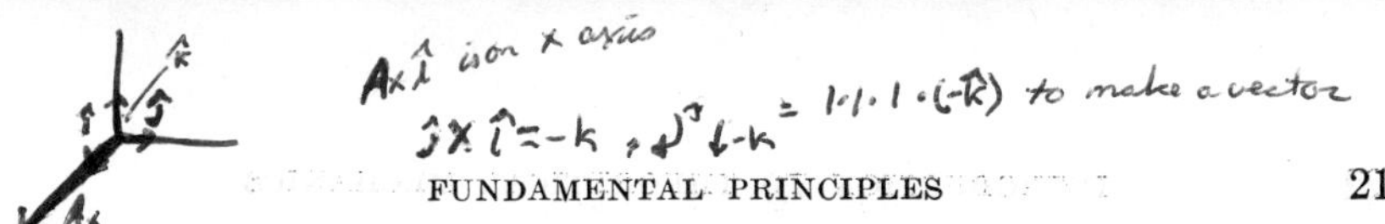

A close inspection of Eq. (1-61) reveals that $\mathbf{a} \times \mathbf{b}$ can be written in the extremely useful determinant form

$$\mathbf{a} \times \mathbf{b} = \begin{vmatrix} \mathbf{i} & \mathbf{j} & \mathbf{k} \\ a_x & a_y & a_z \\ b_x & b_y & b_z \end{vmatrix} \tag{1-62}$$

The form of Eq. (1-62) is convenient to memorize. It can be verified merely by expanding the determinant and comparing the result with Eq. (1-61).

It will be seen later that an important application of the cross product is in writing the vector form of the moment of a force.

1-14. The Gradient Vector. Consider the scalar function $\phi(x,y,z)$ which has a value at each point of the region of space under consideration. Moreover, it has not more than one value at any one point of the region, and the derivatives $\partial\phi/\partial x$, $\partial\phi/\partial y$, and $\partial\phi/\partial z$ exist at all points of the region. A situation of this type is exemplified by the temperature of the air in a room.

Employing the methods of calculus we compute the differential of ϕ to be

$$d\phi = \frac{\partial\phi}{\partial x}\,dx + \frac{\partial\phi}{\partial y}\,dy + \frac{\partial\phi}{\partial z}\,dz \tag{1-63}$$

It is useful to define a vector

$$\mathbf{i}\,\frac{\partial\phi}{\partial x} + \mathbf{j}\,\frac{\partial\phi}{\partial y} + \mathbf{k}\,\frac{\partial\phi}{\partial z} \tag{1-64}$$

which we call the *gradient* of ϕ, employing for it the alternate symbols $\nabla\phi$ or **grad** ϕ. The symbol ∇ is called *del*, and $\nabla\phi$ is read "del ϕ." Since the differentials dx, dy, dz are the components of the differential $d\mathbf{r}$ of the radius vector, we have

$$\begin{aligned} \nabla\phi \cdot d\mathbf{r} &= \left(\mathbf{i}\,\frac{\partial\phi}{\partial x} + \mathbf{j}\,\frac{\partial\phi}{\partial y} + \mathbf{k}\,\frac{\partial\phi}{\partial z}\right) \cdot (\mathbf{i}\,dx + \mathbf{j}\,dy + \mathbf{k}\,dz) \\ &= \frac{\partial\phi}{\partial x}\,dx + \frac{\partial\phi}{\partial y}\,dy + \frac{\partial\phi}{\partial z}\,dz \end{aligned} \tag{1-65}$$

It is possible to give a geometrical interpretation of $\nabla\phi$ through the use of Eq. (1-65). Consider a surface

$$\phi(x,y,z) = c \tag{1-66}$$

in which c is a particular constant. Let us select an infinitesimal displacement $d\mathbf{r}$ of $\mathbf{r}$, and let us confine our attention to only those displacements which are tangent to the surface described by Eq. (1-66). If such a change in $\mathbf{r}$ is sufficiently small, ϕ is not altered. Consequently

$d\phi$ is zero. Hence

$$d\phi = \nabla\phi \cdot d\mathbf{r} = 0 \tag{1-67}$$

Now $d\mathbf{r}$ is not equal to zero, nor in general is $\nabla\phi$. Accordingly, since $d\phi$ is zero, we must have that $\nabla\phi$ is a vector which is perpendicular to the particular $d\mathbf{r}$ chosen. Only if this is so will Eq. (1-67) be true. Consequently, since we have selected $d\mathbf{r}$ such that it is tangent to the surface $\phi(x,y,z) = c$, we see that $\nabla\phi$ is a vector which is normal to the surface $\phi(x,y,z) = c$.

If ϕ can be expressed as a function of a single scalar variable u, a useful theorem follows, for if $\phi = \phi(u)$, then

$$\begin{aligned}\nabla\phi &= \mathbf{i}\frac{\partial\phi}{\partial x} + \mathbf{j}\frac{\partial\phi}{\partial y} + \mathbf{k}\frac{\partial\phi}{\partial z} = \mathbf{i}\frac{d\phi}{du}\frac{\partial u}{\partial x} + \mathbf{j}\frac{d\phi}{du}\frac{\partial u}{\partial y} + \mathbf{k}\frac{d\phi}{du}\frac{\partial u}{\partial z} \\ &= \frac{d\phi}{du}\left(\mathbf{i}\frac{\partial u}{\partial x} + \mathbf{j}\frac{\partial u}{\partial y} + \mathbf{k}\frac{\partial u}{\partial z}\right) = \frac{d\phi}{du}\nabla u \end{aligned} \tag{1-68}$$

The application of Eq. (1-68) to the frequently occurring case in which ϕ is a function of the single scalar variable r yields the result

$$\nabla\phi = \frac{d\phi}{dr}\frac{\mathbf{r}}{r} = \frac{d\phi}{dr}\mathbf{r}_1 \tag{1-69}$$

in which $\mathbf{r}_1$ is a dimensionless unit vector pointing in the direction of $\mathbf{r}$. The proof of (1-69) is left to the student.

1-15. Scalar and Vector Fields. The Line Integral of a Vector. When a continuous scalar function ϕ has a value at every point of a region of space, it is said that ϕ constitutes a *scalar field* in that region. By taking the gradient of such a function a vector function is obtained which likewise possesses a value at every point of the region. This vector is $\nabla\phi$, and $\nabla\phi$ constitutes a *vector field* in the same region. The example which has been employed, of a vector field derivable from a scalar field, is not always the rule. It is possible to have a vector field which cannot be derived from a scalar field. Thus not all vectors are gradients of scalars.

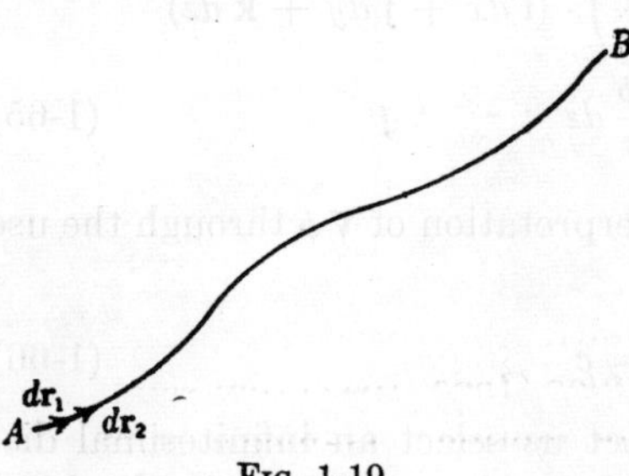

FIG. 1-19

A very useful notion to have, in dealing with a vector field, is that of the *line integral* of the vector along some path connecting a pair of points in the field. In Fig. 1-19, A and B are two points in a region in which the vector $\mathbf{F}$ constitutes a vector field. Select a path, such as is shown, and divide it into tiny elements $d\mathbf{r}_1$, $d\mathbf{r}_2$, etc. The line integral of $\mathbf{F}$ along this path is defined as the limit of the sum of all of the quantities $\mathbf{F} \cdot d\mathbf{r}_1$, $\mathbf{F} \cdot d\mathbf{r}_2$, etc., where in each case the value of $\mathbf{F}$ is its average value within the tiny interval. Thus the line integral of $\mathbf{F}$ along the path is

$$\int_{r_A}^{r_B} \mathbf{F} \cdot d\mathbf{r} \tag{1-70}$$

The integral (1-70) is a scalar since $\mathbf{F} \cdot d\mathbf{r}$ is itself a scalar.

Clearly it may be possible to select another path between A and B such that the line integral will have a different value from that taken along the path of Fig. 1-19. Thus, in general, the value of the line integral between two points in a vector field may depend upon the path taken between the two points.

Of special interest is the case when the vector field is the gradient of a scalar field, for from Eq. (1-67) if the vector $\mathbf{F}$ is the gradient of a scalar function ϕ we have

$$\int_{r_A}^{r_B} \mathbf{F} \cdot d\mathbf{r} = \int_{r_A}^{r_B} \nabla\phi \cdot d\mathbf{r}$$
$$= \int_A^B d\phi = \phi_B - \phi_A \tag{1-71}$$

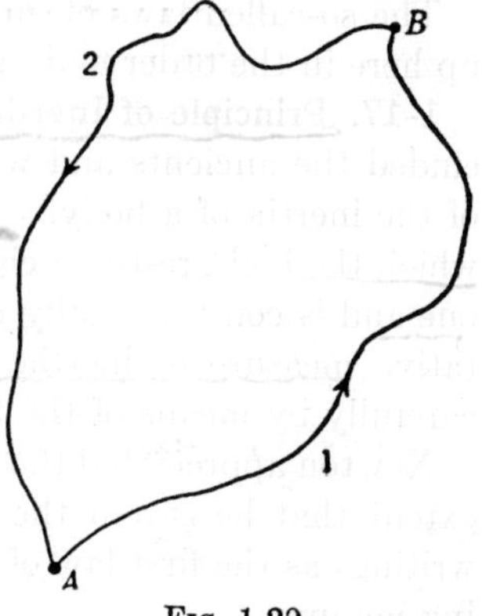

Fig. 1-20

Evidently, if $\mathbf{F}$ is the gradient of a scalar function ϕ, the line integral of $\mathbf{F}$ along a path between the points A and B is just the difference between the values of ϕ at the two points and has nothing to do with the particular path chosen. In particular, if the line integral is taken along a path, such as 1, 2 in Fig. 1-20, which comes back to the starting point A, the result is zero. This follows since

$$\int_{r_A}^{r_A} \mathbf{F} \cdot d\mathbf{r} = \int_{\phi_A}^{\phi_A} d\phi = \phi_A - \phi_A = 0 \tag{1-72}$$

THE LAWS OF MOTION

1-16. The Development of Dynamics. *Dynamics* is that division of mechanics in which the motions of bodies are described. It is a relatively modern science. It was not until the experimental method achieved its great fruit in the hands of Galileo (1564–1642) that significant advances were made in the mechanics of motion. Galileo's chief contributions to dynamics were perhaps contained in the so-called *law of inertia* and in his experiments on accelerated motion. The story (probably apocryphal) of Galileo's experiments on falling bodies dropped from the Leaning Tower of Pisa is an oft-repeated one. It is interesting, however, that in Galileo's mind the law of inertia was apparently not granted the prominence later attributed to it by Newton and others.

We mention next the intellectual giant of the eighteenth century, Sir Isaac Newton (1642–1727), who was born the year Galileo died. Newton was the audacious thinker who placed dynamics on a quantitative foundation. The work of Galileo, Huygens, and others furnished the basis of Newton's formulation of the celebrated laws of motion. Likewise the contributions of Tycho Brahe and Kepler, in Newton's hands, culminated in the law of universal gravitation.

There are other names, such as those of Benedetti and Descartes, which should in fairness be mentioned in connection with the development of dynamics. The serious student of mechanics is urged to consult the excellent book of Mach[1] for greater detail in this respect.

The so-called laws of motion, for reasons of convention, will be taken up here in the order and in phrasing similar to Mach's.

1-17. Principle of Inertia. Newton's First Law. The concept which eluded the ancients and which Galileo was the first to recognize is that of the inertia of a body. *Inertia* is the property of a body by means of which the body resists a change in its motion. The term is a qualitative one and is conventionally employed in a qualitative sense. The quantitative measure of inertia is *mass*, which we shall shortly define more carefully by means of the third law.

Newton appreciated the concept of inertia and continued motion to the extent that he stated the principle, or law, of inertia separately in his writings as the first law of motion. We state the first law in the following manner:

VII. *Every body perseveres in its state of rest or of uniform motion in a straight line, except in so far as it is compelled to change that state by impressed forces.*

It is impossible to outline an experiment in which no forces are acting, since, to do so, an isolated body would be required. For this reason part of the content of the first law must be considered to be intuitive in the sense of an extrapolation from the results of experiments in which small but finite forces are acting. If we can imagine a block sliding on a perfectly smooth horizontal table, the first law states that if a velocity $\mathbf{v}$ is impressed on that block it will continue with that same velocity $\mathbf{v}$, undiminished and with unchanging direction.

1-18. Newton's Second Law. The Equation of Motion. The second law is the quantitative expression of the accelerated motion of a body when it is subjected to forces. It is written:

[1] "The Science of Mechanics," 5th English ed., The Open Court Publishing Company, La Salle, Ill., 1942.

VIII. *The rate of change of the momentum is proportional to the impressed force and takes place in the direction of the straight line along which the force acts.*

Mathematically this can be expressed as a vector equation

$$\frac{d}{dt}(m\dot{\mathbf{r}}) = \mathbf{F} \tag{1-73}$$

When written in this form, Eq. (1-73) states that the vector obtained by taking the time rate of change of the *momentum* $m\dot{\mathbf{r}}$ has the direction of $\mathbf{F}$, and if *consistent units* are employed it has also the same magnitude as $\mathbf{F}$. By *consistent units* is meant a set of units such that, if the mass m, the velocity $\dot{\mathbf{r}}$, the time t, and the force $\mathbf{F}$ are all expressed in terms of these units, all proportionality factors are unity. For example, one set of consistent units for Eq. (1-73) would be $\mathbf{F}$ expressed in *dynes*, m in *grams*, $\dot{\mathbf{r}}$ in *centimeters per second*, and t in *seconds*.

The first law of motion can be regarded as the special case of the second law when the force is zero. If $\mathbf{F}$ is zero in Eq. (1-73) the momentum $m\dot{\mathbf{r}}$ will be a constant. If the mass does not change with time, this is equivalent to saying that the velocity is a constant. Thus Eq. (1-73) can be said to be the analytical expression of the first and second laws of motion. It is the equation of motion and is a vector equation. Equation (1-73) can be expressed in terms of component equations of motion involving the rectangular coordinates x, y, and z. These equations are

$$\frac{d}{dt}(m\dot{x}) = F_x \qquad \frac{d}{dt}(m\dot{y}) = F_y \qquad \frac{d}{dt}(m\dot{z}) = F_z \tag{1-74}$$

and are scalar equations. For the usual case in which m does not change with time, Eqs. (1-74) assume the form

$$m\ddot{x} = F_x \qquad m\ddot{y} = F_y \qquad m\ddot{z} = F_z \tag{1-75}$$

Equations (1-74) are sufficient to treat the motion of a single particle in a field of force. The left side of Eq. (1-73), namely, $d/dt\,(m\dot{\mathbf{r}})$, is often called the *inertial reaction*. Thus the equations of motion are written by merely equating the components of the acting forces to the analogous components of the inertial reaction.

Newton's second law, embodied in Eq. (1-73), can be regarded as the *definition of force*, provided mass is defined elsewhere. The absolute unit of force in the cgs (centimeter-gram-second) system is the *dyne*. This is the force necessary to impart to a one-gram mass an acceleration of one centimeter per second per second.

1-19. Newton's Third Law of Motion. Whenever more than one particle is involved. the first two laws of motion are insufficient to describe the motion. It is necessary to invoke a third law. This can be stated, in the words of Mach, as follows:

IX. *Reaction is always equal and opposite to action; that is to say, the actions of two bodies upon each other are always equal and directly opposite.*

[As employed in the statement of the third law, the term action (and reaction) is synonymous with the term force.]

$\mathbf{F}_{21}$ m_1 m_2 $\mathbf{F}_{12}$

Fig. 1-21

In Fig. 1-21 are shown two mutually interacting particles m_1 and m_2. Particle m_1 exerts a force $\mathbf{F}_{12}$ on m_2, and m_2 in turn exerts a force $\mathbf{F}_{21}$ on m_1. Newton's third law states that $\mathbf{F}_{12}$ is equal in magnitude to $\mathbf{F}_{21}$ and is oppositely directed.

Consider also the example of a block of wood placed on a level floor, assuming the latter to be rigidly attached to the earth. In Fig. 1-22*a* the force $\mathbf{W}$, the weight of the block, is the gravitational force exerted on the block by the earth (as will be shown in Chap. 3, it acts at the

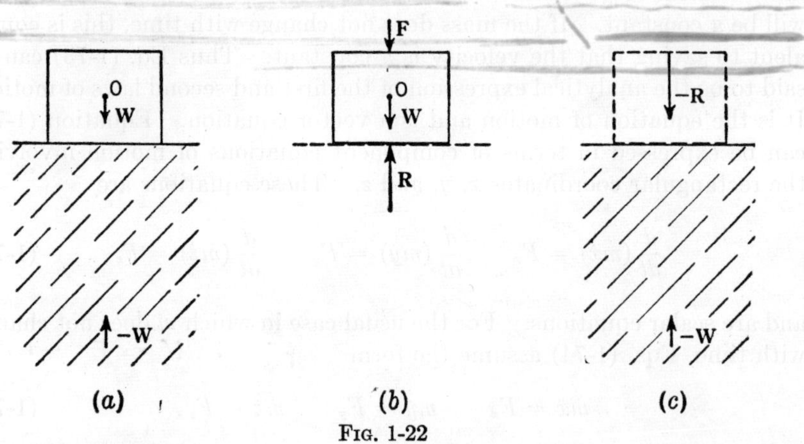

Fig. 1-22

center O of the block), and $-\mathbf{W}$ is the gravitational force exerted on the earth by the block. The two forces $\mathbf{W}$ and $-\mathbf{W}$ are related by Newton's third law and constitute the action and reaction stated there. Such forces are to be contrasted to those which are related through certain auxiliary conditions such as obtain in a state of equilibrium. For example, Fig. 1-22*b* shows only the forces acting on the block. In this figure, as an added feature, a force $\mathbf{F}$ is pushing vertically down upon the block. The presence of $\mathbf{R}$ is deduced in the following manner: Since the forces $\mathbf{W}$ and $\mathbf{F}$ act on the block in a downward direction, the block must experience accelerated motion in that direction in accordance with Newton's second law. The fact that there is no acceleration relative to

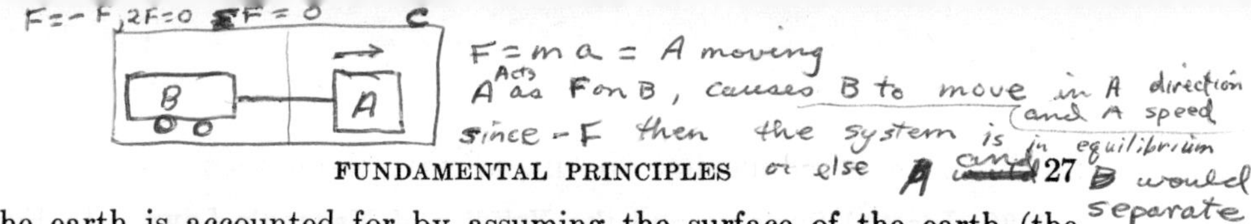

the earth is accounted for by assuming the surface of the earth (the floor) to exert a force **R** on the block which is equal and opposite to **W** + **F**. Thus **R** is related to **W** and **F** through the equilibrium condition.

In a similar vein, part *c* of the figure depicts only the forces acting on the earth. **R** and −**R** are related through Newton's third law. It is apparent that there is a net force equal to **F** acting upon the earth (plus the block), and consequently the earth (plus the block) will experience accelerated motion in the downward direction in accordance with Newton's second law.

Many apparent paradoxes have arisen because of a misunderstanding of the content of the third law. One of these is the following: A horse is attempting to draw a wagon along a level road. The horse is applying a force **F** on the wagon, but by Newton's third law the wagon exerts an equal and opposite force −**F** on the horse! How then is the horse able to start the wagon in motion? The answer is that there is only one force acting on the wagon, and that is **F**. Consequently, in view of Eq. (1-73), the wagon starts to move (assuming no frictional forces). The reaction −**F** of the wagon on the horse is a force that acts not on the wagon but on the horse.

It is possible to formulate a definition of mass in terms of the third law. Suppose two bodies, *A* and *B*, as a result of their mutual interaction receive the accelerations **a** and −**a**′, respectively. The negative sign in the second case is necessary to make the action and reaction opposite in direction. It is stated that *B* has a/a' times the *mass* of *A*. *A* is chosen to be the standard body and is said to have a mass of unity. To any other body is assigned a mass m if this other body imparts to *A* an acceleration which is m times that which *A*, in the reaction, imparts to it. The ratio of the masses is always equal to the inverse ratio of the accelerations. Similarly, experiment reveals that the accelerations always have opposite signs.[1]

Newton performed extensive experiments with the simple pendulum in order to show that the mass of a body can be measured by its weight.[2] By the *weight* of a body is meant the force with which the earth attracts the body at sea level at a specified latitude. If g is the acceleration of gravity, the weight of a body of mass m is simply mg. In other words, the weight of a body is its mass times an acceleration, the acceleration of gravity.

[1] The mass of a body, as defined here, is termed its *inertial mass*.

[2] The mass of a body, as determined by weighing the body, is called its *gravitational mass*. The gravitational force with which another object such as the earth attracts the body is proportional to the body's gravitational mass. The inertia of the body, however, is measured by its *inertial mass*. Experiment shows that the two masses are equal. This equality follows from the *principle of equivalence*, one of the postulates of the *general theory of relativity*.

1-20. Inertial Systems and the Validity of Newton's Laws. In Fig. 1-23, $O_0x_0y_0z_0$ is a reference system in which Newton's laws hold. Thus, in this system Eq. (1-73) provides an accurate description of the motion of a particle on which a force is being applied. $Oxyz$ is a second reference system, which is moving relative to the first. For simplicity we let Ox remain always parallel to O_0x_0, Oy always parallel to O_0y_0, and Oz always parallel to O_0z_0. The position of O with respect to O_0 is defined by the vector **R**, which has the rectangular components X_0, Y_0, Z_0 in the $O_0x_0y_0z_0$ system. These components are changing, since O is moving relative to O_0. The position of a particle of mass m, with reference to O_0 and O, is defined by the vectors $\mathbf{r}_0$ and $\mathbf{r}$ as shown. If, at this point, we limit ourselves to situations in which the mass is independent of time, the component equations of motion in the $O_0x_0y_0z_0$ system for the particle are, from (1-75),

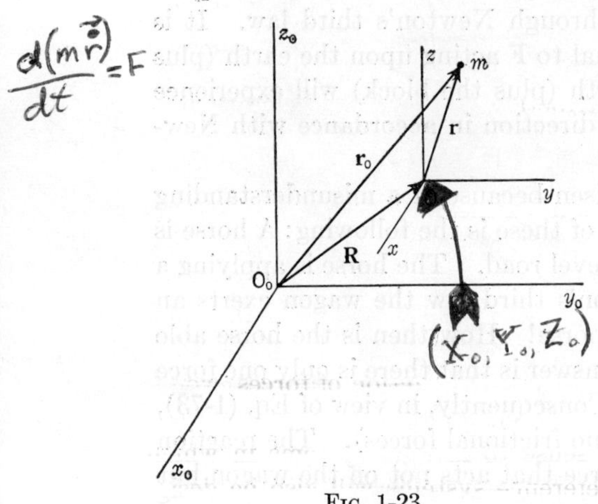

FIG. 1-23

$$m\ddot{x}_0 = F_1 \qquad m\ddot{y}_0 = F_2 \qquad m\ddot{z}_0 = F_3 \tag{1-76}$$

where $\ddot{x}_0$, $\ddot{y}_0$, and $\ddot{z}_0$ are the components of $\ddot{\mathbf{r}}_0$ in the zero system and F_1, F_2, and F_3 are the analogous components of **F**. **F** is assumed to be a function of position only. Now x, y, and z are the components of **r** in the $Oxyz$ system, and are parallel to x_0, y_0, and z_0, respectively. Accordingly

$$x_0 = X_0 + x \qquad y_0 = Y_0 + y \qquad z_0 = Z_0 + z \tag{1-77}$$

from which

$$\ddot{x}_0 = \ddot{X}_0 + \ddot{x} \qquad \ddot{y}_0 = \ddot{Y}_0 + \ddot{y} \qquad \ddot{z}_0 = \ddot{Z}_0 + \ddot{z} \tag{1-78}$$

Substituting these in Eqs. (1-76), we obtain

$$m\ddot{x}_0 = m(\ddot{x} + \ddot{X}_0) = F_1 \tag{1-79}$$

and so on. It is apparent from Eq. (1-79) that the force F_1 gives rise to a motion which, if described in the system $Oxyz$, would be

$$m\ddot{x} = F_1 - m\ddot{X}_0 \tag{1-80}$$

Equation (1-80) is different from the simple equations (1-76), there being an additional inertial reaction term, $-m\ddot{X}_0$, on the right side of (1-80). This arises because of the acceleration of the coordinate system $Oxyz$

relative to $O_0x_0y_0z_0$. The additional term is zero if the origin O is moving with a constant vector velocity with respect to O_0. It is zero also if O is at rest with respect to O_0. In both these cases the second time derivatives of the components of **R** vanish, and the form of the equations of motion is that of Eqs. (1-75). It will be shown later (cf. Chap. 11) that, if $Oxyz$ were rotating with respect to $O_0x_0y_0z_0$, additional reaction terms would be present in the equations of motion referred to $Oxyz$. These would arise because of the centripetal and angular accelerations of $Oxyz$ relative to $O_0x_0y_0z_0$.

In modern parlance, systems in which Newton's laws hold are called *inertial systems*. Thus $O_0x_0y_0z_0$ above is an inertial system, and so also is any other system $Oxyz$, not rotating with respect to $O_0x_0y_0z_0$, traveling with a uniform velocity relative to $O_0x_0y_0z_0$. Strictly speaking, however, the term inertial system originally was applied to a system in which Newton's first law, or the law of inertia, is true. On this basis any system relative to which a particle, not under the action of forces, would describe a straight line is an inertial system. In the present text we shall employ the term in the larger sense to include systems in which all Newton's laws are true. Such reference systems will also be designated as *Newtonian systems*.

It is to be noted that, in the discussion of Eqs. (1-76) to (1-79), no restriction upon the magnitude of the velocity of a particle relative to a coordinate system was explicitly stated. This implies, also, no restriction upon the magnitude of the velocity of one coordinate system relative to another, since if a particle is at rest in one system it necessarily travels with a velocity in the second system which is just the relative velocity of the two reference systems. The *special theory of relativity* does place an upper limit on attainable velocities; this limit is equal to c, the velocity of light. Moreover, it is also shown in special relativity that, if the principle of conservation of linear momentum is to hold at particle velocities approaching c, the mass of the particle must be regarded as varying with the velocity in the manner

$$m = m_0/\sqrt{1 - v^2/c^2} \tag{1-81}$$

in which m_0 is the mass of the particle in a reference system relative to which it is at rest and m is its mass in a system relative to which it is traveling with a velocity v. Since the velocity v in general will depend explicitly upon the time, it follows that m will contain t implicitly in cases where v^2/c^2 is not negligible and thus will have a time rate of change in such instances. (The mass m may have a rate of change also for other reasons; some of these are given in Chap. 8.) Accordingly if m is written as in (1-81), Eq. (1-73) may be assumed to be the correct definition of force even at high velocities. However, the transformation equations involving coordinate systems possessing a high relative velocity do not have the simple form given by (1-77) and (1-78).

It may be said that in Newtonian mechanics the existence of a *preferred system* is implied, in which Eq. (1-73) provides an exact description of the motion of a particle. A system of axes at rest with respect to the earth approximates such a situation for most practical purposes. A reference system at rest with respect to the sun provides an even better approximation. To carry this point of view still further, it is apparent that one arrives eventually at a reference system which is at rest with respect to the average positions of the stars.

The implication of absolute rest contained in the above statements is meaningless from the point of view of relativity. It has been suggested[1] that a better definition of an inertial system, obtained from considerations of general relativity theory, is one in the neighborhood of which the gravitational field vanishes.

1-21. Fundamental Dimensions. The Dimensional Consistency of Equations. The left side of Eq. (1-73) contains three fundamental quantities. These are *mass*, *length*, and *time*. This suggests the possibility of three fundamental *dimensions* corresponding to these quantities. In dimensional notation they are customarily written $[m]$, $[l]$, and $[t]$, respectively. The left side, or inertial reaction, has the dimensions

$$[m][l][t]^{-2}$$

All the other quantities in mechanics have dimensions which can be expressed in terms of these same fundamental dimensions. Thus, for example, the force **F** on the right side of Eq. (1-73) has the dimensions

$$[F] = [m][l][t]^{-2} \tag{1-82}$$

The equation of motion illustrates a very important fact. This is that both sides of a physical equation must have the same dimensions. It makes no physical sense to equate quantities having the dimensions of mass, for instance, to others having the dimensions of length. Consider the equation

$$ac + b = g + \frac{h}{k} + \frac{s}{2 + n} \tag{1-83}$$

The product ac must have the same dimensions as b, g, the ratio h/k, and the ratio $s/(2 + n)$. Furthermore, since the number 2 in the denominator of the last term is a dimensionless number, it is necessary that n be dimensionless also. If the separate groups of terms, such as ac and h/k, have the same dimensions, the equation is said to be *dimensionally*

[1] H. C. Corbin and P. Stehle, "Classical Mechanics," John Wiley & Sons, Inc., New York, 1950.

consistent. This property is often of value in providing a rapid check on the correctness of a physical equation.

Similarly the arguments of functions such as $e^{b/a}$, $\cos \omega t$, and so on, must be dimensionless. (The argument of $e^{b/a}$ is b/a, and ωt is that of $\cos \omega t$.) It is not difficult to see this. When these functions occur in an equation, they are frequently expanded in an infinite series and the expansion must be such that each term of the series has the same dimensions as every other. Consider the function $e^{b/a}$. From elementary calculus if we express e^x in series form, we obtain

$$e^x = 1 + x + \frac{x^2}{2!} + \frac{x^3}{3!} + \cdots$$

Replacing x by b/a, we have

$$e^{b/a} = 1 + \frac{b}{a} + \frac{1}{2!}\left(\frac{b}{a}\right)^2 + \cdots \qquad (1\text{-}84)$$

It is apparent that if b/a were not dimensionless the results would be physically absurd, since each term on the right side of Eq. (1-84) would have different dimensions. If b has the dimensions of length, then a must have the dimensions of length also.

Problems

Problem marked *C* are reprinted by kind permission of the Cambridge University Press.

1-1. Taking the earth to be a sphere of diameter 7,930 miles, find the velocity, in feet per second, of a point at the equator. Consider only motion arising as the result of the daily rotation. Find also the linear velocity of a point at a latitude of 30°.

1-2. Two trains leave the station at an interval of 5 min and are uniformly accelerated in a manner such that they will reach their maximum speeds of 60 mi/hr after having traveled 1 mile. Find the distance the first one has traveled by the time the second starts. Find also the distance between the two trains after both have reached maximum speed.

1-3*C*. A train has a maximum speed of 60 mi/hr. It can accelerate to this speed in 5 min and stop from this speed in $\frac{1}{2}$ mile. Assuming uniform acceleration and deceleration, determine the minimum time taken to travel from one stop to another 10 miles distant if over the central mile the speed must not exceed 10 mi/hr but is unlimited over the remaining portions.

1-4. Two trains of length 75 m and 100 m are running on parallel tracks. The short train moves with twice the speed of the long one. A man sitting in the shorter train observes that it takes the longer train 3 sec to pass by his window. Find the velocities of the two trains if the trains are going in opposite directions.

1-5. A ship is heading due north at 10 mi/hr through a 4 mi/hr current which is directed southeast. Determine the actual course of the ship.

1-6. A ship whose head is pointing due south steams across a current running northwest. At the end of 2 hr it is found that the ship has gone 20 miles in the direction 30° west of south. Find the speed of the ship relative to the water, and the speed of the current.

1-7. A ship is traveling at the rate of 20 knots in a direction 30° north of east. A current of 8 knots has a direction 10° east of south. How far will the ship have gone after 2 hr, and in what direction? (A knot is 1 nautical mile per hour, or 1.1516 land-miles per hour.)

1-8*C*. A ship *A*, steaming due east at 15 knots, sights a second ship *B*, steaming with constant velocity, at a distance of 10 nautical miles and in a direction 30° east of north. Ten minutes later *B* lies in a direction 45° east of north and, after a further 5 min, in a direction 60° east of north. Find the distance *AB* at the last observation, and determine the speed and course of *B* (see Prob. 1-7).

1-9*C*. A cyclist observes that while cycling due north at 15 mi/hr the wind appears to come from due east, but when cycling due east at the same speed the wind appears to come from $22\frac{1}{2}°$ south of east. Determine the velocity of the wind.

1-10. A steamship is traveling at the rate of 15 mi/hr due south. A man on the ship notices that smoke pours out of the funnel in a direction which is due east with respect to the ship. If the speed of the ship is increased to 20 mi/hr, he then observes that the initial direction of the smoke is northeast. Find the velocity of the wind. Assume that as the smoke pours out of the funnel it immediately takes up the velocity of the wind.

1-11. A river has a width b. The water is flowing downstream with a speed that is zero at each bank but increases linearly with distance out from each shore in a manner such that the speed at the center is v_0. A boat capable of a speed v starts at one bank with its prow pointed directly across. Show that by the time the boat reaches the other bank it will have drifted downstream a distance $v_0b/2v$.

1-12. A boat capable of a speed of 5 mi/hr is traveling across a river 1 mile wide. The water in the stream flows with the same speed as the boat. If the boat is pointed in a direction 30° upstream, find the time required for the boat to reach the opposite bank. Find also the distance traveled by the time it reaches the other bank.

1-13*C*. A particle is projected vertically upward and is at a height h after t_1 sec and again after t_2 sec. Prove that

$$h = \tfrac{1}{2}gt_1t_2$$

and that the initial velocity was

$$\tfrac{1}{2}g(t_1 + t_2)$$

1-14*C*. The speed of a train increases at a constant rate A from 0 to V, then remains constant for an interval, and finally decreases to 0 at a constant rate B. If S is the total distance described, prove that the total time occupied is

$$\frac{S}{V} + \frac{1}{2}V\left(\frac{1}{A} + \frac{1}{B}\right)$$

1-15*C*. A bullet, traveling horizontally, pierces in succession three thin screens placed at equal distances a apart. If the time from the first to the second is t_1 and from the second to the third t_2, prove that the retardation (assumed to be consant) is

$$\frac{2a(t_2 - t_1)}{t_1t_2(t_1 + t_2)}$$

and that the velocity at the middle screen is

$$\frac{a(t_1^2 + t_2^2)}{t_1t_2(t_1 + t_2)}$$

1-16. A wheel of radius b is rolling along a muddy road with a speed V. Particles of mud attached to the wheel are being continuously thrown off from all points of the wheel. If $V^2 > bg$, where g is the acceleration of gravity, show that the maximum height above the road attained by the mud will be

$$b + \frac{V^2}{2g} + \frac{b^2 g}{2V^2}$$

1-17. Show that the vectors $\mathbf{a} = 2\mathbf{i} - 4\mathbf{j} - 2\mathbf{k}$ and $\mathbf{b} = 3\mathbf{i} + 4\mathbf{j} - 5\mathbf{k}$ are perpendicular to one another.

1-18. Find the magnitude of the radius vector $\mathbf{r} = 4\mathbf{i} + 3\mathbf{j} + 6\mathbf{k}$. What are its direction cosines?

1-19. Show that the vector $\mathbf{a} = 3\mathbf{i} + 2\mathbf{k}$ lies entirely in a plane perpendicular to the Oy axis.

1-20. The radius vector $\mathbf{r}_1$, to a point P_1, makes angles $\alpha_1, \beta_1, \gamma_1$ with the Ox, Oy, and Oz axes, respectively. Show that

$$\mathbf{r}_1 = r_1(\mathbf{i} \cos \alpha_1 + \mathbf{j} \cos \beta_1 + \mathbf{k} \cos \gamma_1)$$

where r_1 is the scalar magnitude of $\mathbf{r}_1$. Show also, if $\mathbf{r}_2$ is the radius vector to another point P_2, that the angle θ, between $\mathbf{r}_1$ and $\mathbf{r}_2$, can be expressed in the form

$$\cos \theta = \cos \alpha_1 \cos \alpha_2 + \cos \beta_1 \cos \beta_2 + \cos \gamma_1 \cos \gamma_2$$

in which $\alpha_2, \beta_2, \gamma_2$ are the angles which $\mathbf{r}_2$ makes, respectively, with Ox, Oy, and Oz.

1-21. Find the cosine of the angle between the two vectors

$$\mathbf{a} = 3\mathbf{i} + 2\mathbf{j} + \sqrt{3}\,\mathbf{k} \qquad \mathbf{b} = 3\mathbf{i} - \sqrt{2}\,\mathbf{j} + 3\mathbf{k}$$

1-22. A triangle is represented by three vectors $\mathbf{a}$, $\mathbf{b}$, $\mathbf{c}$, joined end to end. If α is the angle opposite $\mathbf{a}$, deduce vectorially that

$$a^2 = b^2 + c^2 - 2bc \cos \alpha$$

1-23. A particle is moving along a path such that its radius vector is

$$\mathbf{r} = (a \sin kt)\mathbf{i} + (b \cos kt)\mathbf{j}$$

where a, b, and k are constants. Show that the path is an ellipse of axes $2a$ and $2b$. Show also, by expressing the acceleration in terms of the vector $\mathbf{r}$, that the acceleration is directed toward the origin at all times.

1-24. Show that the quantities

$$\mathbf{a} = \mathbf{i} \sin \theta + \mathbf{j} \cos \theta \qquad \mathbf{b} = \mathbf{i} \cos \varphi - \mathbf{j} \sin \varphi$$

are unit vectors in the xy plane. Prove also that $\sin (\theta - \varphi) = \cos (\mathbf{a},\mathbf{b})$.

1-25. The vectors $\mathbf{a}$, $\mathbf{b}$, and $\mathbf{c}$ form a closed triangle ABC, where angle A is opposite $\mathbf{a}$, etc. Deduce, by the use of the vector cross product, the sine law of trigonometry.

1-26. Find the locus of points at which the gradient of $ay(x^2 + y^2 + z^2)^{-\frac{3}{2}}$, where a is a constant, is parallel to a line making equal angles with the three coordinate axes.

1-27. Given a force $\mathbf{F} = (x^2 - y^2)\mathbf{i} + 2xy\mathbf{j}$. Find the line integral of $\mathbf{F}$ in the clockwise direction from the origin around the square bounded by the coordinate axes and the lines $x = a$, $y = a$, and back to the origin.

1-28. The vector $\mathbf{b}$ has a constant length b but has its direction in space changing with time. Show that the rate of change of $\mathbf{b}$ is another vector which is always perpendicular to $\mathbf{b}$.

CHAPTER 2

STATICS OF A PARTICLE

2-1. Equilibrium of Forces. It is convenient to divide mechanics into the two branches, *statics* and *dynamics*. Dynamics, as has already been stated, deals with the motion of bodies when forces are acting on them. Statics, on the other hand, is concerned with systems of bodies which are at rest. If forces are to act, the problem of statics is to determine what the configuration of the forces is to be in order that the system of bodies will remain at rest. A system of bodies at rest under the action of forces is said to be in *static equilibrium*. It is this phase of mechanics which will be considered in the present chapter. Moreover we shall be restricted initially to the simplest class of problems, that which deals with the equilibrium of a single particle. Occasionally problems which at first sight seem to be more complicated than this are solvable by the methods of the single-particle case. These will be considered also.

Since the fundamental problem here is the statics of a single particle, it is evident that all forces acting must act at one point, the particle. Newton's second law, which finds quantitative expression in the equation of motion [Eq. (1-73)], states that if a particle is subject to forces it will experience accelerated motion unless the resultant of these forces is zero. Accordingly, the analytical condition for the static equilibrium of a particle subject to forces $\mathbf{F}_1, \mathbf{F}_2, \ldots, \mathbf{F}_n$ can be expressed as

$$\mathbf{F}_1 + \mathbf{F}_2 + \cdots + \mathbf{F}_n = 0 \tag{2-1}$$

Equation (2-1) is equivalent to three scalar equations. If the x, y, and z components of the forces are $X_1, X_2, \ldots, X_n, Y_1, Y_2, \ldots, Y_n$, and $Z_1, Z_2, \ldots, Z_n$, respectively, the three scalar equations are

$$\begin{aligned} X_1 + X_2 + \cdots + X_n &= 0 \\ Y_1 + Y_2 + \cdots + Y_n &= 0 \\ Z_1 + Z_2 + \cdots + Z_n &= 0 \end{aligned} \tag{2-2}$$

These state that:

I. *If a particle is to be in static equilibrium under the action of a number of forces, the algebraic sum of the x components, the y components, and the z components of the forces must separately vanish.*

The types of forces which will be considered in connection with the statics of a particle are the weight of the particle, reactions of constraining surfaces (including frictional forces), and tensions in elastic and inelastic strings. The weight of the particle is assumed to act vertically downward along the plumb line.

It will be found useful to cultivate the point of view of *isolating the body* (or, in the present chapter, of isolating the particle) the equilibrium of which is being investigated. All forces which act upon the body are then to be recognized and appropriately to be taken into account. Accordingly, in Example 2-1 it is apparent that the convenient particle for which to investigate the conditions of static equilibrium is one situated at point C (since the three forces acting all intersect at C). Consequently this particle is isolated, the three adjacent portions of the strings being replaced by the forces T_1, T_2, and W which they transmit to C.

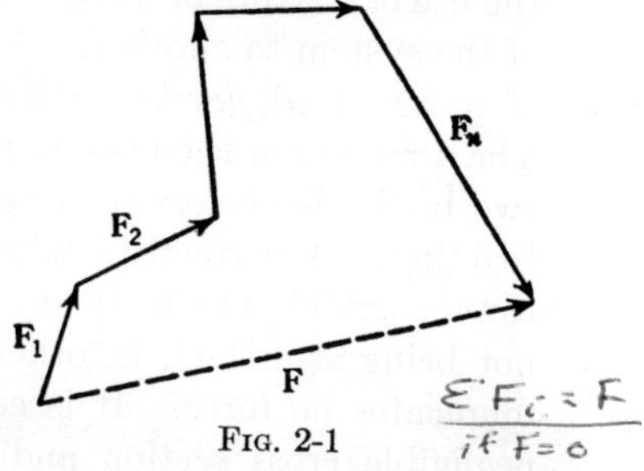

FIG. 2-1

2-2. Polygon of Forces. Triangle of Forces. From Chap. 1 we learned that the sum of a number of vectors $\mathbf{F}_1$, $\mathbf{F}_2$, $\mathbf{F}_3$, . . . , $\mathbf{F}_n$ is found by joining all the vectors end to end so that $\mathbf{F}_2$ has its initial point at the terminal point of $\mathbf{F}_1$, and so on. The sum or resultant of these vectors is a vector $\mathbf{F}$ (dashed in Fig. 2-1) drawn from the initial point of $\mathbf{F}_1$ to the terminal point of $\mathbf{F}_n$. Clearly, if the resultant $\mathbf{F}$ is to be zero, the terminal point of $\mathbf{F}_n$ must join the initial point of $\mathbf{F}_1$, and the system of forces $\mathbf{F}_1$, $\mathbf{F}_2$, . . . , $\mathbf{F}_n$ itself forms a closed polygon.

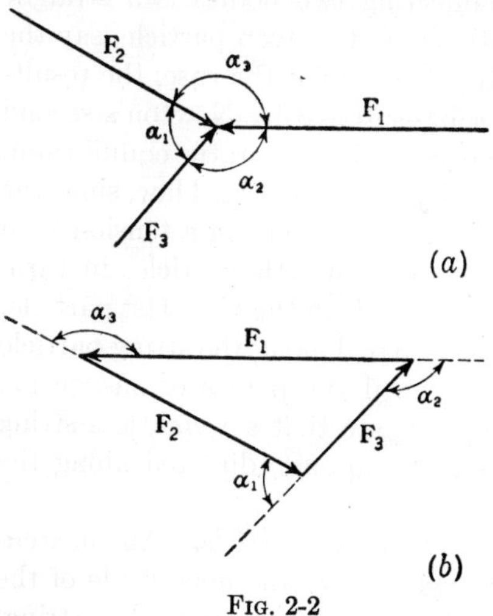

FIG. 2-2

Equilibrium under the action of three forces constitutes a special case. The existence of equilibrium requires that the three vectors form a closed triangle. In order to accomplish this, the lines of action of the three forces must all lie in the same plane. This is to be contrasted to the possibility that, when more than three forces are acting, not all need lie in the same plane. Figure 2-2a shows a particle in equilibrium as the result of the action of three forces $\mathbf{F}_1$, $\mathbf{F}_2$, and $\mathbf{F}_3$, respectively. Figure 2-2b shows the corresponding force triangle, in which the lengths of the sides are proportional to the magnitudes of the forces. The angles designated correspond to the angles between the forces in Fig. 2-2a. Consulting Fig. 2-2b, and from the sine law of elementary trigonometry, we may write

$$\frac{F_1}{\sin \alpha_1} = \frac{F_2}{\sin \alpha_2} = \frac{F_3}{\sin \alpha_3} \tag{2-3}$$

It is useful, also, to state this result verbally:

II. *When a particle is in equilibrium under the action of three forces, the magnitude of each force is proportional to the sine of the angle between the other two forces.*

2-3. The Flexible String. Many systems in static equilibrium involve the use of a string or a chain as a medium to transmit force from one part of the system to another. A force such as that communicated by means of a stretched *flexible* string (or a stretched rod) is called a *tension.* The direction of the tension transmitted by a stretched string connecting two bodies is always such as to tend to draw the two connected bodies together. A perfectly flexible string cannot be under *thrust,* that is, it cannot exhibit the tendency to push the bodies apart. If the string is not being stretched, in other words, if it is not under tension, it communicates no force. It is convenient to idealize the string as being of negligible cross section and consisting of a large number of particles connected in a row.

In many problems only small error is incurred by considering the string to be weightless. If the weight of each particle is negligible, the configuration of a stretched string connecting two bodies is a straight line. Furthermore, the tension on both sides of a given particle, say the ith, will be the same in magnitude. If this were not the case, the resultant force on the ith particle would not be zero and by Newton's second law it would move, thus violating the condition of static equilibrium.

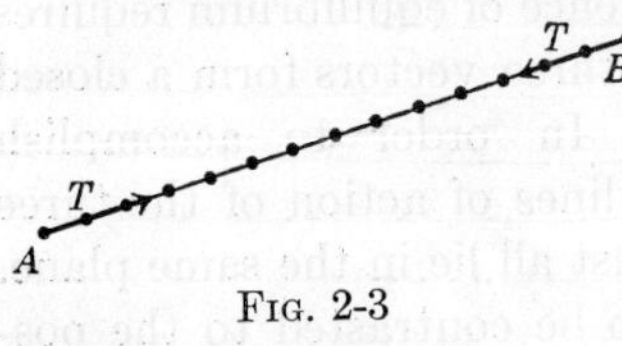

Fig. 2-3

Moreover, by Newton's third law, since the $(i - 1)$st particle is exerting a tension T on the ith particle, the ith particle, in turn, exerts a tension T on the $(i - 1)$st particle. This can be carried along the string particle by particle until the points of suspension (A and B in Fig. 2-3) are reached, and we see that a weightless string under tension T exerts a force T on each support, directed along the string as shown.

Flexible strings may be either *extensible* or *inextensible.* An inextensible string does not change in length regardless of the magnitude of the tension which may be applied to it. The extensible, or *elastic,* string, when subjected to a tension, suffers a change in length which depends

Δl

on the magnitude of the tension. If the tension is zero, the string has its normal length L_0. If the string is under tension T, the change ΔL in its length is related to T through *Hooke's law.* Hooke's law states that:

III. *When an elastic string is subject to a tension, the resulting extension of its length is proportional to the tension.*

We write

$$T = k\,\Delta L \tag{2-4}$$

The constant of proportionality k is called the *elastic constant* (or elastic coefficient) of the string. (Hooke's law is not restricted to flexible elastic strings, but applies equally well to all elastic bodies undergoing stress, whether the stress is a thrust or a tension. The resulting change in a linear dimension of the body is proportional to the stress and is in the direction in which the stress acts. Thus an elastic rod, subject to a thrust along its length, will suffer a decrease in length which is proportional to the thrust.) Occasionally another parameter λ, called the *modulus of elasticity* of the string, is employed in place of k. This quantity λ is just the tension per unit fractional extension of the string. Thus

$$\lambda = \frac{T}{\Delta L/L_0} \tag{2-5}$$

where L_0 is the unstretched length and ΔL is the extension. It is evident that λ is related to k as

$$\lambda = kL_0 \tag{2-6}$$

An examination of Eqs. (2-4) to (2-6) shows that k has the dimensions of force divided by length and that λ has the dimensions of force.

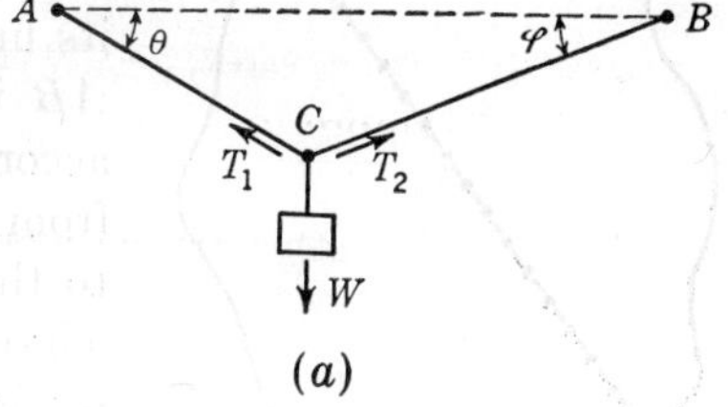

(b)

Fig. 2-4

Example 2-1. A heavy particle of weight W is suspended from two points A and B at the same level by two inextensible weightless strings (see Fig. 2-4*a*), making angles θ and φ with the horizontal. (W is tied at point C.) Find the tensions in the two strings.

The appropriate choice of the particle the equilibrium of which it is desired to investigate is the one situated at point C. Isolating this particle, as in Fig. 2-4*b*, we note that the forces acting may be represented by the tensions $\mathbf{T}_1$ and $\mathbf{T}_2$ in the inclined portions of the string and the tension $\mathbf{W}$ in the vertical portion. From the sine law we have

$$\frac{T_1}{\sin\alpha_1} = \frac{T_2}{\sin\alpha_2} = \frac{W}{\sin\alpha_3} \tag{2-7}$$

Consulting Fig. 2-4*a*, we see that

$$\begin{aligned} \sin \alpha_1 &= \sin (\pi - \alpha_1) = \cos \varphi \\ \sin \alpha_2 &= \cos \theta \\ \sin \alpha_3 &= \sin (\theta + \varphi) \end{aligned} \tag{2-8}$$

and Eq. (2-7) becomes

$$T_1 = \frac{W \cos \varphi}{\sin (\theta + \varphi)} \qquad T_2 = \frac{W \cos \theta}{\sin (\theta + \varphi)} \tag{2-9}$$

An important fact to note is that the weight is tied at point C and is not allowed to slide freely. If the weight had been permitted to slide smoothly, it would have changed its position until the two tensions had become equal in magnitude. (It will be shown later in this chapter that the tension of a weightless string does not vary in the portion of the string which is in contact with a smooth surface.)

2-4. The Rigid Body. Transmissibility of Force. A rigid body is a system of particles in which the distance between any pair of particles remains constant regardless of the forces acting on the system. Such a body is a convenient idealization in many problems. Frequently its use greatly simplifies the mathematical details, and at the same time not much error is incurred by so neglecting the small deformations that actually do occur in real substances when they are being subjected to forces.

A very useful property of rigid bodies lies in the *transmissibility of force* through such bodies. In Fig. 2-5 a force **F** is being applied at point A. The effect will be precisely the same if, instead of being applied at point A, it is applied in the same direction at any other point along its line of action AB. The line of particles AB is a straight line, and the force **F**, in accord with Newton's third law, is transmitted from particle to particle in a manner similar to that for the weightless string of Fig. 2-3. The only difference here is that **F** can be pointing in either direction for the rigid body, that is, the body can be either under thrust or under tension, whereas the flexible string can transmit only a tension. If the body were not rigid, deformations would result from the application of forces and the conditions of equilibrium in general might not be independent of the points of application of the forces.

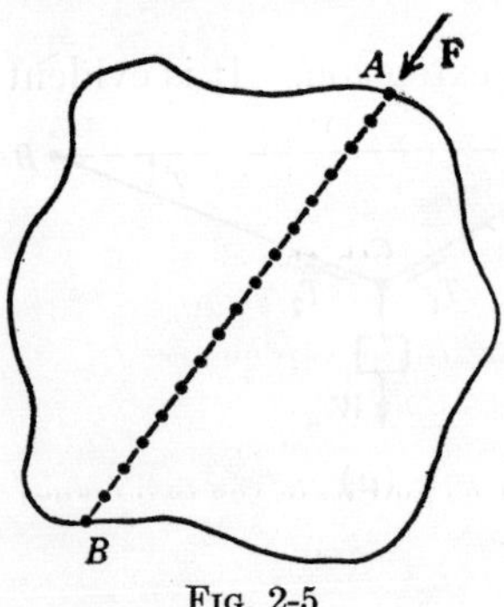

Fig. 2-5

At this early stage the usefulness of the notion of the transmissibility of force along its line of action in a rigid body can be made apparent by considering the case of a rigid sphere. Forces which are acting at a point on the surface of the sphere and in a direction normal to the surface can be considered to be acting at the center of the sphere. In this connection it will be shown later that the weight of the sphere is a force which also can be regarded as acting at the geometrical center provided that the sphere is homogeneous. Consequently in many situations the static equilibrium of a rigid spherical body constitutes a problem identical to that of the statics of a particle.

2-5. Smooth Constraining Surfaces. Many examples in statics involve the contact of bodies with surfaces. Usually such surfaces exert forces on the bodies with which they are in contact. In Fig. 2-6 a block of mass m rests on a flat horizontal table. The weight mg of the block is a force which can be assumed to act vertically downward at the center of the block. However, since the block does not move, the table must exert an opposing force which is equal in magnitude to mg. This is included in R in Fig. 2-6.

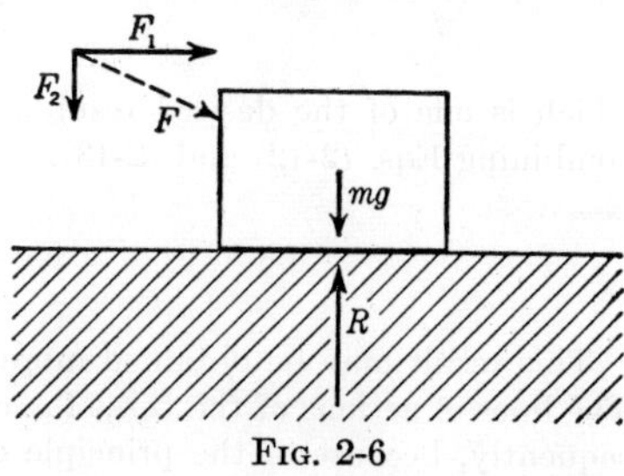

FIG. 2-6

If the table in Fig. 2-6 has a *smooth surface*, any reaction which the surface of the table may exert upon the block must always be normal to the surface of the table. This behavior constitutes the definition of a smooth constraining surface. If another force **F** is applied to the block in a direction such that it has components F_1 and F_2 parallel to and normal to the surface, respectively, the block will experience accelerated motion in the direction of F_1 in accordance with Newton's second law. The only reaction which the table exerts (R in the figure) is normal to the surface and has the magnitude $mg + F_2$.

Example 2-2. A sphere of weight W rests between two smooth inclined planes, of inclination α_1 and α_2 with the horizontal, as shown in Fig. 2-7. Find the forces being exerted on the ball by the two planes.

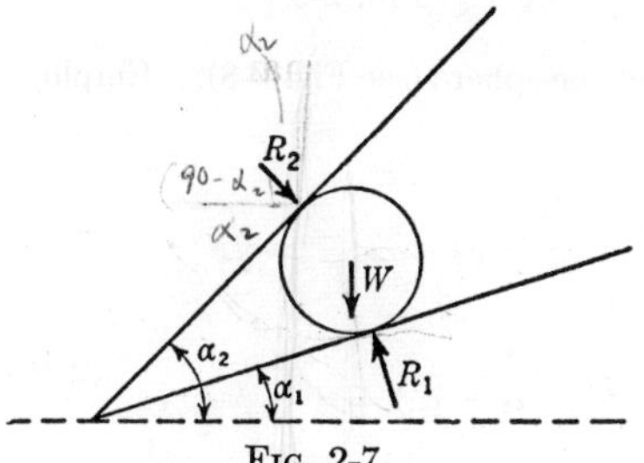

FIG. 2-7

We start by isolating the ball. It is clear that the forces acting on it are its weight W and the reactions R_1 and R_2 of the two planes on the ball. Since the planes are perfectly smooth, these reactions must be normal to their surfaces. The quantities which are known are W, α_1, and α_2, and it remains to determine R_1 and R_2 in terms of these. The ball is in equilibrium under the action of the three forces W, R_1, and R_2, and consequently, taking horizontal and vertical components of these, we have

$$R_2 \cos \alpha_2 + W = R_1 \cos \alpha_1 \tag{2-10}$$

$$R_2 \sin \alpha_2 = R_1 \sin \alpha_1 \tag{2-11}$$

from which

$$R_2 = R_1 \frac{\sin \alpha_1}{\sin \alpha_2} \tag{2-12}$$

Combining Eq. (2-12) with Eq. (2-10), we obtain

$$R_1(\cos \alpha_1 \sin \alpha_2 - \cos \alpha_2 \sin \alpha_1) = W \sin \alpha_2$$

or

$$R_1 = \frac{W \sin \alpha_2}{\sin (\alpha_2 - \alpha_1)} \tag{2-13}$$

which is one of the desired results. The remaining reaction, R_2, can be found by combining Eqs. (2-12) and (2-13). We have

$$R_2 = \frac{W \sin \alpha_1}{\sin (\alpha_2 - \alpha_1)} \tag{2-14}$$

This result may be obtained more directly by using the sine law of trigonometry. The lines of actions of the three forces all intersect at the center of the sphere. Consequently, because of the principle of transmissibility of forces in a rigid body, the

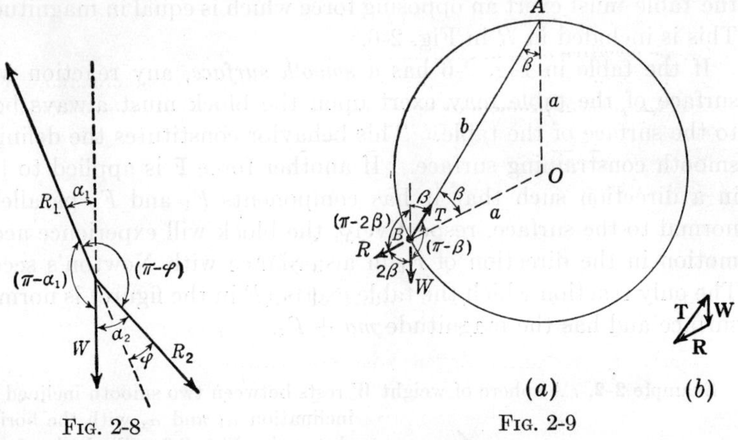

FIG. 2-8

FIG. 2-9

forces all can be regarded as acting at the center of the sphere (see Fig. 2-8). Employing the sine law, we have

$$\frac{R_1}{\sin \alpha_2} = \frac{R_2}{\sin (\pi - \alpha_1)} = \frac{W}{\sin (\pi - \varphi)}$$

or

$$\frac{R_1}{\sin \alpha_2} = \frac{R_2}{\sin \alpha_1} = \frac{W}{\sin \varphi} = \frac{W}{\sin (\alpha_2 - \alpha_1)} \tag{2-15}$$

from which Eqs. (2-13) and (2-14) follow at once.

Example 2-3. A bead of weight W is constrained to slide on a smooth circular loop of wire the plane of which is vertical. The radius of the loop is a. The bead is further connected, as shown in Fig. 2-9a, to the top of the loop by a light, inextensible string of length b. Find the tension in the string and the reaction of the wire on the bead.

Since the quantities given are a, b, and W, it will be necessary to express the tension T and the reaction R in terms of these quantities. In Fig. 2-9a, point O is the center of the circle; the bead is at B and is connected to point A by means of the string. The three forces T, R, and W are seen to act at the point B. Consequently we may isolate the bead and employ the sine law. This yields

$$\frac{T}{\sin 2\beta} = \frac{R}{\sin (\pi - \beta)} = \frac{W}{\sin (\pi - \beta)}$$

or

$$\frac{T}{\sin 2\beta} = \frac{R}{\sin \beta} = \frac{W}{\sin \beta}$$

Hence we have the results

$$T = \frac{W \sin 2\beta}{\sin \beta} = 2W \cos \beta \tag{2-16}$$

and

$$R = W \tag{2-17}$$

Since β is not given directly, it is desirable to express β in terms of the quantities a and b. Thus $\cos \beta = b/2a$.

It is observed that the direction of R was correctly chosen initially to point outward. If it had been arbitrarily chosen to point inward, the result (2-17) would have turned out to be negative in sign, thus showing that R is oppositely directed. However, it is not difficult in most cases to guess the correct directions beforehand. This can be seen in the present problem by consulting the vector triangle shown in Fig. 2-9b. The particle is in equilibrium; so the force triangle (similar to OAB) must consist of the three vectors **T**, **W**, and **R** joined end to end. The terminal point of **R**, in Fig. 2-9b, must intersect **T** at its initial point.

2-6. Rough Constraining Surfaces. Static Friction. If a weight is resting against a *rough surface*, it is possible for the reaction of the surface to have a component in the plane of contact, in addition to the normal component mentioned in Sec. 2-5. The practical importance of this is very easy to see since the smooth constraint of the previous section is a convenient idealization and in reality does not exist. In nature there are no perfectly smooth bodies. The reaction component, which is tangent to the contact surfaces, is called the *frictional reaction* of the surface, or force of friction exerted upon the body by the surface.

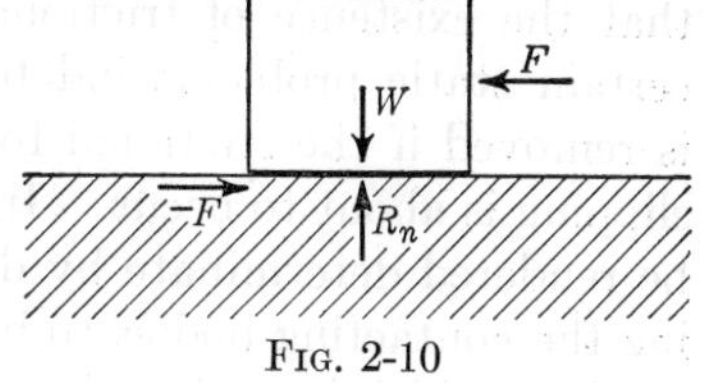

Fig. 2-10

The simple experiment shown in Fig. 2-10 serves to illustrate the frictional force. In Fig. 2-10, a block of weight W is resting on a rough, horizontal table. For the simple case in which no forces other than the weight W are applied, we know that the block will remain at rest. Consequently, the reaction of the surface will be a force R_n, equal in magnitude but opposite in direction to W.

Suppose now a small force F, as shown, is applied horizontally. We know from experiment that the block will remain at rest. The only explanation of the fact that there is no motion is to state that the rough surface exerts an equal and opposite force $-F$ on the block. If this were not the case, then by Newton's second law there would be motion in the direction of the applied force. We also know from experiment that, as F is further increased, the block will not move until a certain critical value of F is reached. This means that the opposing frictional reaction varies over the same range in a way such as always to be equal in magnitude to F.

However, the fact that motion eventually does occur shows that the frictional force finally attains a value beyond which it cannot further increase (see Fig. 2-11). The classical interpretation of experiments with frictional forces is that the critical value F_L for a given block is very nearly independent of the area of contact of the block with the table and essentially depends only on the nature of the contacting substances. In addition it is concluded that the critical value F_L which is necessary to move the block, is directly proportional to the force pressing the block against the table (in this case the force is just the weight of the block). A *coefficient of static friction* can be defined which is approximately constant for a given pair of contact surfaces and approximately independent of the total normal force. This coefficient is defined by the equation

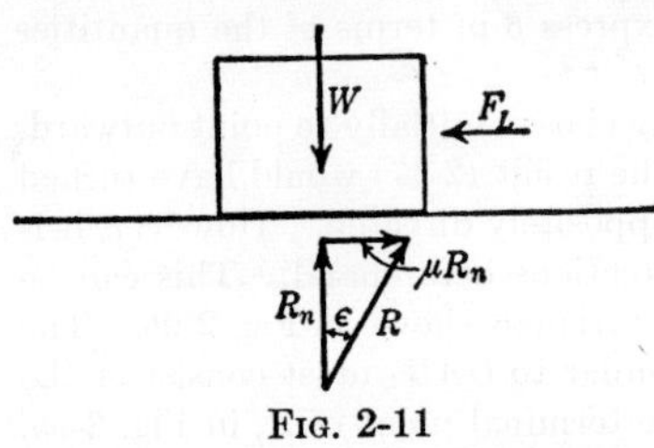

FIG. 2-11

$$F_L = \mu R_n \tag{2-18}$$

where R_n is the normal component of the reaction (here equal to the weight W) and μ is the coefficient of static friction. It is to be emphasized that μ is defined only when slipping is about to occur. This means not that the frictional force occurs only in this limiting case but that only here is it equal to μR_n. It will be pointed out in a subsequent chapter that the existence of frictional forces is sometimes sufficient to render certain static problems indeterminate. Usually this indeterminateness is removed if the frictional forces have their limiting values, namely, if slipping is about to occur. In the nonlimiting case the problem can also be rendered determinate by departing from the idealized case of assuming the contacting bodies to be perfectly rigid and considering the deformations which in reality do occur.

Equation (2-18) states that the ratio of the critical value of the tangential applied force to the normal reaction is a constant, characteristic of the two surfaces and equal to μ the coefficient of static friction. The actual reaction R is the resultant of R_n and μR_n, the normal and frictional reactions, respectively. The *angle of friction*, ϵ, is defined as

$$\tan \epsilon = \frac{\mu R_n}{R_n} = \mu \tag{2-19}$$

and is a constant for a given pair of surfaces. ϵ and μ can be determined for two given surfaces by the action of an inclined plane. In Fig. 2-12a, α is the angle of inclination (less than ϵ) of the plane, and W is the weight of the block, acting vertically downward. The force which tends to make the block slide down the plane is just the component of W along the plane, or $W \sin \alpha$. The frictional reaction which prevents this is R_t (equal in magnitude to $W \sin \alpha$). If we increase the angle α, we find that the critical value of α is ϵ, the angle of friction. From the classical viewpoint this angle is independent of the weight of the block and of the area of contact. The frictional reaction, in the limiting case, is equal to μR_n, that is, in magnitude it is equal to $\mu W \cos \epsilon$ (Fig. 2-12b).

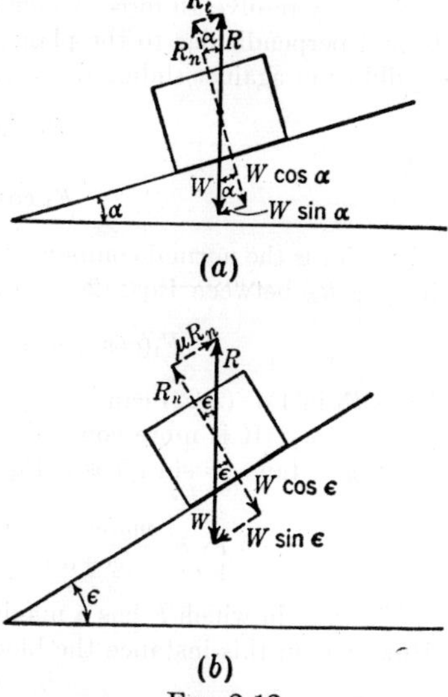

FIG. 2-12

The classical ideas concerning static friction which are expressed above and in part contained in Eq. (2-18) will be adopted in the present text for the solution of problems.[1] Before considering applications in which frictional forces occur, it will be well to express a word of caution. Although, in the above case of the block about to slide down the plane, the frictional force was directed up the plane, this is not always the case. The frictional force will always be such as to *oppose* the resultant of the tangentially applied forces[2] and thus may point down the plane, up the plane, or in any other direction which is parallel to the plane.

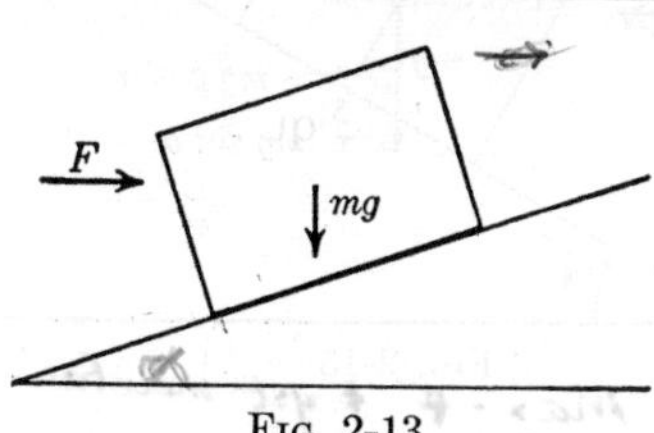

FIG. 2-13

Example 2-4. A cubical block of mass m rests on a rough plane inclined at an angle α to the horizontal, under the action of a force F applied horizontally as shown in Fig. 2-13 and whose line of action passes through the geometrical center of the block. The coefficient of static friction is μ. Find the minimum and maximum values of F consistent with static equilibrium.

We shall first investigate the case of minimum F. F will have a minimum value F_1

[1] The views here stated comprise the so-called classical interpretation of experiments involving frictional forces. The accuracy of some of the classical laws has been questioned recently. See, in this regard, Palmer's article, *Am. J. Phys.*, **17**: 181, 327, 336 (1949).

[2] See Sec. 5-10 for the definition of an *applied force* in the sense in which the term is employed here.

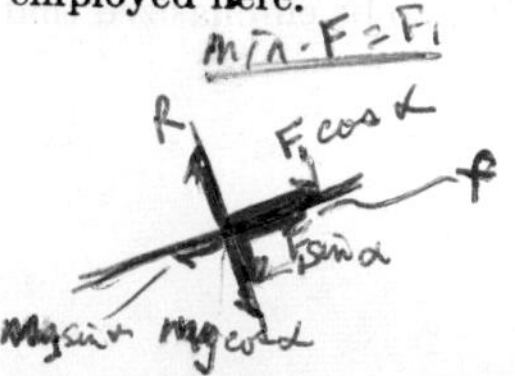

if the block is at the point of slipping down the plane. The forces acting upon the block are F_1 (to be found), the weight mg, and the reaction (not shown) of the plane on the block.

We may resolve all forces which are acting upon the block into components parallel to and perpendicular to the plane, as shown in Fig. 2-14. The condition of limiting equilibrium against sliding down the plane requires that

$$F_1 \sin \alpha + mg \cos \alpha = R_N \tag{2-20}$$

and

$$F_1 \cos \alpha + \mu R_N = mg \sin \alpha \tag{2-21}$$

where R_N is the normal component of the reaction R of the plane on the block. Eliminating R_N between Eqs. (2-20) and (2-21), we obtain

$$F_1(\cos \alpha + \mu \sin \alpha) = mg(\sin \alpha - \mu \cos \alpha) \tag{2-22}$$

Thus F_1 in Eq. (2-22) can be expressed as mg times a quantity which is a function of μ and of α. It is more convenient, however, to make use of the angle ϵ of friction. Since $\mu = \tan \epsilon = \sin \epsilon / \cos \epsilon$, Eq. (2-22) becomes

$$F_1 = \frac{mg(\cos \epsilon \sin \alpha - \sin \epsilon \cos \alpha)}{(\cos \alpha \cos \epsilon + \sin \epsilon \sin \alpha)} = mg \tan (\alpha - \epsilon) \tag{2-23}$$

The case in which F has a maximum value F_2 can be obtained in a similar fashion. However, in this instance the block is at the point of sliding up the plane, and con-

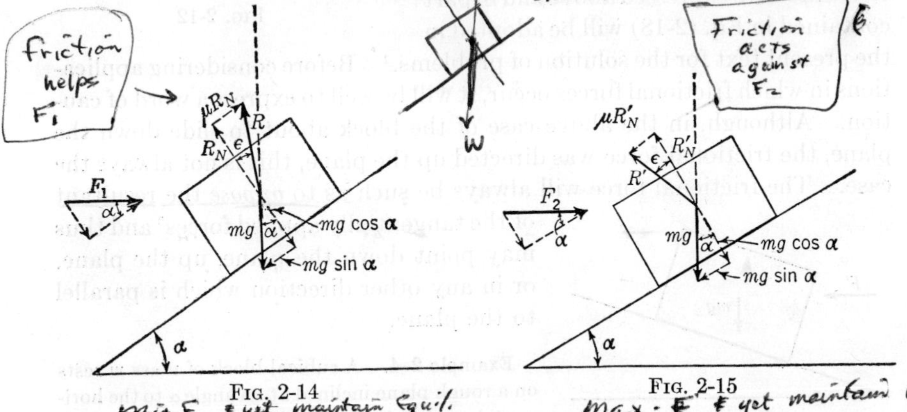

Fig. 2-14 Fig. 2-15

sequently the frictional force is directed down the plane in the direction opposite to that in which motion is about to occur. The forces are shown in Fig. 2-15. Again taking components parallel and perpendicular to the plane, we obtain

$$F_2 \sin \alpha + mg \cos \alpha = R'_N \tag{2-24}$$

and

$$F_2 \cos \alpha = \mu R'_N + mg \sin \alpha \tag{2-25}$$

In the same manner as before we obtain, finally,

$$F_2 = mg \tan (\alpha + \epsilon) \tag{2-26}$$

It is worthwhile to notice the difference in the directions of the resultant reaction R in the two cases. This is evident in Figs. 2-14 and 2-15. It is to be emphasized that the angle between R and R_N, in the case of limiting equilibrium, is always the angle ϵ of static friction.

The sine law can also be employed to solve this problem. We shall consider only the case of minimum F. The lines of action of the three forces $\mathbf{F}_1$, mg, and $\mathbf{R}$ pass through the geometrical center of the block, and the problem is that of the equilibrium of a particle under the action of three forces. From Fig. 2-16 we have

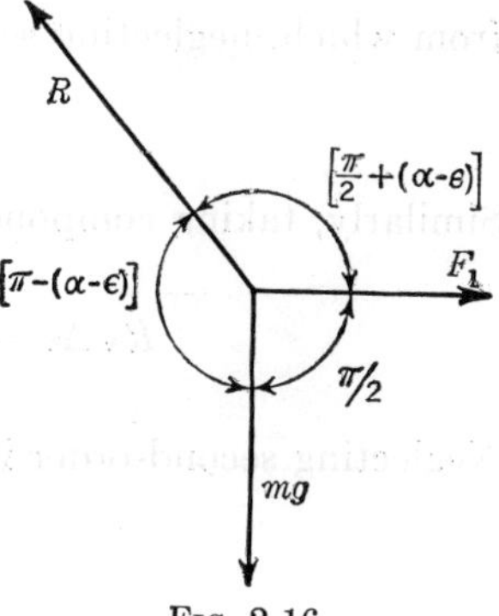

FIG. 2-16

$$\frac{F_1}{\sin[\pi - (\alpha - \epsilon)]} = \frac{mg}{\sin[\pi/2 + (\alpha - \epsilon)]}$$

$$\frac{F_1}{\sin(\alpha - \epsilon)} = \frac{mg}{\cos(\alpha - \epsilon)} \tag{2-27}$$

from which

$$F_1 = mg \tan(\alpha - \epsilon) \tag{2-28}$$

2-7. String in Contact with a Rough Curved Surface. It is common to everyone's experience that a relatively small force F_1, applied to a rope, can often be made to support a much larger force F_2 by merely wrapping the rope a few times about a post (see Fig. 2-17). The only limiting factor is that the maximum tension capable of being sustained by the rope must be greater than F_2.

Figure 2-18 shows, in exaggerated fashion, a small increment Δs (curved line AB in the figure) of the string in contact with an arbitrary

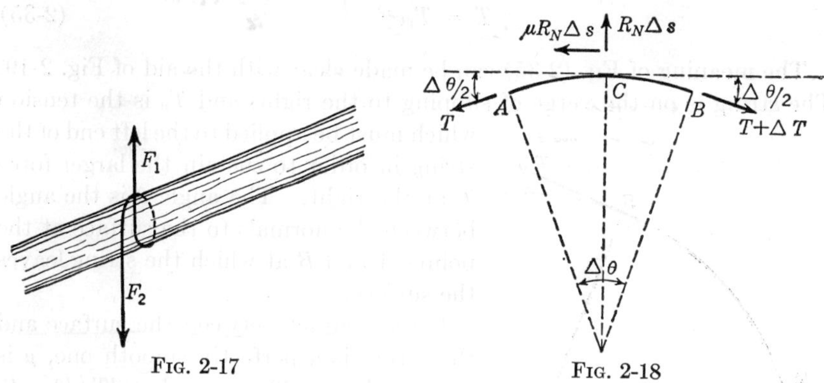

FIG. 2-17 FIG. 2-18

rough surface. The coefficient of static friction is μ. The string is on the verge of slipping to the right. If the normal reaction of the surface is R_N per unit length of the string, there will be a frictional force on Δs, directed to the left, of magnitude $\mu R_N \Delta s$. It is sufficiently accurate to assume both these to act at the mid-point C of the small increment Δs as shown. Furthermore, Δs is in static equilibrium, and consequently the sum of the components, in any direction, of all the forces must be zero.

Thus, considering components tangent to the surface, we have

$$T \cos \frac{\Delta\theta}{2} + \mu R_N \, \Delta s = (T + \Delta T) \cos \frac{\Delta\theta}{2} \tag{2-29}$$

from which, neglecting second-order infinitesimals,

$$\mu R_N \, \Delta s = \Delta T \tag{2-30}$$

Similarly, taking components normal to the surface at C, we have

$$R_N \, \Delta s = T \sin \frac{\Delta\theta}{2} + (T + \Delta T) \sin \frac{\Delta\theta}{2} \tag{2-31}$$

Neglecting second-order infinitesimals, we have

$$R_N \, \Delta s = T \, \Delta\theta \tag{2-32}$$

Combining Eqs. (2-30) and (2-32), and passing to the limit as $\Delta\theta$ becomes very small, we obtain, finally,

$$\frac{dT}{T} = \mu \, d\theta \tag{2-33}$$

This integrates to

$$\log_e T = \mu\theta + \log_e T_0 \tag{2-34}$$

to

$$T = T_0 e^{\mu\theta} \tag{2-35}$$

The meaning of Eq. (2-35) can be made clear with the aid of Fig. 2-19. The string is on the verge of slipping to the right, and T_0 is the tension which must be applied to the left end of the string in order to sustain the larger force T at the right. The angle θ is the angle between the normals to the surface at the points A and B at which the string leaves the surface.

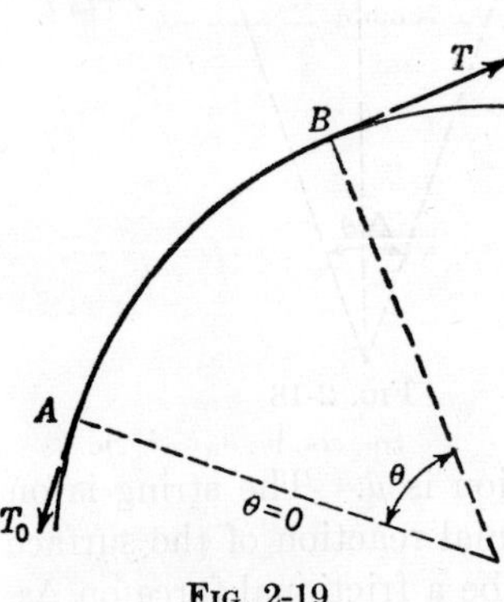

FIG. 2-19

If the contact between the surface and the string is a perfectly smooth one, μ is zero and Eq. (2-33) states that $dT/d\theta = 0$. This means that the tension is not changed by a contact with a smooth surface. The same result can be seen with the aid of Eq. (2-35).

It is sometimes useful to know the normal reaction of the surface in terms of the radius of curvature of the surface. It can be found from Eq. (2-31). On passing to the limit where $\Delta\theta$ becomes small, Eq. (2-31) becomes

$$R_N = \frac{T}{ds/d\theta} \tag{2-36}$$

or

$$R_N = \frac{T}{r} \tag{2-37}$$

where r is the radius of curvature ($= ds/d\theta$) of the surface.

Example 2-5. A string with weights W_2 and W_1 ($W_2 > W_1$) connected to the ends as shown in Fig. 2-20 rests in limiting equilibrium across a rough log in a plane normal to the axis of the log. The coefficient of static friction between the log and the string is $\frac{1}{2}$. Find the ratio of W_2 to W_1.

The angle between the normals at the points at which the strings leave the log is π. Consequently, from Eq. (2-35)

$$\frac{W_2}{W_1} = e^{\pi/2} = e^{1.57} = 4.81$$

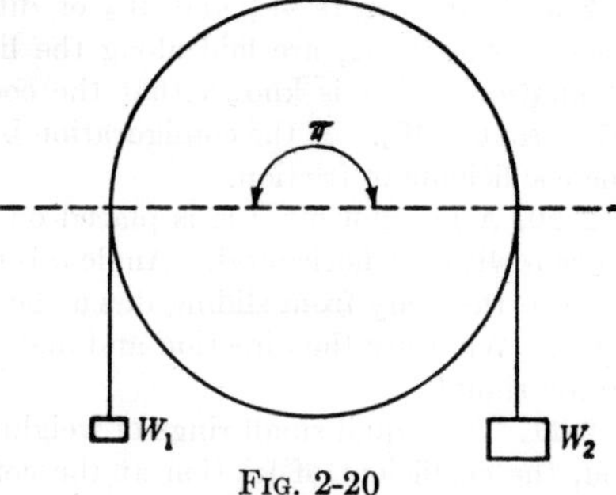

FIG. 2-20

If the string is wrapped completely around an additional time, namely, if $\theta = 3\pi$, the ratio W_2/W_1 assumes the much larger value of approximately 111. From this it is clear that considerable advantage may be obtained by even a small angle of contact (θ in Fig. 2-19) of a rope with a rough curved surface.

Problems

Problems marked C are reprinted by kind permission of the Cambridge University Press.

2-1. A weight of 200 lb is suspended by two weightless inextensible strings, each of which makes an angle of 30° with the horizontal. Find the tension in each string.

2-2. A weight of 50 lb is tied to two points A and B, 5 ft apart at the same level. The natural length of each string is 2.5 ft, and the modulus of elasticity is 100 lb. What is the distance below the level AB at which the hanging weight will be in equilibrium? HINT: Use an approximate method such as Newton's rule.

2-3. A horse capable of exerting a pull of 800 lb attempts to drag a weight of 10,000 lb along a rough road. The coefficient of friction is 0.4. A crane is available to help the horse. If a vertical cable from the crane is attached to the weight, find the tension in the cable such that the horse will just be able to start the weight in motion.

2-4. An ant falls into a bowl which has the shape of an inverted spherical cap of radius b. How deep is the bowl if the ant can just crawl out, the coefficient of friction between the bowl and its feet being $\frac{1}{3}$?

2-5. A body of weight W rests on a horizontal plane, the angle of static friction between the contact surfaces being ϵ. Find the direction and magnitude of the least force that will move the body.

2-6. A smooth inextensible string of length b is attached to two points which are at the same level and which are a distance a ($a < b$) apart. A heavy bead of mass m is free to slide on the string. If a horizontal force F is applied such that the bead is in static equilibrium directly beneath one of the points of attachment of the string, show that the tension in the string is

$$\frac{mg(a^2 + b^2)}{2b^2}$$

Find also the magnitude of F in terms of a, b, m, g.

2-7. A weight W is hanging by three light inextensible strings of equal length b. The strings are hanging from three points which comprise the vertices of an equilateral triangle of side a. Find the tension in each string.

2-8. Two equal uniform spheres of weight W and radius a rest in a smooth hemispherical bowl of radius b. Find the force between the two spheres and also the reaction of the bowl on each sphere.

2-9. Two weights W_1 and W_2 of different material, connected by a weightless inextensible string, are laid along the line of greatest slope of an inclined plane of inclination α. It is known that the coefficient of friction for the upper one W_2 is twice that of W_1. If the configuration is one of limiting static equilibrium, determine the coefficients of friction.

2-10. A body of mass m is placed on a rough plane which is inclined at an acute angle α with the horizontal. Angle α is greater than the angle ϵ of friction. In order to keep the body from sliding down the plane, it is necessary to apply an additional force. What are the direction and magnitude of the least force required to maintain equilibrium?

2-11. Two equal small rings of weight W are free to slide along a rough horizontal rod, the coefficient of friction at the contacts being μ. The weights are connected by a smooth weightless inextensible string of length b. A weight W_1 is free to slide along the string. Find the maximum distance between the two rings consistent with static equilibrium.

2-12. Two weights W_1 and W_2 are connected by a light inextensible string which passes over a smooth peg ($W_2 > W_1$). W_2 is in contact with a rough inclined plane, below the peg, of inclination α to the horizontal, and α is less than the angle ϵ of friction. Show that the extreme values of the angle β which the string from W_2 to the peg may make with the normal to the plane are determined by

$$\sin(\beta + \epsilon) = \frac{W_2}{W_1}\sin(\epsilon \pm \alpha)$$

2-13. The two opposite sides of a hill are rough planes which are inclined at angles α and β to the horizontal. A weight W on one side is connected by a weightless string over a smooth pulley at the top to another unequal weight on the other side. The configuration is one of limiting equilibrium, and W is about to slide up the hill. Find the maximum weight which can be added to W without disturbing the equilibrium. The angle of friction between each weight and the hill is ϵ.

$$\frac{W \sin 2\epsilon \sin(\alpha + \beta)}{\sin(\alpha - \epsilon)\sin(\beta - \epsilon)}$$

2-14C. Two rings of equal weight, connected by a weightless inextensible string, can slide on two fixed rough rods which are in the same vertical plane and are inclined toward each other at equal angles α to the horizontal. Deduce that, if μ is the coefficient of friction, the extreme angle θ which the string can make with the horizontal is given by

$$\tan\theta = \frac{\mu}{\sin^2\alpha - \mu^2\cos^2\alpha}$$

2-15C. Two beads of weights W_1 and W_2 can slide on a smooth wire bent into the form of a circle in a vertical plane. They are connected by a light inextensible string which subtends an angle 2β at the center of the circle when the beads are in equilibrium on the upper half of the wire. Prove that the inclination α of the string to the horizontal is given by

$$\tan \alpha = \frac{W_1 - W_2}{W_1 + W_2} \tan \beta$$

2-16. Two weights W_1 and W_2 hang in limiting equilibrium from a weightless string which passes over a rough circular cylinder the axis of which is in a horizontal position. The string is in a plane perpendicular to the axis of the cylinder. If W_1 is on the point of descending, what weight, expressed in terms of W_1 and W_2, may be added to W_2 without disturbing the equilibrium.

2-17. A mountain climber of mass m is roped to a guide of equal mass on the side of a mountain which may be taken to be an inverted hemisphere of radius b. If the length of the rope is $b\alpha$, where α is a constant, how far down the mountain (measured along the circumference) can the mountain climber descend before both he and the guide fall to the bottom? The angle of friction between each man and the mountain and between the rope and the mountain is ϵ. Assume the rope to be massless and in contact with the mountain along its entire length.

CHAPTER 3

STATICS OF RIGID BODIES

3-1. Introduction. The material of the preceding chapter was concerned with the static equilibrium of a single particle and with certain examples involving extended bodies all of which were reducible to the single-particle case. It will be remembered that the criterion for the applicability of the single-particle methods was that the lines of action of the forces applied to one body must intersect in a single point. It is true that certain of the problems at the end of Chap. 2 involved such a situation as a pair of bodies connected by a string and that the forces acting on the entire system did not all intersect in a single point. However, each body of the system, when considered separately, was indeed acted upon by forces the lines of action of which all intersected in a single point, and equations of static equilibrium could be written for each body separately, each being considered as a single particle (see, for example, Probs. 2-8, 2-9, 2-11).

In the present chapter, the statics of rigid bodies will be considered for cases in which the lines of action of the forces applied to one body do not all intersect at one point. The additional feature which is introduced by such a set of forces is a tendency to cause rotation, as well as translation, of the body. Consequently, if we are interested in the conditions for the static equilibrium of a body which is subjected to an arbitrary set of forces, it is necessary to be concerned not only with equilibrium against translation but with equilibrium against rotation as well.

MASS CENTERS

3-2. Center of Mass of a System of Particles. Of great usefulness in the solution of problems in both statics and dynamics is the concept of the *center of mass*, or *centroid*, of a body. We define first the center of mass of a system of n particles and then determine its coordinates with respect to an arbitrary cartesian coordinate system $Oxyz$ (see Fig. 3-1). The radius vectors to each of a system of n particles with reference to these axes are $\mathbf{r}_1, \mathbf{r}_2, \ldots, \mathbf{r}_n$, respectively. Let the masses of these particles be $m_1, m_2, \ldots, m_n$. The vector $\mathbf{r}_i$ is the radius vector to the

ith particle. Selecting some other point C, draw the radius vector $\mathbf{r}_c$ from O to C and the radius vector $\mathbf{r}_{ci}$ from C to the ith particle. The three vectors form a closed triangle the equation of which is

$$\boxed{\mathbf{r}_i = \mathbf{r}_c + \mathbf{r}_{ci}}$$

Multiplying through by m_i, we have

$$m_i\mathbf{r}_i = m_i\mathbf{r}_c + m_i\mathbf{r}_{ci} \tag{3-1}$$

A similar equation may be written for each of the remaining particles, and, adding them all together, we obtain

$$m_1\mathbf{r}_1 + m_2\mathbf{r}_2 + \cdots + m_n\mathbf{r}_n = (m_1 + m_2 + \cdots + m_n)\mathbf{r}_c + m_1\mathbf{r}_{c1} + m_2\mathbf{r}_{c2} + \cdots + m_n\mathbf{r}_{cn}$$

or, in a much simpler notation,

$$\sum_{i=1}^{n} m_i\mathbf{r}_i = \left(\sum_{i=1}^{n} m_i\right)\mathbf{r}_c + \sum_{i=1}^{n} m_i\mathbf{r}_{ci} \tag{3-2}$$

The vector $\mathbf{r}_c$ is not being summed over since it is the same no matter which particle is being discussed, in other words, it does not possess any subscripts over which a summation is being taken. Such summations over vectors are vector sums the elements of which are combined according to the parallelogram law of addition.

If the second summation on the right side of Eq. (3-2) is zero, that is, if

$$\boxed{\sum_{i=1}^{n} m_i\mathbf{r}_{ci} = 0} \tag{3-3}$$

then point C is defined to be the center of mass. If this is the case, it is customary to replace $\mathbf{r}_c$ by the symbol $\bar{\mathbf{r}}$, a vector $\mathbf{r}$ with a bar over it. Thus, if Eq. (3-3) is true, we have, from Eq. (3-2),

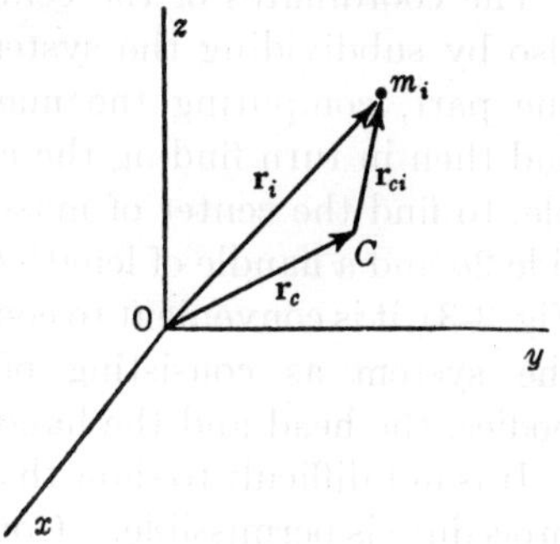

Fig. 3-1

$$M\bar{\mathbf{r}} = \sum_{i=1}^{n} m_i\mathbf{r}_i \tag{3-4}$$

in which M is the total mass of all of the particles $\left(= \sum_{i=1}^{n} m_i\right)$. The quantity $\bar{\mathbf{r}}$ is the radius vector from the arbitrary origin O to the center of mass.

Usually it is more useful to employ the rectangular components $\bar{x}$, $\bar{y}$, and $\bar{z}$ of $\bar{\mathbf{r}}$. These are

$$\bar{x} = \frac{\sum_{i=1}^{n} m_i x_i}{M} \qquad \bar{y} = \frac{\sum_{i=1}^{n} m_i y_i}{M} \qquad \bar{z} = \frac{\sum_{i=1}^{n} m_i z_i}{M} \tag{3-5}$$

A simple example is to find the center of mass of two particles having masses m_1 and m_2. Let the line joining them be the x axis (see Fig. 3-2). Since the y and z coordinates of the two particles are both zero, it is clear from Eqs. (3-5) that the center of mass will lie on the x axis along the line joining them. The x coordinate of the center of mass is

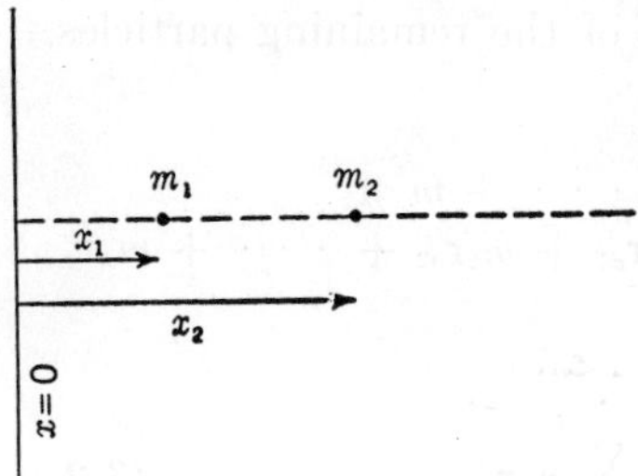

FIG. 3-2

$$\bar{x} = \frac{m_1 x_1 + m_2 x_2}{m_1 + m_2} \tag{3-6}$$

For the particular case in which the two particles have the same mass, the center of mass lies at the mid-point of the line joining the two particles.

The coordinates of the center of mass of a system can be determined also by subdividing the system, in an arbitrary manner, into more than one part, computing the mass center for each of the parts separately, and then in turn finding the center of mass of these centers. For example, to find the center of mass of a mallet consisting of a cubical head of side $2a$ and a handle of length b (see Fig. 3-3), it is convenient to consider the system as consisting of two bodies, the head and the handle.

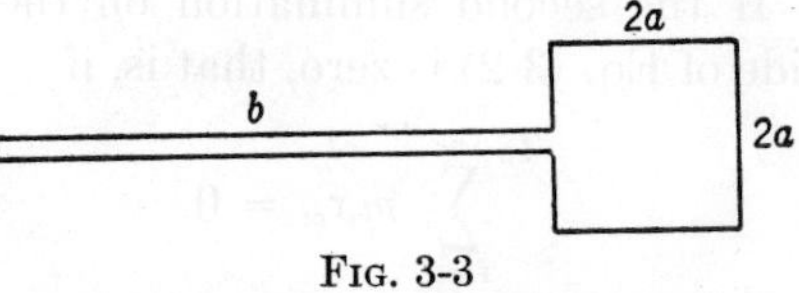

FIG. 3-3

It is not difficult to show that this procedure is permissible. Consider again the above case of n particles. The system is divided into two parts, the first of which contains the first k particles and the second the remaining $n - k$ particles. The masses of the two groups are

$$M' = \sum_{i=1}^{k} m_i \qquad M'' = \sum_{i=k+1}^{n} m_i \tag{3-7}$$

where $M = M' + M''$. The radius vectors to the center of mass of each group are

$$M'\bar{\mathbf{r}}' = \sum_{i=1}^{k} m_i \mathbf{r}_i \qquad M''\bar{\mathbf{r}}'' = \sum_{i=k+1}^{n} m_i \mathbf{r}_i$$

Clearly

$$\sum_{i=1}^{n} m_i\mathbf{r}_i = \sum_{i=1}^{k} m_i\mathbf{r}_i + \sum_{i=k+1}^{n} m_i\mathbf{r}_i$$

and Eq. (3-4) becomes

$$M\bar{\mathbf{r}} = \sum_{i=1}^{k} m_i\mathbf{r}_i + \sum_{i=k+1}^{n} m_i\mathbf{r}_i = M'\bar{\mathbf{r}}' + M''\bar{\mathbf{r}}'' \tag{3-8}$$

The x component of Eq. (3-8) is

$$M\bar{x} = M'\bar{x}' + M''\bar{x}'' \tag{3-9}$$

The similarity to the case of two particles [Eq. (3-6)] is apparent. Equation (3-9) states that, so far as finding the coordinates $\bar{x}$, $\bar{y}$, and $\bar{z}$ of the center of mass is concerned, we may consider all the mass M' of the first system to be located at its center of mass $\bar{x}'$, $\bar{y}'$, $\bar{z}'$ and all the mass M'' of the second system to be located at its mass center $\bar{x}''$, $\bar{y}''$, $\bar{z}''$.

3-3. Mass Centers of Solid Bodies. In order to determine the coordinates of the center of mass of a solid body, the integral notation is convenient. In place of m_i, the mass of the ith particle, the mass element dm is employed. The summations over the particles are replaced by integrals over the volume of the body. The mass element dm is equivalent to $\rho\, dV$, where ρ is the density (which may vary from point to point) and dV is the element of volume. In rectangular coordinates $dV = dx\, dy\, dz.$ Referring to Eq. (3-5) for comparison, the rectangular coordinates of the center of mass of a solid body of mass M are

$$\bar{x} = \frac{\int_V \rho x\, dV}{\int_V \rho\, dV} \qquad \bar{y} = \frac{\int_V \rho y\, dV}{\int_V \rho\, dV} \qquad \bar{z} = \frac{\int_V \rho z\, dV}{\int_V \rho\, dV} \tag{3-10}$$

Example 3-1. Show that the center of mass of a triangular lamina of uniform density is located at the intersection of the medians of the triangle.

In Fig. 3-4 construct the median AE. Choosing as a mass element the strip GH parallel to BC, we see that, in effect, GH is a uniform rod having its mass center at its mid-point K. But K is also the point of intersection of GH with AE. Similarly the centers of mass of all other parallel mass elements will be at their intersections with AE. Thus the center of mass of the entire triangle must lie somewhere on AE.

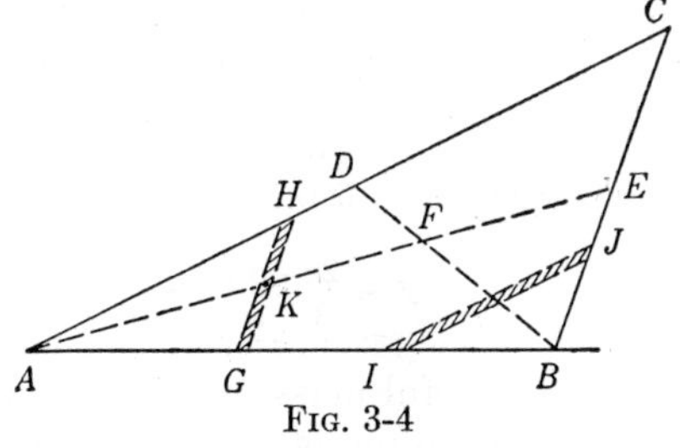

Fig. 3-4

Similarly we may draw a second median BD, and selecting mass elements, such as IJ parallel to AC, we see that the mass center of the lamina must also lie along BD.

However, the only way this may be true, while at the same time we require the mass center to lie along AE, is for it to be at their intersection, F. From plane geometry we remember that all three medians of a triangle intersect in one point, and therefore the proposition of Example 3-1 is demonstrated. The point F lies two-thirds of the distance along any median from the vertex to the opposite side.

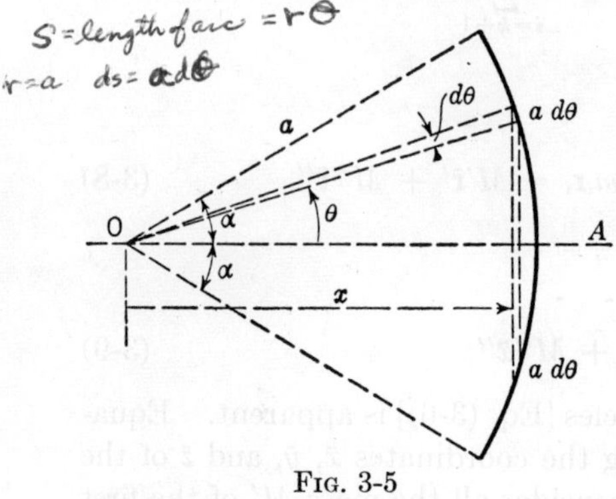

Fig. 3-5

Example 3-2. Find the center of mass of a uniform wire bent into the form of an arc of a circle of radius a, subtending an angle 2α at the center of the circle.

In Fig. 3-5 take the mass per unit length of the wire to be σ. Clearly the center of mass will lie on the axis of symmetry OA, which we take to be the axis of x. Choosing symmetrical elements $a\,d\theta$ as shown, their mass center will lie midway between them on line OA at a distance x from O, with an effective mass $2\sigma a\,d\theta$. Thus, the center of mass $\bar{x}$ of the entire wire lies on the x axis. Since $x = a\cos\theta$, we have

$$\bar{x} = \frac{\int x\,dm}{\int dm} = \frac{\int_0^\alpha 2\sigma a^2 \cos\theta\,d\theta}{\int_0^\alpha 2\sigma a\,d\theta} = \frac{a\sin\alpha}{\alpha} \tag{3-11}$$

Example 3-3. Employ the results of Examples 3-1 and 3-2 to find the center of mass of a thin lamina which is the sector of a circle.

In Fig. 3-6 the sector OAB has a vertex angle 2α and a radius a. We may take as an element of integration a small triangle ODE. From Example 3-1 the mass center will lie at C, two-thirds of the way along the median from point O. We can subdivide the entire lamina into many elements, all of the same size. All the mass centers of these will lie, as shown, along the arc FG, which has a radius $\frac{2}{3}a$. After we permit the elements to become very small, the line FG may be considered to be, in effect, a uniform wire bent into the form of a circular arc. Consequently the result (3-11) of Example 3-2 is useful, and the mass center of GF, and therefore of OAB, lies along the bisector, OH, at a distance from O which is

$$\bar{x} = \frac{2a}{3}\frac{\sin\alpha}{\alpha} \tag{3-12}$$

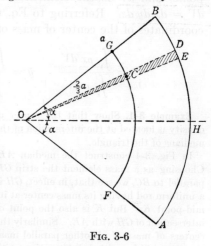

Fig. 3-6

Other useful mass centers are stated as follows without proof:

1. The center of mass of a homogeneous pyramid with a flat base and having any number of sides is three-fourths the distance along a straight line drawn from the apex to the centroid of the base.

2. Two parallel planes, normal to the x axis, intersect a thin homogeneous spherical shell of radius a. The zone of the sphere included between the two planes has a mass center given by

$$\bar{x} = \frac{a}{2} (\cos \alpha_1 + \cos \alpha_2) \tag{3-13}$$

where $\bar{x}$ is measured from the center of curvature. α_1 and α_2 are the plane semiangles of the cones having as bases the circular areas intercepted in the planes by the shell and having a common vertex at the center of curvature.

3. The center of mass of a sector of a homogeneous sphere of radius a lies along the axis of symmetry at a distance

$$\bar{x} = \frac{3a}{8} (1 + \cos \alpha) \tag{3-14}$$

from the vertex. The vertex angle of the cone of slant height a bounding the figure is 2α. In particular the mass center of a homogeneous hemisphere is on the line perpendicular to the flat side and passing through the center of curvature, at a distance $3a/8$ from the flat side.

3-4. Center of Mass of a Body Containing a Cavity. Occasionally it is necessary to determine the mass center of a body possessing some sort of cavity. If the cavity is absent, it may be that the centroid is either already known or can easily be found. An example of this is a thick homogeneous hemispherical shell of radii a and b ($b > a$). Here the solid is bounded by two hemispherical surfaces of radii a and b, respectively, and the cavity is a hemispherical hollow of radius a.

From Eq. (3-14) we know that, if the hollow were filled with material of the same density as the rest of the shell, the center of mass would lie (on the axis of symmetry) at a distance $\frac{3}{8}b$ from the center of curvature. However, it is not immediately clear where the mass center will be located in the case of the shell. This may be found either by integration or much more simply by a short method which we proceed to outline.

To do this, we shall employ the notation of a finite number of particles employed in Sec. 3-2. Suppose that a complete system comprises n particles; imagine, however, that all the particles from $k + 1$ to n are absent. It is desired to find the center of mass of the system consisting of particles 1 to k if at the same time we know the centers of mass of the two systems, 1 to n and $k + 1$ to n. In other words, if all n particles were present, the mass center would be known and also the mass center of the portion $k + 1$ to n would be known. In the example of the hemisphere cited above, this amounts to knowing the location of the center of mass of the solid hemisphere of radius b and of the absent hemisphere of radius a.

Accordingly, as before we designate the radius vector to the center of mass of all n particles as $\bar{\mathbf{r}}$, and that to the mass center of the particles $k+1$ to n as $\bar{\mathbf{r}}''$, both of which are given. The unknown, to be determined in terms of $\bar{\mathbf{r}}$ and $\bar{\mathbf{r}}''$, is $\bar{\mathbf{r}}'$, the mass center of the first k particles. By definition

$$M'\bar{\mathbf{r}}' = \sum_{i=1}^{k} m_i\mathbf{r}_i \tag{3-15}$$

However, the summation may clearly be rewritten

$$\sum_{i=1}^{k} m_i\mathbf{r}_i = \sum_{i=1}^{n} m_i\mathbf{r}_i - \sum_{i=k+1}^{n} m_i\mathbf{r}_i$$

from which Eq. (3-15) becomes

$$M'\bar{\mathbf{r}}' = \sum_{i=1}^{n} m_i\mathbf{r}_i - \sum_{i=k+1}^{n} m_i\mathbf{r}_i = M\bar{\mathbf{r}} - M''\bar{\mathbf{r}}''$$

The x component of $\bar{\mathbf{r}}'$ is

$$\bar{x}' = \frac{M\bar{x} - M''\bar{x}''}{M - M''} \tag{3-16}$$

This yields the interesting result that the center of mass of the system 1 to k can be regarded as being the centroid of two centroids, the first of which is that of the system of n particles having a total mass M and the second of which is the centroid of the system $k+1$ to n having a negative effective mass of magnitude M''.

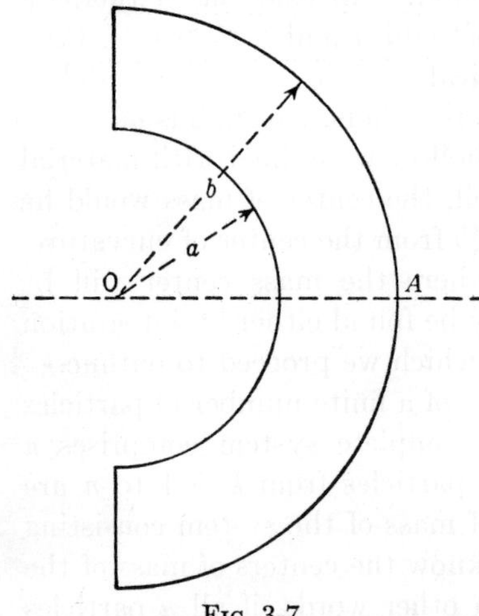

Fig. 3-7

Example 3-4. Find the center of mass of the above-mentioned example of the homogeneous hemispherical shell of radii a and b (see Fig. 3-7).

Clearly the center of mass lies on the axis of symmetry OA, which may be taken as the x axis, with origin at O. Following the procedure outlined above, consider the problem in the light of two hemispheres, of radii a and b, respectively, of which the first possesses a negative mass. Accordingly the coordinate of the center of mass of the shell is, if ρ is the density of the shell,

$$\begin{aligned}\bar{x} &= \frac{(3b/8)(2\pi\rho b^3/3) - (3a/8)(2\pi\rho a^3/3)}{(2\pi\rho/3)(b^3 - a^3)} = \frac{3(b^4 - a^4)}{8(b^3 - a^3)} \\ &= \frac{3(a + b)(b^2 + a^2)}{8(a^2 + ab + b^2)}\end{aligned} \tag{3-17}$$

3-5. Use of Arbitrary Coordinates. In addition to the applicability of the methods given in Secs. 3-2 to 3-4, the center of mass of a body may always be found by a straightforward integration. It is necessary only to know the volume element or surface element, as the case may be, in the coordinates chosen. Several commonly employed systems of coordinates, other than the cartesian system, are defined in Appendix 1. However, in making use of coordinates other than rectangular coordinates, attention should be paid to the fact that it is under the integral sign of Eqs. (3-10) that the change is made. What we arrive at is still the rectangular coordinates $\bar{x}$, $\bar{y}$, $\bar{z}$ of the mass center. Thus, employing spherical polar coordinates for the integration, we have

$$\begin{aligned} M\bar{x} &= \iiint \rho(r \sin\theta \cos\varphi)(r^2 \sin\theta \, dr \, d\theta \, d\varphi) \\ M\bar{y} &= \iiint \rho(r \sin\theta \sin\varphi)(r^2 \sin\theta \, dr \, d\theta \, d\varphi) \\ M\bar{z} &= \iiint \rho(r \cos\theta)(r^2 \sin\theta \, dr \, d\theta \, d\varphi) \end{aligned} \tag{3-18}$$

where $M = \iiint \rho r^2 \sin\theta \, dr \, d\theta \, d\varphi$ and in all cases the integration is over the entire volume of the body.

If it is desired, for example, to write the spherical coordinates of the center of mass $\bar{r}$, $\bar{\theta}$, $\bar{\varphi}$ rather than the rectangular coordinates $\bar{x}$, $\bar{y}$, $\bar{z}$, the latter must first be found in the usual manner [cf. Eqs. (3-10)]. Following this, $\bar{r}$, $\bar{\theta}$, and $\bar{\varphi}$ may be determined with the aid of Eqs. (3), Appendix 1. We have

$$\begin{aligned} \bar{r} \sin\bar{\theta} \cos\bar{\varphi} &= \bar{x} \\ \bar{r} \sin\bar{\theta} \sin\bar{\varphi} &= \bar{y} \\ \bar{r} \cos\bar{\theta} &= \bar{z} \end{aligned} \tag{3-19}$$

from which

$$\begin{aligned} \bar{r}^2 &= \bar{x}^2 + \bar{y}^2 + \bar{z}^2 \\ \bar{\theta} &= \tan^{-1} \frac{\sqrt{\bar{x}^2 + \bar{y}^2}}{\bar{z}} \\ \bar{\varphi} &= \tan^{-1} \frac{\bar{y}}{\bar{x}} \end{aligned} \tag{3-20}$$

EQUILIBRIUM OF RIGID BODIES

3-6. Extent of a System. Internal and External Forces. Before examining the conditions which are required in order to ensure the static equilibrium of an extended body, it is necessary to recognize the geometrical boundaries of the body. As an example let us consider the equilibrium of two equal uniform ladders OP and PQ (Fig. 3-8a) joined together by a frictionless pivot at P and by a horizontal, weightless, rigid strut AB smoothly joined to the ladders at A and B. The whole rests under gravity on a smooth floor. Let the weight of each ladder be W. First we examine the system consisting of the two ladders OPQ and the strut AB. The forces acting upon this system, owing to external agencies, are the total weight $2W$, assumed to act at the center of mass C, of the entire system and the two reactions N_1 and N_2 ($N_1 = N_2$, by symmetry here) of the floor upon the ladders (cf. Example 3-8 and also Probs. 3-31 and 3-32). The forces $2W$, N_1, and N_2 are therefore termed *external forces*. In addition to the external forces, there are *internal forces* acting. Such forces are those which might exist between adjacent portions such as a and b of the ladder or between members meeting at a joint such as P. (Internal forces are not necessarily transmitted by contact, however. If a system consists of n particles, the gravitational, electrostatic, or other forces acting between pairs of particles are all forces which are internal to the system and are therefore called internal forces.)

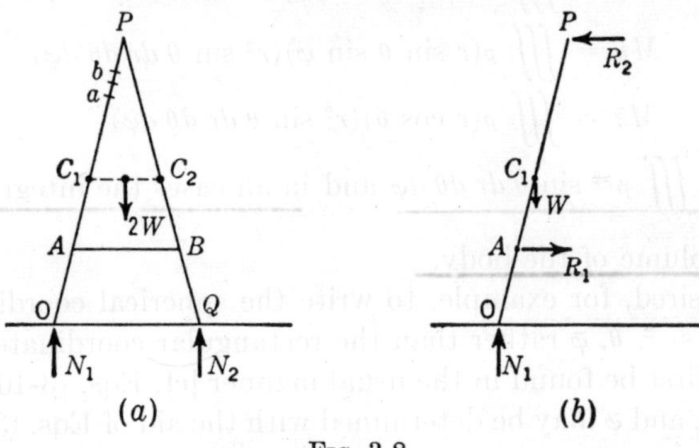

FIG. 3-8

In Fig. 3-8b we have selected one ladder, OP, as the new system of interest. Thus the external forces acting on OP are N_1, as before, W acting at the center of mass C_1, and R_1 and R_2, which are the reactions (horizontal; cf. Example 3-8 and also Prob. 3-31), respectively, of the strut AB and of the other ladder PQ upon the ladder OP. R_1 and R_2 are not considered in part a of the figure, since for that system they are internal forces.

The size or extent of a system is entirely arbitrary and is chosen according to the dictates of convenience for the solution of the problem at hand. It is evident that, if a system is made sufficiently large in extent, *all* forces become internal forces.

3-7. Moment of a Force. Figure 3-9 shows a body able to rotate about an axis through O perpendicular to the plane of the paper. A force $\mathbf{F}$ acts at point P of the body. The *moment* of $\mathbf{F}$ is a vector $\mathbf{L}$

defined by

$$\mathbf{L} = \mathbf{r} \times \mathbf{F}_f \tag{3-21}$$

where $\mathbf{r}$ is the radius vector from O to P. Point P and the line of action of $\mathbf{F}$, for convenience, are taken to be in the plane of the paper. Thus the angles θ and φ are also measured in that plane, and the magnitude of $\mathbf{L}$ may be written

$$L = rF \cos \theta = rF \sin \varphi \tag{3-22}$$

where r is the magnitude of $\mathbf{r}$, θ is the angle which $\mathbf{r}$ makes with the perpendicular from O to the line of action of $\mathbf{F}$, and φ is the complement of θ. As is evident from Eq. (3-22) and the figure, L can be obtained either by multiplying r by the component of $\mathbf{F}$ normal to $\mathbf{r}$ or by multiplying F by the component of $\mathbf{r}$ normal to the line of action of $\mathbf{F}$. The direction of $\mathbf{L}$ is that of the advance of a right-handed screw which carries $\mathbf{r}$ into $\mathbf{F}$ (through the angle φ). In Fig. 3-9 the direction of $\mathbf{L}$ is perpendicular to the plane of the paper and is directed toward the reader.

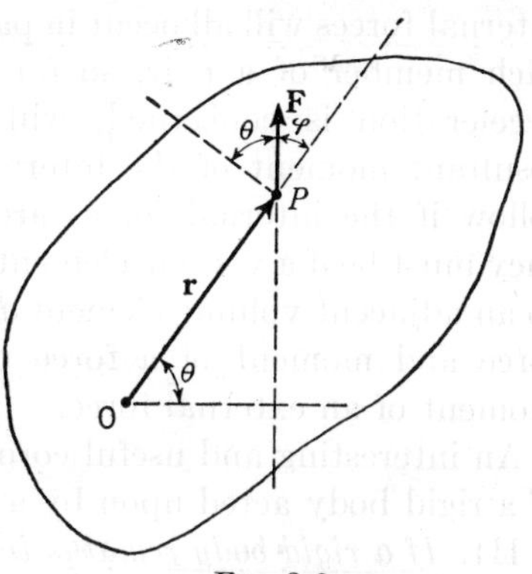

Fig. 3-9

If a number of forces $\mathbf{F}_1, \mathbf{F}_2, \ldots, \mathbf{F}_n$ are acting at point P, the vector sum of their moments about point O is equal to the moment of their resultant. This may be demonstrated quite readily, for the moment $\mathbf{L}$ of their resultant $\mathbf{F}$ can be written

$$\begin{aligned}\mathbf{L} &= \mathbf{r} \times \mathbf{F} = \mathbf{r} \times (\mathbf{F}_1 + \mathbf{F}_2 + \cdots + \mathbf{F}_n) \\ &= \mathbf{r} \times \mathbf{F}_1 + \mathbf{r} \times \mathbf{F}_2 + \cdots + \mathbf{r} \times \mathbf{F}_n \\ &= \mathbf{L}_1 + \mathbf{L}_2 + \cdots + \mathbf{L}_n\end{aligned} \tag{3-23}$$

3-8. General Conditions of Equilibrium for a Rigid Body Acted upon by a System of Coplanar Forces. It will be shown in Chap. 8 that, when an array of forces acts on a system of particles, the center of mass of the system will experience a translational acceleration as if all the forces were applied at the mass center. Accordingly, in order to ensure static equilibrium against translation, the vector sum of all the forces must be zero. Likewise in Chap. 8 it will be shown that the system of particles will experience an angular acceleration about any selected axis of rotation, which is itself unaccelerated, if there is a moment of force about that axis. Hence, if there is to be no angular acceleration about such an axis, arbitrary in orientation, passing through any selected point of the system, the vector sum of the moments of the forces with respect to that point must vanish.

In the present chapter the system of particles comprises a rigid body. In addition we shall, for simplicity, restrict ourselves to a coplanar system of forces being applied to the body. Accordingly the conditions for the complete equilibrium of a rigid body in one plane are:

I. *The algebraic sum of the components of all the forces in each of two mutually perpendicular directions in the plane of the forces must separately vanish.*

II. *The algebraic sum of the moments of the forces about any point in the plane of the forces must be zero.*

In this discussion no differentiation has been made between internal and external forces. Clearly, however, by Newton's third law the internal forces will all occur in pairs of oppositely directed forces. Hence each member of a pair, so far as its effect in producing translational acceleration is concerned, will cancel the other member. That the resultant moment of the internal forces about any point vanishes will follow if the internal forces are central (cf. Secs. 8-1 and 8-3), which they must be if a volume element dV_1 of the body is to be rigidly attached to an adjacent volume element dV_2. Accordingly, in I and II, the terms force and moment of a force may be replaced by external force and moment of an external force.

An interesting and useful corollary to I and II is provided by the case of a rigid body acted upon by a set of three forces:

III. *If a rigid body remains in equilibrium in one plane under the action of three forces, the three forces all lie in the same plane and intersect in one point.*

This is not difficult to see. First, if the forces were not all in the same plane, they would not form a closed triangle and consequently they would possess a resultant different from zero. Furthermore, if we suppose that two of the forces, $\mathbf{F}_1$ and $\mathbf{F}_2$, intersect in a point P, we know that $\mathbf{F}_1$ and $\mathbf{F}_2$ will each possess zero moment with respect to P. Thus if $\mathbf{F}_3$ is to possess zero moment with respect to P, it must also pass through this same point.

It is to be noticed that conditions I and II provide three equations for determining the conditions of equilibrium of a body. This means that as many as three unknowns may be determined by means of these conditions. If there are more than three unknowns, the problem is statically indeterminate so far as the methods here considered are concerned. It is possible to render such problems determinate by taking into account the minute deformations which actually occur when all real bodies are subjected to forces. We shall consider a simple example of a statically indeterminate problem in Example 3-7.

3-9. Composition of Parallel Forces. It is possible to replace a system of forces the lines of action of which are all parallel by a *single* force. In

Fig. 3-10 is shown the case in which two parallel forces $\mathbf{F}_1$ and $\mathbf{F}_2$ are acting at points A_1 and A_2, respectively, and it is postulated that the force $\mathbf{F}$, equal to $\mathbf{F}_1 + \mathbf{F}_2$ and acting through O, is the resultant of $\mathbf{F}_1$ and $\mathbf{F}_2$ and consequently may replace them. Now if this is true, $\mathbf{F}$, in addition to being equal to $\mathbf{F}_1 + \mathbf{F}_2$, must have, about any point in the plane of Fig. 3-10, a moment which is equal to the vector sum of the moments

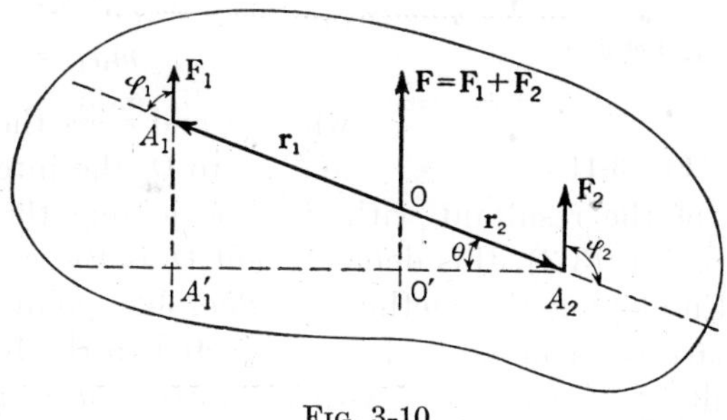

FIG. 3-10

of $\mathbf{F}_1$ and $\mathbf{F}_2$ taken about the same point. This follows since, if $\mathbf{F}_1$ and $\mathbf{F}_2$ are replaceable by $\mathbf{F}$, $\mathbf{F}$ must have the same tendency to produce rotation (as well as translation) as the combination of $\mathbf{F}_1$ and $\mathbf{F}_2$. In particular, since $\mathbf{F}$ passes through O and therefore possesses zero moment with respect to O, the vector sum of the moments of $\mathbf{F}_1$ and $\mathbf{F}_2$ about O must also vanish. It is evident from the figure that this requires the magnitudes of their moments about O to be equal, since the two moments tend to produce rotations in opposite directions about an axis through O normal to the paper. Thus we have

$$F_1(O'A'_1) = F_2(O'A_2)$$

or

$$F_1 r_1 \cos\theta = F_2 r_2 \cos\theta \tag{3-24}$$

In vector notation this result is expressed as

$$\boxed{\mathbf{r}_1 \times \mathbf{F}_1 + \mathbf{r}_2 \times \mathbf{F}_2 = 0} \tag{3-25}$$

In the same way the resultant of two antiparallel forces may be found provided that the magnitudes of the two forces are not equal. It is *not* possible to replace two antiparallel forces of equal magnitude the lines of action of which do not coincide by a single force. This latter case will be considered in Sec. 3-13.

3-10. Center of Gravity of a System of Particles. The center of mass of a system of particles assumes special importance in the light of the composition of parallel forces. If the particles $m_1, m_2, \ldots, m_n$ are all close to the surface of the earth, the gravitational force on each is directed vertically downward. These are of magnitude m_1g, m_2g, . . . , m_ng, and so on, as shown in Fig. 3-11 for the ith and jth particles. From Eq. (3-24) the resultant of m_ig and m_jg is a force of magnitude $m_ig + m_jg$, directed vertically downward, such that

$$m_igr_i = m_jgr_j \tag{3-26}$$

or

$$m_ir_i = m_jr_j$$

Fig. 3-11

where r_i and r_j are the distances of m_i and m_j to O, the intersection of the line of action of the resultant with the line joining the two particles. But, in view of Eq. (3-3), this defines point O as the center of mass of m_i and m_j. Employing the methods of Sec. 3-2, point O can be considered as a particle of mass $m_i + m_j$ subjected to the force $m_ig + m_jg$. Since the resultant of $m_ig + m_jg$ with the force m_kg on a third particle m_k may be obtained, and so on, it is evident that the final result, so far as the gravitational forces are concerned, will be that all the masses $m_1, m_2, \ldots, m_n$ may be considered to be located at the center of mass of the entire system and a force $m_1g + m_2g + \cdots + m_ng$ to be acting vertically downward through that point. Hence the center of mass becomes the *center of gravity* for the special case where the gravitational forces on the particles of the system are all parallel. For this reason the term center of gravity is often employed synonymously with the term center of mass. However, it is preferable here to use the latter term exclusively since, if the system of particles is so large that the gravitational acceleration (g at the surface of the earth) varies from one point to another in the system, the center of mass may no longer be a center of gravity. In such a case the center of mass is the useful parameter. Indeed, a center of gravity may not even exist, since the concept, as developed above, is predicated upon the replacement of the gravitational forces on all the particles by a single force. It will be pointed out in Sec. 3-14 that it is not always possible to replace an arbitrary system of forces by a single force.

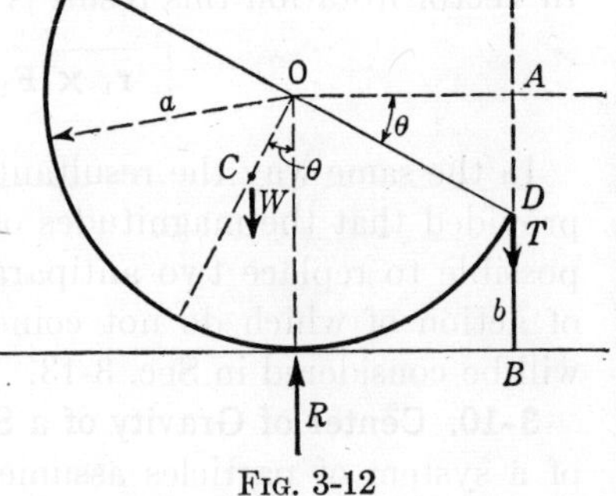

Fig. 3-12

3-11. Miscellaneous Examples

Example 3-5. A homogeneous hemisphere of weight W and radius a rests on a perfectly smooth horizontal table. As shown in Fig. 3-12, one edge, D, is tied to a point B on the table by means of a light inextensible string of length b ($b < a$). Find the tension in the string.

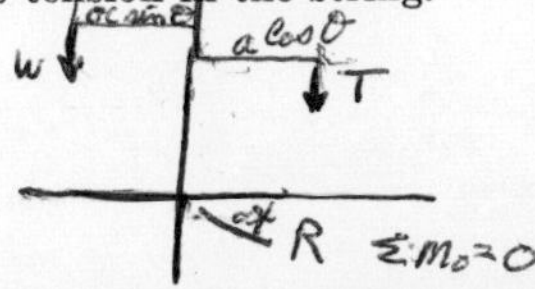

We isolate the hemisphere. The forces acting on it are its weight W, acting at C, a distance $\frac{3}{8}a$ from O, the reaction R of the table, and the tension T in the string. Since the contact between the hemisphere and the table is smooth, R must be normal to the surface, passing through point O. Thus, since W and R are vertical, T must also have no horizontal component. Consequently the string DB is vertical.

The moments of the forces about point O must be zero. We select O since it is on the line of action of R, and we therefore immediately eliminate the unknown R, in which we have no interest in this problem. For equilibrium to be possible, we must have

$$W \cdot \overline{OC} \sin \theta - Ta \cos \theta = 0 \tag{3-27}$$

But $\sin \theta = (a - b)/a$, $\cos \theta = (2ab - b^2)^{\frac{1}{2}}/a$, and $\overline{CO} = 3a/8$. Hence

$$T = \frac{3}{8} \frac{W(a - b)}{\sqrt{2ab - b^2}} \tag{3-28}$$

Example 3-6. A uniform heavy ladder is in limiting equilibrium, standing on a rough floor, and leaning against an equally rough wall. The angle of friction is ϵ. Find the angle which the ladder makes with the horizontal.

The forces acting on the ladder are shown in Fig. 3-13a. The weight W of the ladder acts vertically downward at the middle B of the ladder. Since point A is about to slide, so also must point C; the forces at these points may be represented by

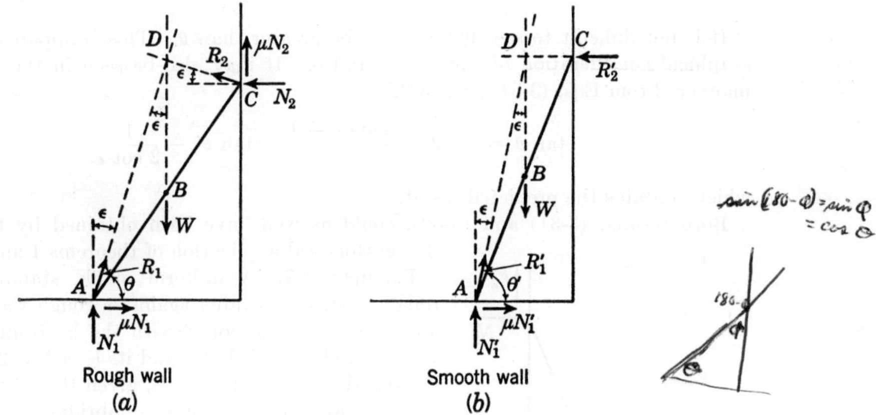

FIG. 3-13

N_1, μN_1, N_2, and μN_2, as shown. These have the resultants R_1 at A and R_2 at C, each making an angle ϵ with the normal since the system is in limiting equilibrium. The problem can be reduced to one involving the three forces R_1, R_2, and W, and consequently their lines of action must intersect at a single point D. In triangle ABD, employing the sine law,

$$\frac{\overline{AB}}{\sin \epsilon} = \frac{\overline{AD}}{\cos \theta} \tag{3-29}$$

But, by construction, angle ADC is $\pi/2$, and therefore

$$\overline{AD} = \overline{AC} \cos \left(\frac{\pi}{2} - \epsilon - \theta \right) \tag{3-30}$$

Employing Eq. (3-30) and also the fact that $\overline{AC} = 2\overline{AB}$, Eq. (3-29) yields

$$\cos\theta = 2\sin\epsilon\sin(\epsilon+\theta) = 2\sin^2\epsilon\cos\theta + 2\sin\epsilon\cos\epsilon\sin\theta$$

This reduces to

$$\tan\theta = \frac{1-2\sin^2\epsilon}{2\sin\epsilon\cos\epsilon} = \frac{\cos 2\epsilon}{\sin 2\epsilon} = \cot 2\epsilon \tag{3-31}$$

and we see that, in limiting equilibrium, $\theta = (\pi/2) - 2\epsilon$.

It is interesting to consider the case in which the wall is perfectly smooth. This is shown in Fig. 3-13*b*. Since the wall is now smooth, the reaction R_2' is normal to the surface and consequently the three forces must intersect at point D, at the same level as C. We have

$$\frac{\overline{AB}}{\sin\epsilon} = \frac{\overline{AD}}{\cos\theta'}$$

and

$$\overline{AC}\sin\theta' = \overline{AD}\cos\epsilon$$

Hence

$$\overline{AB} = \frac{\overline{AC}\sin\epsilon\sin\theta'}{\cos\theta'\cos\epsilon} = 2\overline{AB}\tan\epsilon\tan\theta'$$

from which

$$\tan\theta' = \tfrac{1}{2}\cot\epsilon \tag{3-32}$$

It is not difficult to see that θ' must be greater than θ. This is apparent from a graphical consideration of Fig. 3-13*a* and *b*. It may also be seen in the following manner: From Eqs. (3-31) and (3-32),

$$\tan\theta = \cot 2\epsilon = \frac{\cot^2\epsilon - 1}{2\cot\epsilon} = \tan\theta' - \frac{1}{2\cot\epsilon}$$

which contains the predicted result.

Both results, (3-31) and (3-32), could as well have been obtained by the more straightforward application of theorems I and II.

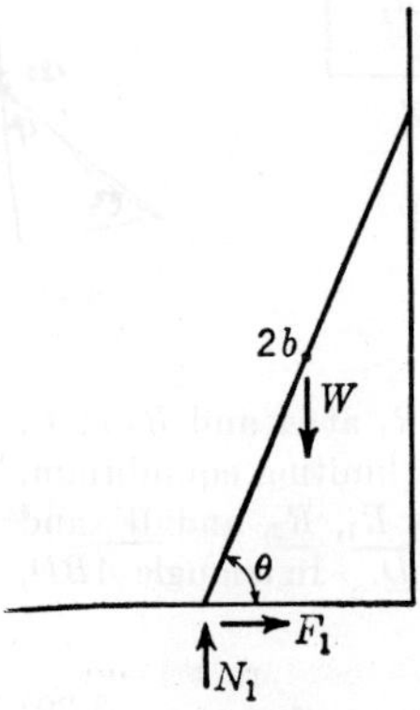

Example 3-7. A uniform plank, standing on a rough floor, and leaning against a rough wall, makes an angle of inclination θ with the horizontal. The weight of the plank is W, and its length is $2b$. Find the reactions at the two ends, given that the configuration is not one of limiting equilibrium.

The situation is shown in Fig. 3-14. It is an equilibrium configuration, but it is not a limiting one. F_1 and F_2 are not known directly in terms of N_2 and N_1. Consequently we have four unknowns, F_1, N_1, F_2, and N_2, and but three equations with which to determine them. These are

FIG. 3-14

$$F_1 - N_2 = 0$$
$$N_1 + F_2 - W = 0$$
$$Wb\cos\theta - 2N_2b\sin\theta - 2F_2b\cos\theta = 0$$

in which moments were taken about the lower end of the ladder. Consequently, with the methods which we have at our command, the problem is indeterminate.

3-12. Reactions at Smooth Joints. These are present in so many static problems that some emphasis on them is justified at this point.

When two beams are connected by means of a *smooth joint* (pivot), the action of one on the other can be represented by a single force acting at the junction. Usually the direction, as well as the magnitude, of this force is not apparent, and both must be deduced from the equations of equilibrium.

In two cases, however, the direction of the force is immediately apparent. The first of these is when the geometry of the body and the external forces are both symmetrical about a straight line passing through the joint. In this event the action and reaction between the two members must also be symmetrical with respect to this line. Since the action and reaction are equal and opposite, this requires that each be perpendicular to the line of symmetry (cf. the case of R_2 in Example 3-8).

The second example in which the direction is readily evident is provided by a member which is smoothly hinged at two points. If no forces other than the reactions at the joints are acting upon the member, then, in order to ensure the static equilibrium of the member, both reactions must lie along the straight line joining the two joints and must be equal and opposite (cf. the reactions at the two ends of the strut in Example 3-8).

Example 3-8. An A frame consists of two uniform bars, each of length b and weight W, standing in a vertical plane on a smooth floor, as shown in Fig. 3-15a, and

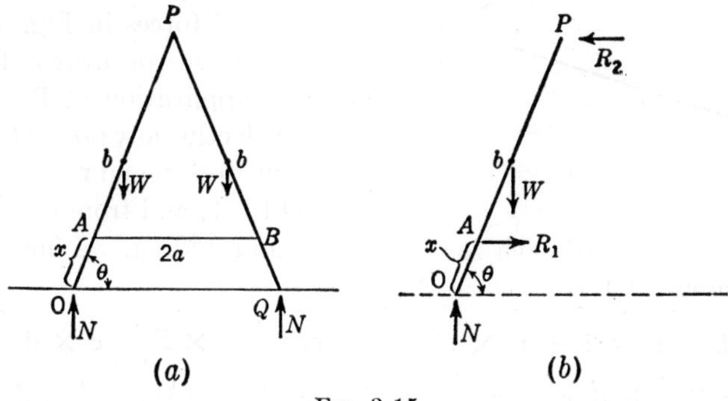

FIG. 3-15

smoothly hinged at the top. At a distance x from the floor (measured along a bar) is a light horizontal strut AB of length $2a$ smoothly hinged to the two bars. Find the stress in the strut.

The external forces acting on the system consisting of both bars are the normal reactions N of the smooth floor and the weights W acting vertically at the mid-points of the bars. From symmetry considerations, the reaction N is the same for both bars. Considering the system as a whole, the stresses in the rod and the reactions at P are internal forces and thus cancel in pairs. Accordingly, taking vertical components of the external forces,

$$2N = 2W \tag{3-33}$$

We next consider the forces acting on one bar, OP, for example. The forces acting on OP, indicated in Fig. 3-15b, are W, N, the reaction R_1 of the strut, and the reaction R_2 at the joint P. Since the two bars are equal, R_2 must be horizontal, as shown. If it were to possess a vertical component, upward, for example, this would mean that the reaction of OP on QP would likewise have a vertical component, but in the opposite sense, in this case downward. This is not consistent with the symmetry of the figure, in which OP is equal to QP. Thus the line of action of R_2 can only be horizontal. In view of this, and of Eq. (3-33), R_1 must also be horizontal and must be equal and opposite to R_2.

Taking moments about point P, we have

$$Nb \cos \theta - R_1(b - x) \sin \theta - W \frac{b}{2} \cos \theta = 0 \tag{3-34}$$

and since $\tan \theta = [(b - x)^2 - a^2]^{\frac{1}{2}}/a$, we have, making use of (3-33),

$$R_1 = \frac{Wab}{2(b - x)[(b - x)^2 - a^2]^{\frac{1}{2}}} \tag{3-35}$$

3-13. The Couple. The configuration which was not considered in Sec. 3-9 is that of two equal and opposite forces the lines of action of which are parallel but which do not coincide. Such a pair of forces constitutes a *couple* and cannot be reduced to a single force. It is characterized by a tendency to produce a rotation, but not a translation, of the body upon which it acts.

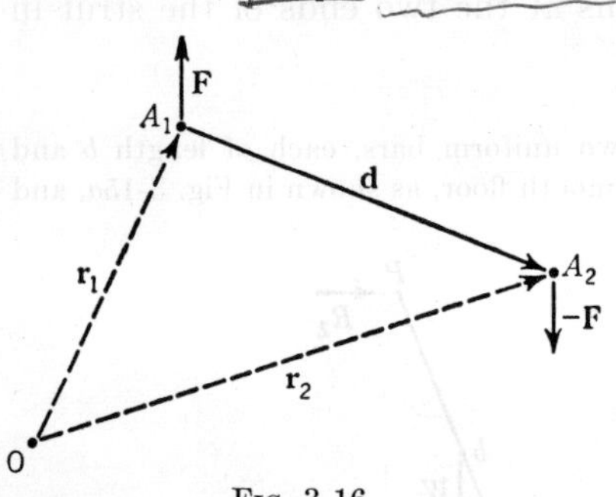

FIG. 3-16

It is useful to determine the moment of a couple. Let $\mathbf{F}$ and $-\mathbf{F}$ be two parallel forces in Fig. 3-16; $\mathbf{d}$ is the radius vector drawn from the point of application of $\mathbf{F}$ to that of $-\mathbf{F}$. Selecting any point O, draw the radius vectors $\mathbf{r}_1$ and $\mathbf{r}_2$, respectively, from O to A_1 and from O to A_2. Taking moments of each force about O and adding, we find that the resultant moment $\mathbf{L}$ is

$$\mathbf{L} = \mathbf{r}_1 \times \mathbf{F} + \mathbf{r}_2 \times (-\mathbf{F}) = (\mathbf{r}_1 - \mathbf{r}_2) \times \mathbf{F} = \mathbf{F} \times \mathbf{d} \tag{3-36}$$

since, from Fig. 3-16, $\mathbf{r}_1 - \mathbf{r}_2$ is just the negative of the vector $\mathbf{d}$. At the same time the explicit dependence on $\mathbf{r}_1$ and $\mathbf{r}_2$ disappears, and we immediately perceive the further fact that $\mathbf{L}$ is independent of the point with respect to which the moment is taken. It is important to notice, also, that it is not necessary for O to be in the plane of the forces.

The couple, as applied to a rigid body, is an example of what is called a *free, or nonlocalized, vector*, since its moment is the same with respect to all points of the rigid body. A single force applied to a body, however, is a *localized* vector since its total effect on the body, including rotation, depends on the location of its line of action.

The definition of the couple facilitates an alternate way of stating the conditions I and II of equilibrium: This is:

I *and* II (*alternate*). *A rigid body will be in equilibrium in one plane under the action of a system of forces confined to this same plane if the sums of the moments of all of the forces about any three noncollinear points are each zero.*

This can be seen since, if the moments about one of the points A_1 vanish, this guarantees only that there is no couple, but not that the forces are zero. Carrying this further, the vanishing of the sum of the moments about a second point A_2 ensures only that all forces are zero except those whose lines of action are coincident with A_1A_2. Consequently, to establish that the resultant force is zero, the sum of the moments about a third point A_3, not along the line which passes through A_1A_2, must also vanish.

3-14. Reduction of an Arbitrary System of Forces to a Single Force Plus a Couple. Any system of forces acting on a rigid body may be replaced by a single force acting at an arbitrary point plus a suitable couple. In Fig. 3-17 consider a force $\mathbf{F}_1$ to be acting at point P of a rigid body. Selecting any other point O, apply two forces $\mathbf{F}_2$ and $-\mathbf{F}_2$, equal in magnitude to $\mathbf{F}_1$, acting through O with a common line cf action parallel to that of $\mathbf{F}_1$. Since the combined effect of $\mathbf{F}_2$ and $-\mathbf{F}_2$ is zero, the behavior of the body is precisely that when $\mathbf{F}_1$ is acting by itself. However, another way of interpreting the figure is to consider the system of forces as made up of the force $\mathbf{F}_2$ plus the couple resulting from the forces $\mathbf{F}_1$ and $-\mathbf{F}_2$ (the moment of the couple being $\mathbf{F}_1 \times \mathbf{d}$). Clearly then, we may regard the single force $\mathbf{F}_1$ to be replaced by the couple $\mathbf{F}_1 \times \mathbf{d}$, plus the force $\mathbf{F}_2$ acting through the point O.

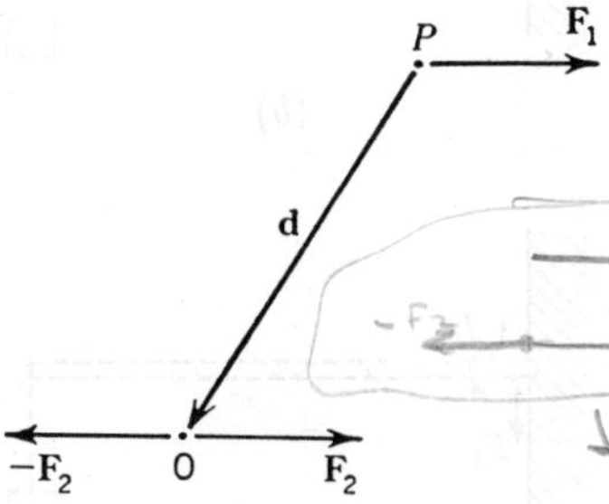

Fig. 3-17

Now if forces other than $\mathbf{F}_1$ are acting on the body, we are able to replace each of them by a single force passing through O, plus a couple All these forces passing through O may be combined vectorially into a single force passing through O, and all the couples may be combined vectorially into a single couple. Consequently any system of forces acting on a rigid body may be replaced by a single force passing through any convenient point, plus a suitable couple. (It is evident that the moment of the couple will vary depending on the point chosen at which the force is to act.)

3-15. Reactions at Rigid Joints. This case differs from that of a smooth joint. A smooth joint connecting two members of a system may transmit a force, but not a couple, from one member to the other. A *rigid joint*, on the other hand, may transmit both a force and a couple.

For purposes of discussion let us consider (see Fig. 3-18a) a uniform beam AB, of length b and weight W_0, rigidly attached to a wall at A, and with a weight W suspended at B. Let us inquire into the stresses acting at a point D of the beam, a distance x from A, and also those acting on the wall at the point of attachment A.

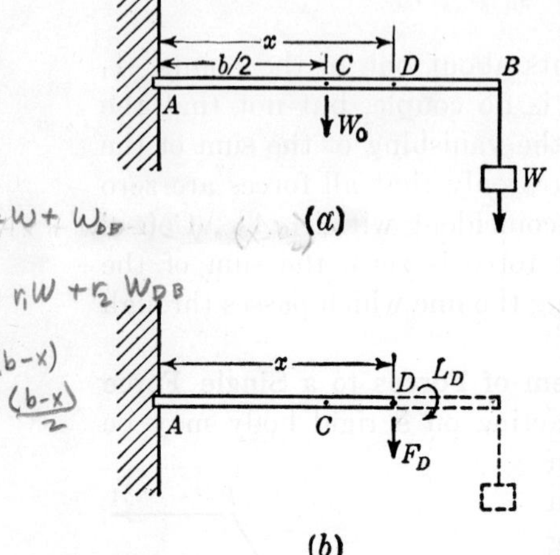

In part b of the figure, the section DB of the beam and the hanging weight W are replaced (by virtue of Sec. 3-14) by a force

$$F_D = W + \frac{W_0}{b}(b - x) \tag{3-37}$$

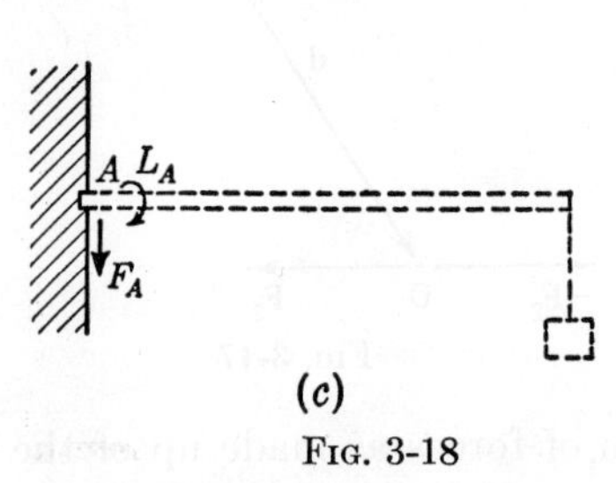

acting vertically downward and a couple, or bending moment,

$$L_D = W(b - x) + \frac{W_0}{b}(b - x)\frac{(b - x)}{2} \tag{3-38}$$

tending to rotate the beam in the clockwise sense. (The bending moment is usually the important factor in determining whether or not the member will break.) Thus the stresses acting at D are a shearing force and a couple, given by Eqs. (3-37) and (3-38).

(c)

Fig. 3-18

In part c of the figure we examine the stresses acting at A. Evidently the beam AB may be replaced by a shearing force

$$F_A = W + W_0 \tag{3-39}$$

and a bending moment

$$L_A = Wb + W_0\frac{b}{2} \tag{3-40}$$

both acting upon the wall at point A. The couple L_A has the sense of inducing a rotation of the wall in the clockwise direction. From Newton's third law, the wall exerts a force upon the beam which is equal in magnitude to F_A but which is directed upward and, in addition, a couple, equal in magnitude to L_A tending to rotate the beam in a counterclockwise sense.

Problems

Problems marked C are reprinted by kind permission of the Cambridge University Press.

3-1. Find the center of mass of each of the following homogeneous laminae:

a. One quadrant of an ellipse, of major and minor axes $2a$ and $2b$.

b. A sheet bounded by the parabolas $y^2 = 4ax$ and $x^2 = 4ay$.

c. One arch of the cycloid $x = a(\theta - \sin\theta)$, $y = a(1 - \cos\theta)$.

d. Half of a circular ring of inner and outer radii a and b.

e. The lamina bounded by the semicubical parabola $x^3 = ay^2$, the axis of x, and the line $x = a$.

3-2. Find the center of mass of each of the following homogeneous solids:

a. A paraboloid of revolution bounded by a right section through the focus.

b. One octant of the volume common to two equal cylinders the axes of which intersect at right angles.

c. The volume consisting of a cone and hemisphere placed base to base, where the common radius is a and the distance from the vertex to the base of the cone is h.

3-3. A homogeneous rod of radius a is bent into the form of a semicircle of mean radius b. Find the center of mass of the figure.

3-4. A flat plate, of variable density, is shaped in the form of a quadrant of a circle of radius a. The density has the value m_0 at the center of curvature of the circular edge and at other points increases in such manner that it is proportional to the square of the distance of the point from the vertex. At the circular edge the density is $2m_0$. Find the center of mass.

3-5. Find the center of mass of that portion of a homogeneous solid sphere of radius a situated between the conical surfaces $\theta = \alpha_1$ and $\theta = \alpha_2$, where θ is the angle between the radius vector passing through the center of the sphere and the z axis.

$$\frac{3a(\cos\alpha_1 + \cos\alpha_2)}{8}$$

3-6. An elliptical plate, of constant thickness, has major and minor axes $2a$ and $2b$. Find the center of mass of that part of the positive quadrant bounded by the x axis, the elliptical edge, and the line $y = bx/a$.

3-7. A thin circular lamina of radius a has a small circle of radius $a/2$ cut from it in a manner such that the periphery of the smaller circle passes through the center of, and is tangent to the edge of, the larger circle. Find the center of mass of the remainder.

3-8. A hemisphere of radius a has a conical sector of semivertical angle 45° cut from it. The axis of the conical hole is the z axis. Obtain the center of mass of the remainder by (a) integration employing spherical coordinates and (b) integration by slicing the figure into circular laminae by planes perpendicular to the z axis.

3-9. A uniform beam of weight W stands on a smooth floor and leans against a smooth wall the angle of inclination with the horizontal being α. Find what horizontal force applied at the foot of the ladder will maintain equilibrium.

3-10. A cardboard square of side a has cut out of the middle of one side a notch in the form of an isosceles triangle of base b, reaching to the center of the square. Find the center of mass of the remainder.

3-11. Two homogeneous right circular cones of similar material and altitudes of 6 in. and 2 in. are placed base to base. Find the center of mass of the combination.

3-12. A right circular cone is divided into two equal parts by a plane through the axis. Find the center of mass of either half.

3-13. Two parallel forces of magnitude 5 lb and 15 lb act, respectively, at one end and the mid-point of a rod 3 ft in length. Their lines of action are perpendicular to the rod. A force of magnitude 10 lb is acting at the other end in a direction opposite to the other two forces. Find the magnitude, direction, and line of action of the resultant.

3-14. Along the sides AB, BC, and CD of the square $ABCD$ are acting the forces F, $2F$, and $3F$, respectively. Find the magnitude and direction of the resultant.

3-15. A uniform rod rests with one end on a rough horizontal plane (coefficient μ of friction) and the other on a smooth plane inclined at an angle θ to the horizontal. Find the least angle which the rod can make with the horizontal.

3-16. A cord passing over a frictionless pulley bears two weights W_1 and W_2 at its ends. Another cord of length a connects W_1 to a point at the same level of, and at a distance b from, the pulley. (Take $a > b$.) Write the equations determining the position of equilibrium.

3-17. A uniform ladder of weight W_1 and length b stands on a rough horizontal plane and leans against a rough vertical wall, the coefficients of friction at the two contacts being the same. Find how far a man of weight W_2 can ascend the ladder without its slipping. The ladder makes an angle θ with the horizontal, and is not initially in limiting equilibrium.

3-18. A uniform rod of weight W and length b is supported from a single point by means of two strings, each of length b, connected to the ends of the rod. A weight W is hung from one of the ends of the rod. Find the angle which the rod makes with the horizontal and the tension in each string.

3-19. A long plank has its upper end resting against a smooth vertical wall and its lower end on a rough floor. The coefficient of static friction at the lower end is μ. The upper half of the plank is twice as thick as the lower half. Find the inclination to the horizontal at which sliding will just occur.

3-20. A uniform plank of negligible thickness rests in equilibrium horizontally across a rough fixed circular cylinder of radius a. If b is the length of the plank, ϵ the angle of friction, and W the weight of the plank, find the greatest weight which can be attached to one end without causing the plank to slip.

3-21. Prove in a general manner, employing the methods of vector analysis, that a system of forces acting along and represented by the sides of a triangle, taken in order, is equivalent to a couple the moment of which is represented by twice the area of the triangle.

3-22C. Two uniform rods AB, BC, rigidly joined at B so that ABC is a right angle, hang freely in equilibrium from a fixed point at A. The lengths of the rods are a and b and their weights are wa and wb. Prove that, if AB makes an angle θ with the vertical,

$$\tan \theta = \frac{b^2}{a^2 + 2ab}$$

3-23C. Two smooth spheres of radius r and weight W are placed inside a hollow cylinder of radius a, open at both ends, which rests on a horizontal plane with its axis vertical. Prove that, in order that the cylinder may not upset, its weight must at least be

$$2W\left(1 - \frac{r}{a}\right)$$

(where $r > a/2$).

3-24C. A uniform bar of weight W and length $2a$ is suspended from two points in a horizontal plane by two equal strings of length b, which are originally vertical. The strings are connected to the ends of the bar. Show that the couple which must be applied to the bar in a horizontal plane to keep it at rest at right angles to its former direction is

$$\frac{Wa^2}{\sqrt{b^2 - 2a^2}}$$

3-25*C*. A uniform beam of length c hangs from a single point by two light inextensible strings of lengths a and b. Show that the tensions in the strings are

$$\frac{Wa}{\sqrt{2a^2 + 2b^2 - c^2}} \quad \text{and} \quad \frac{Wb}{\sqrt{2a^2 + 2b^2 - c^2}}$$

3-26*C*. A beam of weight W rests against a smooth horizontal rail, with its lower end on a smooth horizontal plane. Find what horizontal force must be applied to the lower end in order that the beam may be in equilibrium at a given inclination θ to the horizontal, having given the height h of the rail above the horizontal plane and the distance a of the center of mass of the beam from the lower end.

$$\frac{Wa}{h} \sin^2 \theta \cos \theta$$

3-27. Two identical uniform ladders of length b and weight W_1 are leaned together, being smoothly hinged at the top and standing on a smooth floor. They are prevented from slipping by a light inextensible cord of length $2a$ connecting the other two ends of the ladders ($2a < 2b$). A man of weight W_2 is at a distance x, measured along one ladder from the end of the ladder in contact with the floor. Find the tension in the cord.

3-28. Two smooth planes, inclined in the opposite directions to the horizontal at angles α and β, have a common line of intersection with a horizontal plane. A uniform rod rests on the two planes. Show that its inclination θ to the horizontal can be expressed in the form

$$\tan \theta = \tfrac{1}{2}(\cot \alpha - \cot \beta)$$

3-29. Two identical uniform ladders, each of weight W and length b, are smoothly hinged at one end and stand on a rough horizontal plane. A man, also of weight W, ascends one of the ladders to a distance x from the foot of the ladder. Find the normal components of the reactions of the floor on the ladders. Which end will slip first?

3-30. In Example 3-8, let bar OP have a weight W_1 and QP a different weight W_2. By considering the structure as a whole, determine the reactions N_1 and N_2 of the floor on OP and QP, respectively. Then by examining the equilibrium of each bar and the strut separately, and making no initial assumptions as to the directions of the reactions R_1 and R_2, show that R_1 has only a horizontal component, and find the horizontal and vertical components of R_2.

3-31. In Example 3-8, let the strut AB have a weight $2W$. Determine the reactions at A and P. Let $x = b/4$.

3-32*C*. Two equal beams AA', BB', without weight, are smoothly hinged at their common middle point C and placed in a vertical plane on a smooth horizontal table. The upper ends A, B of the rods are connected by a light string ADB, on which a small heavy ring can slide freely. Show that, in equilibrium, a horizontal line through the ring D will bisect AC and CB. (See Fig 3-19.)

3-33. A stiff plank, of negligible weight, is laid across a small stream. A man of weight W walks slowly across the plank. Show that the instantaneous bending couple acting on the plank is greatest at the point of the plank on which the man stands. Show also that the probability that the plank will break is greatest when the man is at the middle of the plank.

3-34. A uniform heavy rod of length b is supported at each end. If W is the weight per unit length, show that the bending couple at a point a distance x from one end is $(Wx/2)(b - x)$.

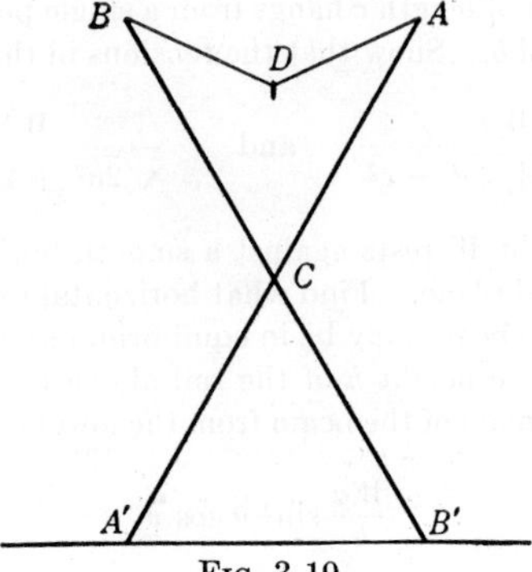

FIG. 3-19

3-35C. A light uniform horizontal beam, which is to be equally loaded at all points of its length, rests smoothly upon supports at one end and at some other point along the beam. Find where the second support should be situated in order that the greatest possible total load, uniformly distributed, may be placed upon the beam without breaking it, and show that the second point of support will divide the beam in the ratio $1/(\sqrt{2} - 1)$.

3-36. A heavy ladder of length b is standing on a rough floor and leans against a smooth wall, the whole being in limiting equilibrium. A man begins to ascend the ladder. How high along the ladder can he climb without the ladder slipping?

CHAPTER 4

STATICS OF THE SUSPENDED STRING OR CABLE

4-1. Introduction. No bodies exist in nature which are perfectly rigid. All are *deformable* to a greater or lesser degree. In many cases, such as those considered in previous chapters, it is a sufficient approximation to treat the bodies as being perfectly rigid. This is possible since many bodies are only very slightly deformed under the action of forces of the magnitudes usually encountered. In some situations, however, such an assumption is not possible. For example, if a large force is applied normal to a beam at one end, the beam will be deformed to an extent depending on the magnitude of the force. If the deformation is not too great, the beam will assume its former shape when the force is removed. Other deformable bodies, such as a string lying loosely on a table, may be deformed by very small forces and moreover will not assume their previous shapes when the force is removed. Such bodies are said to be *freely deformable*, as contrasted to the former case, in which the bodies are called elastic bodies. In the present chapter we shall limit ourselves to the consideration of a single freely deformable body, a string hanging in static equilibrium in a uniform gravitational field under various conditions. Moreover we shall treat only those cases in which the string hangs in a curve in a single plane.

In treating deformable bodies the assumption is made that they consist of a large number of rigid elements joined together and that the forces acting on each element are such external forces as have already been considered, including the effect of the adjacent elements on the element under consideration. The latter force is the tension in the string and is assumed to act along the line joining the two elements.

Before proceeding further, the student is advised to consult Appendix 3. The elementary identities there listed, of the class of functions known as hyperbolic functions, will have wide application in the present chapter.

4-2. Light Cable Supporting a Horizontal Roadway. The Suspension Bridge. We treat first the case in which the weight of the cable is neglected and the load supported is distributed uniformly along a horizontal direction. Such is the situation which obtains for the ordinary suspension bridge (see Fig. 4-1). The points of support, C and C', are at the same level. The problem is analyzed most simply by considering a section of the cable one end of which is the lowest point and the other end is any other point of the string. Such a pair of points are A and B of Fig. 4-2. The forces acting on AB are the tension T_0 at A, which is in a horizontal direction, the tension T tangent to the cable at B, and

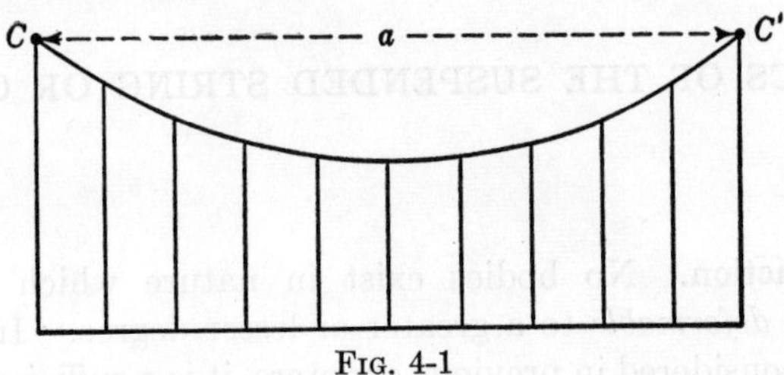

FIG. 4-1

the total weight W of the section $A'B'$ of the roadway supported by AB. The latter force may be regarded as acting through the mid-point of $A'B'$. (It is worthy of note that, since there are but three external forces acting on the segment AB and since the configuration is one of static equilibrium, the lines of action of the three forces must intersect in a single point E.) If w is the weight per unit length of the roadway (the length measured horizontally) and θ is the angle which the tangent to the cable at B makes with the horizontal, the conditions of equilibrium of AB are

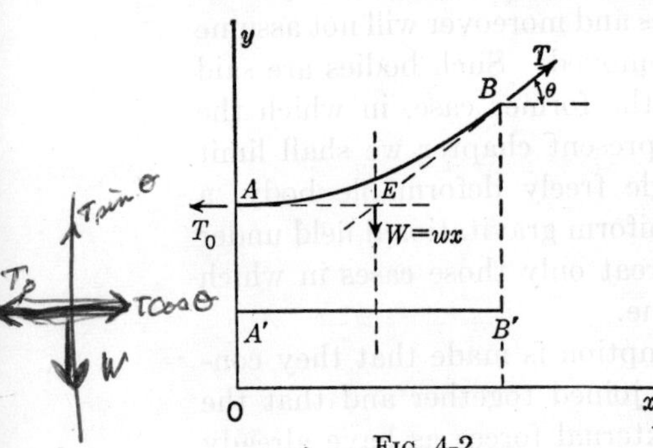

FIG. 4-2

$$T \sin \theta = wx \tag{4-1}$$

$$T \cos \theta = T_0 \tag{4-2}$$

Dividing Eq. (4-1) by Eq. (4-2), and since $\tan \theta = dy/dx$, we obtain

$$\frac{dy}{dx} = \frac{wx}{T_0}$$

from which

$$y = \frac{wx^2}{2T_0} + c_1 \tag{4-3}$$

the equation of a *parabola*, with c_1 an arbitrary constant of integration. The meaning of c_1 is clear. It is the height OA of the mid-point of the cable above the origin O. The axis of the parabola is Oy. The tension T at any point of the cable can be obtained quite readily from Eqs. (4-1) and (4-2) by squaring the two and adding them. We obtain finally

$$\boxed{T = \sqrt{T_0{}^2 + w^2x^2}} \tag{4-4}$$

Clearly the tension will increase with increasing x^2, the maximum being at the points of support. If we designate the *span*, or horizontal distance between the points of support, as a, the maximum tension T_m will be

$$T_m = \frac{1}{2}\sqrt{4T_0^2 + w^2a^2} \tag{4-5}$$

4-3. Uniform Cable Supporting Its Own Weight. The Uniform Catenary. In the present section, we consider a uniform heavy string or cable hanging under its own weight. Here the situation is different from the previous case, in which the load was uniformly distributed along a horizontal direction. We now have a load which is uniformly distributed along the cable itself. This load is the weight w per unit length of the cable. In Fig. 4-3 we have a segment AB of a cable which is suspended from two points (not shown) at the same level. Point A is the lowest point of the cable, and B is any other point having coordinates x, y with respect to the axes shown. The axis Oy passes through the point A. It is convenient, also, to designate a coordinate s, having the dimensions of length, measured along the cable from point A as origin (not to be confused with the origin, O, of x and y). The equations of static equilibrium of the segment AB are

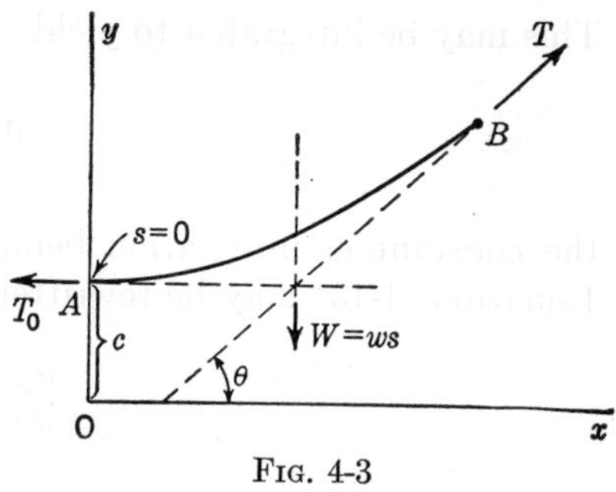

FIG. 4-3

$$T \cos \theta = T_0 \tag{4-6}$$
$$T \sin \theta = ws \tag{4-7}$$

in which T_0 is the tension at A and T is the tension at B. As before, we divide Eq. (4-7) by Eq. (4-6), obtaining

$$s = c \tan \theta \tag{4-8}$$

where c is defined by

$$c = \frac{T_0}{w} \tag{4-9}$$

and has the dimensions of length. Equation (4-8) is called the *intrinsic equation of the catenary.* The form is a bit more familiar, however, if we express it in terms of the rectangular coordinates x and y. To do this, we note that $\tan \theta = dy/dx$, whence Eq. (4-8) becomes

$$\frac{dy}{dx} = \frac{s}{c} \tag{4-10}$$

from which

$$\frac{d^2y}{dx^2} = \frac{1}{c}\frac{ds}{dx} = \frac{1}{c}\frac{\sqrt{dx^2 + dy^2}}{dx} = \frac{1}{c}\sqrt{1 + \left(\frac{dy}{dx}\right)^2} \tag{4-11}$$

an equation which contains neither y nor x explicitly. We make the substitution $p = dy/dx$. Equation (4-11) becomes a first-order equation in p and x:

$$\frac{dp}{dx} = \frac{1}{c}\sqrt{1 + p^2}$$

Separating the variables, we have

$$\frac{dp}{\sqrt{1 + p^2}} = \frac{1}{c}\,dx \tag{4-12}$$

This may be integrated to yield

$$\sinh^{-1} p = \frac{x}{c} \tag{4-13}$$

the constant of integration being zero, since $p = dy/dx = 0$ at $x = 0$. Equation (4-13) may be rewritten

$$\frac{dy}{dx} = \sinh \frac{x}{c} \tag{4-14}$$

Equation (4-14) may be integrated to

$$\boxed{y = c \cosh \frac{x}{c}} \tag{4-15}$$

in which the constant of integration is zero if $y = c$ at $x = 0$. Equation (4-15) is the familiar form of the equation of the catenary with vertex at $x = 0$, $y = c$. The axis Oy is called the *axis* of the catenary. The quantity c is called the *parameter* of the catenary and characterizes it. Catenaries which have the same parameter c are termed *equal*. The Ox axis lies a distance c below the vertex and as such is called the *directrix* of the catenary.

It is sometimes convenient to express the variable s as a function of x. This may be done by combining Eqs. (4-10) and (4-15). Differentiating Eq. (4-15) and substituting in Eq. (4-10), we obtain

$$\boxed{s = c \sinh \frac{x}{c}} \tag{4-16}$$

It is possible to obtain other relationships in a similar manner. We write only one of them, the relation involving y and s, with x absent. This is obtained by squaring Eqs. (4-15) and (4-16) and subtracting (4-16) from (4-15). We have

$$\boxed{y^2 = s^2 + c^2} \tag{4-17}$$

The equation [Eq. (15)] of the catenary is represented by the solid line shown plotted in Fig. 4-4. It is to be noted that the curve is symmetrical about Oy, the axis of the catenary. The x axis in the figure is the directrix. The solid curve is merely the sum of two exponentials, as may be seen from the definition of the hyperbolic cosine (cf. Appendix 3). We may rewrite Eq. (4-15) as

$$y = \frac{c}{2}\left(e^{x/c} + e^{-x/c}\right) \tag{4-18}$$

The tension at any point of the cable may be found from Eqs. (4-6), (4-15), and (4-16). Since $\cos\theta = dx/ds$, we have

$$T = T_0 \frac{ds}{dx} = T_0 \cosh\frac{x}{c} = \frac{T_0}{c}\, y = wy \tag{4-19}$$

Hence the tension at any point of the string is proportional to the height

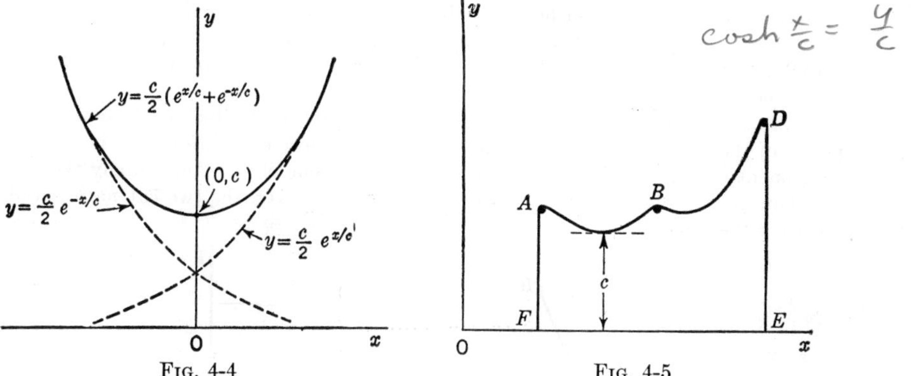

Fig. 4-4

Fig. 4-5

of that point above the directrix. The maximum value of T occurs at the points of support, at which points

$$|x| = \frac{a}{2} \qquad y = h + c \tag{4-20}$$

where a is the span and h is the *sag*, or maximum distance at which the cable hangs below the straight line joining the points of support. The maximum value T_m of the tension is

$$T_m = T_0 \cosh\frac{a}{2c} = wc \cosh\frac{a}{2c} = w(h + c) \tag{4-21}$$

4-4. Points of Support Not at the Same Level. In Fig. 4-5, a uniform heavy string hangs in static equilibrium over three smooth fixed pegs A, B, and D, as shown, with its ends hanging freely. A and B are at the same level. Let us consider first the portion FAB of the string. We know that the portion AB comprises a catenary with the lowest point hanging a distance c above the directrix. (The x axis may be chosen to be the directrix, as shown.) Equation (4-19) states that the end F of the hanging portion must just touch the directrix if equilibrium is to be maintained at point A. Similarly, if equilibrium is to exist at points B and D, the portion BD of the string must have the same directrix as AB and the end E must also touch the directrix. The fact that the two segments AB and BD have the same directrix does not necessarily mean that the segments are arcs of equal catenaries. This follows because the peg B is fixed in space, and, even though it is smooth, the resultant horizontal force applied by the two segments of the string need not be zero. This means that the horizontal component of the tension on the two sides of the peg need not be the same. Accordingly, if it is different on the two sides, each side will possess a different parameter c.

Example 4-1. A uniform heavy chain hangs between fixed points A and B, not necessarily at the same level. A weight is attached to the chain at a position D between points A and B (see Fig. 4-6). Show that the segments AD and DB form arcs of equal catenaries.

We are given that the chain has a weight w per unit length which is the same at all points of the chain. Furthermore, if W is to be in static equilibrium, the horizontal component of the tension must be the same on each side of D. Accordingly the two segments possess the same parameter $c = T_0/w$. However, the two catenaries need not have the same directrix.

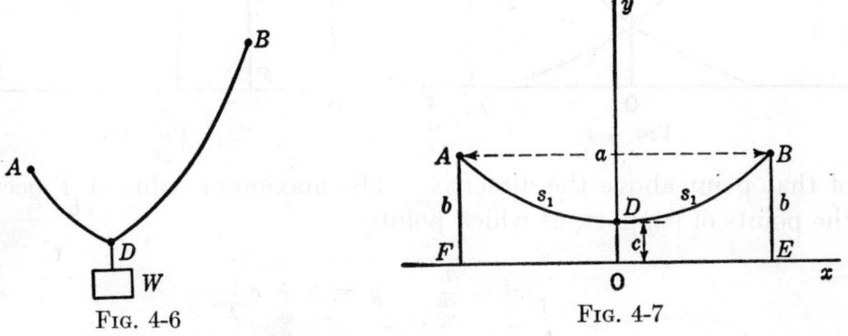

FIG. 4-6

FIG. 4-7

Example 4-2. A heavy string hangs over two smooth pegs at the same level a distance a apart. Show that equilibrium is impossible unless the length of the string is equal to or greater than ae, where e is the base of the system of natural logarithms.

The situation is as shown in Fig. 4-7, where Ox is the directrix. The total length L of the string is the sum of its parts, or $L = 2s_1 + 2b$. The parameter of the catenary ADB is c. For equilibrium to obtain at point B, we must have

$$T_m = bw = wc \cosh \frac{a}{2c}$$

Also

$$b = c \cosh \frac{a}{2c} \qquad \text{and} \qquad s_1 = c \sinh \frac{a}{2c}$$

Hence the total length of the string may be written

$$L = 2c\left(\cosh\frac{a}{2c} + \sinh\frac{a}{2c}\right) = 2ce^{a/2c} \tag{4-22}$$

If the length L of the string is altered, the only other quantity in this equation which changes is the parameter c. Hence we investigate stationary values of L by computing the derivative of L with respect to c. We have

$$\frac{dL}{dc} = 2e^{a/2c} - 2c\frac{a}{2c^2}e^{a/2c} = 0 \tag{4-23}$$

from which $a = 2c$ and thus, from Eq. (4-22), $L = ae$. In order to ascertain whether the stationary value of L, thus found, is a maximum or a minimum, it is necessary to differentiate Eq. (4-23) again. It is easily found that, if $a = 2c$, $d^2L/dc^2 = 2e/c$, a positive quantity. Accordingly ae is the minimum length for which equilibrium is realized. That equilibrium obtains for lengths greater than this is clear on physical grounds.

4-5. Determination of the Parameter c of the Catenary. It is perhaps apparent that many problems involving suspended strings or cables require the determination of the parameter c of the catenary. We first derive a rigorous expression for c. Consider the uniform string AB hanging from two points A and B, not necessarily at the same level, in Fig. 4-8. Let us take, as coordinate axes, the axis and directrix of the catenary of which AB is an arc. Point A has rectangular coordinates x_1 and y_1 with respect to the axes shown and also the coordinate s_1 measured along the catenary from point O' a distance c above O (c is the parameter of the catenary). Similarly B has coordinates x_2, y_2, s_2. It is convenient to employ the symbols

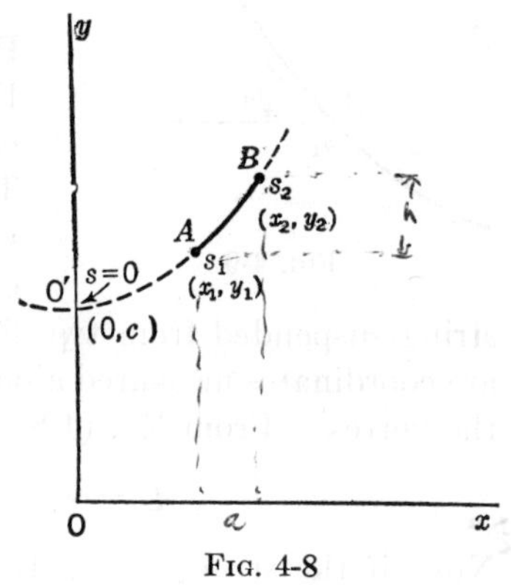

FIG. 4-8

$$L = s_2 - s_1 \qquad a = x_2 - x_1 \qquad h = y_2 - y_1 \tag{4-24}$$

where the analogies with the previous meanings of these symbols are clear. From Eqs. (4-15) and (4-16), we may write

$$L = c\left(\sinh\frac{x_2}{c} - \sinh\frac{x_1}{c}\right) \quad \text{and} \quad h = c\left(\cosh\frac{x_2}{c} - \cosh\frac{x_1}{c}\right)$$

Squaring these and subtracting, we have

$$(L^2 - h^2) = c^2\left(\sinh^2\frac{x_2}{c} + \sinh^2\frac{x_1}{c} - 2\sinh\frac{x_2}{c}\sinh\frac{x_1}{c}\right) - c^2\left(\cosh^2\frac{x_2}{c} + \cosh^2\frac{x_1}{c} - 2\cosh\frac{x_2}{c}\cosh\frac{x_1}{c}\right)$$

This may be simplified (see Appendix 3) to become

$$(L^2 - h^2) = 2c^2\left(\cosh\frac{a}{c} - 1\right) \tag{4-25}$$

from which

$$(L^2 - h^2) = 4c^2 \sinh^2\frac{a}{2c} \tag{4-26}$$

This is an equation from which c may be determined numerically. It is pointed out by Routh[1] that Eq. (4-26) furnishes two real finite values of c, one positive and one negative, of which only the positive value corresponds to a catenary with vertex downward. Accordingly, the positive root is to be selected. It is to be emphasized that, since only one positive value of c is obtained from Eq. (4-26), a uniform heavy string suspended from two fixed points can assume but one position of equilibrium.

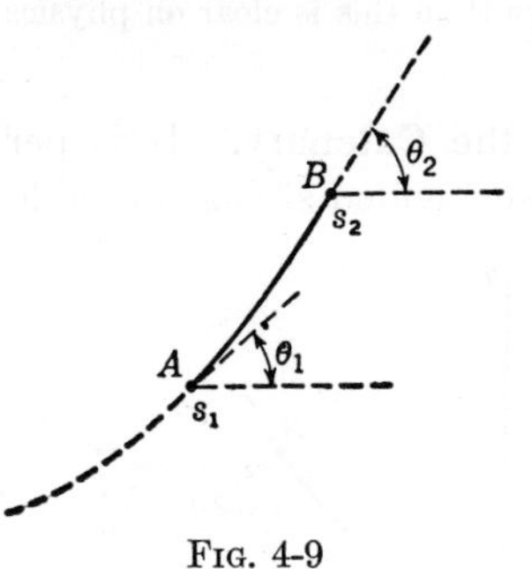

FIG. 4-9

4-6. Approximate Determination of the Parameter c for a Tightly Stretched String. For a tightly stretched string it is not difficult to obtain an approximate value for c. This rests upon the fact that in this case c is very large. This may be seen by a consideration of Fig. 4-9, which shows a length L of string suspended from two fixed points A and B. In Fig. 4-9, s_1 and s_2 are coordinates measured along the catenary (of which AB is an arc) from the vertex. From Eq. (4-8) we have

$$L = s_2 - s_1 = c(\tan\theta_2 - \tan\theta_1)$$

Now, if the string is tightly stretched, θ_1 must be approximately equal to θ_2 and thus, to preserve the equality, c must be very large compared with L. For any heavy string of length L suspended from two fixed points at the same level a distance a apart, we may write [from Eq. (4-16)]

$$\frac{L}{2} = c\sinh\frac{a}{2c} \tag{4-27}$$

For the case of a tightly stretched string we make use of the fact that c is large. Expanding the quantity on the right of Eq. (4-27), we have

$$\frac{L}{2} = c\left(\frac{a}{2c} + \frac{1}{3!}\frac{a^3}{8c^3} + \cdots\right) \simeq \frac{a}{2} + \frac{a^3}{48c^2} \tag{4-28}$$

neglecting powers of a/c higher than the second. Solving for c, we

[1] "Analytical Statics," Vol. I, Sec. 447, Cambridge University Press, London, 1896.

obtain

$$c = \left[\frac{a^3}{24(L - a)}\right]^{\frac{1}{2}} \tag{4-29}$$

as an approximate expression for the parameter c, in terms of the length L and the span a, of a string hanging from two fixed points at the same level.

It is an easy matter to obtain the maximum tension T_m to the first approximation for such a case. Employing Eq. (4-21), we write, neglecting higher powers of a/c than the first,

$$T_m = wc \cosh \frac{a}{2c}$$

$$= wc\left(1 + \frac{1}{2!}\frac{a^2}{4c^2} + \frac{1}{4!}\frac{a^4}{16c^4} + \cdots\right) \simeq w\left(c + \frac{a^2}{8c}\right) \tag{4-30}$$

Example 4-3. It is required to determine the sag of a tightly stretched string, of given length L, suspended between two fixed points at the same level a distance a apart.

If h is the sag and y_1 is the height of a point of support above the directrix, we may write

$$h = y_1 - c = c\left(\cosh \frac{a}{2c} - 1\right) \tag{4-31}$$

Expanding the hyperbolic cosine in Eq. (4-31), we have

$$h = c\left(1 + \frac{1}{2!}\frac{a^2}{4c^2} + \frac{1}{4!}\frac{a^4}{16c^4} + \cdots\right) - c \simeq \frac{a^2}{8c} \tag{4-32}$$

expressed in terms of the span a and the parameter c. By employing Eq. (4-29), this becomes

$$h \simeq [\tfrac{3}{8}a(L - a)]^{\frac{1}{2}} \tag{4-33}$$

4-7. Parabolic Catenary. It is of interest to examine Eq. (4-15) for the case of c large (tightly stretched string). We have

$$y = c \cosh \frac{x}{c} = c\left(1 + \frac{1}{2!}\frac{x^2}{c^2} + \frac{1}{4!}\frac{x^4}{c^4} + \cdots\right) \simeq c + \frac{x^2}{2c} \tag{4-34}$$

to a first approximation. Equation (4-34) is the equation of a parabola which, if c_1 in Eq. (4-3) is properly defined, is exactly that of a light cable supporting a uniform horizontal load. This is not surprising since the weight w per unit length of the heavy string approximates a uniform horizontal load if the string is tightly stretched.

4-8. Catenary of Uniform Strength. An interesting variation of the suspended string or cable problem is that in which the cross section of the cable is not uniform but varies in a manner such that the tension per unit area of the cross section is constant along the cable. If this is the case, the cable will have an equal probability of breaking at all points. A little reflection will enable the student to see that the cable should be heaviest at the points of support and that the thinnest portion should be at the mid-point (we consider only the case in which the points of suspension are at the same level).

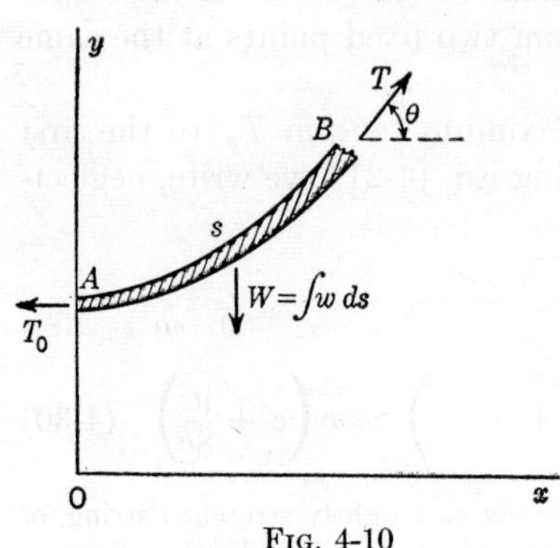

FIG. 4-10

As before, we consider the equilibrium of a segment AB of the cable (see Fig. 4-10), in which A is the lowest point and B is any other point a distance s away from A measured along the cable. If w is the weight per unit length of the cable (w is a variable here), the conditions of equilibrium of AB are

$$T \cos \theta = T_0 \tag{4-35}$$

$$T \sin \theta = \int_0^s w \, ds \tag{4-36}$$

where T is the tension. Dividing Eq. (4-36) by Eq. (4-35), we obtain

$$\tan \theta = \frac{dy}{dx} = \frac{1}{T_0} \int_0^s w \, ds \tag{4-37}$$

It is desirable to eliminate w from Eq. (4-37). This may be done by noting that, since the tension is proportional to the cross-sectional area, it is also proportional to the weight w per unit length. Denoting k as the constant of proportionality, we may put $T = kw$, and from Eq. (4-35)

$$w = \frac{T_0}{k} \frac{1}{\cos \theta} = \frac{T_0}{k} \frac{ds}{dx} \tag{4-38}$$

It is convenient to make the substitution $p = dy/dx$ in Eq. (4-37). If this is done, we may take the differential of p to be

$$dp = \frac{w}{T_0} ds = \frac{1}{k} \frac{ds}{dx} ds = \frac{1}{k} \left(\frac{ds}{dx}\right)^2 dx \tag{4-39}$$

in which we have made use of Eq. (4-38). Now since

$$ds^2 = dx^2 + dy^2 = \left[1 + \left(\frac{dy}{dx}\right)^2\right] dx^2 = (1 + p^2)\, dx^2$$

Eq. (4-39) becomes

$$\frac{dp}{1 + p^2} = \frac{1}{k} dx$$

This integrates readily to

$$\tan^{-1} p = \frac{x}{k}$$

where the constant of integration is zero since $p = 0$ at $x = 0$. We have therefore

$$p = \frac{dy}{dx} = \tan\frac{x}{k}$$

Separating the variables and integrating, we have

$$y = k \log \sec \frac{x}{k} + c_1 \tag{4-40}$$

where c_1 is a constant of integration. This equation possesses vertical asymptotes at $x = \pm\pi k/2$. We therefore have the interesting fact that the maximum possible span for the catenary of uniform strength is πk.

4-9. Heavy String or Cable Subject to Smooth Constraint. We now consider a heavy string which is in contact with a smooth curve in a single vertical plane. This is best treated by examining an element ds of the string (AB in Fig. 4-11); the procedure is very similar to that of Sec. 2-7. The forces acting on ds are the tension T at A, $T + dT$ at B, the weight $w\,ds$ of the element, and the normal reaction $N\,ds$ of the surface on the string.

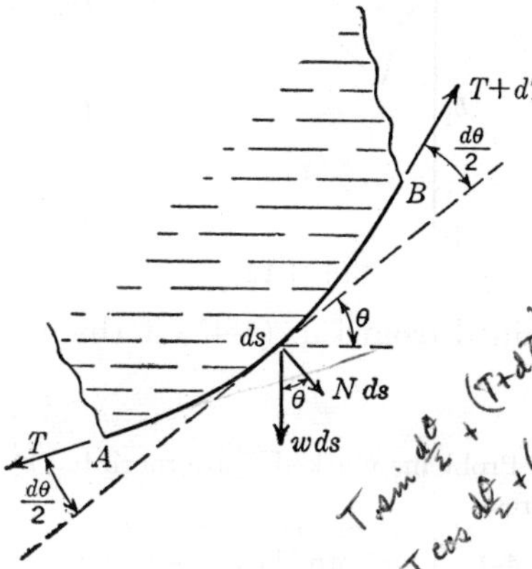

FIG. 4-11

Neglecting infinitesimals of the second order, the equations of equilibrium for ds are (resolving in directions tangent to and normal to the surface)

$$T \sin\frac{d\theta}{2} + (T + dT)\sin\frac{d\theta}{2} \simeq T\,d\theta = N\,ds + w\,ds\cos\theta \tag{4-41}$$

$$dT\cos\frac{d\theta}{2} \simeq dT = w\,ds\sin\theta \tag{4-42}$$

But $\sin\theta = dy/ds$, and therefore Eq. (4-42) becomes

$$dT = w\,ds\frac{dy}{ds} = w\,dy$$

and

$$T = wy + \text{const} \tag{4-43}$$

the same equation which governs the tension in the uniform catenary [see Eq. (4-19)]. In the interests of further clarification, Eq. (4-43) may be applied profitably to a simple example. Consider a uniform string hanging in a vertical plane in contact with a smooth surface (AB

in Fig. 4-12), it being required to determine the equilibrium configuration. This follows at once from Eq. (4-43), which states that, if the string is to be in equilibrium, the ends D and E must be at the same level, namely,

$$b_1 + h = b_2$$

If it is desired, the normal reaction per unit length N may be found from Eq. (4-41). Equation (4-41) may be rewritten

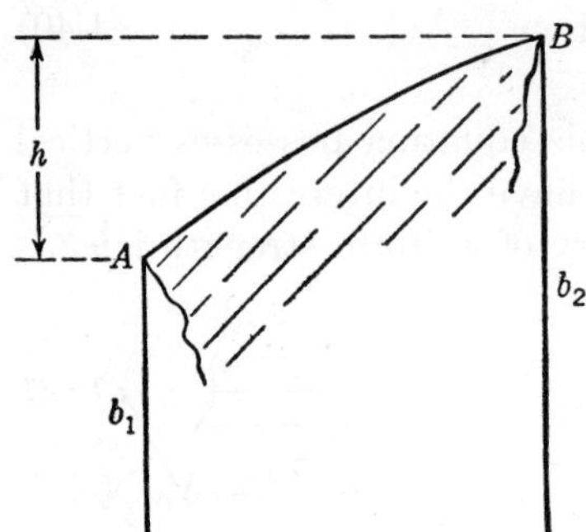

Fig. 4-12

$$T\frac{d\theta}{ds} = N + w \cos \theta \qquad (4\text{-}44)$$

But $d\theta/ds = 1/\rho$, where ρ is the radius of curvature of ds. Accordingly Eq. (4-44) becomes

$$N = \frac{T}{\rho} - w \cos \theta \qquad (4\text{-}45)$$

The normal reaction N may be determined from Eq. (4-45) if the equation of the curve is known.

Problems

Problems marked C are reprinted by kind permission of the Cambridge University Press.

4-1. A uniform heavy string hangs between two fixed points situated a certain distance apart at the same level. Show that

$$y = c \sec \theta$$

where c is the parameter of the catenary and θ is the angle the string makes with the horizontal at any point.

4-2. A uniform heavy wire 100 ft long is suspended between two points at the same level 75 ft apart. Find, by approximate means, the numerical value of the sag in the middle.

4-3. A uniform chain 100 ft long hangs between two points at the same level. If the sag is 15 ft and the maximum tension is 75 lb, find the weight per unit length of the chain. See Page 9 of Notes

4-4*C*. The span of a suspension bridge is 100 ft, and the sag at the middle is 10 ft. If the total load on each chain is 25 tons, find the greatest tension in each chain and the tension at the lowest point.

4-5. A cable composed of two parts of different densities joined end to end hangs between two fixed points. Show that the curvatures on the two sides of the junction are directly proportional to the two linear densities, respectively.

4-6. A string of total weight W and length b is suspended from two points at the same level, and a weight W_1 is attached to the lowest point of the string. If the string on either side of the lowest point makes an angle β with the horizontal, find the parameter of the catenary assumed by either branch of the string. Find also the maximum tension in the string.

4-7. The end links of a uniform chain of length b can slide on two smooth rods in the same vertical plane which are inclined toward each other at equal angles α to the vertical. Find the span.

4-8. In Prob. 4-7 show that the sag in the middle is $(b/2)\tan(\alpha/2)$.

4-9*C*. A uniform heavy cable of length L is tightly stretched over a river, the middle point just touching the surface of the water, while each of the extremities has an elevation h above the surface. Show that the difference between the length of the cable and the width of the river is nearly $8h^2/3L$.

4-10*C*. The end links of a uniform chain can slide on a fixed rough horizontal rod of coefficient of friction μ. Show that the ratio of the extreme span to the length of the chain is

$$\mu \log \frac{1 + \sqrt{1 + \mu^2}}{\mu}$$

4-11. A uniform, tightly stretched string of length $2b$ hangs between two points at the same level, the sag being a small quantity h. Show that, neglecting quantities of the order of $(h/b)^4$, the span is

$$2b\left(1 - \frac{2h^2}{3b^2}\right)$$

4-12*C*. A box kite is flying at a height h, with a length b of the string paid out, and with the vertex of the catenary on the ground. Show that, at the kite, $\tan\theta/2 = h/b$ and that the tension there is equal to the weight of a length $(b^2 + h^2)/(2h)$ of the string.

4-13*C*. Show that the maximum horizontal span for a uniform wire of given material is $1.325b$, where the maximum permissible tension is equal to the weight of a length b of the wire. Show also that the length of the wire under these conditions is $1.667b$.

4-14*C*. A heavy string of length $2b$ is suspended from two fixed points A and B in the same horizontal line at a distance apart equal to $2a$. A ring of weight W can slide freely on the string and is in equilibrium at the lowest point. Find the parameter of the catenary and the position of the weight.

4-15*C*. A chain of length b is stretched nearly straight between two points at different levels. If W is the weight and T the tension, show that the sag, measured vertically from the middle point of the chord, is $Wb/8T$.

4-16. A nonuniform string of total weight W hangs in the form of a circular arc, of radius a and semiangle α, between two points at the same level. Find the weight per unit length as a function of W, a, α, and the variable angle θ, where θ is the angle which the tangent at any point makes with the horizontal direction.

$$(W/2a)\cot\alpha \sec^2\theta$$

4-17. Prove that, in the catenary of uniform strength, $\rho = k\sec\theta = k\cosh(s/k)$.

4-18*C*. A heavy string just fits around a smooth vertical circle. Show that the tension at the highest point is three times that at the lowest.

4-19*C*. A heavy string is laid on a rough catenary with its vertex upward and its axis vertical, so that one extremity of the string is at the vertex. The length of the string is just equal to the parameter of the catenary. If the string rests in limiting equilibrium under gravity, show that the coefficient of static friction is $(2\log 2)/\pi$.

4-20*C*. A heavy string resting on a rough vertical circle with one extremity at the highest point is on the point of motion. If the length of string is equal to a quadrant, prove that $(\pi/2)\tan\epsilon = \log\tan 2\epsilon$, where ϵ is the angle of friction.

CHAPTER 5

WORK AND THE STABILITY OF EQUILIBRIUM

In this chapter the concept of energy is developed. Many problems in statics are more conveniently treated by energy methods than by utilizing the force components following the procedures outlined in Chaps. 2 and 3. The energy methods are particularly useful when it is desired to investigate the stability of equilibrium.

WORK AND POTENTIAL ENERGY

5-1. Work Done by a Force. If a particle experiences a force **F** at any point in a region of space, it is said to be in a *field of force* in that region. Suppose the particle is initially at some point A and is transported along some path to another point B. If the force is variable and the path an arbitrary curve, it is necessary to define the *element of work* dW as the amount of work done by the force on the particle while the latter is traversing an element $d\mathbf{r}$ along that curve. This is written

$$dW = \mathbf{F} \cdot d\mathbf{r} \tag{5-1}$$

The total work done by the force when the particle passes from A to B along the chosen path is the line integral of **F** along the path. It is

$$W_{A \to B} = \int_{\mathbf{r}_A}^{\mathbf{r}_B} \mathbf{F} \cdot d\mathbf{r} \tag{5-2}$$

and in general its value depends upon the particular path chosen. The work may also be written in terms of the components of the force along the coordinate axes. Equation (5-1) then becomes

$$\begin{aligned} dW &= (\mathbf{i}F_x + \mathbf{j}F_y + \mathbf{k}F_z) \cdot (\mathbf{i}\,dx + \mathbf{j}\,dy + \mathbf{k}\,dz) \\ &= F_x\,dx + F_y\,dy + F_z\,dz \end{aligned} \tag{5-3}$$

Integrating Eq. (5-3), we have

$$W_{A \to B} = \int_{x_A}^{x_B} F_x\,dx + \int_{y_A}^{y_B} F_y\,dy + \int_{z_A}^{z_B} F_z\,dz = W_x + W_y + W_z$$

which shows that the total work done by the force is equal to the sum of the work done by each of the components of the force.

The work done by a moment of force, or torque, acting through an elementary angle $d\theta$ can be obtained by considering a particle to move along a circular path of radius r. (We omit the use of vector notation for this demonstration.) The force F is assumed, for simplicity, to act in a direction tangent to the path at all points of the path.

$$dW = F(r\,d\theta) = L\,d\theta \tag{5-4}$$

since rF is the moment of the force, or torque, L. Thus the work done by F during the passage along the circular path from $\theta = \theta_1$ to $\theta = \theta_2$ is

$$W_{\theta_1 \to \theta_2} = \int_{\theta_1}^{\theta_2} L\,d\theta \tag{5-5}$$

It should be apparent from Eq. (5-1) that work is done in a given displacement only by those forces which have components in the direction of the displacement. W, in Eq. (5-2), may at times turn out to be a positive quantity and at other times a negative quantity, depending upon the point of view. Thus when a force does negative work, for example on a particle, this can also be regarded as the particle doing positive work on the system giving rise to the forces.

5-2. Potential Energy and the Conservative Field. Of special interest is the class of force fields in which the value of the line integral of the force between two points is independent of the particular path chosen (see Chap. 1). This means that, if a mechanical system being acted upon by such a force is changed from one configuration to another, the work done on the system by the force is independent of the manner in which the change was made. In particular, the work done on the system by the force is zero if the system is changed through a complete cycle of configurations back to the original one. A force field having this property is called a *conservative field*, and the force which acts at each point of the field is called a *conservative force*. Examples of conservative forces are the gravitational force and the force of electrostatic attraction and repulsion. Examples of nonconservative forces are frictional forces, forces of air resistance, and the like.

For a mechanical system being acted upon by conservative forces it is convenient to define the *potential energy* of the system as follows:

I. *Selecting any standard configuration O for a mechanical system being acted upon by conservative forces, the work done on the system by the forces in the passage from a configuration P to the configuration O is called the potential energy of the system when it is in the configuration P.*

It follows as a consequence of the definition of a conservative force that the potential energy of the system in the configuration P has a unique value with respect to the standard configuration O, regardless of the manner in which the change from O to P was made. The potential energy V_P, in the configuration P, may be written

$$V_P = \int_P^O \mathbf{F} \cdot d\mathbf{r} = -\int_O^P \mathbf{F} \cdot d\mathbf{r} \tag{5-6}$$

where $\mathbf{F}$ is the resultant of the conservative forces acting and the limits of the integral signify that the appropriate limits of the variable of integration $\mathbf{r}$ are employed. The work done upon the system by the forces when the system is changed from the configuration P to another configuration Q is found to be

$$W_{P\to Q} = \int_P^Q \mathbf{F} \cdot d\mathbf{r} = \int_P^O \mathbf{F} \cdot d\mathbf{r} + \int_O^Q \mathbf{F} \cdot d\mathbf{r} = V_P - V_Q \tag{5-7}$$

Thus we have:

II. *The work done by conservative forces upon a mechanical system, in the passage of the system from the configuration P to the configuration Q, is equal to the potential energy of P minus the potential energy of Q, both being taken with respect to an arbitrary standard configuration O.*

In particular, let us apply Eq. (5-7) to the case of a single particle being acted upon by a conservative force. Let the particle be moved a small distance $d\mathbf{r}$ from a point P to another point Q. The work performed by the force on the particle is, in view of Eq. (5-7),

$$dW = \mathbf{F} \cdot d\mathbf{r} = -dV \tag{5-8}$$

Now

$$dV = \frac{\partial V}{\partial x} dx + \frac{\partial V}{\partial y} dy + \frac{\partial V}{\partial z} dz$$

and

$$d\mathbf{r} = \mathbf{i}\, dx + \mathbf{j}\, dy + \mathbf{k}\, dz$$

Thus Eq. (5-8) becomes

$$\begin{aligned}(\mathbf{i}F_x + \mathbf{j}F_y + \mathbf{k}F_z) \cdot (\mathbf{i}\, dx + \mathbf{j}\, dy + \mathbf{k}\, dz) &= F_x\, dx + F_y\, dy + F_z\, dz \\ &= -\frac{\partial V}{\partial x} dx - \frac{\partial V}{\partial y} dy - \frac{\partial V}{\partial z} dz \end{aligned} \tag{5-9}$$

If Eq. (5-9) is to be true for any $d\mathbf{r}$, that is, for any elements dx, dy, and dz, we must equate the coefficients of the infinitesimals term by term. Accordingly

$$F_x = -\frac{\partial V}{\partial x} \qquad F_y = -\frac{\partial V}{\partial y} \qquad F_z = -\frac{\partial V}{\partial z} \tag{5-10}$$

Hence:

III. *The components of the force at any point in a conservative field are the negative of the corresponding components of the gradient of the potential energy at that point.*

Written down analytically, this becomes

$$\mathbf{F} = -\nabla V \tag{5-11}$$

An important point to be remembered is that, if a new fixed point O' is chosen as a reference point of potential energy, the potential energy V'_Q, relative to O', of a particle at Q differs from V_Q, its potential energy relative to O, only by an additive constant. This follows since

$$V_Q = \int_Q^O \mathbf{F} \cdot d\mathbf{r} = \int_Q^{O'} \mathbf{F} \cdot d\mathbf{r} + \int_{O'}^{O} \mathbf{F} \cdot d\mathbf{r} = V'_Q + \text{const} \tag{5-12}$$

both O and O' being fixed points. Hence from Eq. (5-11), the force at point Q is

$$\mathbf{F}_Q = -\nabla V_Q \equiv -\nabla V'_Q \tag{5-13}$$

5-3. Work Required to Raise a System of Particles at the Earth's Surface. The Uniform Field. A point of frequent application is the amount of work necessary to raise a body from one height to another in the neighborhood of the earth's surface. We confine ourselves only to those cases in which the dimensions of the bodies, and the heights through which they are raised, are very small compared with the radius of the earth. Thus we consider, to a very good approximation, that the force mg of gravity on a particle of mass m is acting vertically downward and that it will have the same value regardless of where the particle may be located. Such a force field, in which the force has the same magnitude and direction at all points of the field, is called a *uniform field*.

We consider a system of n particles. The initial height of the ith particle above the ground is h_i, and its final height is h'_i. Accordingly, the work done by the gravitational force on the ith particle when it is elevated through the distance $h'_i - h_i$ is $-m_i g(h'_i - h_i)$, where the minus sign follows since the force is oppositely directed to the displacement from h_i to h'_i. Hence the amount of work which must be done against gravity in the elevation of m_i is

$$W_i = m_i g(h'_i - h_i) \tag{5-14}$$

The total work which must be done against gravity in order to raise all n particles is found by merely summing Eq. (5-14) over the n particles or, what is equivalent, by adding together the n equations of the type (5-14). We obtain

$$W = \sum_{i=1}^{n} W_i = \sum_{i=1}^{n} [m_i g(h'_i - h_i)] = g\left[\sum_{i=1}^{n} m_i h'_i - \sum_{i=1}^{n} m_i h_i\right] \tag{5-15}$$

But the final sum within the brackets in Eq. (5-15), in view of Eqs. (3-5), is equal to the total mass M of the system of particles multiplied by the initial height h_c of the center of mass. Similarly, if h'_c is the final height of the center of mass, the first sum within the brackets is equal to the total mass times the quantity h'_c. Consequently Eq. (5-15) takes the form

$$W = g(Mh_c' - Mh_c) = Mg(h_c' - h_c) \tag{5-16}$$

This states that the work which must be done against gravity in order to raise the system of particles can be computed by imagining the mass to be concentrated at the center of mass and then elevating this mass from the initial height of the mass center to its final height. Also, since Eq. (5-16) is identical to the work done by gravity in lowering the system from the height h_c' to h_c, expression (5-16) is just the potential energy of the system at height h_c' relative to the configuration at h_c. If the lower position is the ground, the potential energy of the center of mass at height h relative to the ground is therefore merely mgh.

5-4. Conservation of Energy. Kinetic Energy. The importance of the concept of potential energy may be appreciated by an examination of a simple problem, that of the motion of a particle under the action of a force. The equation of motion in vector form is

$$\frac{d}{dt}(m\dot{\mathbf{r}}) = \mathbf{F}$$

Multiplying both sides by $\dot{\mathbf{r}}\cdot$, namely, taking the dot product of $\dot{\mathbf{r}}$ with both sides of the equation, we have

$$\dot{\mathbf{r}} \cdot \frac{d}{dt}(m\dot{\mathbf{r}}) = \mathbf{F} \cdot \dot{\mathbf{r}} \tag{5-17}$$

where the order of the factors in the dot product is immaterial. Assuming a constant mass, the left side may be rewritten as

$$\frac{m}{2}\frac{d}{dt}(\dot{\mathbf{r}} \cdot \dot{\mathbf{r}}) \equiv \frac{m}{2}\frac{d}{dt}|\dot{\mathbf{r}}|^2$$

That this is permissible is easily shown. We have

$$\frac{1}{2}\frac{d}{dt}(\dot{\mathbf{r}} \cdot \dot{\mathbf{r}}) = \frac{1}{2}(\ddot{\mathbf{r}} \cdot \dot{\mathbf{r}} + \dot{\mathbf{r}} \cdot \ddot{\mathbf{r}}) = \dot{\mathbf{r}} \cdot \frac{d}{dt}(\dot{\mathbf{r}})$$

Now $|\dot{\mathbf{r}}|^2 \equiv v^2$, where v is the magnitude of $\dot{\mathbf{r}}$. Accordingly, Eq. (5-17) becomes

$$\frac{m}{2}\frac{d(v^2)}{dt} = \mathbf{F} \cdot \frac{d\mathbf{r}}{dt}$$

whence

$$mv\,dv = \mathbf{F} \cdot d\mathbf{r}$$

Hence

$$\int_{v_1}^{v_2} mv\,dv = \int_{r_1}^{r_2} \mathbf{F} \cdot d\mathbf{r} \tag{5-18}$$

For the case in which $\mathbf{F}$ is a conservative force, it is the negative of the gradient of a potential energy V. Hence Eq. (5-18) becomes

$$\int_{v_1}^{v_2} mv\,dv = -\int_{r_1}^{r_2} \nabla V \cdot d\mathbf{r} = -\int_{V_1}^{V_2} dV \tag{5-19}$$

or

$$\frac{m}{2}v_1^2 + V_1 = \frac{m}{2}v_2^2 + V_2 \tag{5-20}$$

The second equality in (5-19) follows from Sec. 1-13. Equation (5-20) states the very important fact that:

IV. *In a motion of a particle which is acted upon only by conservative forces, the sum of the kinetic and potential energies remains constant.*

This is not difficult to see, since point 2 is arbitrary, Eq. (5-20) being true for any choice of point. In more general form, if T is the kinetic energy,

$$T + V = \text{const} \tag{5-21}$$

A simple application of Eq. (5-21) is the case of a particle sliding down a smooth inclined plane from a position of rest at a vertical height h above the floor. It is required to find its final velocity along the floor at the foot of the plane. Initially the particle has a potential energy mgh (with respect to the floor) and zero kinetic energy. At the foot of the plane the potential energy is zero. Consequently, then, the energy of the particle must be entirely kinetic, and the velocity is the familiar expression $\sqrt{2gh}$.

5-5. Work Required to Stretch an Elastic String. If a stretched string is perfectly elastic, it will return to its original length when the stretching force is removed. Hence the restoring force in the string, which is present when the string is stretched, can be thought of as a conservative force. In view of this we shall compute the potential energy acquired by a mass m (see Fig. 5-1) if the string is stretched from its natural length x_0 to a new length x. We consider also that the force constant of the string is k and that the string is not to be stretched beyond its *elastic limit*. We assume, moreover, that the string is stretched infinitely slowly, so that no appreciable kinetic energy is acquired by the system. If the stretched length is x, the restoring force F will be in the direction so as to restore the string to its original length. Consequently it will point in the $-x$ direction, and we may write

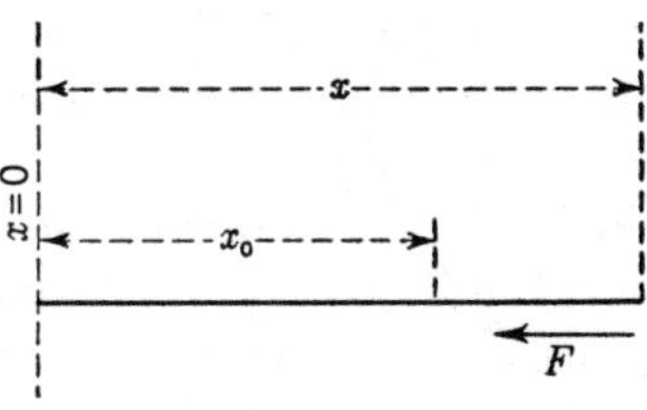

FIG. 5-1

$$F = -k(x - x_0)$$

Taking the zero of potential energy to be when $x = x_0$, we have, from

Eq. (5-6),

$$V = -\int_{x_0}^{x} F\,dx = +k\int_{x_0}^{x} (x - x_0)\,dx = \frac{k}{2}(x - x_0)^2 \qquad (5\text{-}22)$$

In terms of the modulus of elasticity λ, defined in Chap. 2, Eq. (5-22) takes the form

$$V = \frac{\lambda}{2x_0}(x - x_0)^2 \qquad (5\text{-}23)$$

since $\lambda = kx_0$. It is to be noticed that the potential energy *stored* in such a stretched elastic string is a positive quantity. This is the amount of work which must be done against the restoring force in order to stretch the string to its new length x, provided that no kinetic energy is imparted to the system during the process.

THE LAW OF GRAVITATION

5-6. Origin and Statement of the Law. Gravitational Field Strength. Possibly Newton's greatest single contribution to our present-day knowledge is the celebrated law of gravitation. The law may be stated in the form:

V. *Every body in the universe attracts every other body with a force, directed along the line joining the two bodies, of magnitude directly proportional to the masses of each of the bodies and inversely proportional to the squares of the distances separating the bodies.*

In equation form the force of attraction experienced by a mass m in the field of, and at a distance $\mathbf{r}$ from, a mass M is

$$\mathbf{F} = -\frac{GMm}{r^2}\mathbf{r}_1 \equiv -\frac{GMm}{r^3}\mathbf{r} \qquad (5\text{-}24)$$

in which the origin is placed at M, and $\mathbf{r}_1$ $(= \mathbf{r}/r)$ is a unit vector pointing along the radius vector from the origin. The minus sign indicates that the force is directed oppositely to $\mathbf{r}_1$, that is, it is directed always toward the origin. The quantity G is the so-called gravitational constant, which in cgs units has the numerical value 6.66×10^{-8}. The components of $\mathbf{F}$ along a set of rectangular axes with origin at M, from Eq. (5-24), are

$$F_x = -\frac{GMmx}{r^3} \qquad F_y = -\frac{GMmy}{r^3} \qquad F_z = -\frac{GMmz}{r^3} \qquad (5\text{-}25)$$

It is often convenient to use the term *field strength*, or *field intensity*, in connection with a force field. In the case of the gravitational field the field strength at a given point is the force per unit mass at that point and is a vector quantity. If the origin of force is considered to be the mass

M, the gravitational field strength at the given point, because of M, is

$$\mathbf{f} = \frac{\mathbf{F}}{m} = -\frac{GM}{r^2}\mathbf{r}_1 \tag{5-26}$$

In cgs units the field strength of a gravitational field is in dynes per gram. Clearly this has the dimensions of acceleration in the case of the gravitational field and may be stated as so many centimeters per second per second. Thus, for the uniform field at the earth's surface, the field intensity $\mathbf{f}$ is simply $\mathbf{g}$, the acceleration of gravity.

5-7. Potential Energy in a Gravitational Field. Potential. Since a gravitational field is an example of a conservative field, it is possible to define a potential energy in connection with it. We shall take as an example a mass M situated at a point O, attracting a particle of mass m situated at some point P the position of which with respect to the origin is defined by the radius vector $\mathbf{r}$. Since the force (5-24) vanishes at $r = \infty$, it is customary to consider ∞ as the zero of potential energy. In fact this is a suitable procedure for all forces of the form $1/r^n$, provided that n is greater than 1. Accordingly the potential energy of m situated at P, a distance r from O, may be found by computing the negative of the work done by the force while m is brought from $r = \infty$ to $r = r$. A straight-line path extending through the origin is convenient. We have

$$V = -\int_\infty^r \mathbf{F} \cdot d\mathbf{r} = -\int_\infty^r F\,dr = GMm \int_\infty^r \frac{dr}{r^2} = -\frac{GMm}{r} \tag{5-27}$$

which is a negative quantity. It may be remarked that, provided that the zero of potential energy is placed at infinity, the potential energy will be negative for all attractive forces varying as $1/r^n$, where $n > 1$, and positive for all repulsive forces varying in the same fashion.

In a manner similar to the relationship between the field strength $\mathbf{f}$ and the force $\mathbf{F}$, it is possible to define a *potential* ϕ which is just the potential energy per unit mass at a point in the field. Thus the potential at a distance r from M is

$$\phi = -\frac{GM}{r} \tag{5-28}$$

5-8. Field and Potential of an Extended Body. As it stands, Eq. (5-24) expresses the gravitational force existing between two particles. In general it applies to larger bodies as well only if the distances separating them are very large compared with the dimensions of the bodies (compare, however, in Sec. 5-9, with the field due to a uniform sphere, at an internal point). Thus, to compute the force between two extended bodies, it is necessary to consider the contributions of all the elements of mass of each body. The procedure for finding the force or the field strength by a direct integration is difficult except for symmetrically shaped bodies. Consider, for example, the body in Fig. 5-2. It is desired to find the field at point P produced by the body shown. Two mass elements dm_i and dm_j are shown, each of which gives a tiny contribution $d\mathbf{f}_i$ and $d\mathbf{f}_j$, respectively, to the field at P. Clearly these must be added vectorially. Then, in turn, the contributions of the other mass elements of the body must be vectorially compounded with this result. This procedure, which is often a laborious one, will be applied below to the relatively simple problem of finding the field of a homogeneous circular disk at a point on its axis of symmetry.

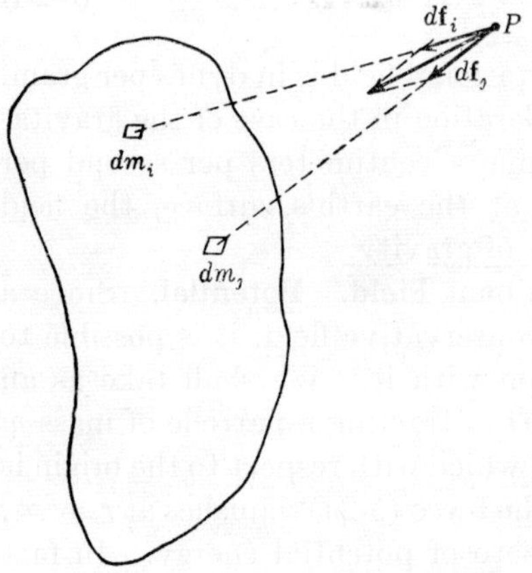

FIG. 5-2

A better method, for many instances, is first to compute the potential of the body. This is usually a much simpler procedure since the contribution of each mass element is a scalar and can thus be added to the contributions of the rest of the body by simple algebraic addition (see, for example, the case of electrostatics, in which two signs of charge afford the possibility of two signs of potential; in the gravitational case we have only positive mass). Following this, the field strength may be obtained by computing the gradient of the potential.

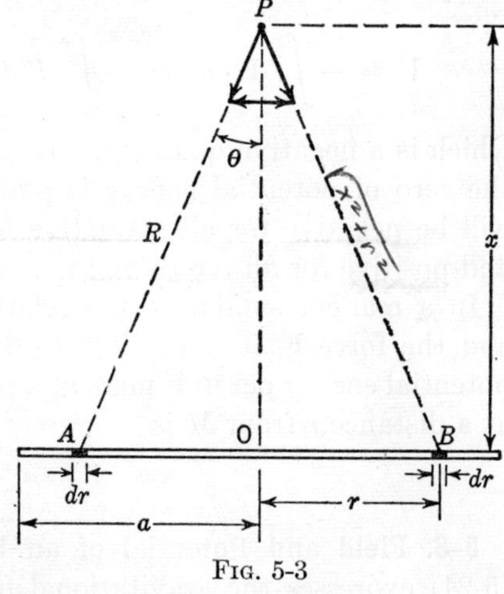

FIG. 5-3

Example 5-1. Find the potential at a point outside of, and on the axis of symmetry of, a homogeneous circular disk of radius a, small thickness t, and mass σ per unit volume.

A vertical section of the disk is pictured in Fig. 5-3. We select a ring element a distance r from the center. Since all parts of this ring are at the same distance from the axis, the potential at point P, a distance x from the center, due to this ring is simply

$$d\phi = -\frac{G\,dm}{R} = -\frac{G(2\pi r\sigma t\,dr)}{R} = -\frac{2\pi\sigma Gtr\,dr}{(x^2 + r^2)^{\frac{1}{2}}}$$

Integrating, we have

$$\begin{aligned}\phi &= \int_0^a \frac{2\pi\sigma Gtr\,dr}{(x^2 + r^2)^{\frac{1}{2}}} = -2\pi\sigma Gt(x^2 + r^2)^{\frac{1}{2}}\Big|_0^a \\ &= -2\pi\sigma Gt[(x^2 + a^2)^{\frac{1}{2}} - x] \qquad (5\text{-}29)\end{aligned}$$

Example 5-2. Find the field **f**, for the disk of Example 5-1, at the same point.

The simplest way of achieving this result is merely to take the gradient of the expression in Eq. (5-29). It is clear, from Fig. 5-3, that the symmetry of the body requires that **f** will point vertically downward at P, and thus we need only consider $-\partial\phi/\partial x$. Accordingly, the magnitude of **f** is given by

$$f = -\frac{\partial\phi}{\partial x} = -2\pi\sigma Gt\left[1 - \frac{x}{(x^2 + a^2)^{\frac{1}{2}}}\right] \tag{5-30}$$

The same result may be achieved by a direct integration. Consider Fig. 5-3 again. Again the ring element of mass is selected. However, we first consider two small segments of this ring, one at A and one at B, in the plane of the paper. Clearly the contribution of each to the field at P is represented, respectively, by the arrow directed from P toward each segment. Moreover it is evident that these small contributions have horizontal components which mutually cancel. Accordingly, the vector sum of the two contributions is the sum of the projections along the x axis. This same procedure may be applied to symmetrical pairs of segments for the remainder of the ring element. Consequently the magnitude df of the field at P due to the entire ring element is

$$df = -\frac{G\,dm}{R^2}\cos\theta = -\frac{2\pi\sigma Gtxr\,dr}{(x^2 + r^2)^{\frac{3}{2}}} \tag{5-31}$$

We employ scalar notation since the resultant field contributed by all concentric ring elements lies along the x axis. The negative sign states that the direction of **df** is in the direction of decreasing x, as is to be expected. Integrating Eq. (5-31), we obtain a result identical to (5-30).

5-9. Field and Potential of a Homogeneous Spherical Body. Because of its importance this case merits special attention. Consider first the problem of finding by direct integration the potential of a spherical shell. In Fig. 5-4 the origin is at the center of the shell of radius a, and it is required to find the potential at an external point P, distant r from O. The shell has a mass σ per unit area. A ring element is selected the axis of which is OP. Accordingly the potential at P due to this element is

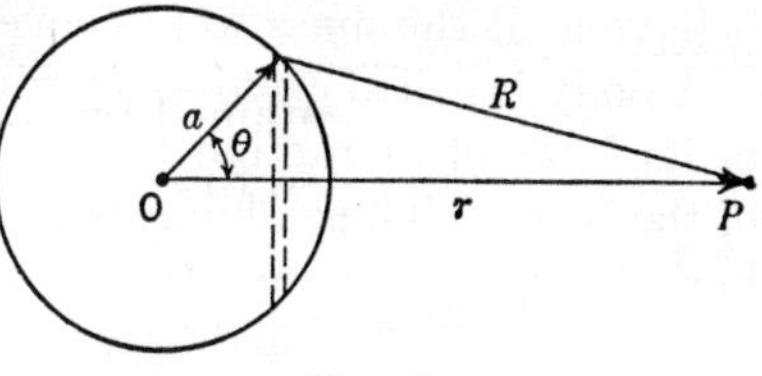

FIG. 5-4

$$d\phi = -\frac{2\pi Ga^2\sigma\sin\theta\,d\theta}{R} \tag{5-32}$$

Now

$$R^2 = a^2 + r^2 - 2ar\cos\theta$$

and

$$2R\,dR = 2ar\sin\theta\,d\theta$$

Substituting for $\sin\theta\,d\theta$ in Eq. (5-32), we have

$$d\phi = -\frac{2\pi Ga\sigma}{r}\,dR$$

and integrating with respect to the variable R, between the limits $r - a$ and $r + a$, we obtain

$$\phi = -\frac{2\pi G a \sigma}{r}\int_{r-a}^{r+a} dR = -\frac{2\pi G a \sigma}{r}[(r + a) - (r - a)] = -\frac{MG}{r} \quad (5\text{-}33)$$

where $M\ (= 4\pi a^2 \sigma)$ is the total mass of the shell. Equation (5-33) states that the potential of a homogeneous spherical shell at an external point is the same as if all the mass were concentrated at the center, a very important result indeed. It is a simple matter to extend this result to the case of a homogeneous solid sphere, since such a body may be regarded as a large number of concentric spherical shells. The effect of each of these, so far as an external point is concerned, is as if all its mass were concentrated at the center. Accordingly, Eq. (5-33) is also the potential of a homogeneous solid sphere at an external point, in which M is the total mass of the sphere.

The field can be computed at once by employing Eq. (1-69). We have

$$\mathbf{f} = -\frac{d\phi}{dr}\mathbf{r}_1 = -\frac{MG}{r^2}\mathbf{r}_1 \quad (5\text{-}34)$$

as is to be expected. Here $\mathbf{r}_1$ is again a unit vector pointing in the direction of increasing r.

It is worthwhile repeating that the results (5-33) and (5-34) are true only if the mass distribution of the spherical body possesses radial symmetry. If this is not the case, the field and potential do not in general behave as if the mass were concentrated at the center.

A body is called *centrobaric* if it has the property that the line of action of its force of attraction for a point mass passes through a single point of the body. Thus a homogeneous sphere (or spherical shell) is a centrobaric body with respect to any external point.

Equation (5-33), slightly altered, may be employed to find the potential at an internal point. In this instance the only change is that the limits of integration are from $R = a - r$ to $R = a + r$. For the potential at an internal point we find, therefore, that

$$\phi = -4\pi a \sigma G = -\frac{Mg}{a} \quad (5\text{-}35)$$

which states that the potential at any point inside a homogeneous spherical shell is a constant, equal to the potential at the surface of the shell. Thus the field inside a homogeneous spherical shell is zero. Accordingly, a particle can be moved around inside the shell without any work being done.

In passing, it is of interest to find the field $\mathbf{f}$ at any point, a distance r from the center, inside a homogeneous solid sphere of radius a. In view of the considerations given immediately above, the field at a distance r from the center must arise only from the portion of the sphere interior to r. Accordingly, at an interior point of a homogeneous solid sphere of radius a and density σ, we have

$$\mathbf{f} = -\frac{4}{3}\pi\frac{r^3\sigma G}{r^2}\mathbf{r}_1 = -\frac{4}{3}\pi\sigma G r\mathbf{r}_1 \tag{5-36}$$

and the magnitude of $\mathbf{f}$ is directly proportional to r.

PRINCIPLE OF VIRTUAL WORK

5-10. Applied Forces and Forces of Constraint. Virtual Displacements. In Sec. 1-19, the case of a block of weight W resting on a level floor was considered. W is the gravitational force exerted by the earth on the block. That the block does not move is accounted for by assuming the floor to exercise on the block a reaction R equal and opposite to W. The procedure has been followed in Chaps. 2 and 3 of designating such a force as R a *reaction of constraint*. The constraint in this case is the floor.

It is often useful to employ a terminology which in a given system distinguishes the reactions of constraint from other forces which are acting. The procedure will be to call all forces which are not reactions of constraints *applied forces*. Thus W, the gravitational force exerted on the block by the earth, is an applied force.

In the present chapter, the concept of a *virtual displacement* will be introduced. In the example above, of the block on the level table, let the surface of the table be the xy plane. Let us imagine the block to suffer a small displacement in the x direction. We call this imaginary, or virtual, displacement δx, in contrast to the differential dx for a *real displacement*. The virtual displacement δx is identical to dx geometrically, the difference being that *time is held constant* during the virtual displacement δx. Thus, even if the surface of the table is moving or changing in shape, the dependence of the position and shape of the surface upon the time will not affect the virtual displacement δx. This is not in general the case if the block is compelled to execute a real displacement dx during a time dt.

If the top of the table is smooth, there is no frictional reaction of constraint. Consequently no work is done during a virtual displacement such as δx, which is parallel to the surface (and hence which does not violate the constraint). (However, if a displacement δz, in a direction normal to the surface and thus violating the constraint, is considered, work will be done by R, the reaction of constraint. This will amount to $R\ \delta z$.) A constraint which admits of at least one virtual displacement in which the constraining forces do no work is termed a *workless constraint*. (Note that, since time is held fixed during a virtual displacement, a smooth surface the form of which depends upon the time is still a workless constraint.)

In general, since in a static problem forces other than constraining forces will also be acting, work will be done by the applied forces during a virtual displacement. The work done in this manner is called *virtual work*.

Although in the above discussion infinitestimal displacements were considered, it will be instructive in the next section first to examine several very simple mechanical systems in which finite displacements are conceived to occur.

5-11. Equilibrium and Finite Displacements. The Pulleys of Stevinus and the Inclined Plane of Galileo. Thus far, the foundations of statics have been presented on the basis of the equilibrium of forces and of moments of forces. Instead of selecting this approach we could as well have based statics on a single principle, and this we shall now consider. From this single principle it is possible, in turn, to derive the principles of the equilibrium of forces and of moments of forces. In Fig. 5-5, a weight W_1, hanging vertically, is connected, by means of a light inextensible string which passes over a smooth pulley, to a weight W_2. W_2 is constrained to slide on a perfectly smooth plane inclined at an angle α to the horizontal. Galileo recognized a very important feature of this arrangement. This was that the equilibrium could be interpreted in terms of displacements of the weights toward and away from the center of the earth. Galileo stated that the weights would be in static equilibrium provided that, in a displacement of the system, the product of W_1 and the vertical height through which it moves is equal to the product of W_2 and the vertical height through which W_2 moves. Moreover, the two products must be in the opposite sense: if W_1 is lowered, W_2 must be raised. Therefore, if we imagine W_1 to be lowered a distance h and if the system were initially in equilibrium, we must have

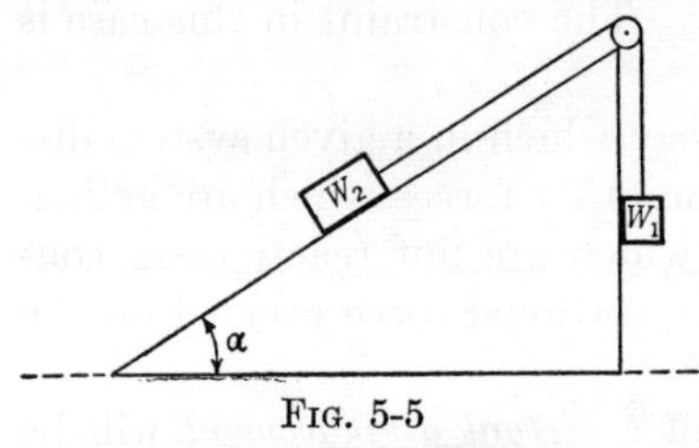

FIG. 5-5

$$W_1 h = W_2 h \sin \alpha$$

or

$$W_1 h - W_2 h \sin \alpha = 0 \tag{5-37}$$

In modern terms this states simply that the total work done by the applied forces (W_1 and W_2 here) during the displacement is zero. Clearly the result (5-37) is identical, save for the presence of h, to that obtained on the basis of Chap. 2. The perception of the content of Eq. (5-37) by Galileo is further evidence of his intellectual greatness.

The same result was apparent in the pulley systems of Stevinus, a contemporary of Galileo. In Fig. 5-6, we have systems of smooth, massless pulleys arranged in equilibrium. On the basis of the principles of statics already stated in previous chapters, we see that, in Fig. 5-6*a*, if the two weights are to be in equilibrium, they must be equal. In Fig. 5-6*b*, we may consider the situation one part at a time. Accordingly, if pulley A is to remain in equilibrium, we see that twice the tension T_1 must be equal to W_2. Further, since T_1 is equal to W_1, we have that $W_1 = W_2/2$. In Fig. 5-6*c*, a similar analysis shows that $W_1 = W_3/4$.

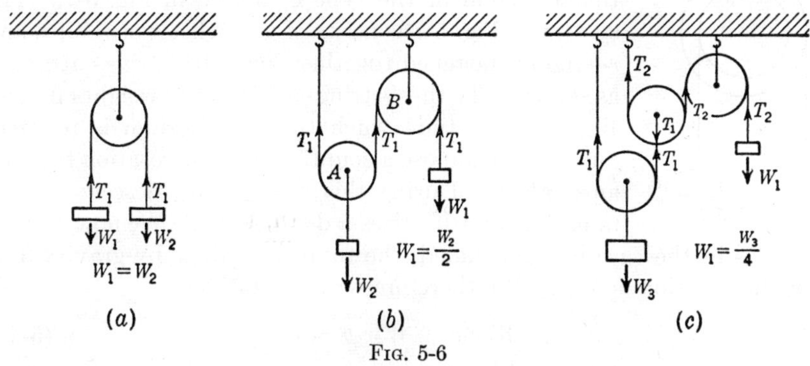

FIG. 5-6

However, these are just the results we obtain if we require that, during a virtual displacement of one of the weights of the system in a vertical direction, the remaining weights in the system will be displaced in a manner such that the total work done by the gravitational force is zero. This is easily verified, since, if we imagine W_1 in each system to be lowered a distance s_1, it is apparent from the geometry that in Fig. 5-6*a*, for example, we have

$$W_1 s_1 - W_2 s_2 = 0 \tag{5-38}$$

where s_2 is the distance W_2 is raised. But $s_1 = s_2$; hence

$$W_1 = W_2$$

In Fig. 5-6*b*, we have, if s_2 is again the constrained displacement of W_2 while W_1 is lowered a distance s_1,

$$W_1 s_1 - W_2 s_2 = W_1 s_1 - W_2 \frac{s_1}{2} = 0 \tag{5-39}$$

or

$$W_1 = \frac{W_2}{2}$$

In Fig. 5-6c, if W_3 is constrained to be elevated a distance s_3 while W_1 is lowered a distance s_1, the corresponding equation is

$$W_1 s_1 - W_3 s_3 = W_1 s_1 - W_3 \frac{s_1}{4} = 0 \tag{5-40}$$

or

$$W_1 = \frac{W_3}{4}$$

Clearly the notion of vanishing work in a displacement of the above systems provides us with the criteria of equilibrium. It is interesting to note that the same result is obtained for a rotational system of the type exhibited in Fig. 5-7. The system in Fig. 5-7 consists of two wheels of radii r_1 and r_2, rigidly fastened together, and with strings attached as shown. To these strings are fastened weights W_1 and W_2, of magnitudes such that equilibrium is realized. Suppose we impose a counterclockwise rotation through an angle θ. During this rotation W_1 descends a distance $r_1\theta$, and W_2 rises a distance $r_2\theta$. We now suspect that, as in the case just concluded, the total work done by gravity during the rotation is zero. We therefore write tentatively

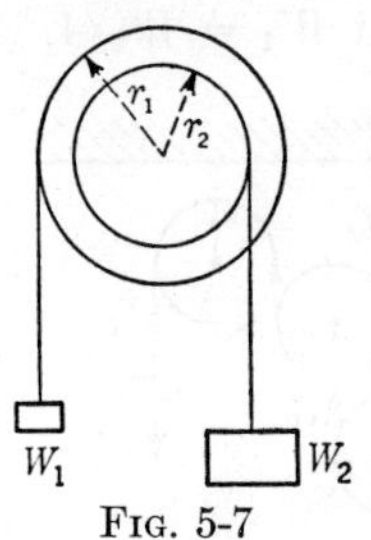

FIG. 5-7

$$W_1 r_1 \theta - W_2 r_2 \theta = 0 \tag{5-41}$$

or

$$W_1 r_1 = W_2 r_2$$

But this is just the condition of the equilibrium of moments for this case. We therefore have that the vanishing of the virtual work during the virtual displacements of these simple systems in equilibrium provides us simultaneously with the two conditions (equilibrium of forces and equilibrium of moments of forces) for static equilibrium which were previously stated separately.

5-12. Infinitesimal Virtual Displacements. Statement of the Principle of Virtual Work. It is necessary to see clearly one thing concerning each of the cases treated in Figs. 5-5 and 5-6. There, no limitations were made on the magnitudes of the displacements, and we notice that in each case the displaced system was also in an equilibrium configuration. We considered, therefore, special types of systems in which the components could be given finite virtual displacements, during any portion of which the system was capable of existing in static equilibrium. Only certain types of systems exhibit this quality. We may assume, in fact, that the general rule will be for a finite displacement to carry the system into a neighboring configuration which is no longer an equilibrium one. The suggestion is made that, instead, we should employ vanishingly small, or *infinitesimal, displacements.* A basis for this suggestion can be detected in Fig. 5-8. This pictures the same system which was shown in Fig. 5-5 except that portions of the plane adjacent to a tiny region at point A are distorted in the manner indicated by the dotted line. In spite of this, however, the equilibrium condition at point A,

$$W_1 - W_2 \sin \alpha = 0$$

again holds provided that the tangent at A is inclined at the angle α to the horizontal. We postulate that the condition of vanishing work must in this case be written, if W_1 is lowered an amount δs_1,

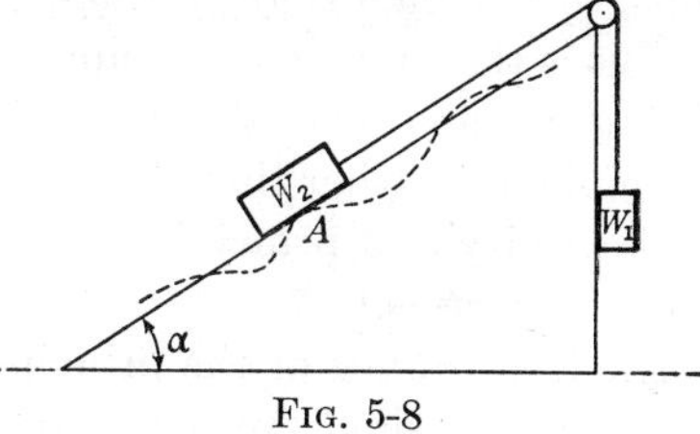

Fig. 5-8

$$W_1\, \delta s_1 - W_2\, \delta s_2 \sin \alpha = 0 \tag{5-42}$$

where the quantities δs_1 and δs_2 are vanishingly small, δs_2 being tangent to the plane at point A. Since δs_1 and δs_2 are equal in magnitude, the result is the same as before. Evidently the previous cases, in which the displacements were finite, are satisfied as well by infinitesimal displacements.

The successful application of an equation of the general type of (5-42), in which infinitesimal displacements are employed, verifies the correctness of the assumption. Johann Bernoulli, in 1717, was the first to recognize the universal applicability of this principle to all systems which are in static equilibrium. The principle, which is variously called the *principle of virtual work*, *virtual velocities*, or *virtual displacements*, can be stated in general terms for a system of any number of parts acted upon by a number of forces $\mathbf{F}_1, \mathbf{F}_2, \ldots, \mathbf{F}_n$, in which each is acting through an infinitesimal displacement $\delta\mathbf{s}_1, \delta\mathbf{s}_2, \ldots, \delta\mathbf{s}_n$, respectively. We have

$$\mathbf{F}_1 \cdot \delta\mathbf{s}_1 + \mathbf{F}_2 \cdot \delta\mathbf{s}_2 + \cdots + \mathbf{F}_n \cdot \delta\mathbf{s}_n = 0$$

or, more compactly,

$$\sum_{i=1}^{n} \mathbf{F}_i \cdot \delta\mathbf{s}_i = 0 \tag{5-43}$$

It is to be emphasized that the displacements $\delta\mathbf{s}_i$ are virtual in the sense that in reality the system is in static equilibrium and is not moving, the displacement being only a conceptual one. The displacements, however, must be ones which are compatible with the geometrical constraints. Thus they are geometrically related. These statements will be made more clear below, when examples are considered. Equation (5-43) states that:

VI. *The total work (virtual), performed in a set of virtual displacements by all the forces acting on a system in static equilibrium is zero. The displacements in general must be vanishingly small ones in the neighborhood of the equilibrium configuration.*

This is the so-called principle of virtual work. It should be made clear that the above considerations are not intended as a derivation of the principle but rather that they lead us to suspect and recognize the general applicability of a principle which we perceive to be valid for a number of particular cases.[1]

The importance of the principle of virtual work lies in the fact that it constitutes a single principle upon which all statics can be based. All the conditions of static equilibrium which we have considered in the foregoing chapters can be derived directly from the principle of virtual work. Conversely the same situation is true, since the individual problems concerning static equilibrium which have been encountered previously can also be solved as well by the direct application of the principle of virtual work. However, it is frequently found that the student is able to apply the principle with ease and confidence only after having had a fair amount of experience with it. The fact is comforting, therefore, that all problems solvable by the method of virtual work are readily solved also by the straightforward methods presented in previous chapters.

5-13. Work Done by Internal Forces. In the direct application of the principle of virtual work to specific problems, the manner in which the virtual displacement is taken depends upon which of the forces it is desired to determine. For example, it may be that smooth surfaces comprise a part of the system which is in static equilibrium. If we are not interested in the reactions of these surfaces upon the bodies with which they are in contact, appropriate displacements are those which are tangent to the smooth surfaces. This is not difficult to see, since the reactions are normal to the (smooth) surfaces, and no work will be done by them during displacements which are tangent to the surfaces. Consequently they do not appear in Eq. (5-43). However, if it is desired to determine these reactions, the displacements should be conceived in a manner which involves work being done by the reactions.

Frequently a system involves weightless rigid rods, weightless inextensible strings, and so on, connecting various parts of a system which is in equilibrium. Usually these components transmit forces between the parts of the system they connect. In the case of the rods, it has been seen (Chap. 3) that they transmit either tension or thrust, and similarly it has been seen that light inextensible strings transmit tension only. If

[1] See Mach's "The Science of Mechanics," 5th English ed., The Open Court Publishing Company, LaSalle, Ill., 1942.

it is not desired to determine internal forces such as these thrusts and tensions, the virtual displacements are so conceived that the members transmitting these forces are neither stretched nor, in the case of the rod, compressed. The notion of stretching a rigid rod or inextensible string seems contradictory. This further emphasizes the virtual character of the displacements; they are conceptual ones.

It is not difficult to see that the work done by the tension in a weightless string vanishes identically in any set of displacements not involving the stretching of the string. In Fig. 5-9, two particles i and j, respectively at A and B, are connected by a light string which, in turn, is in contact with a smooth surface CD. Since the reaction of the surface is normal to the surface, any sliding of the string over the surface may be performed without work having been done by the reaction of the surface. Imagine the string to be very slightly displaced to the new configuration $A'CDB'$, point A having been displaced through $\delta\mathbf{s}_1$ to A' and B through $\delta\mathbf{s}_2$ to B'. Let us consider only the effect of the tension. If we fix our attention on the string as a whole, the tension at A can be regarded as a force $\mathbf{T}_i$ acting on i and directed along the string produced through A and as a force $\mathbf{T}_j$ acting along the string produced through B. The work done by the forces during the displacement is

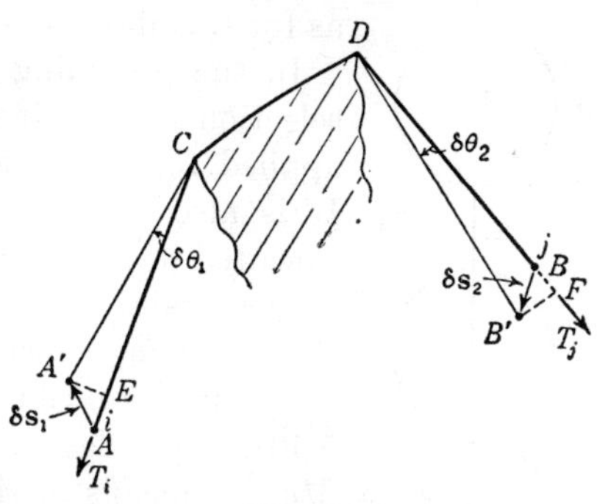

FIG. 5-9

$$\delta W = \mathbf{T}_i \cdot \delta\mathbf{s}_1 + \mathbf{T}_j \cdot \delta\mathbf{s}_2 \tag{5-44}$$

However, the displacement can be considered as composed of two parts, a linear translation along the string of $ACDB$ to $ECDF$, followed by a rotation $\delta\theta_1$ of EC about C to $A'C$ and a rotation $\delta\theta_2$ of DF about D to DB'. Accordingly, Eq. (5-44) becomes

$$\delta W = -T_i \cdot \overline{AE} + T_j \cdot \overline{BF} \tag{5-45}$$

which follows since the displacements EA' and FB' are normal to the tension, and thus no work is done by the tension during these displacements. Furthermore, the magnitude of the tension is constant along a weightless string, provided that the contacting surfaces are not rough. Consequently $T_i = T_j$, and since the string is not changed in length, we have also that $AE = BF$. Therefore the right side of Eq. (5-45) vanishes. Consequently we see that no work is done by the tension in a weightless string in contact with a smooth surface during a displacement which does not alter the length of the string.

A quite similar situation may be demonstrated with respect to the internal forces of a rigid body during a displacement which does not alter the shape of the body. In this instance the forces acting between any two adjacent particles of the body lie along the straight line joining the particles and, by Newton's third law, are equal and opposite. During a displacement of the body which leaves the distances between all pairs of particles unaltered, the work done by the internal forces cancels out in precisely the same manner as for the above case of the tension in the string.

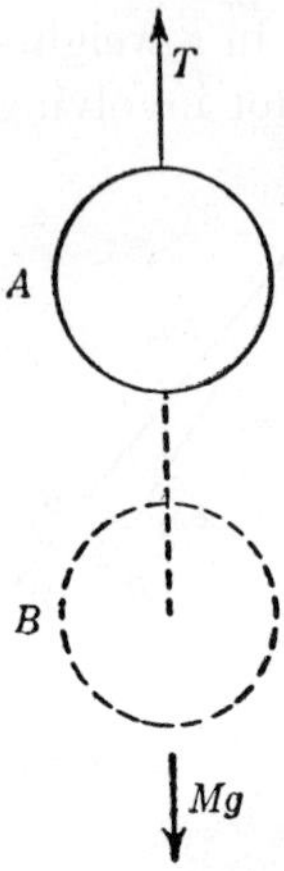

In the preceding sections we were concerned with the selection of the virtual displacements such that only the applied forces would do work during the displacements. It is also possible to conceive the displacements to be executed in a manner such that work is done by the forces of constraint as well (see Sec. 5-10). To select a very simple example, consider the case of a mass M suspended vertically by means of a light inextensible string (see Fig. 5-10). It is desired to find the tension (which of course is Mg) by means of the principle of virtual work. We imagine a vertical displacement of the particle from A to B. The work done by the gravitational force is just $+Mg \cdot \overline{AB}$, and since the displacement is opposite to the direction of T, the latter does negative work. Thus the total work done by the forces is

T

A

B

Mg

FIG. 5-10

$$\delta W = Mg \cdot \overline{AB} - T \cdot \overline{AB} = (Mg - T)\overline{AB}$$

This must vanish because of the principle of virtual work, and we have simply that $T = Mg$.

5-14. Miscellaneous Examples

Example 5-3. A uniform plank of weight W is resting on a smooth floor and is leaning against a smooth wall. The lower end of the plank is connected by means of a weightless inextensible string to the base of the wall. The angle of inclination of the plank to the horizontal is α. It is required to find the tension in the string.

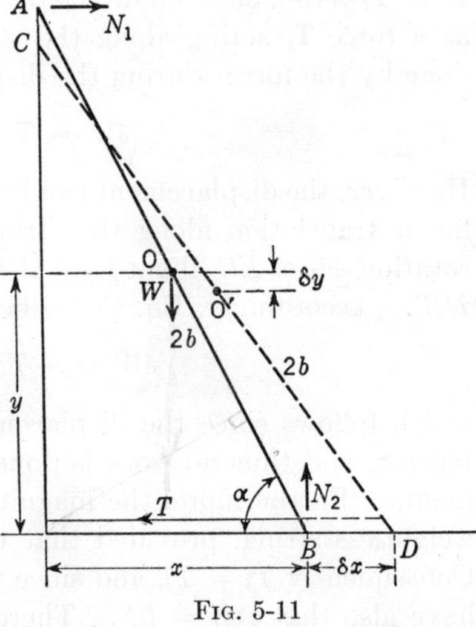

FIG. 5-11

The arrangement is shown in Fig. 5-11. We take for convenience the length of the plank as $2b$. AB is the initial position of the plank. The mid-point of the plank is at O, a distance y above the floor, and the base B is at a distance x from the wall, where

$$y = b \sin \alpha \qquad x = 2b \cos \alpha \tag{5-46}$$

The forces acting on the plank are the tension T in the string, the weight W, and the reactions N_1 and N_2 normal to the respective surfaces. Let us take the displacement to the configuration shown by the dotted line CD. (If α were $\pi/4$, we should have $AC = BD$.) We desire the displacements for arbitrary α. In particular we need to know the change δy in the level of O and the shift δx in the horizontal position of B. From Eqs. (5-46) these are

$$|\delta y| = b \cos \alpha \; \delta\alpha \qquad |\delta x| = 2b \sin \alpha \; \delta\alpha \tag{5-47}$$

Now, since the displacements of the ends of the plank are tangent to the surfaces, the normal reactions do no work. Accordingly, from the principle of virtual work we must have that the work done on the plank by the remaining forces during the displacement is zero. Hence we have

$$-T \cdot |\delta x| + W \cdot |\delta y| = 0 \tag{5-48}$$

where the sign convention is that, since δx is in a direction opposite to T, T does negative work and, since δy is in the direction of W, W does positive work. Substituting from Eqs. (5-47), this becomes

$$-T \cdot 2b \sin \alpha \; \delta\alpha + W \cdot b \cos \alpha \; \delta\alpha = 0$$

from which $T = (W/2) \cot \alpha$, a result readily checked by the methods of Chaps. 2 and 3.

Example 5-4. A framework consists of four equal rods, of weight W_1, joined by smooth hinges, and suspended in a vertical plane at point A as shown. A light rigid rod forms the horizontal diagonal. Suspended from point C is a weight W_2. It is desired to find the stress in the light rod (tension or thrust) in terms of W_1, W_2, and the angle θ (equal to $\angle ABO$ in the figure).

The initial configuration of the framework is the rhombus $ABCD$ in Fig. 5-12. The external forces acting upon the system are the four weights, each W_1, the weight W_2, and the reaction at the point of support A. Since we wish to determine the stress in the rod BD, it is necessary to select a virtual displacement such that the stress will do work during the displacement. Consider the slightly altered configuration $AB'C'D'$. We let the symbol x designate BD and y designate AC, where

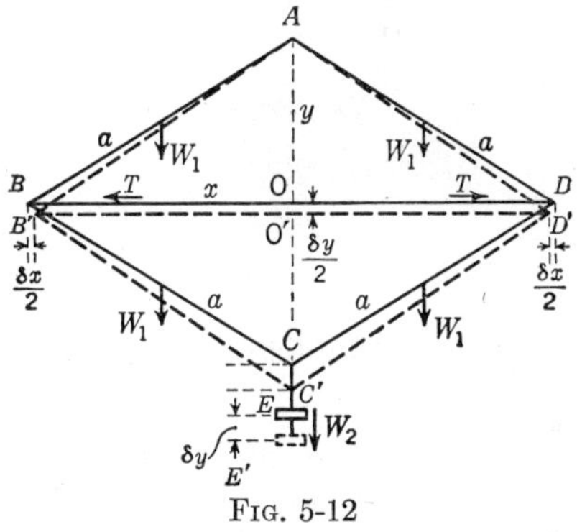

FIG. 5-12

$$x = 2a \cos \theta \qquad y = 2a \sin \theta$$

During the displacement, rod BD is compressed an amount $\delta x/2$ at each end, or a total of δx, and the diagonal distance AC is increased a distance δy. Since the joints are all smooth, the reactions at the joints do no work, and since point A remains fixed, the reaction of the support at that point also does no work. However, the center of mass of the four heavy rods, originally at O, is lowered a distance $\delta y/2$ to point O', and the weight W_2 is lowered a distance δy from E to E'. Consequently the gravitational force does work on the weights of the system. This, combined with the fact that work is done against the stress T in the rod (assuming for the moment that the stress is a thrust; the arrows indicate that the stress is assumed to be a thrust, that is, the arrows indicate forces against the joints B and D), enables us to write the equation of the total work done by all the forces during the set of virtual displacements δx and δy. Since

$$|\delta x| = 2a \sin\theta\,\delta\theta \qquad |\delta y| = 2a\cos\theta\,\delta\theta$$

and

$$4W_1 \cdot \left|\frac{\delta y}{2}\right| + W_2 \cdot |\delta y| - 2T \cdot \left|\frac{\delta x}{2}\right| = 0 \tag{5-49}$$

we have, therefore,

$$4W_1 \cdot a\cos\theta\,\delta\theta + W_2 \cdot 2a\cos\theta\,\delta\theta - 2T \cdot a\sin\theta\,\delta\theta = 0$$

This simplifies to $T = (2W_1 + W_2)\cot\theta$.

It is to be noted that we conceive of such a displacement as the above even though the rod is perfectly rigid. Also, we could just as well have assumed the rod to be under tension. In that event the third term in Eq. (5-49) would have been positive. Consequently the end result would have been a negative quantity, demonstrating that we had made the wrong choice.

Example 5-5. Given a smooth string of length b, attached to point A as shown in Fig. 5-13, and free to slide over a smooth pulley at point B. B is at the same level as A and is a fixed distance $2a$ from it. A mass m is attached to the smooth, light

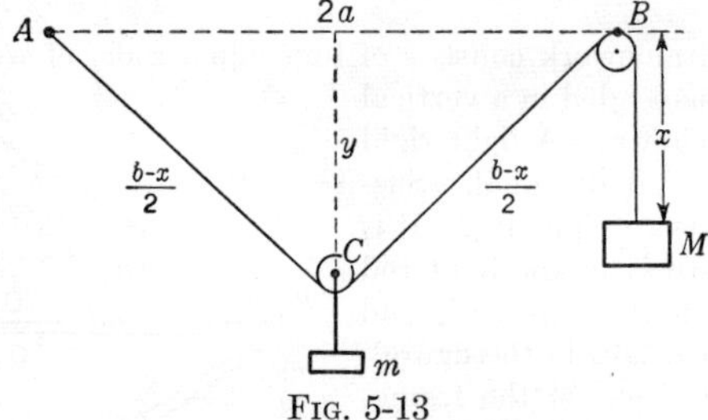

FIG. 5-13

pulley at C, and at the end of the string which is hanging from B is attached a mass M. It is desired to find, by use of the principle of virtual work, the equilibrium distance x of M below B.

Let us imagine a displacement of M a distance δx downward. Thus m will be displaced a distance δy upward. Clearly

$$y^2 + a^2 = \frac{(b-x)^2}{4} \qquad \text{and} \qquad \delta y = \frac{-(b-x)}{[4(b-x)^2 - 16a^2]^{\frac{1}{2}}}\,\delta x \tag{5-50}$$

The principle of virtual work enables us to write

$$-mg \cdot |\delta y| + Mg \cdot |\delta x| = 0$$

Eliminating δy by means of Eq. (5-50), we obtain, finally,

$$x = b - \left(\frac{16M^2a^2}{4M^2 - m^2}\right)^{\frac{1}{2}}$$

Example 5-6. Two masses m_1 and m_2 are connected by a light inextensible string, the whole being arranged as in Fig. 5-14. The pulley at B is smooth, and the contacts between m_1 and the plane and m_2 and the plane are equally rough. The angles of inclination of the planes are α and β. If the masses are about to slip in the counterclockwise direction, find the coefficient of static friction μ by means of the principle of virtual work.

We take a small displacement δx along the planes in a counterclockwise direction. The total work done by the applied forces during this virtual displacement will be zero if the system is in equilibrium. We have

$$(m_1 g \sin \alpha - \mu m_1 g \cos \alpha) \cdot |\delta x| - (m_2 g \sin \beta + \mu m_2 g \cos \beta) \cdot |\delta x| = 0$$

from which

$$\mu = \frac{m_1 \sin \alpha - m_2 \sin \beta}{m_1 \cos \alpha + m_2 \cos \beta}$$

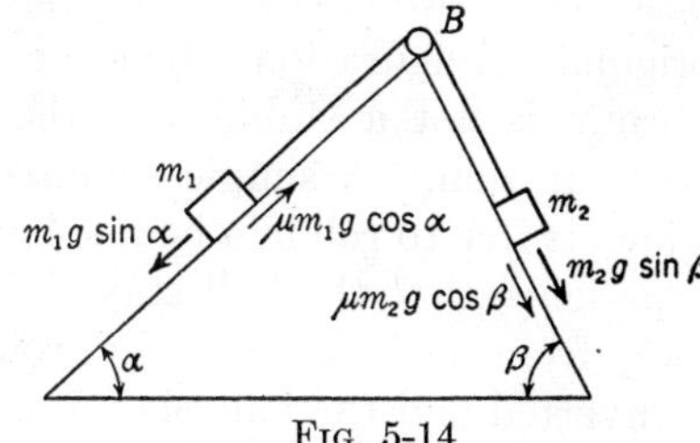

Fig. 5-14

a result which may be readily verified by the methods of Chaps. 2 and 3. This example emphasizes again the virtual character of the displacement δx, since in any real displacement we should be concerned with the coefficient of sliding friction rather than that of static friction. Another way to look at the situation is that the frictional force concerned with the state of limiting static equilibrium involves the coefficient of static friction, and not kinetic friction.

STABILITY OF EQUILIBRIUM

5-15. Equilibrium of Conservative Forces. The principle of virtual work tells us that, for any system which is in static equilibrium, the total work performed by all the forces during any virtual displacement of the system is zero. In equation form, this states that

$$\delta W = 0 \tag{5-51}$$

Now, if the system of forces is conservative, it is possible to speak of a potential energy V of the system which is a function of the coordinates. Furthermore, the change in the potential energy during the displacement, by Eq. (5-8), is related to δW by the equation

$$\delta W = -\delta V = -\left(\frac{\partial V}{\partial x}\,\delta x + \frac{\partial V}{\partial y}\,\delta y + \frac{\partial V}{\partial z}\,\delta z\right) \tag{5-52}$$

Since δx, δy, and δz are independent, the only way Eqs. (5-51) and (5-52) will be satisfied, for any δx, δy, and δz, is to equate each term to zero separately. Accordingly

$$\frac{\partial V}{\partial x} = 0 \qquad \frac{\partial V}{\partial y} = 0 \qquad \frac{\partial V}{\partial z} = 0 \tag{5-53}$$

These are just the necessary conditions that the potential energy will have a stationary value in the equilibrium configuration. [They are also the negatives of the components of the forces; thus $\partial V/\partial x = -F_x$, and Eq. (5-53) makes the familiar statement that in equilibrium the resultant force is zero.] However, they do not tell us whether or not the equilibrium is a stable one. Experiment tells us that in stable equilibrium the potential energy is always a *minimum*. Thus, if a slight disturbance from a position of stable equilibrium is experienced, the system returns to the original configuration when the disturbance is removed. If the equilibrium is not a stable one, the system does not return to the initial configuration. A simple example of the first situation is that of a ball which is free to roll in a hemispherical bowl. It is common to everyone's experience that the ball stays in the bottom of the bowl and, if pushed away from the bottom, always rolls back. On the other hand, if the bowl is inverted and the ball placed on top of the convex surface, a slight disturbance causes the ball to roll off and there is no tendency for the ball to return to the original position at the top.

In the first instance, of the ball in the bottom of the bowl, the equilibrium is a *stable* one; in the second case, it is *unstable*. Still a third case is possible, that in which a small disturbance produces neither an increase nor a decrease of potential energy. Such, for example, is the case of a ball free to move on a flat horizontal table. This type of equilibrium is called *neutral equilibrium*.

5-16. Potential Energy a Function of a Single Scalar Variable. For simplicity we consider a system possessing a potential energy which is a function of a single scalar independent variable—the situation when the system has but one degree of freedom (see Sec. 13-1). Designate this variable as x, taking the origin to be at an equilibrium point. We wish to investigate the behavior of the potential energy $V = V(x)$ in the neighborhood of the equilibrium point $x = 0$. In order to do this we expand V in a Taylor series in one variable about the point $x = 0$. We have

$$V(x) = V_0 + x\left(\frac{dV}{dx}\right)_0 + \frac{x^2}{2!}\left(\frac{d^2V}{dx^2}\right)_0 + \frac{x^3}{3!}\left(\frac{d^3V}{dx^3}\right)_0 + \frac{x^4}{4!}\left(\frac{d^4V}{dx^4}\right)_0 + \cdots \tag{5-54}$$

where the zero subscript indicates that the particular quantity is evaluated at the origin. In Eq. (5-54), the potential energy V_0, at the origin, is a constant, and no loss of generality is incurred by calling this constant zero. (An additive constant in the potential energy has no effect on the forces.) Furthermore, since $x = 0$ is a point of equilibrium, the forces must be zero there and therefore

$$\left(\frac{dV}{dx}\right)_0 = 0$$

Accordingly $V(x)$ becomes

$$V(x) = \frac{x^2}{2!}\left(\frac{d^2V}{dx^2}\right)_0 + \frac{x^3}{3!}\left(\frac{d^3V}{dx^3}\right)_0 + \frac{x^4}{4!}\left(\frac{d^4V}{dx^4}\right)_0 + \cdots \qquad (5\text{-}55)$$

Since the expansion is in a neighborhood very close to $x = 0$, x is in general a very small quantity. Therefore each of the terms of Eq. (5-55) will be an infinitesimal, one order higher than the preceding term in each case. Consequently, if $(d^2V/dx^2)_0$ is not zero, a good approximation in most cases is to neglect all the higher terms, that is, to write approximately

$$V(x) = \frac{x^2}{2!}\left(\frac{d^2V}{dx^2}\right)_0 \qquad (5\text{-}56)$$

Now if $V(x)$ is to be a minimum at $x = 0$, and since it is defined to be zero there, it must be greater than zero in the immediate neighborhood of $x = 0$. In Eq. (5-56), x^2 is positive whether x is positive or negative. Thus $V(x)$ has the same sign as $(d^2V/dx^2)_0$, and therefore the equilibrium is stable or unstable according as $(d^2V/dx^2)_0$ is positive or negative.

If $(d^2V/dx^2)_0 = 0$, it is necessary to investigate the higher-order terms. In this case the next approximation to $V(x)$ is

$$V(x) = \frac{x^3}{3!}\left(\frac{d^3V}{dx^3}\right)_0 \qquad (5\text{-}57)$$

If $(d^3V/dx^3)_0$ is positive, then, by virtue of Eq. (5-57), $V(x)$ is positive or negative according as x is positive or negative. Also, if $(d^3V/dx^3)_0$ is negative, $V(x)$ is positive or negative according as x is negative or positive, respectively. Hence if the first nonvanishing term is the one containing $(d^3V/dx^3)_0$, the equilibrium is unstable whatever the sign of $(d^3V/dx^3)_0$. If $V(x)$ can be written in the form of Eq. (5-57) the appearance of $V(x)$ is that of Fig. 5-15. It is seen that $V(x)$ has a point of inflection at the origin. The first case considered, of positive $(d^3V/dx^3)_0$, is shown in Fig. 5-15a, while the case of negative $(d^3V/dx^3)_0$ is shown in Fig. 5-15b.

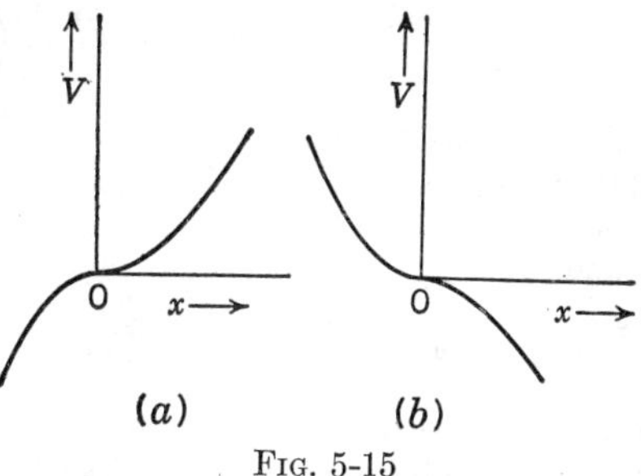

Fig. 5-15

If $(d^3V/dx^3)_0$ is zero, we must investigate the next higher derivative. We may write approximately

$$V(x) = \frac{x^4}{4!}\left(\frac{d^4V}{dx^4}\right)_0 \qquad (5\text{-}58)$$

Here, as for the second derivative, $V(x)$ is positive or negative according as $(d^4V/dx^4)_0$ is positive or negative, and the equilibrium is stable or unstable, in the same order.

The above considerations can be carried to any number of higher derivatives, and the pattern of the results can easily be predicted. We summarize the criteria of stability as follows:

VII. *If the first nonvanishing derivative of V is one of even order, the equilibrium is stable or unstable according as this derivative, evaluated at the equilibrium point, is positive or negative.*

VIII. *If the first nonvanishing derivative of V is one of odd order, the equilibrium is unstable, regardless of the sign of the derivative.*

If the potential energy V is a function of more than one independent variable, the situation becomes more complex. If it is a function of two variables, it is necessary to examine the Taylor expansion in the two variables. It is possible to carry out the analysis in the same manner as for a single variable, save that the situation has greater complexity.

Example 5-7. Find the equilibrium configuration in Example 5-5, and determine whether or not it is stable (see Fig. 5-13).

Aside from an additive constant the potential energy may be written, if both x and y are measured positively downward,

$$V = -mgy - Mgx = -\frac{mg}{2}\sqrt{(b-x)^2 - 4a^2} - Mgx$$

whence

$$\frac{dV}{dx} = \frac{mg}{2}\frac{b-x}{[(b-x)^2 - 4a^2]^{\frac{1}{2}}} - Mg$$

The equilibrium value of x occurs at $dV/dx = 0$. This yields

$$x = b - \left(\frac{16M^2a^2}{4M^2 - m^2}\right)^{\frac{1}{2}} \tag{5-59}$$

as before. It is interesting to note that for x to be a real quantity we must have $4M^2 > m^2$. In order to ascertain whether or not the equilibrium is stable, we examine the next higher derivative of V. We have, after simplifying and making use of Eq. (5-59),

$$\frac{d^2V}{dx^2} = \frac{g(4M^2 - m^2)^{\frac{3}{2}}}{4m^2a} \tag{5-60}$$

which is a positive quantity provided that $4M^2 > m^2$. Thus we will have no equilibrium unless $4M^2 > m^2$, and moreover the equilibrium, in that case, will be a stable one.

Example 5-8. A uniform solid cube of side $2a$ rests in the equilibrium position on the top of a cylindrical log of radius r. The plane of one side of the cube is normal to the axis of the log. Determine whether or not the equilibrium is stable. Assume perfectly rough contact.

In Fig. 5-16 the cube is slightly displaced (by rolling) from its equilibrium position, in which Q was in contact with the log at C. The new point of contact is C', where OC' makes an angle θ with OC and PC' makes an angle φ with PQ. If there is to be no sliding, we must have that

$$\overline{QC'} = r\theta \qquad \text{or} \qquad r\theta = a\tan\varphi \tag{5-61}$$

However, we assume a very small displacement, in which case Eq. (5-61) becomes approximately

$$r\theta = a\varphi \tag{5-62}$$

Since the zero point of potential energy is arbitrary, we choose the zero to be at the level of the axis of the log. Hence the potential energy of the cube is $+Mgh$, where h

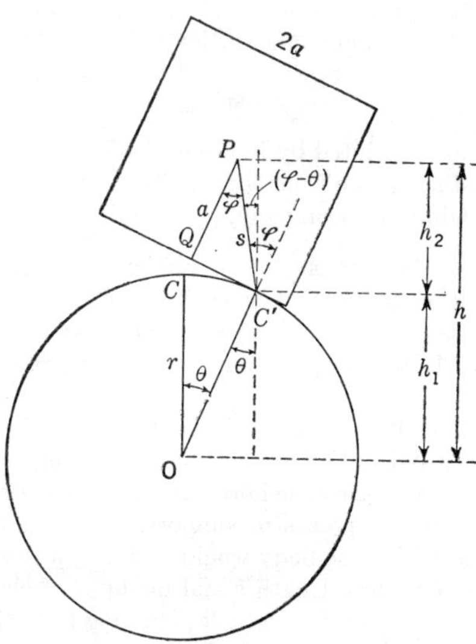

FIG. 5-16

is the height above O of the center P of the cube. But we see from Fig. 5-16 that

$$h = h_1 + h_2 = r\cos\theta + s\cos(\varphi - \theta)$$

where

$$s = \overline{PC'} = (a^2 + r^2\theta^2)^{\frac{1}{2}}$$

Therefore

$$\begin{aligned} V = mgh &= mgr\cos\theta + mg(a^2 + r^2\theta^2)^{\frac{1}{2}}\cos(\varphi - \theta) \\ &= mgr\cos\theta + mg(a^2 + r^2\theta^2)^{\frac{1}{2}}\cos\left[\left(\frac{r}{a} - 1\right)\theta\right] \end{aligned} \tag{5-63}$$

in which use has been made of Eq. (5-62). Differentiating Eq. (5-63) with respect to θ, and equating to zero, we have

$$\begin{aligned} \frac{dV}{d\theta} = 0 = -mgr\sin\theta &+ \frac{mgr^2\theta}{(a^2 + r^2\theta^2)^{\frac{1}{2}}}\cos\left[\left(\frac{r}{a} - 1\right)\theta\right] \\ &- mg(a^2 + r^2\theta^2)^{\frac{1}{2}}\left(\frac{r}{a} - 1\right)\sin\left[\left(\frac{r}{a} - 1\right)\theta\right] \end{aligned} \tag{5-64}$$

This is satisfied by $\theta = 0$, for arbitrary values of a and r. Now

$$\begin{aligned} \frac{d^2V}{d\theta^2} &= -mgr\cos\theta + \frac{mgr^2}{(a^2 + r^2\theta^2)^{\frac{1}{2}}}\cos\left[\left(\frac{r}{a} - 1\right)\theta\right] - \frac{mgr^4\theta^2}{(a^2 + r^2\theta^2)^{\frac{3}{2}}}\cos\left[\left(\frac{r}{a} - 1\right)\theta\right] \\ &- \frac{mgr^2\theta}{(a^2 + r^2\theta^2)^{\frac{1}{2}}}\left(\frac{r}{a} - 1\right)\sin\left[\left(\frac{r}{a} - 1\right)\theta\right] - \frac{mgr^2\theta}{(a^2 + r^2\theta^2)^{\frac{1}{2}}}\left(\frac{r}{a} - 1\right)\sin\left[\left(\frac{r}{a} - 1\right)\theta\right] \\ &- mg(a^2 + r^2\theta^2)^{\frac{1}{2}}\left(\frac{r}{a} - 1\right)^2\cos\left[\left(\frac{r}{a} - 1\right)\theta\right] \end{aligned}$$

At $\theta = 0$ this takes the value

$$\left(\frac{d^2V}{d\theta^2}\right)_0 = mg\left[-r + \frac{r^2}{a} - a\left(\frac{r}{a} - 1\right)^2\right] = mg(r - a) \tag{5-65}$$

Consequently, if $r > a$, the equilibrium is stable. If $r < a$, the equilibrium is unstable, and finally if $r = a$, higher derivatives must be examined in order to ascertain the nature of the equilibrium configuration.

Problems

Problems marked *C* are reprinted by kind permission of the Cambridge University Press. Those marked *J* are taken from Jeans, "Theoretical Mechanics" and are used by kind permission of Ginn and Company.

5-1. A ring of mass m slides on a smooth vertical rod. Attached to the ring is a light inextensible string passing over a smooth peg distant a from the rod. At the other end of the string is a mass M $(M > m)$. The ring is released from rest at the same level as the peg. Determine, in terms of M, a, m, the maximum distance the ring will fall.

5-2. A body of mass M falls from a height h and engages the center of a light elastic string, of modulus λ and natural length b, which is strung without any appreciable tension between two points in the same horizontal plane a distance b apart. Find the maximum distance, below the points of support, to which the body will descend. Find also the position at which the body would rest in equilibrium.

5-3. An elastic rope of natural length b and modulus of elasticity W is suspended vertically from one end. A man of weight W, hanging from the lower end, wishes to climb the rope. Find the ratio of the work the man must do to climb to the top to the work he would have had to do if the rope had been inextensible and of length b.

5-4. Find the potential at an external point on the axis of a homogeneous right-circular cylinder of mass M, altitude h, and radius a, distant R from the center of the cylinder.

5-5. Find the potential energy of a particle of mass m situated at a point on the axis of a homogeneous right circular cone, with vertex upward, of mass M, altitude h, and radius a of the base. The particle m is external to the cone and at a distance R above the vertex.

5-6. A uniform rod of length b and mass m_1 per unit length lies on the axis of symmetry of a circular loop of wire of mass m_2 per unit length. The radius of the loop is a, and the nearer end of the rod is located a distance h from the center of the loop. Find the work you would have to do in order to move the rod to an infinite distance away. Find also the force on the rod while in its original position.

5-7. The center of a uniform spherical shell of mass M and radius a is at a distance b $(b > a)$ from an infinite thin sheet having a mass σ per unit area. Find the resultant force on the sheet due to the sphere.

5-8. A uniform hemispherical shell of mass M and radius a is oriented so that it is concave upward, and the bottom of the shell is at a vertical distance h above an infinite thin sheet having a mass σ per unit area. The shell is then shifted in orientation so that its axis of symmetry is now parallel to the sheet and at a height $h + a$ above it. How much work was done?

5-9. In Fig. 5-17, an infinite plane sheet joins symmetrically, as shown, to a uniform spherical shell of radius a. O is the center of the shell. Both the sheet and the shell have a mass σ per unit area. How much work is required to move a mass m from O to P, a distance h above O?

5-10. Two concentric spherical shells of masses m_1 and m_2, have radii a and b, respectively ($b > a$). What is the total gravitational potential energy of the system? What is the pressure on the outer sphere?

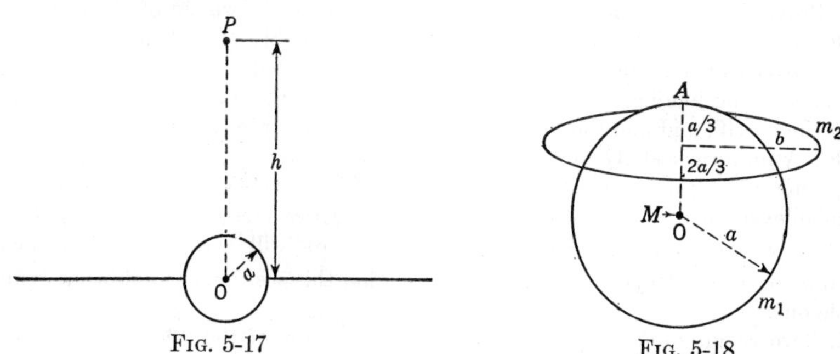

FIG. 5-17 FIG. 5-18

5-11. Point O is the center of a uniform spherical shell of radius a and mass m_1. A uniform wire of mass m_2 is bent into the form of a circle of radius b ($b > a$) and is oriented (Fig. 5-18) so that its plane lies a distance $2a/3$ above O. AO is the axis of symmetry of the wire. A particle of mass M is situated at O.

a. If M were originally situated at A, how much work would you have to do in order to move M from A to O?

b. What is the resultant force on M if it is located at O?

c. If M and m_2 remain fixed, as in the figure, determine the amount of work you would have to do in order to remove sphere m_1 a large distance away from M and m_2.

5-12. Two ladders, of weights W each, are joined smoothly together at the top and are placed on a smooth floor. A light chain tied at a distance two-thirds of the way up each ladder keeps each ladder at an angle α with the vertical. A man of weight $2W$ stands on a step halfway up one of the ladders. What is the tension in the chain? Solve by means of the principle of virtual work. Solve also by the methods of Chap. 3.

5-13. A ladder of weight W and length b rests on a smooth floor and leans against a smooth vertical wall, the lower end of the ladder being attached by a string to a point on the wall at a height h above the floor. If the ladder makes an angle θ with the wall, find the tension in the string by use of the principle of virtual work.

5-14. A weight W can slide without friction along a weightless rod which is suspended from a frictionless pivot at one end O, and W is attached to O by an elastic string of modulus λ and natural length a. Find the potential energy of the system when the rod makes an angle α with the downward vertical ($\alpha < 90°$), and from this find the couple which must be applied to the rod to keep the system in equilibrium in this position.

5-15*C*. A rhombus $ABCD$ is formed of four equal uniform rods smoothly hinged together and suspended from the point A; it is kept in position by a light rod joining the mid-points of BC and CD; prove that if T be the thrust in this rod and W the weight of the rhombus, $T = W \tan \frac{1}{2}A$.

5-16*C*. Four equal uniform rods, each of weight W, are smoothly hinged so as to form a square $ABCD$; the side AB is fixed in a vertical position with A uppermost, and the figure is kept in shape by a string joining the middle points of AD, DC. Show that the tension is $5.66W$.

5-17*C*. Six uniform bars, smoothly hinged together, hang from a fixed point A and form a regular hexagon $ABCDEF$ which is kept in shape by light horizontal struts BF, CE. Prove that the thrusts in these are as $5:1$.

5-18*C*. Six equal bars are smoothly hinged at their extremities, forming a regular hexagon *ABCDEF* which is kept in shape by vertical strings joining the middle points of *BC*, *CD*, and *AF*, *FE*, respectively, the side *AB* being held horizontal and uppermost. Prove that the tension of each string is three times the weight of a bar.

5-19*C*. Four equal uniform rods of weight W are smoothly hinged so as to form a square *ABCD* which is suspended from *A* and is prevented from collapsing by an inextensible string joining the middle points of *AB* and *BC*. Prove that the tension of the string is $4W$, and find the magnitude and direction of the reaction at *B*.

5-20*C*. A uniform rod *AB* of weight W can turn freely around a horizontal axis through one end *A*; a fine cord is attached to a point *C* vertically above *A*, passes through an eyelet fixed on the rod at the end *B*, and carries a hanging weight W_1 at its other end. Prove that, in the absence of friction, the rod will be in equilibrium when BC is equal to $W_1 \cdot \overline{AC}/(\frac{1}{2}W + W_1)$. Prove also that this position of equilibrium is an unstable one.

5-21. Two rough stones rest in equilibrium, one on top of the other. The lower one is fixed. At the point of contact the radii of curvature of the two surfaces are r_1 and r_2, and the common normal there is vertical. If the height of the center of mass of the upper stone is a distance h above the point of contact, show that the equilibrium is stable or unstable according as

$$\frac{1}{h} > \text{or} < \frac{1}{r_1} + \frac{1}{r_2}$$

5-22. A uniform plank of thickness $2a$ rests in the equilibrium configuration horizontally across the top of a perfectly rough horizontal cylinder of radius a. Determine whether or not the equilibrium is stable.

5-23. A uniform plank of thickness $2a$ lies in the equilibrium position across a cylinder of diameter $2b$, where $a < b$. The plank is in a plane normal to the axis of the cylinder, and the contact is perfectly rough. Find the maximum displacement of the plank consistent with stable equilibrium.

5-24. A cylindrical can, open at one end, whose height is $2a$ and whose diameter is $2a$, rests with its closed end on the top of a fixed sphere. Find the least diameter of the sphere consistent with stable equilibrium of the can.

5-25. A solid, of uniform density, is made up of a right circular cone and a hemisphere, both having a common base. Find the maximum value of the ratio of the height of the cone to the radius of the hemisphere such that the solid will stand upright in stable equilibrium on the table.

5-26. A rod of length b is composed of two pieces, of equal length and cross section, having densities of 9 and 1, respectively, and placed end to end. The rod is placed in a smooth hemispherical bowl of radius R. Find the angle the rod makes with the horizontal when it is in equilibrium. Show analytically that the equilibrium is stable.

5-27. A bowl, of radius of curvature R, rests on a rough horizontal table. Show that the bowl is in stable equilibrium about its bottom center point as long as the center of mass of the material stacked in the bowl is at a height h above the bottom of the bowl which is less than R.

5-28*J*. The radii of curvature at the blunt and pointed ends of a hard-boiled egg are a and b, respectively, and the egg can just be made to balance on its blunt end when stood on a rough surface. Show that it can be made to balance on its pointed end when stood inside a hemispherical basin of radius less than $b(c - a)/(c - b - a)$, where c is the longest axis of the egg. If the radius of the basin is just equal to the critical length, would the equilibrium be stable or unstable? Take $b > (c - a)/2$.

5-29*J*. A ladder of length h and weight W stands in a vertical position on a rough floor, an elastic string being tied to its topmost point and to a point in the ceiling at a height b above the floor, its tension being T. (T may vary.) Show that the equilibrium is stable or unstable according as

$$T > \text{or} < \frac{W(b-h)}{2b}$$

If the tension is equal to $W(b-h)/2b$, determine whether the equilibrium is stable or unstable.

5-30. A uniform heavy bar AB, of length $2b$, rests against a smooth vertical wall at A and over a smooth fixed rod C, as shown in Fig. 5-19, which is situated at a distance a from the wall. Take $b > a$. Find the value of θ, the inclination of the bar to the wall, at which equilibrium occurs. Determine whether or not the equilibrium is stable.

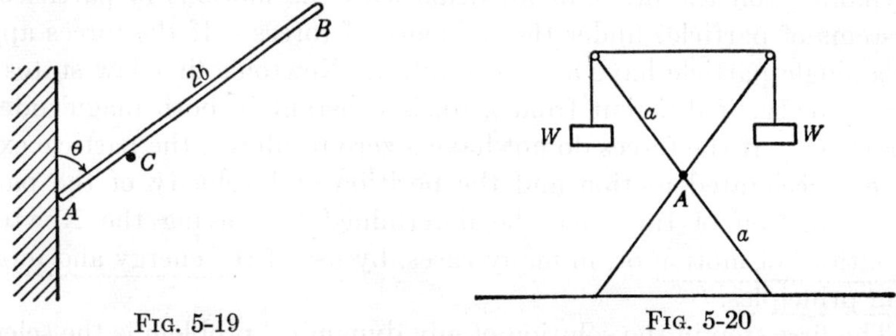

FIG. 5-19

FIG. 5-20

5-31. Two equal light rods of length $2a$ are joined at their mid-points A by a smooth pivot (see Fig. 5-20). The rods are constrained always to be in a vertical plane with their lower ends resting on a smooth floor. Determine the angle either rod makes with the vertical when the system is in equilibrium. Show whether the equilibrium is stable or unstable.

5-32. A particle moves in one direction and has a potential energy

$$V = V_0(e^{-ax} + bx)$$

where x is the distance from the origin and V_0, a, and b are positive constants. Find the force on the particle at any point x. Determine the equilibrium position and whether or not it is stable.

5-33. Two smooth solid cylinders, each of mass m and radius a, rest in a smooth cylindrical trough of radius b. Determine whether the equilibrium configuration is stable.

CHAPTER 6

MOTION OF A PARTICLE IN A UNIFORM FIELD

Statics deals with particles and systems of particles which are at rest. Dynamics, on the other hand, deals with the motions of particles and systems of particles under the influence of forces. If the forces applied to a single particle have a zero resultant, Newton's first law states that the velocity, if different from zero, is constant in both magnitude and direction. If the forces do not have a zero resultant, the particle experiences accelerated motion and the position and velocity of the particle as a function of time may be determined by solving the Newtonian equations of motion or, in many cases, by use of the energy and momentum principles.

The first step in the solution of any dynamical problem is the selection of an appropriate coordinate system. The equations of motion are then stated in terms of the coordinate system selected, or the equations of energy and momentum may be written, following which the solution is carried out in either case. In many instances the problem may be made easy or difficult depending upon the particular choice of coordinates. It will become apparent, as we go further into the study of dynamics, that it is possible with practice to acquire a fair amount of skill in the selection of the coordinate system which is most suitable for a given problem.

ONE-DIMENSIONAL MOTION OF A PARTICLE ACTED UPON BY A CONSTANT FORCE

6-1. Falling Body. For the present, we limit ourselves to a very simple exercise in one dimension, that of a body falling in a uniform gravitational field. It is convenient to employ a single coordinate y, positive upward, with origin on the ground. The equation of motion for a particle of mass m falling under the influence of a constant gravitational force mg is

$$m\ddot{y} = -mg \tag{6-1}$$

where the negative sign on the right follows since the force points in the direction of negative y. Equation (6-1) may be integrated successively

with respect to time, as

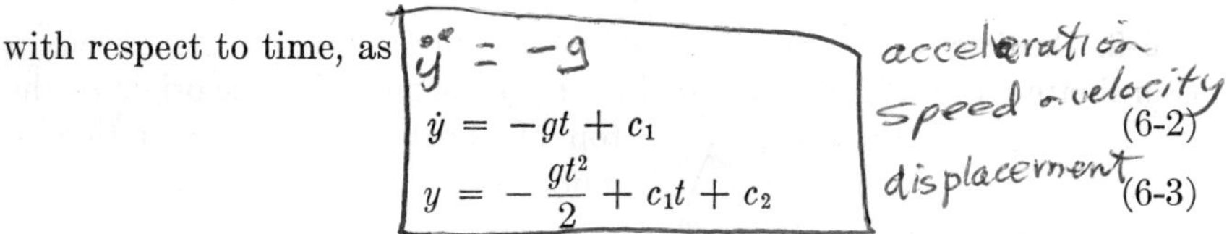

$$\dot{y} = -gt + c_1 \tag{6-2}$$

$$y = -\frac{gt^2}{2} + c_1 t + c_2 \tag{6-3}$$

in which c_1 and c_2 are arbitrary constants of integration. If, for example, the initial velocity were directed downward and of magnitude v_0, Eq. (6-2) would become

$$\dot{y} = -gt - v_0$$

If, in addition, the initial position were a height h above the ground, Eq. (6-3) would be

$$y = \frac{gt^2}{2} - v_0 t + h$$

From this it is clear how the boundary conditions of any particular problem enter into the final results. It will be noticed that in Eq. (6-2) we have obtained the speed of the particle as a function of time and in Eq. (6-3) the position of the particle as a function of time. It may be desired, instead, to express the speed $\dot{y}$ as a function of position y. In order to do this, we multiply both sides of Eq. (6-1) by $2\dot{y}$. We have, with the m canceled,

$$2\dot{y}\ddot{y} = -2g\dot{y}$$

or, rewriting,

$$\frac{d}{dt}(\dot{y}^2) = -2g\frac{dy}{dt} \tag{6-4}$$

The transformation of the left side may be verified by merely carrying out the operation of taking the derivative. Equation (6-4) may be integrated at once to yield

$$\dot{y}^2 = -2gy + c_3 \tag{6-5}$$

which expresses the speed $\dot{y}$ as a function of position. The quantity c_3 may be determined in terms of the boundary conditions. It is the square of the speed at $y = 0$ (or the value of $2gy$ at $\dot{y} = 0$).

Equation (6-5) may also be obtained by use of the principle of conservation of energy. If the origin is taken to be the point of zero potential energy and if the particle is at rest at a height h, its total energy at that point is all potential energy, of magnitude mgh. Accordingly, if it drops from rest at point h, we should have, at any later position y,

$$\frac{m}{2}\dot{y}^2 + mgy = mgh$$

6-2. Particle on a Smooth Inclined Plane. Suppose we select a coordinate x (see Fig. 6-1) parallel to the plane and with the origin at the top of the plane. The equation of motion is

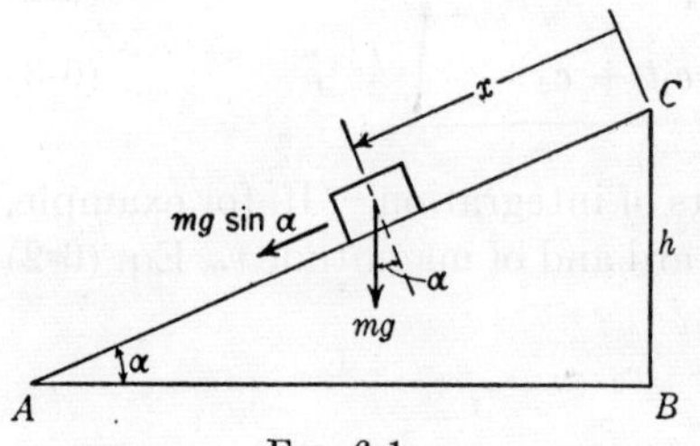

Fig. 6-1

$$m\ddot{x} = mg \sin \alpha \tag{6-6}$$

where α is the angle of inclination of the plane. The positive sign on the right follows since the force is in the direction of x increasing. Equation (6-6) states that the acceleration $\ddot{x}$ is a constant, equal to $g \sin \alpha$. Suppose we determine the velocity at A, the foot of the plane. Multiplying both sides of Eq. (6-6) by $2\dot{x}$, we have

$$2\dot{x}\ddot{x} = 2\dot{x}g \sin \alpha \qquad \text{or} \qquad \frac{d}{dt}(\dot{x}^2) = 2g \sin \alpha \frac{dx}{dt}$$

from which

$$\dot{x}^2 = 2gx \sin \alpha \tag{6-7}$$

where the constant of integration is zero if we take $\dot{x} = 0$ at $x = 0$. From Eq. (6-7), if v_1 is the value of $\dot{x}$ when the particle is at point A, and since $\overline{AC} \sin \alpha = h$, we have $v_1{}^2 = 2gh$, a result which again could have been written at once from energy considerations. It is interesting to note that the speed of the particle at the foot of the plane is the same as it would have been had it dropped directly from C to B.

6-3. Atwood's Machine. A slightly more complicated situation is that of Atwood's machine, shown in Fig. 6-2. Here two masses m_1 and m_2, connected by a light inextensible string which passes over a smooth pulley, are hanging in a vertical plane. We require the acceleration of the system and the tension in the cord.

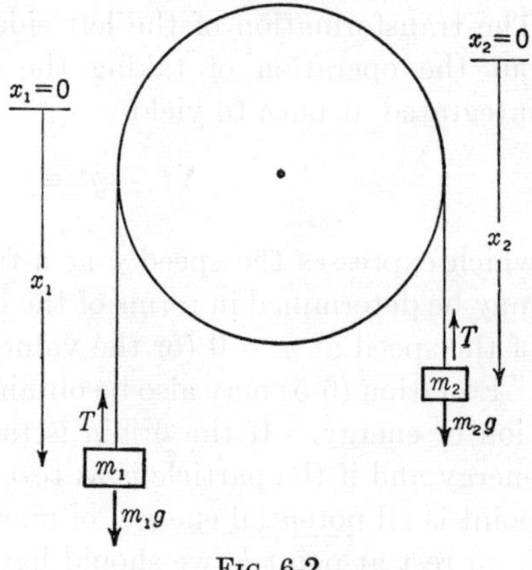

Fig. 6-2

The tension T is the same at all points of the cord, for two reasons. First, the pulley is smooth, and, second, the mass of the cord is neglected. If either of these conditions were not met, the tension T would not be uniform over the length of the cord. The other forces acting are the weights m_1g and m_2g of the two masses. Selecting coordinates x_1 and x_2, positive downward as shown, we may write the equation of motion of each particle. They are

$$m_1\ddot{x}_1 = m_1g - T \tag{6-8}$$

$$m_2\ddot{x}_2 = m_2g - T \tag{6-9}$$

where the signs on the right sides follow from the fact that the weights are forces which are in the direction of increasing coordinates and T is, in each case, in the direction of the coordinate decreasing. Now, since the string is inextensible, we must have $\ddot{x}_2 = -\ddot{x}_1$. Consequently Eq. (6-9) becomes

$$-m_2\ddot{x}_1 = m_2g - T \tag{6-10}$$

Equations (6-8) and (6-10) comprise two simultaneous equations in two unknowns, $\ddot{x}_1$ and T. We may first eliminate T between the two, obtaining

$$\ddot{x}_1 = \frac{(m_1 - m_2)g}{m_1 + m_2}$$

The quantity g is a positive number, and so, if m_1 is greater than m_2, $\ddot{x}_1$ is positive. This means that in this case the motion is in the counterclockwise sense, with m_1 descending. The tension T, obtained by combining this result with Eq. (6-8), is $2m_1m_2g/(m_1 + m_2)$.

Example 6-1. Consider the simple Atwood's machine when mounted in an elevator which is (a) descending with a constant speed and (b) descending with a uniform acceleration a; find the acceleration of the masses and the tension in the cord in both cases.

The result of part a of the problem can be perceived at once. Since the Atwood's machine is moving with a uniform velocity in a straight line, a system of coordinates which is moving along with the pulley, as in Sec. 6-3, is an inertial system. Consequently the same equations of motion, (6-8) and (6-9), are valid in the moving system (see Sec. 1-20). Thus the acceleration $\ddot{x}_1$ and the tension T are not affected so long as the elevator itself is not being accelerated.

In part b the elevator is being accelerated in the direction of increasing coordinates (x_1 and x_2 are both positive downward). Consequently the equations of motion, in view of Eq. (1-80), will be

$$m_1\ddot{x}_1 = m_1g - T - m_1a \qquad m_2\ddot{x}_2 = m_2g - T - m_2a$$

For the same reason as before, $\ddot{x}_2 = -\ddot{x}_1$, and we have the results

$$\ddot{x}_1 = -\ddot{x}_2 = \frac{(m_1 - m_2)(g - a)}{m_1 + m_2} \tag{6-11}$$

and

$$T = \frac{2m_1m_2(g - a)}{m_1 + m_2} \tag{6-12}$$

The results are precisely those which would obtain if the acceleration g of gravity were diminished by an amount equal to the acceleration a of the elevator. Evidently, if the elevator were being accelerated in the upward direction, the sign of a in Eqs. (6-11) and (6-12) would be changed.

6-4. Kinetic Friction. The concept of *sliding* or *kinetic friction* can be introduced in much the same way as was static friction (see Sec. 2-6). It is observed experimentally that when one body is sliding over another a frictional force opposes this motion. Let us consider the case of a block sliding across a rough table (Fig. 6-3). Suppose it is being accelerated by a force F. We observe that there is present a frictional force opposing this motion, which force we denote by the symbol f. Clearly the resultant acceleration of the block is $(F - f)/m$, where m is the mass of the block. We may find the value of f by the simple expedient of reducing the magnitude of F until the acceleration is zero. The magnitude F_0 of F at which this occurs is just equal to the frictional force f.

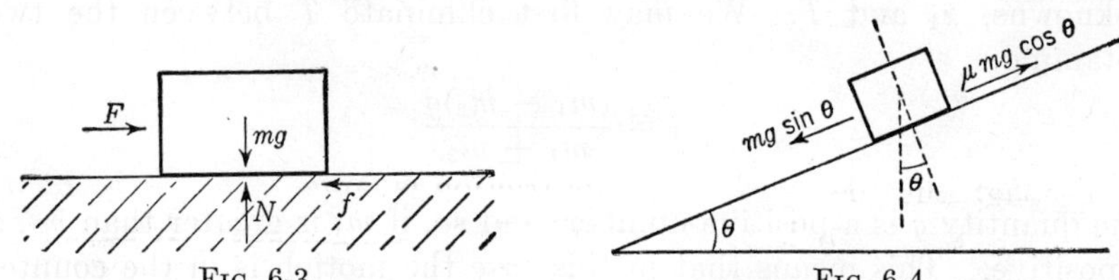

FIG. 6-3 FIG. 6-4

The forces of kinetic friction are found to depend on the nature of the contact surfaces and are very nearly proportional to the normal force pressing the two surfaces together.[1] In the present case, the normal force is N $(= mg)$, and the frictional force may therefore be written

$$f = \mu N \tag{6-13}$$

The constant of proportionality μ is called the *coefficient of sliding* or *kinetic friction*. The coefficient μ is very nearly independent of the area of contact and for moderate pressures very nearly independent of the velocity. There need be no confusion about the use of the same symbol μ to denote both the coefficients of sliding and of kinetic friction. It will be clear from the context which is the appropriate meaning. It is possible to measure the coefficient of kinetic friction in much the same way as that in which the coefficient of static friction is measured. In Fig. 6-4, a block is free to slide on a rough plane inclined at an angle θ to the horizontal. If the block is in motion, the forces acting upon it parallel to the plane are the force $mg \sin \theta$ down the plane and the force $\mu mg \cos \theta$ up the plane. The quantity μ is the coefficient of kinetic friction for the pair of surfaces in contact. Now the value of μ is found by merely adjusting the angle θ until the block slides down the plane with a

[1] The views here stated comprise the so-called *classical* interpretation of experiments involving frictional forces. The accuracy of some of the *classical laws* has been questioned recently. See, in this regard, Palmer's article, *Am. J. Phys.*, **17**, 181, 327, 336 (1949).

constant speed. Let this be true at an angle $\theta = \epsilon$. If there is no acceleration we must have

$$mg \sin \epsilon = \mu mg \cos \epsilon$$

or

$$\boxed{\mu = \tan \epsilon}$$

Thus, in analogy to the angle of static friction, it is also possible to define an *angle of kinetic friction.* It is found experimentally, for a given pair of contacting surfaces, that the coefficient of sliding friction is always less than the coefficient of static friction. If the angle θ in Fig. 6-4 is at first sufficiently small so that the block is initially at rest, it is found that θ must be increased to a value ϵ_0 in order to initiate the motion, after which it must be reduced to the value ϵ in order to prevent further acceleration of the block. Here ϵ_0 is the angle of static friction, and ϵ is the angle of kinetic friction.

Although the coefficient of static friction is larger than the coefficient of kinetic friction, it is of the same order of magnitude. It should not be inferred, however, that the frictional force undergoes a discontinuous change as the motion begins. When the relative velocity between the surfaces in contact is very small, the coefficient of kinetic friction changes rapidly and approaches in magnitude the coefficient of static friction for the two surfaces (see Fig. 6-5) as the relative velocity approaches zero. We may, for the moment, label the coefficient of static friction μ_0. For larger velocities μ is smaller in magnitude and changes much less rapidly. For many practical purposes the coefficient of kinetic friction varies sufficiently slowly over a wide enough range of velocities (excluding the region of low velocities just mentioned) to treat μ as being approximately constant over this range.

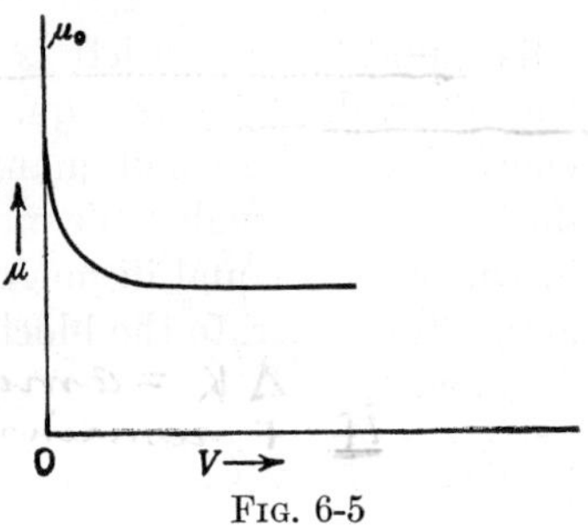

FIG. 6-5

In the case of the block sliding on the rough table, it is interesting to note that, if it requires a given force F_1 to move the block at a constant speed v_1, the same force F_1 is required to move the block at a constant speed v_2, where v_2 is greater than v_1 (to the approximation in which μ is assumed to be the same in both cases). Of course a force F, greater than F_1, must be applied for the time during which the block is accelerated from v_1 to v_2. However, as soon as the speed reaches v_2 and the force is reduced to F_1, the acceleration is again zero. It is useful to compute the work done in increasing the speed of the block from v_1 to v_2 in this manner. The equation of motion during the period of acceleration is

$$m\ddot{x} = F - \mu mg \tag{6-14}$$

in which μ is the coefficient of sliding friction, assumed constant. Multiplying both sides by $\dot{x}$, we have

$$m\dot{x}\ddot{x} = F\dot{x} - \mu mg\dot{x}$$

This may be rewritten as

$$F\frac{dx}{dt} = m\dot{x}\frac{d\dot{x}}{dt} + \mu mg\frac{dx}{dt}$$

Multiplying through by dt and integrating, we have

$$\int_{x_1}^{x_2} F\,dx = \int_{v_1}^{v_2} m\dot{x}\,d\dot{x} + \int_{x_1}^{x_2} \mu mg\,dx \tag{6-15}$$

in which x_1 is the point at which the acceleration begins and x_2 is the point at which it is again zero. If F is kept constant during this interval, Eq. (6-15) becomes simply

$$F(x_2 - x_1) = (\tfrac{1}{2}mv_2^2 - \tfrac{1}{2}mv_1^2) + \mu mg(x_2 - x_1) \tag{6-16}$$

The quantity on the left is the total work done by the force F during the interval. The first quantity on the right is the change in kinetic energy, and the second quantity is the energy *dissipated* through friction during this interval. We may regard F as the sum of two parts $F_1 + F_2$, in which F_1 is equal in magnitude to μmg and F_2 is the additional force acting to accelerate the block. Thus, if we substitute this in Eq. (6-16), we have

$$F_1(x_2 - x_1) + F_2(x_2 - x_1) = (\tfrac{1}{2}mv_2^2 - \tfrac{1}{2}mv_1^2) + \mu mg(x_2 - x_1) \tag{6-17}$$

Since $F_1 = \mu mg$, we see that the amount of work done to change the speed from v_1 to v_2 is just $F_2(x_2 - x_1)$, the result which should be expected.

Example 6-2. A block of mass m is sliding down a rough plane inclined at an angle θ to the horizontal. The coefficient of sliding friction is μ. It is desired to find the acceleration.

We may employ Fig. 6-4 for the present situation. The equation of motion is

$$m\ddot{x} = mg \sin\theta - \mu mg \cos\theta \tag{6-18}$$

where x is positive down the plane. By defining an angle ϵ of sliding friction, where $\mu = \tan\epsilon$, Eq. (6-18) becomes

$$\ddot{x} = g \sec\epsilon(\sin\theta\cos\epsilon - \sin\epsilon\cos\theta) = g \sec\epsilon \sin(\theta - \epsilon)$$

Example 6-3. A block of mass m_1 constrained to slide across a rough table is connected by means of a rough light string to a mass m_2 hanging over the edge of the table. Assuming that the string makes contact with the table only at the edge, it is

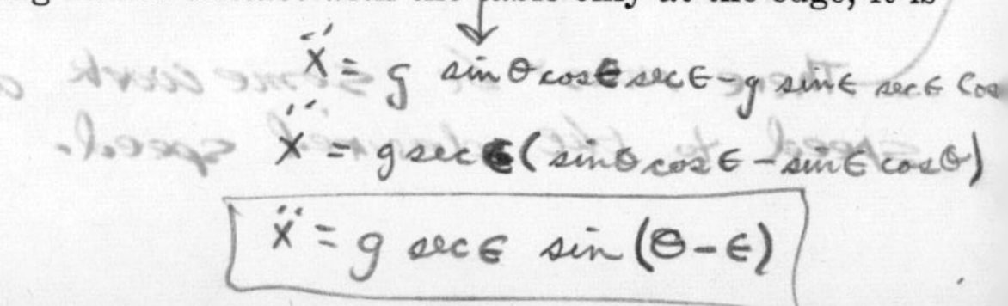

desired to determine the acceleration of m_1 and m_2. The coefficient of sliding friction between m_1 and the table is μ_1, and that between the string and the table is μ_2.

The situation is shown in Fig. 6-6. The forces acting on m_1 along the line of motion are the tension T_1 in the part of the string along the table top and the frictional force $\mu_1 m_1 g$. Similarly the forces acting on m_2 along the line of motion are T_2, the tension in the vertical portion of the string, and the weight $m_2 g$. Selecting coordinates x and y, positive in the direction of motion, we may write the equations of motion as

$$m_1\ddot{x} = T_1 - \mu_1 m_1 g \tag{6-19}$$
$$m_2\ddot{y} = m_2 g - T_2 \tag{6-20}$$

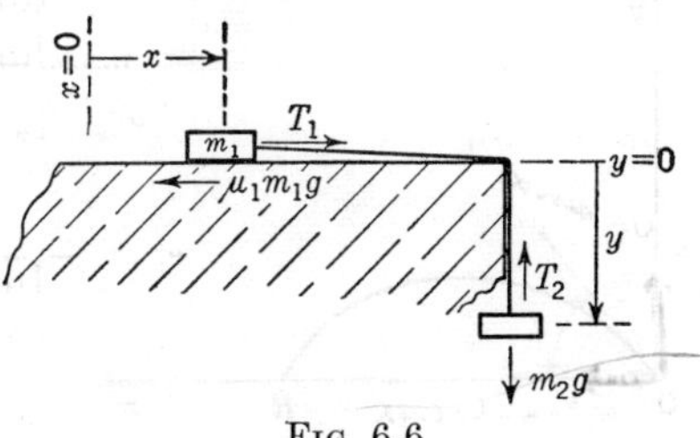

FIG. 6-6

But if the string is inextensible, $\ddot{y} = \ddot{x}$. Furthermore we may assume the string to be in contact with the edge of the table through an angle $\pi/2$. Since the conditions at this contact are precisely the same for kinetic friction as for static friction (save that μ is different), we may employ the method of Sec. 2-7 in order to determine the relation of T_1 to T_2. Clearly $T_2 = T_1 e^{\mu_2 \pi/2}$, and Eq. (6-20) becomes

$$m_2\ddot{x} = m_2 g - T_1 e^{\mu_2 \pi/2} \tag{6-21}$$

Eliminating T_1 between Eqs. (6-19) and (6-21), we have

$$\ddot{x} = \frac{g(m_2 - \mu_1 m_1 e^{\mu_2 \pi/2})}{m_2 + m_1 e^{\mu_2 \pi/2}}$$

Since $\ddot{x}$ has been found, the tensions T_1 and T_2 may be found readily with the aid of Eqs. (6-19) and (6-20).

FLIGHT OF A PROJECTILE

We now consider the motion of a projectile, regarding the problem to be that of a particle moving in two dimensions in a uniform gravitational field. It is evident that this may not be a good approximation for the consideration of many ballistic problems which might be expected to occur in practice. For example, a projectile fired from a rifled gun possesses a *spinning* motion about its axis of symmetry and, moreover, may also *yaw* (a deviation of the axis of symmetry from the instantaneous flight direction). Such details, in reality, bring the discussion into the category of the three-dimensional motions of a rigid body, taken up in Chap. 12. Complications of this nature will not be considered in this text; the student is referred to texts on exterior ballistics for the treatment of these.

Other possible factors which may be encountered in practice are changes in the gravitational acceleration, owing to the extreme height of the trajectory, and the influence of air resistance upon the motion. The latter may be rendered complicated by the effects of the shape of the projectile. A few of the simplest cases of air resistance, treating the moving body as a particle, will be examined briefly below in Secs. 6-7 to 6-9.

6-5. Projectile in a Vacuum. Equation of the Path. In Fig. 6-7, a particle of mass m is projected from the origin O with a velocity v_0 in a direction making an angle α with the horizontal. The equations of motion are

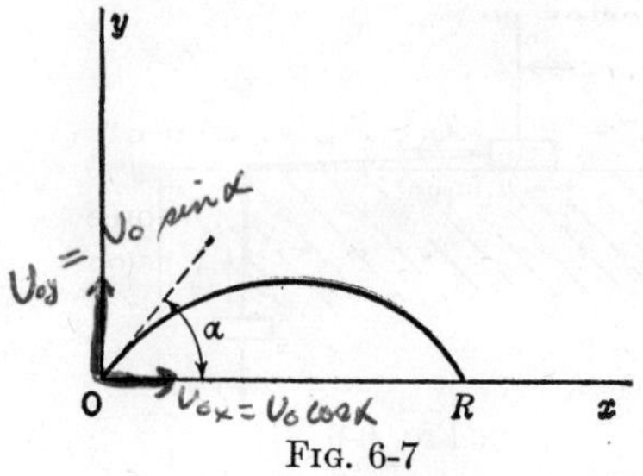

$$m\ddot{y} = -mg \tag{6-22}$$

$$m\ddot{x} = 0 \tag{6-23}$$

Fig. 6-7

Integrating at once with respect to time, we have

$$\dot{y} = -gt + v_0 \sin \alpha \tag{6-24}$$

$$\dot{x} = v_0 \cos \alpha \tag{6-25}$$

Integrating a second time, we obtain

$$y = \frac{-gt^2}{2} + v_0 t \sin \alpha \tag{6-26}$$

$$x = v_0 t \cos \alpha \tag{6-27}$$

where, at $t = 0$, $x = y = 0$, $\dot{x} = v_0 \cos \alpha$, and $\dot{y} = v_0 \sin \alpha$. At the end of the flight $y = 0$. Hence, from Eq. (6-26), we find the time of flight to be

$$t_0 = \frac{2v_0}{g} \sin \alpha \tag{6-28}$$

Eliminating t between Eqs. (6-26) and (6-27), we find the equation of the path to be

$$y = -\frac{g}{2v_0^2 \cos^2 \alpha} x^2 + x \tan \alpha \tag{6-29}$$

The horizontal range R may be determined from Eq. (6-29) by putting $y = 0$. We obtain (since, at $y = 0$, $x = R$)

$$R = \frac{2v_0^2}{g} \sin \alpha \cos \alpha = \frac{v_0^2}{g} \sin 2\alpha \tag{6-30}$$

This is a maximum at $\sin 2\alpha = 1$, or $\alpha = \pi/4$. Thus the maximum range R_m is equal to v_0^2/g. The curve described by Eq. (6-29) is a *parabola*. It will help the student to see this if, for the moment, we select a different origin. Suppose we choose the origin at the top of the path, that is, where $\dot{y} = 0$. Thus we take $x = y = \dot{y} = 0$ at $t = 0$. Accordingly, Eq. (6-26) takes the form

$$y = -\frac{gt^2}{2} \tag{6-31}$$

which, combined with Eq. (6-27), yields

$$x^2 = -\frac{2v_0^2 \cos^2 \alpha}{g} y$$

This is easily seen to be a parabola with vertex convex upward at the origin and in which the latus rectum is $(2v_0^2 \cos^2 \alpha)/g$. The origin being temporarily retained at the top of the path, it is interesting to examine the kinetic energy of the particle at any point of the trajectory. If the speed is v, and since $v^2 = \dot{x}^2 + \dot{y}^2$, we have, employing Eqs. (6-24), (6-25), and (6-31),

$$\frac{1}{2} mv^2 = \frac{m}{2} (v_0^2 \cos^2 \alpha + g^2t^2) = \frac{m}{2} (v_0^2 \cos^2 \alpha - 2gy) \qquad (6\text{-}32)$$

Now $(v_0^2 \cos^2 \alpha)/2g$ is just one-fourth of the latus rectum, which, in turn, is the distance of the directrix above the x axis. Upon introducing a symbol p for this quantity, Eq. (6-32) becomes

$$\tfrac{1}{2}mv^2 = mgp - mgy = mg(p - y) \qquad (6\text{-}33)$$

Now, since p is the distance from the vertex to the directrix of the parabola and since the vertex is at the origin, with y positive upward, the quantity $p - y$ must be the depth, below the directrix, of the point on the trajectory at which the speed is v. Hence Eq. (6-33) states that the kinetic energy at any point of the trajectory is the same as if the particle had fallen from a position of rest on the directrix.

6-6. Miscellaneous Examples

Example 6-4. A gun fires two shots, the muzzle velocity in each case being v_0. In the first case the shot is fired at an angle of elevation α, and in the second case the shot is fired at a smaller angle of elevation α'. Find the time interval between the two firings such that the two shots will collide in mid-air.

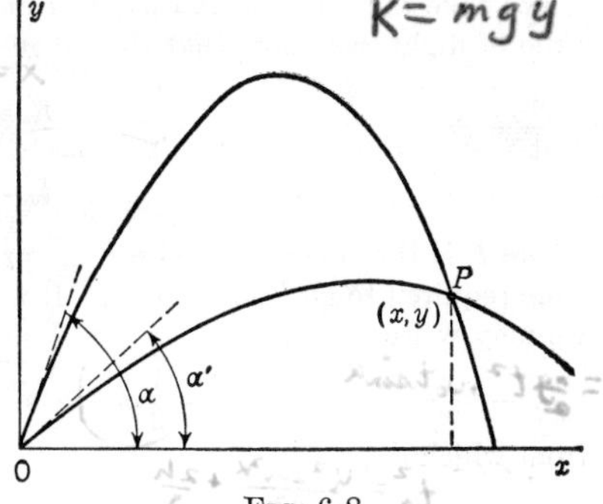

FIG. 6-8

Denoting x_1, y_1 as the coordinates of the point P (see Fig. 6-8) at which the collision occurs, we may employ Eq. (6-27) to find the time of flight of each body from O to P. We have

$$t_1 = \frac{x_1}{v_0 \cos \alpha} \qquad t_1' = \frac{x_1}{v_0 \cos \alpha'}$$

from which the interval Δt between the time of firings must be

$$\Delta t = t_1 - t_1' = \frac{x_1}{v_0} \frac{\cos \alpha' - \cos \alpha}{\cos \alpha \cos \alpha'} \qquad (6\text{-}34)$$

In order to eliminate x_1, we make use of Eq. (6-29). We have, since y_1 is the same for both trajectories,

$$x_1 \tan \alpha - \frac{g x_1^2}{2 v_0^2 \cos^2 \alpha} = x_1 \tan \alpha' - \frac{g x_1^2}{2 v_0^2 \cos^2 \alpha'}$$

from which

$$x_1 = \frac{2 v_0^2}{g} \frac{(\cos^2 \alpha \cos^2 \alpha')(\tan \alpha - \tan \alpha')}{\cos^2 \alpha' - \cos^2 \alpha}$$

Substituting this in Eq. (6-34), we obtain

$$\Delta t = \frac{2 v_0}{g} \frac{(\cos \alpha \cos \alpha')(\tan \alpha - \tan \alpha')}{\cos \alpha' + \cos \alpha} = \frac{2 v_0}{g} \frac{\sin (\alpha - \alpha')}{\cos \alpha' + \cos \alpha} \tag{6-35}$$

Example 6-5. A gun is mounted on a cliff at a height h above a level plain. It fires a projectile with a muzzle velocity v_0 at an angle of elevation α. Find the horizontal range R (the distance along the level plain from the foot of the cliff).

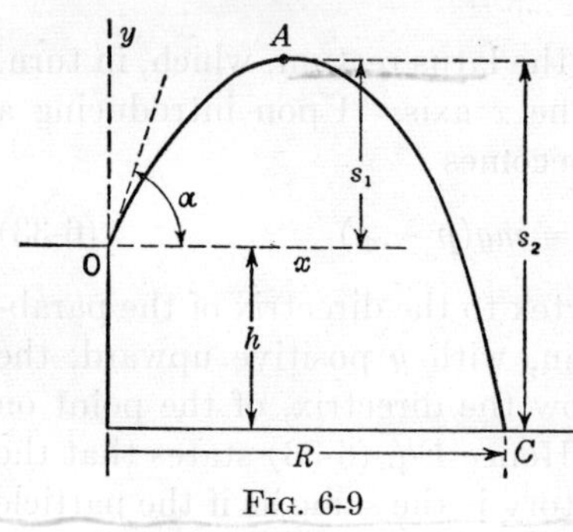

FIG. 6-9

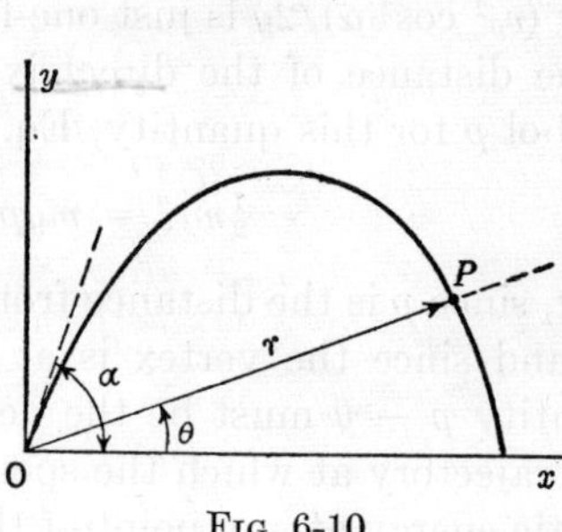

FIG. 6-10

We take the origin O at the point of firing, as shown in Fig. 6-9. If t_0 is the total time of flight, we know that the range R will be

$$R = v_0 t_0 \cos \alpha$$

But

$$t_0 = t_1 + t_2$$

where t_1 is the time required for the projectile to get from O to point A and t_2 is the time required to go from A to C. If we denote the height of A above the plain as s_2, we may write

$$s_2 = \frac{1}{2} g t_2^2 = s_1 + h = \frac{1}{2} \frac{v_0^2 \sin^2 \alpha}{g} + h$$

from which

$$t_2 = \frac{1}{g} \sqrt{v_0^2 \sin^2 \alpha + 2gh}$$

Thus, since $t_1 = (v_0 \sin \alpha)/g$, we have finally

$$R = \frac{v_0 \cos \alpha}{g} [v_0 \sin \alpha + (v_0^2 \sin^2 \alpha + 2gh)^{\frac{1}{2}}] \tag{6-36}$$

Example 6-6. A gun fires a projectile with a muzzle velocity v_0. Determine the maximum range in any direction, given that the angle of elevation of the gun may be freely varied.

The picture is as shown in Fig. 6-10. The projectile, fired from the origin with a velocity v_0 at an angle of elevation α, will pass through a point P. The polar coordinates of P are r, θ as shown. Evidently the range in the direction OP is just the coordinate r, of P. Since

$$x = r \cos \theta \quad \text{and} \quad y = r \sin \theta$$

Eq. (6-29) becomes

$$r \sin \theta = -\frac{g}{2v_0^2 \cos^2 \alpha} r^2 \cos^2 \theta + r \cos \theta \tan \alpha$$

Simplifying, we have

$$r = \frac{2v_0^2}{g \cos \theta} (\sin \alpha \cos \alpha - \cos^2 \alpha \tan \theta) = \frac{2v_0^2}{g \cos^2 \theta} \cos \alpha \sin (\alpha - \theta) \tag{6-37}$$

which is the range, expressed as a function of the variable α, in a given direction making an angle θ with the horizontal. The maximum value R_θ of the range must satisfy the condition that $dr/d\alpha = 0$. Carrying out this operation, we have

$$\begin{aligned} \frac{dr}{d\alpha} &= \frac{2v_0^2}{g \cos^2 \theta} [\cos \alpha \cos (\alpha - \theta) - \sin \alpha \sin (\alpha - \theta)] \\ &= \frac{2v_0^2}{g \cos^2 \theta} \cos (2\alpha - \theta) = 0 \end{aligned} \tag{6-38}$$

from which $\alpha = (\pi/4) + (\theta/2)$. Upon substituting this in Eq. (6-37), the maximum range R_θ in a direction making an angle θ with the horizontal is

$$\begin{aligned} R_\theta &= \frac{2v_0^2}{g \cos^2 \theta} \cos \left[\frac{1}{2}\left(\frac{\pi}{2} + \theta\right)\right] \sin \left[\frac{1}{2}\left(\frac{\pi}{2} - \theta\right)\right] \\ &= \frac{2v_0^2}{g \cos^2 \theta} \left[\frac{1 + \cos (\pi/2 + \theta)}{2}\right]^{\frac{1}{2}} \left[\frac{1 - \cos (\pi/2 - \theta)}{2}\right]^{\frac{1}{2}} \\ &= \frac{v_0^2}{g \cos^2 \theta} (1 - \sin \theta) = \frac{v_0^2}{g(1 + \sin \theta)} \end{aligned} \tag{6-39}$$

where the last step follows since

$$\cos^2 \theta = (1 - \sin^2 \theta) = (1 - \sin \theta)(1 + \sin \theta)$$

MOTION WITH AIR RESISTANCE

The problem of the motion of a body in a viscous medium, such as air, for a wide range of velocities is a complicated one. Although it is usually assumed that the force is proportional to a power n of the velocity, in reality n varies from low values at speeds small compared with the speed of sound to high values as the speed of sound is approached. As the speed of the body exceeds that of sound, the value of n again decreases. A further complication is that the factor of proportionality does not remain a constant but varies if the speed of the body is varied over a sufficiently wide range.

The above remarks indicate the difficulty attending any attempt to treat most practical cases analytically. In order to consider such problems accurately, extensive use often must be made of the methods of numerical integration, tables of measured data for various speeds and shapes of moving bodies being employed.

In the present discussion, we limit ourselves to the treatment of one or two simple cases in which we assume the moving body to be a particle and consider both n and the proportionality factor to be constant.

6-7. Falling Body. Resistance Proportional to the First Power of the Velocity. We consider the body to be released from rest at an initial height above the ground. It is convenient to choose the coordinate y to be positive downward, with y and t both zero initially. If the proportionality constant is mk, where m is the mass of the body, the equation of motion is

$$\ddot{y} = g - k\dot{y} \tag{6-40}$$

This may be rewritten as

$$\frac{d\dot{y}}{dt} = g - k\dot{y} \tag{6-41}$$

from which, separating the variables, we have

$$\frac{d\dot{y}}{(g/k) - \dot{y}} = k\,dt \tag{6-42}$$

Equation (6-42) may be integrated, to yield

$$\log\left(\frac{g}{k} - \dot{y}\right) = -kt + \log c_1 \tag{6-43}$$

where c_1 is a constant. Taking the antilogarithms, and noting, since $\dot{y} = 0$ at $t = 0$, that $c_1 = g/k$, we have finally

$$\dot{y} = \frac{g}{k}(1 - e^{-kt}) \tag{6-44}$$

We see that the body cannot exceed a limiting velocity g/k, a fact which was also evident from Eq. (6-40) since, for the limiting case, the acceleration is zero. Integrating a second time, we obtain

$$y = \frac{g}{k}t + \frac{g}{k^2}e^{-kt} + c_2 \tag{6-45}$$

Since $y = 0$ at $t = 0$, $c_2 = -g/k^2$. Therefore Eq. (6-45) assumes the form

$$y = \frac{g}{k}t - \frac{g}{k^2}(1 - e^{-kt}) \tag{6-46}$$

It is interesting to consider an approximate solution which obtains when the air resistance is small (namely, when k is very small). Expanding the exponential term in Eq. (6-46),

$$y = \frac{g}{k}t - \frac{g}{k^2}\left[1 - \left(1 - kt + \frac{k^2t^2}{2!} - \frac{k^3t^3}{3!} + \cdots\right)\right] \tag{6-47}$$

Performing the obvious cancellations, and neglecting powers of k higher than the first, we have left

$$y \simeq \frac{1}{2}gt^2 - \frac{gk}{6}t^3 \tag{6-48}$$

The first term on the right is that obtained alone if there is no air resistance. The presence of a small force proportional to the first power of the velocity introduces the second term on the right.

6-8. Resistance Proportional to the Second Power of the Velocity. If we consider motion in the direction of increasing y (such as would be the case if the body had fallen from an initial height, with y measured positively downward), the equation of motion reduces to

$$\ddot{y} = g - k\dot{y}^2 \tag{6-49}$$

where k has the same significance as before. Upon rewriting $\ddot{y}$ as $d\dot{y}/dt$ and separating the variables, Eq. (6-49) becomes

$$\frac{d\dot{y}}{(g/k) - \dot{y}^2} = k\,dt \tag{6-50}$$

Integrating, we obtain

$$\sqrt{\frac{k}{g}}\tanh^{-1}\left(\sqrt{\frac{k}{g}}\,\dot{y}\right) = kt + c_1 \tag{6-51}$$

If $\dot{y} = 0$ at $t = 0$, $c_1 = 0$ and Eq. (6-51) becomes, finally,

$$\dot{y} = \sqrt{\frac{g}{k}}\tanh(\sqrt{gk}\,t) \tag{6-52}$$

Since $\tanh \infty = 1$, Eq. (6-52) shows that the limiting velocity is $\sqrt{g/k}$, a result obtained, as well, from Eq. (6-49).

Equation (6-49) can be integrated in another way, if desired, to find $\dot{y}$ as a function of y. To do this, it is necessary only to note that $\ddot{y} = \dot{y}\,d\dot{y}/dy$. Equation (6-50) becomes, in this case,

$$\frac{\dot{y}\,d\dot{y}}{(g/k) - \dot{y}^2} = k\,dy \tag{6-53}$$

This may be integrated, as

$$-\frac{1}{2}\log\left(\frac{g}{k} - \dot{y}^2\right) = ky + c_2 \tag{6-54}$$

If we take $\dot{y} = 0$ at $t = 0$, $c_2 = -\frac{1}{2} \log (g/k)$ and Eq. (6-54) finally becomes

$$\dot{y}^2 = \frac{g}{k} (1 - e^{-2ky}) \tag{6-55}$$

which gives $\dot{y}$ as a function of y, with time eliminated. It is possible to eliminate $\dot{y}$ between Eqs. (6-52) and (6-55), yielding y as a function of t. We have, from (6-52) and (6-55),

$$\tanh^2 (\sqrt{gk}\, t) = 1 - e^{-2ky}$$

from which, upon consulting Appendix 3,

$$\text{sech}^2 (\sqrt{gk}\, t) = e^{-2ky}$$

The displacement y may be found from this to be

$$y = \frac{1}{k} \log \cosh (\sqrt{gk}\, t) \tag{6-56}$$

It is of interest to note that, by retaining y positive downward, Eq. (6-49) becomes, for upward motion,

$$\ddot{y} = g + k\dot{y}^2 \tag{6-57}$$

The difference between (6-49) and (6-57) arises since in Eq. (6-57) the resistance term is in the positive y direction and the sign of $\dot{y}^2$ is positive even though $\dot{y}$ is negative. Such a change in sign of the resistance term with change in the direction of motion is to be expected whenever the force of resistance is proportional to an *even* power of the velocity. (This is in contrast to the situation when the resistance term is proportional to an *odd* power of the velocity.) An alternative way to take the change of sign into account is to change the sign of the coordinate for the two cases (for example, by taking y positive first for the upward motion and then again for the downward motion).

6-9. Projectile with Air Resistance. We limit ourselves to the simple case of air resistance proportional to the first power of the velocity, a greatly idealized situation from what is the practical case. The projectile is fired with velocity v_0 from the origin at an angle of elevation α. The equations of motion are, with m canceled,

$$\ddot{x} = -k\dot{x} \tag{6-58}$$

$$\ddot{y} = -g - k\dot{y} \tag{6-59}$$

Equation (6-58) becomes, noting that $\ddot{x} = d\dot{x}/dt$ and separating the variables,

$$\frac{d\dot{x}}{\dot{x}} = -k\, dt \tag{6-60}$$

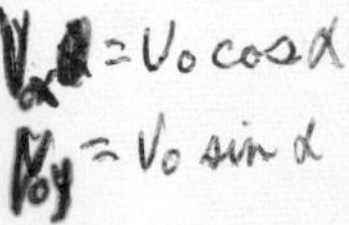

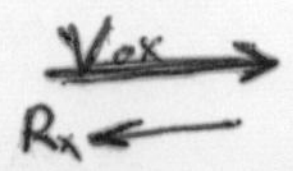

from which

$$\log \dot{x} = -kt + c_1$$

This becomes, since at $t = 0$, $\dot{x} = v_0 \cos \alpha$

$$\dot{x} = v_0 \cos \alpha \, e^{-kt} \tag{6-61}$$

Integrating a second time,

$$x = \frac{v_0 \cos \alpha}{k}(1 - e^{-kt}) \tag{6-62}$$

where the constant of integration has been evaluated by means of the condition that $x = 0$ at $t = 0$. The integration of the y equation is quite similar (cf. also Sec. 6-7). We have

$$\frac{d\dot{y}}{(g/k) + \dot{y}} = -k \, dt \tag{6-63}$$

from which

$$\dot{y} = -\frac{g}{k} + \left(v_0 \sin \alpha + \frac{g}{k}\right) e^{-kt} \tag{6-64}$$

where we have made use of the condition $\dot{y} = v_0 \sin \alpha$ at $t = 0$. Integrating a second time, and since $y = 0$ at $t = 0$, we have finally

$$y = -\frac{g}{k}t + \frac{1}{k}\left(v_0 \sin \alpha + \frac{g}{k}\right)(1 - e^{-kt}) \tag{6-65}$$

The equation of the path may be found by eliminating t between Eqs. (6-62) and (6-65).

It is instructive to examine the appearance of the above equations after a long time has elapsed. Equations (6-61), (6-62), (6-64), and (6-65) yield

$$\dot{x} \xrightarrow[t \to \infty]{} 0 \qquad \dot{y} \xrightarrow[t \to \infty]{} -\frac{g}{k}$$

$$x \xrightarrow[t \to \infty]{} \frac{v_0 \cos \alpha}{k} \qquad y \xrightarrow[t \to \infty]{} -\infty$$

The interpretation follows directly. The trajectory approaches, asymptotically, the vertical line $x = (v_0 \cos \alpha)/k$. The terminal speed (in the absence of an impact with the ground) is $-g/k$.

Viscous forces involving even powers of the velocity introduce complications into the equations of motion if rectangular coordinates are employed. These arise because of the difficulty with the sign after the top of the trajectory has been passed. Consequently, projectile motion with viscous forces is usually treated by resolving the vector equation of motion into one tangent to the path and one normal to the path. In this way a coordinate s can be employed which is continually increasing along the trajectory. This procedure will not be considered here. Such coordinates are introduced in Chap. 11 for systems involving constraints.

Problems

Problems marked C are reprinted by kind permission of the Cambridge University Press.

6-1. A mass of 5 lb resting on a smooth plane inclined at 30° to the horizontal is connected by a fine thread, which passes over a pulley at the summit of the plane, to a mass of 3 lb hanging vertically. The system starts from rest, and after 8 sec the thread is severed. Find how far the 5-lb mass will rise up the inclined plane before falling back.

6-2. A particle of weight W is resting on a rough inclined plane and is being moved up the plane with a uniform velocity by a force F acting at an angle θ with the plane. Draw the triangle of forces, and calculate F and the reaction R of the plane in terms of W, θ, α, and ϵ, where α is the angle of inclination of the plane with the horizontal and ϵ is the angle of friction.

6-3. A ship of 2,000 tons moving 30 ft/min is brought to rest by a hawser in a distance of 2 ft. Find, in tons, the average pull sustained by the hawser.

6-4. A bullet traveling 1,000 ft/sec penetrates a block of wood to a depth of 12 in. Supposing that it had been fired against a block of the same wood 6 in. thick, find the speed it would have had on emergence. (Assume the resistance of the wood to the bullet to be constant.)

6-5. In a system of pulleys with one fixed and one movable block, the cord is attached to the axis of the movable block, then passes over the fixed one, then under the movable block, then over the fixed one, and has a weight P attached to its other end. A weight W is suspended from the movable block. Find the acceleration of W when the system is released, the weight of the cord being neglected and that of the movable block being included in W. (It is assumed that the parts of the cord between the blocks are parallel.)

6-6. A rope bearing a weight P passes three times around a fixed pulley and twice around a movable pulley bearing a weight W, the other end of the rope being attached to the movable pulley. Find the acceleration of W, neglecting friction.

6-7*C*. A bucket of mass M is raised from the bottom of a shaft of depth h by means of a light cord which is wound on a wheel of mass m. The wheel is driven by a constant force which is applied tangentially to its rim for a certain time and then ceases. Show that, if the bucket just comes to rest at the top of the shaft t_0 sec after the beginning of the motion, the greatest rate at which work is being done by this force is $2M^2g^2ht_0/[Mgt_0^2 - 2h(m + M)]$, where the mass of the wheel is condensed in its rim.

6-8*C*. An elevator is lowered for the first third of the shaft with a constant acceleration, for the next third it descends with uniform velocity, and then a constant retarding force just brings it to rest as it reaches the bottom of the shaft. If the time of descent is equal to that taken by a particle in falling four times the whole depth, show that the force of the man inside, on the bottom of the cage, was, at the beginning, $\frac{23}{48}$ of his weight.

6-9. A body of mass m falls with an initial velocity v_0 from a height above the ground. Assuming a force of air resistance proportional to the square of the velocity (proportionality constant mk^2), find the velocity at any time. Find also the limiting velocity.

6-10. The total weight of a paratrooper (including the parachute) is 200 lb. Assume that, when the parachute is open, the retarding force being applied by the air is proportional to the square of his velocity of fall. Also it is given that, if his velocity were 10 ft/sec, the retarding force would be equal to 2 lb/sq ft of projected area of the parachute. (*a*) How large must the projected area of the parachute be if 4 ft/sec is a safe speed at which to reach the ground? (*b*) If the man is initially at rest, how far will he fall before reaching 95 per cent of the safe terminal velocity?

6-11*C*. A cubical box slides down a rough inclined plane, the coefficient of friction of the contact surfaces being μ. The two sides of the base of the box are horizontal during the motion. If the box contains sufficient water to just cover the base of the box during the motion, show that the volume of water is $\mu/2$ times the internal volume of the vessel.

6-12. A bomb is dropped from a height h above the ground. Assuming a force of air resistance proportional to the velocity (constant of proportionality mk, where m is the mass of the bomb), find the approximate time of fall, neglecting quantities of the order of k^2.

6-13. A particle of mass m is projected vertically upward with an initial velocity v_0. The motion of the particle is opposed by a force which is proportional to the velocity (proportionality constant mk). Determine the time required for the particle to arrive at the top of its path. Find also the energy which has been dissipated in the viscous medium during this time. Consider k to be small so that terms in k^2 may be neglected. Also $v_0k/g \ll 1$.

6-14*C*. A particle is projected with a given velocity up a line of greatest slope of a rough inclined plane. Find the height, above the point of projection, of the point at which it comes to rest. Supposing the inclination of the plane to be greater than the angle of static friction, find the velocity with which the particle returns to the point of projection.

6-15. A gun is mounted at the top of a cliff a height h above a level plain. If the muzzle velocity is v_0, show that the maximum horizontal range along the plain will occur when the gun has an elevation

$$\alpha = \sin^{-1} \frac{v_0}{\sqrt{2(v_0^2 + gh)}}$$

6-16. Show that the range for any elevation α of the gun in the previous problem is increased by raising it to the top of the cliff by an amount

$$\frac{R_1}{2}\left[\left(1 + \frac{2gh}{v_0^2 \sin^2 \alpha}\right)^{\frac{1}{2}} - 1\right]$$

where R_1 is the range when the gun is mounted at the foot of the cliff.

6-17. A gun mounted at the foot of a hill is directed up the hill in such a manner that the projectile fired by the gun eventually strikes the hill at right angles. If the hill makes an angle β with the horizontal, determine the angle that the gun barrel makes with the slope of the hill.

6-18. A particle rests initially in equilibrium at the highest point of a smooth horizontal cylinder of radius a. It is disturbed slightly so as to slide down in a plane normal to the axis of the cylinder. Find the latus rectum of the path of the particle which it describes after leaving the cylinder.

6-19*C*. A particle is projected so as to have a range R on a horizontal plane through the point of projection, and the greatest height attained by it is h. Prove that the maximum horizontal range with the same velocity of projection is

$$2h + \frac{R^2}{8h}$$

6-20*C*. Two particles are projected from the same point with velocities v_1 and v_2, at elevations α_1 and α_2, respectively ($\alpha_1 > \alpha_2$). Show that if they are to collide in mid-air the interval between the firings must be

$$\frac{2v_1v_2 \sin (\alpha_1 - \alpha_2)}{g(v_1 \cos \alpha_1 + v_2 \cos \alpha_2)}$$

6-21. Suppose the motion of a falling spherical drop of liquid is opposed by a force proportional to the surface area of the drop, to the density ρ of the medium through which it is falling, and to the nth power of its velocity. The density of the droplet is ρ_0, and it is assumed that the shape remains spherical. Show that the limiting velocity of the drop varies as the nth root of the radius of the drop. Gravity g is acting vertically downward.

6-22. A block slides, under gravity, down a smooth plane which is inclined at an angle θ to the horizontal. The block starts from rest at the top of the incline, a distance b from the foot, the whole being immersed in a medium which resists the motion of the block with a force proportional to the square of its velocity (constant of proportionality mk, where m is the mass of the block). Show that the time required for the block to get to the bottom of the incline is $(\cosh^{-1} e^{kb})/\sqrt{kg \sin \theta}$.

6-23. A body falls, under gravity, in a medium which opposes its motion with a force proportional to the velocity of the body. The limiting velocity acquired by the body is such that it would be attained in a time T if the body were falling in a vacuum. Show that, the medium being present, the body acquires half its limiting velocity in a time $0.693T$.

6-24. A particle is fired upward with a velocity v in a medium which opposes its motion with a force proportional to the square of its velocity. Gravity is acting vertically downward. Show that when the particle returns to the point of projection its speed is $vV/\sqrt{v^2 + V^2}$, where V is the limiting velocity of the particle in the medium.

6-25. A projectile is fired with a muzzle velocity v_1 from a gun which is elevated at an angle of 45° to the horizontal. At the top of the trajectory the velocity is v_2 (horizontal). The medium resists the motion with a force proportional to the square of the velocity of the projectile. Show that, on the descending portion of the trajectory, when the direction of motion of the missile makes an angle of 45° with the horizontal, its speed is given by $v_1v_2/\sqrt{v_1^2 - v_2^2}$.

6-26*C*. A particle moves under gravity in a medium the resistance of which is proportional to the velocity. Prove that the range on a horizontal plane is a maximum, for given velocity of projection, when the angles of elevation at the beginning and end of the trajectory are complementary.

6-27. A block of mass m slides over a smooth horizontal surface, the whole being immersed in a medium which opposes the motion with a force proportional to the velocity (proportionality constant mk). If the initial velocity is V, find the distance traversed and the time elapsed before the block comes to rest.

CHAPTER 7

OSCILLATORY MOTION OF A PARTICLE IN ONE DIMENSION

One of the most important mechanical systems in nature is the *harmonic oscillator*, or system which executes simple harmonic motion. Examples of this motion are found in the behavior of a mass driven by a spring which is being alternately compressed and extended, the small oscillations of a pendulum, and the movements of charges in certain types of electrical circuit. Indeed, this type of oscillatory behavior is approximated by all systems in stable equilibrium, whether the equilibrium is a static or a dynamic one, in carrying out *small motions* about the equilibrium configurations.

It is a characteristic of systems demonstrating simple harmonic motion that the differential equation of motion is *linear* with *constant coefficients*. As a consequence of the linearity of the equation of motion the so-called *superposition principle* holds. This means, for example, that, if two particular solutions of the equation of motion are found, their sum also is a solution of the equation.

A further property of an oscillation which is simple harmonic is that the *period*, or time required to execute one complete oscillation, is independent of the maximum displacement. For most vibratory systems in nature, however, the motion is simple harmonic only for a limited range of conditions. The most common limitation is that the amplitude must be small, and, if this condition is not met, the oscillation departs from the simple harmonic type. This departure exhibits itself in the equation of motion, in general, by the appearance of *nonlinear* terms. The detailed treatment of such cases is above the level of this text. However, brief mention will be made at the end of this chapter of one or two examples of nonharmonic systems, and a few of the physical properties of such nonlinear systems will be pointed out.

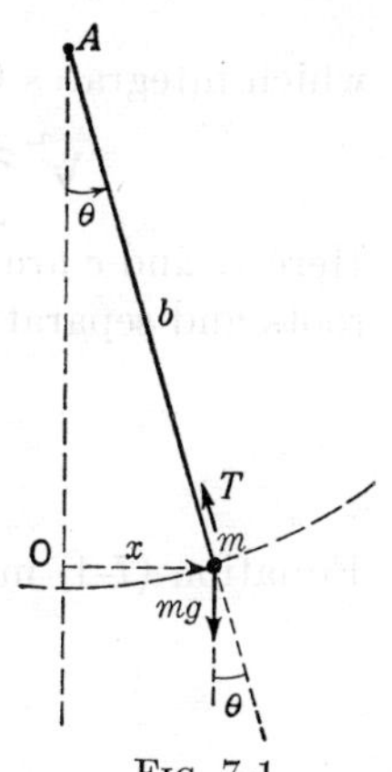

FIG. 7-1

7-1. Motion of a Simple Pendulum. The Oscillator Equation. We treat first the case of a simple pendulum moving in a vertical plane. This consists (see Fig. 7-1) of a mass m suspended by a long massless string of length b, free to perform small

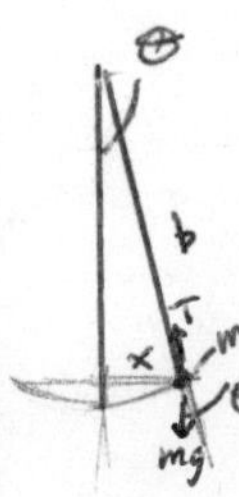

motions in a plane about the equilibrium position. The maximum displacement $b\theta_{\max}$ in this motion is very small compared with b. The forces acting on m are the tension T in the string and the weight mg. If x is a coordinate measured positively to the right, the equation of motion along the x direction is

$$m\ddot{x} = -T \sin \theta \tag{7-1}$$

where the minus sign follows from the fact that the x component of T is in the direction of $-x$. If θ is small throughout the motion, we may assume $\cos \theta \simeq 1$ and $T \simeq mg$. Hence Eq. (7-1) becomes, since $\sin \theta = x/b$,

$$\ddot{x} + \frac{g}{b} x = 0 \tag{7-2}$$

or

$$\ddot{x} + \omega_0{}^2 x = 0 \tag{7-3}$$

where $\omega_0{}^2 = g/b$. Equation (7-3) is known as the equation of the *undamped harmonic oscillator*. As we have just seen, this is the motion followed by a simple pendulum when the displacements are small. Equation (7-3) applies equally well to the small motions of a mass m attached to a massless spring, as shown in Fig. 7-2. The force constant of the spring is k, and the tube is considered to be perfectly smooth. In this instance $\omega_0{}^2 = k/m$.

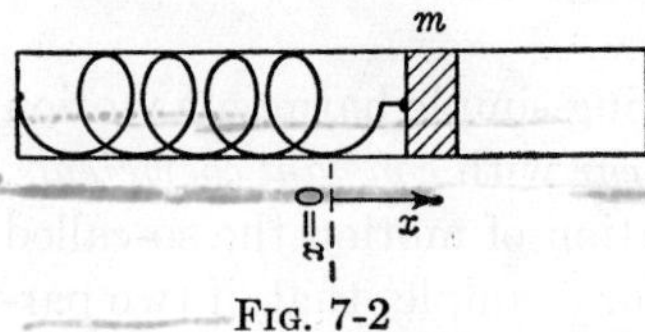

FIG. 7-2

To solve Eq. (7-3), multiply by $2\dot{x}$. We obtain

$$2\dot{x}\ddot{x} = -2\omega_0{}^2\dot{x}x$$

which integrates to

$$\dot{x}^2 = -\omega_0{}^2x^2 + c = \omega_0{}^2(A^2 - x^2)$$

Here A and c are constants, being related as $c = \omega_0{}^2A^2$. Taking square roots and separating variables, we have

$$\frac{dx}{(A^2 - x^2)^{\frac{1}{2}}} = \omega_0 \, dt \tag{7-4}$$

Equation (7-4) may be integrated directly, to become

$$\sin^{-1} \frac{x}{A} = \omega_0 t + B$$

This may be rewritten in the form

$$\boxed{x = A \sin (\omega_0 t + B)} \tag{7-5}$$

and is the general solution of Eq. (7-3), containing the two arbitrary constants A and B.

7-2. Physical Interpretation of Terms. The physical meaning of the quantities A, B, and ω_0 can be clarified by a consideration of Fig. 7-3. The line SS' represents the x component of the motion of the pendulum bob, A being the maximum value of x measured from the mid-point O' of SS'. Construct the circle of radius A, tangent to SS' at $x = 0$. At any time the position x of the particle can be represented by the projection, on SS', of $A \sin \varphi$, the distance from the line OO' of a point traveling along the circular path. Suppose we let $\varphi = \omega_0 t$, that is, the angle φ is increasing uniformly at a rate ω_0, (ω_0 = *angular velocity* or *angular frequency*). Thus

$$x = A \sin \omega_0 t \tag{7-6}$$

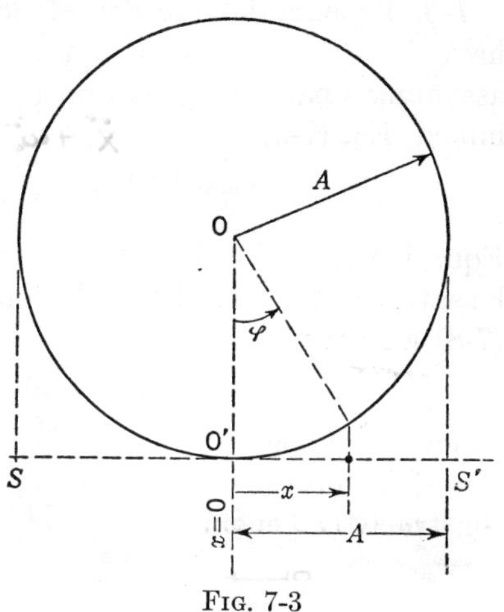

FIG. 7-3

is represented by the uniform motion of the point along the circular path, the angular velocity of the point being ω_0. Referring to Eq. (7-6), A is the *amplitude*, or maximum value of the *displacement* x. The angular frequency ω_0 is related to the actual *frequency* ν_0 by the relation

$$\omega_0 = 2\pi\nu_0$$

The quantity $\omega_0 t$ in Eq. (7-6) or $\omega_0 t + B$ in Eq. (7-5) is called the *phase angle*, or simply the *phase* of the motion. The meaning of the *phase constant*, or *initial phase* B, in Eq. (7-5) is now apparent. It is evident that in the circle of reference, the angle would have the value B at zero time.

Further information is obtained if we expand Eq. (7-5) by a well-known theorem of trigonometry as

$$\begin{aligned} x &= A \sin (\omega_0 t + B) = A \sin \omega_0 t \cos B + A \cos \omega_0 t \sin B \\ &= E \sin \omega_0 t + F \cos \omega_0 t \end{aligned} \tag{7-7}$$

where $E = A \cos B$ and $F = A \sin B$. Accordingly

$$B = \tan^{-1} \frac{F}{E}$$

From Eq. (7-7) we have that, at $t = 0$, $x = F$. Consequently F is the initial value of the displacement. Differentiating Eq. (7-7) with respect to time, we have

$$\frac{dx}{dt} = \dot{x} = E\omega_0 \cos \omega_0 t - F\omega_0 \sin \omega_0 t$$

from which the physical meaning of E is clear. $E\omega_0$ is the initial value of the velocity $\dot{x}$.

7-3. Exponential Method of Solution. Equation (7-3) could as well have been solved by a trial substitution of an exponential. Thus, assuming a particular solution $x = e^{\lambda t}$, where λ is a quantity to be determined, Eq. (7-3) becomes

$$\lambda^2 e^{\lambda t} + \omega_0^2 e^{\lambda t} = 0 \quad \text{or} \quad \lambda^2 + \omega_0^2 = 0 \tag{7-8}$$

Equation (7-8) is called the *characteristic*, or *auxiliary, equation* and must be satisfied if $e^{\lambda t}$ is to be a solution of Eq. (7-3) (see Appendix 2). From (7-8) we have

$$\lambda = \pm i\omega_0 \tag{7-9}$$

and we find that $e^{\lambda t}$ is a particular solution of Eq. (7-3) provided that $\lambda = +i\omega_0$ or $-i\omega_0$. Employing these two roots, we are then able to construct the general solution of Eq. (7-3) to be

$$x = Ge^{i\omega_0 t} + He^{-i\omega_0 t} \tag{7-10}$$

where G and H are arbitrary constants. Equation (7-10) is entirely equivalent to Eq. (7-5). Equation (7-10) may be put into the form of Eq. (7-5) by use of the *Euler formulas*. These are

$$\begin{aligned} e^{i\theta} &= \cos \theta + i \sin \theta \\ e^{-i\theta} &= \cos \theta - i \sin \theta \end{aligned} \tag{7-11}$$

It will be found that, although for some purposes the exponential solution is convenient, it has some disadvantages, since the constants G and H have the undesirable feature of being complex quantities. This is contrasted with the case of Eq. (7-5), in which the constants A and B are real.

7-4. Energy of the Oscillator. In order to examine the energy of a system performing a simple harmonic motion, it is convenient to employ the form of Eq. (7-5). Since the kinetic energy T is $m\dot{x}^2/2$ at any time, with the aid of Eq. (7-5) it may be written in the form

$$T = \frac{mA^2\omega_0^2}{2} \cos^2 (\omega_0 t + B) \tag{7-12}$$

The potential energy (taken zero at $x = 0$) is

$$V = -\int_0^x F\,dx$$

The restoring force F is just $-m\omega_0^2x$ [from Eq. (7-3)], and therefore we have

$$V = \int_0^x m\omega_0^2x\,dx = \frac{1}{2}m\omega_0^2x^2$$
$$= \tfrac{1}{2}m\omega_0^2A^2\sin^2(\omega_0t + B) \qquad (7\text{-}13)$$

Accordingly, the total energy W is

$$W = T + V = \tfrac{1}{2}mA^2\omega_0^2[\cos^2(\omega_0t + B) + \sin^2(\omega_0t + B)]$$
$$= \tfrac{1}{2}mA^2\omega_0^2 \qquad (7\text{-}14)$$

Thus we see that the total energy W is proportional both to the square of the amplitude and to the square of the frequency. Furthermore, since the total energy remains constant, it is clear that the time average W_{av} of the total energy is just equal to W. It is also of interest to look at the time averages of the potential and kinetic energies. If we designate τ_0 as the period of the oscillator, the time average of the potential energy is defined to be

$$V_{av} = \frac{\frac{1}{2}m\omega_0^2A^2}{\tau_0}\int_0^{\tau_0}\sin^2(\omega_0t + B)\,dt = \frac{m\omega_0^2A^2}{4} \qquad (7\text{-}15)$$

Similarly the time average of the kinetic energy is

$$T_{av} = \frac{\frac{1}{2}m\omega_0^2A^2}{\tau_0}\int_0^{\tau_0}\cos^2(\omega_0t + B)\,dt = \frac{m\omega_0^2A^2}{4} \qquad (7\text{-}16)$$

so that the average value of the kinetic energy is equal to the average value of the potential energy. Also, by considering Eqs. (7-12) and (7-13), we see that, when T is a maximum, V is zero, and vice versa.

It is helpful to plot the potential energy V as a function of the displacement x. From Eq. (7-13) we have

$$V = \frac{m\omega_0^2x^2}{2} \qquad (7\text{-}17)$$

For the common case of a stretched spring of force constant k, this expression becomes

$$V = \frac{kx^2}{2} \qquad (7\text{-}18)$$

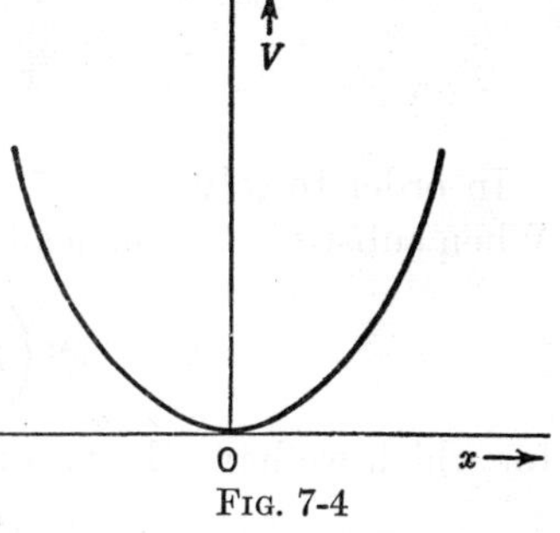

FIG. 7-4

The potential-energy curve is parabolic in shape (see Fig 7-4), with the vertex of the parabola at the equilibrium value $x = 0$.

THE DAMPED HARMONIC OSCILLATOR

7-5. The General Solution. The Underdamped Case. Of more practical interest than the undamped motion discussed above is the problem

of oscillatory motion in which there exists a retarding force proportional to the first power of the velocity.

Consider the free (no driving force) motion of a massless spring attached at one end to an immovable wall (A in Fig. 7-5) and at the other end to a mass m, the whole being immersed in a fluid which affords a retarding force on m proportional to the velocity of m. Gravity is neglected. Let x be the displacement of m from the position B, at which point the spring is unstretched and of length a. The coordinate x is positive to the right. If k is the force constant of the spring, then, for the displacement shown, there is a force of magnitude kx in the direction of x decreasing. The retarding force of the fluid is $R\dot{x}$, where R is a constant of proportionality, and for the direction of motion shown to the right is directed toward x decreasing. Hence the equation of motion is

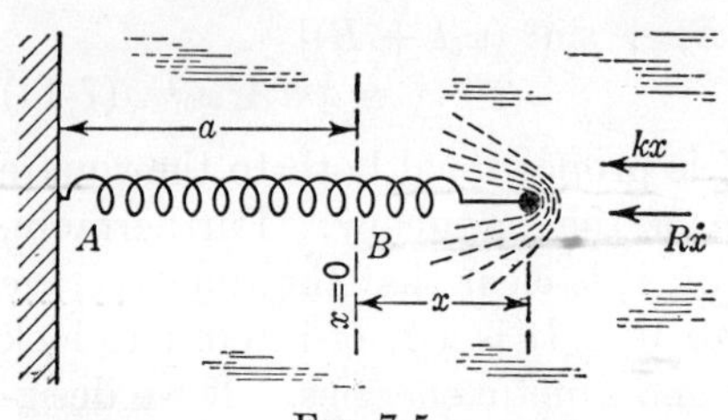

FIG. 7-5

$$m\ddot{x} = -kx - R\dot{x} \tag{7-19}$$

where the negative signs arise because both forces are in the direction of x decreasing for the conditions indicated. The same equation is true for motion in the opposite direction since, in that case, $\dot{x}$ itself is negative, and the sense of $R\dot{x}$ is in the direction of x increasing, as it should be from physical considerations. Since, from Sec. 7-1, the angular frequency ω_0 of the undamped motion is $\sqrt{k/m}$, Eq. (7-19) may be rewritten as

$$\ddot{x} + \frac{R}{m}\dot{x} + \omega_0^2 x = 0 \tag{7-20}$$

In order to solve Eq. (7-20), we assume a particular solution $x = e^{\lambda t}$. When substituted, this leads to the auxiliary equation

$$e^{\lambda t}\left(\lambda^2 + \frac{R}{m}\lambda + \omega_0^2\right) = 0$$

for which we have the two roots

$$\begin{aligned} \lambda_1 &= -\frac{R}{2m} + \sqrt{\frac{R^2}{4m^2} - \omega_0^2} \\ \lambda_2 &= -\frac{R}{2m} - \sqrt{\frac{R^2}{4m^2} - \omega_0^2} \end{aligned} \tag{7-21}$$

We are then able to write the general solution as

$$x = e^{-(R/2m)t}\left[Ae^{+\left(\frac{R^2}{4m^2} - \omega_0^2\right)^{\frac{1}{2}}t} + Be^{-\left(\frac{R^2}{4m^2} - \omega_0^2\right)^{\frac{1}{2}}t}\right] \tag{7-22}$$

Now m, R, and ω_0 are positive numbers, and therefore there are three general cases of interest. These are the following:

$\omega_0{}^2 > \dfrac{R^2}{4m^2}$	*underdamped* case (oscillatory motion)
$\omega_0{}^2 = \dfrac{R^2}{4m^2}$	*critically damped* case (not oscillatory)
$\omega_0{}^2 < \dfrac{R^2}{4m^2}$	*overdamped* case (not oscillatory)

For the underdamped case we put

$$\omega_1{}^2 = \omega_0{}^2 - \frac{R^2}{4m^2} \tag{7-23}$$

and we see that, since $\omega_0{}^2 > R^2/4m^2$, $\omega_1{}^2$ is positive and the exponents of Eq. (7-22) are imaginary, giving us the form

$$x = e^{-(R/2m)t}(Ae^{i\omega_1 t} + Be^{-i\omega_1 t}) \tag{7-24}$$

As was mentioned in Sec. 7-3, the quantity within the parentheses in Eq. (7-24), by the use of Euler's formulas, may be put into the form $C \cos(\omega_1 t + D)$. Therefore Eq. (7-24) may now be written as

$$x = Ce^{-(R/2m)t} \cos(\omega_1 t + D) \tag{7-25}$$

This is the general solution since the constants C and D are arbitrary. The quantity ω_1 is the *natural frequency* of the system. Remembering that ω_1 is related to ω_0, the natural frequency of the undamped oscillator, by Eq. (7-23), we note that the natural frequency of the damped system is less than that for the undamped system.

Equation (7-25) states that displacements will lie between the two curves

$$Ce^{-(R/2m)t} \text{ and } -Ce^{-(R/2m)t} \tag{7-26}$$

In Fig. 7-6 the full solid line is the curve of Eq. (7-25) for the case $D = 0$, or

$$x = Ce^{-(R/2m)t} \cos \omega_1 t \tag{7-27}$$

The dotted lines represent the two quantities in (7-26). It is clear from Eq. (7-27) that the solid curve will touch the envelopes at points where $\cos \omega_1 t = \pm 1$, or at times $t_n = n\pi/\omega_1$, where n is an integer. Thus we see that the points of contact are just a half period $\frac{1}{2}\tau_1$ apart. The maxima and minima of Eq. (7-27) do not occur quite at these points, but they also are separated by the same time interval. We may see this by

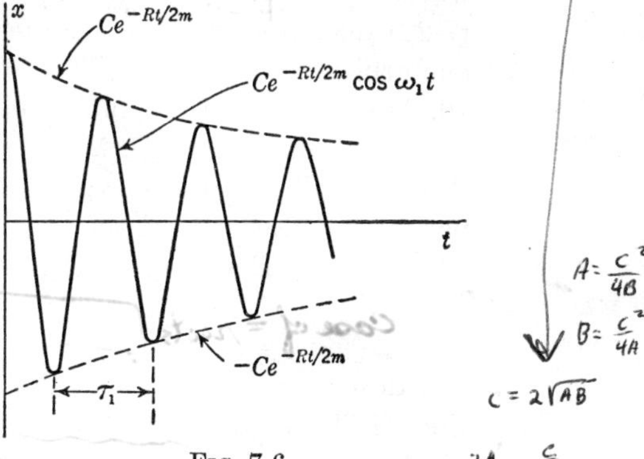

Fig. 7-6

differentiating Eq. (7-27). We have

$$\dot{x} = -\frac{C}{2m} Re^{-(R/2m)t} \cos \omega_1 t - C\omega_1 e^{-(R/2m)t} \sin \omega_1 t = 0$$

from which

$$\tan \omega_1 t = -\frac{R}{2m\omega_1} \tag{7-28}$$

a function which is periodic with a period π/ω_1.

The ratio between the amplitudes of two successive maxima is

$$\frac{Ce^{-(R/2m)t_n'}}{Ce^{-(R/2m)(t_n'+\tau_1)}} = e^{R\tau_1/2m} = e^{R\pi/m\omega_1}$$

The logarithm of this ratio is known as the *logarithmic decrement* of the motion and has the value

$$\delta = \frac{R\tau_1}{2m} = \frac{R\pi}{m\omega_1} \tag{7-29}$$

7-6. Critically Damped and Overdamped Cases. If the damping, or retarding, force is sufficiently large compared with the restoring force in the system, the latter no longer oscillates but instead, from the initial displacement, the system monotonically approaches the equilibrium configuration. One such possibility occurs if R is just large enough so that $\omega_0^2 = R^2/4m^2$. In this event the method of solution leading to Eq. (7-24) fails, since then $\omega_1 = 0$, and Eq. (7-24) takes the form

$$x = (A + B)e^{-(R/2m)t} = Ge^{-(R/2m)t}$$

where G is another constant $(= A + B)$. Thus, for $\omega_0^2 = R^2/4m^2$, Eq. (7-22) does not represent the general solution, since it would yield but one arbitrary constant. This is the *case of equal roots* discussed in Sec. 10, Appendix 2. It is pointed out there that the substitution $x = y(t)e^{\lambda t}$ should be made. Since λ has already been determined to be $-R/2m$ [Eq. (7-21)], the procedure is one for the determination of $y(t)$. Comparing with Eq. (61), Appendix 2, we see that $y(t)$ has the form $A + Bt$, where A and B are constants. Therefore the general solution is

$$x = (A + Bt)e^{-(R/2m)t} \tag{7-30}$$

and describes the motion if the system is critically damped. The solution (7-30) clearly does not represent an oscillatory motion, and the displacement approaches zero asymptotically at large values of the time. Equation (7-30) contains three cases of interest, depending on the magnitudes of A and B. These can be seen by differentiating Eq. (7-30) with respect to time and setting the resulting equation equal to zero, as

$$\frac{dx}{dt} = \lambda e^{\lambda t}(A + Bt) + e^{\lambda t}B = 0$$

yielding

$$t = -\frac{A}{B} + \frac{2m}{R} \tag{7-31}$$

so that the curve of Eq. (7-30) has a maximum at $t = 0$ or at some positive value of t, respectively, if $A/B = 2m/R$ or $< 2m/R$. The remaining case, where $A/B > 2m/R$, does not have a maximum for positive t. These conditions are displayed in the three curves of Fig. 7-7. All three are possible examples of the critically damped motion.

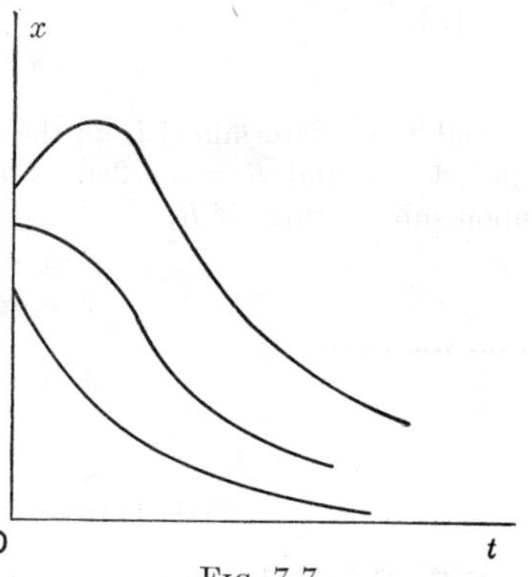

FIG. 7-7

If the damping constant is still larger, such that $\omega_0^2 < R^2/4m^2$, we have the third case, that of overdamped motion. For this situation, Eq. (7-22) again represents the general solution except that, instead of Eq. (7-24), the solution now consists of real exponentials. The solution may be written, again from Eq. (7-22),

$$x = e^{-(R/2m)t}(Ae^{\omega_1' t} + Be^{-\omega_1' t}) \tag{7-32}$$

in which A and B are arbitrary and may be determined to fit the boundary conditions of a particular problem. The quantity ω_1' is a real quantity and is defined as $(R^2/4m^2) - \omega_0^2$.

Example 7-1. A pendulum consists of a mass m suspended from a fixed point by means of a string of length b and is immersed in a viscous medium. The medium provides a retarding force proportional to the velocity, the constant of proportionality being $2m\sqrt{g/b}$, where g is the acceleration of gravity. Initially m is released from rest at a small angular displacement α from the vertical. Find its velocity and angular displacement at a later time.

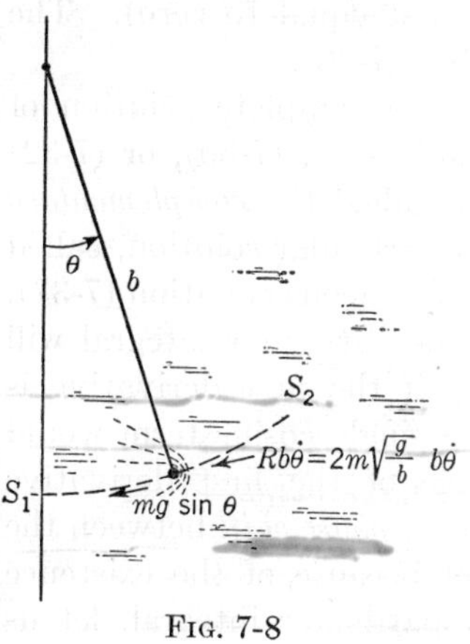

FIG. 7-8

It is convenient to employ the angle coordinate θ measured positively as shown in Fig. 7-8. The equation of motion along the circular path S_1S_2 is

$$mb\ddot{\theta} = -mg\sin\theta - Rb\dot{\theta} \tag{7-33}$$

where we employ the symbol R temporarily for $2m\sqrt{g/b}$. Since the initial displacement is small, the subsequent motion may be approximated by replacing $\sin\theta$ by θ in Eq. (7-33). Equation (7-33) may be rewritten

$$\ddot{\theta} + \frac{R}{m}\dot{\theta} + \frac{g}{b}\theta = 0 \tag{7-34}$$

which is the equation of motion of a damped oscillator (with the variable θ replacing x). In order to determine which class of motions is exhibited by the system, we note that, here, $\omega_0^2 = g/b$, and that $R/m = 2\sqrt{g/b}$. Accordingly

$$\frac{R^2}{4m^2} = \frac{g}{b} = \omega_0^2$$

Thus the motion is critically damped, and, following Eq. (7-30), the solution to (7-34) is

$$\theta = e^{-Rt/2m}(A + Bt) \tag{7-35}$$

A and B are determined from the conditions that, at $t = 0$, $\theta = \alpha$ and $\dot{\theta} = 0$. We find $A = \alpha$ and $B = R\alpha/2m$. Consequently the expressions for θ and $\dot{\theta}$ become, upon substituting for R,

$$\begin{aligned} \theta &= \alpha e^{-\sqrt{g/b}\,t}\left(1 + \sqrt{\frac{g}{b}}\,t\right) \\ \dot{\theta} &= -\frac{g\alpha t}{b}\,e^{-\sqrt{g/b}\,t} \end{aligned} \tag{7-36}$$

THE FORCED HARMONIC OSCILLATOR

7-7. General Solution. For this problem we consider only a harmonic driving force $F \cos \omega t$, although it is clear that we need not be thus limited. The differential equation may be written merely by adding $(F/m) \cos \omega t$ to the right side of Eq. (7-20). We have

$$\ddot{x} + \frac{R}{m}\dot{x} + \omega_0^2 x = \frac{F}{m} \cos \omega t \tag{7-37}$$

where the quantities are defined as before with the exception of F and ω. Here F is the amplitude of the driving force of angular frequency ω. Note that Eq. (7-37) is a linear equation with constant coefficients, but it is no longer homogeneous (right side here not equal to zero). The left side, or reduced equation, is identical to Eq. (7-20).

From the theory of differential equations the complete solution of Eq. (7-37) consists in the general solution [Eq. (7-25), (7-30), or (7-32) as the case requires] of the reduced equation, called the *complementary function*, which we have already found, plus a particular solution, called the *particular integral*, which satisfies the inhomogeneous equation (7-37). We can guess, with a little experience, that the particular integral will have harmonic terms of frequency ω. Now, if the first derivative is absent (no damping), we can easily see that a single cosine term would satisfy the equation. However, the presence of the first derivative necessitates the addition of a sine term, that is, a *phase shift* between the driving force and the displacement is present because of the existence of the damping term. In order to find the particular integral, let us assume a trial solution of the form

$$x = b \cos \omega t + c \sin \omega t \tag{7-38}$$

where b and c are quantities which will be determined by substituting (7-38) in Eq. (7-37). Equation (7-38) is entirely equivalent to

$$x = \sqrt{b^2 + c^2} \cos (\omega t - \varphi) \tag{7-39}$$

where $\varphi = \tan^{-1} (c/b)$, as can easily be shown by expanding the cosine term in Eq. (7-39). Choosing for the moment the form of Eq. (7-38), and substituting in Eq. (7-37), we obtain

$$-b\omega^2 \cos \omega t - c\omega^2 \sin \omega t - b\omega \frac{R}{m} \sin \omega t + c\omega \frac{R}{m} \cos \omega t$$
$$+ \omega_0{}^2 b \cos \omega t + \omega_0{}^2 c \sin \omega t = \frac{F}{m} \cos \omega t$$

An equation of this type can be satisfied for all values of time only by equating separately the coefficients of the sine terms and the cosine terms. This provides us with a pair of simultaneous equations for the determination of b and c. The equations are

$$b(\omega_0{}^2 - \omega^2) + \frac{c\omega R}{m} = \frac{F}{m}$$
$$-\frac{b\omega R}{m} + c(\omega_0{}^2 - \omega^2) = 0$$

from which we find

$$c = \frac{F\omega R}{\omega^2 R^2 + m^2(\omega_0{}^2 - \omega^2)^2}$$
$$b = \frac{Fm(\omega_0{}^2 - \omega^2)}{\omega^2 R^2 + m^2(\omega_0{}^2 - \omega^2)^2}$$

The particular integral (7-39) may now be written

$$x = \frac{F}{[m^2(\omega_0{}^2 - \omega^2)^2 + \omega^2 R^2]^{\frac{1}{2}}} \cos (\omega t - \varphi) \tag{7-40}$$

in which

$$\tan \varphi = \frac{\omega R}{m(\omega_0{}^2 - \omega^2)} \tag{7-41}$$

In order to find the general solution of Eq. (7-37), we add (7-40) to the general solution of the homogeneous or reduced equation. For our purposes we shall consider only the form of the complementary function which represents the underdamped case [that is, the oscillating solution (7-25)]. The general solution of (7-37) may then be written

$$x = Ce^{-(R/2m)t} \cos (\omega_1 t + B)$$
$$+ \frac{F}{[m^2(\omega_0{}^2 - \omega^2)^2 + \omega^2 R^2]^{\frac{1}{2}}} \cos (\omega t - \varphi) \tag{7-42}$$

where C and B are arbitrary constants of integration. The first part of

(7-42) oscillates with the natural frequency of the system and, because of the damping term, dies out at large values of the time. This part of the solution is called the *transient* solution. For large values of the time the second part of the solution is the only part remaining. This is the so-called *steady-state* solution, and it shows that, in the steady state, the system oscillates with the frequency ω of the driving force. However, the presence of the phase constant φ shows that the displacement x is in general shifted in phase from the driving force. The steady-state solution is usually the one of major interest, and so we shall consider only the situation at large values of t, that is, when the solution is represented by Eq. (7-40) alone.

7-8. Resonance. Of great physical interest are the situations in which the frequency of the driving force is such that the largest possible amplitude of the oscillating system in the steady state occurs. The frequency of the driving force for which this takes place is said to be the *resonance* frequency for maximum displacement of the system. In most phenomena involving oscillating systems in nature, the quantity which is observed experimentally is closely related to the energy [for example, (1) absorption of energy from a light wave by an atomic oscillator; (2) resonance behavior of a cavity when excited by an electromagnetic field]. Consequently, since the average potential energy of the harmonic oscillator is proportional to the square of the amplitude, we shall focus our attention mainly upon the behavior of the square of the amplitude near resonance. Moreover, we confine ourselves to the situation in which ω_0 is held fixed and the driving angular frequency ω is permitted to vary.

Let us denote the quantity $m^2(\omega_0{}^2 - \omega^2)^2 + \omega^2R^2$ by the symbol Z (not to be confused with the complex impedance of the circuit theory), so that the steady-state solution may be written

$$x = \frac{F}{Z^{\frac{1}{2}}} \cos (\omega t - \varphi) \tag{7-43}$$

Thus the square of the displacement is given by

$$x^2 = \frac{F^2}{Z} \cos^2 (\omega t - \varphi) = a^2 \cos^2 (\omega t - \varphi) \tag{7-44}$$

where $F^2/Z \equiv a^2$. Since F is fixed, a^2 is a maximum for the particular value of ω^2 at which Z is a minimum. We denote this value by $\omega_2{}^2$. This occurs at $dZ/d\omega^2 = 0$. We obtain

$$\omega_2{}^2 = \omega_0{}^2 - \frac{R^2}{2m^2} \tag{7-45}$$

This is the resonance frequency for fixed ω_0 and variable ω. Note that it is not the same as either the natural frequency of the damped oscillator

$[\omega_1 = \sqrt{\omega_0^2 - R^2/4m^2}]$ or that of the undamped oscillator (ω_0). Let us examine a^2 further. We have

$$a^2 = \frac{F^2}{Z} = \frac{F^2}{m^2(\omega_0^2 - \omega^2)^2 + \omega^2 R^2} \tag{7-46}$$

Let

$$y^2 = \omega^2 - \omega_2^2 \qquad \text{or} \qquad \omega^2 = y^2 + \omega_0^2 - \frac{R^2}{2m^2} \tag{7-47}$$

Substituting (7-47) in (7-46), we obtain

$$a^2 = \frac{F^2}{m^2[(R^2/2m^2) - y^2]^2 + R^2[y^2 + \omega_0^2 - (R^2/2m^2)]}$$

$$= \frac{F^2}{m^2(y^2)^2 + R^2\omega_1^2} \tag{7-48}$$

Now

$$y^2 = \omega^2 - \omega_2^2 = (\omega - \omega_2)(\omega + \omega_2) \simeq 2\omega(\omega - \omega_2) \simeq 2\omega_1(\omega - \omega_2)$$

provided that we are interested only in frequencies very near ω_2 and with not too large damping. Substituting this in Eq. (7-48), we have

$$a^2 = \frac{F^2}{4m^2\omega_1^2(\omega - \omega_2)^2 + R^2\omega_1^2} = \frac{F^2/\omega_1^2}{4m^2(\omega - \omega_2)^2 + R^2} \tag{7-49}$$

The maximum value a_0^2 of a^2 occurs at $y^2 = 0$. Therefore from Eq. (7-48) we have $a_0^2 = F^2/R^2\omega_1^2$. Employing this, together with Eq. (7-49), we may write the ratio a^2/a_0^2 as

$$\frac{a^2}{a_0^2} = \frac{R^2}{4m^2(\omega - \omega_2)^2 + R^2} \tag{7-50}$$

Equation (7-50) is symmetrical in $\omega - \omega_2$, and a plot of a^2/a_0^2 against $\omega - \omega_2$, shown in Fig. 7-9, gives a curve which is symmetrical about $\omega = \omega_2$. It is also of interest to notice the location of the points corresponding to $\omega = \omega_0$ and $\omega = \omega_1$.

Consider the case where $a^2/a_0^2 = \frac{1}{2}$. At this ordinate, $\omega - \omega_2$ has a value Δ which is called the resonance *half width* at half maximum, or, for short, the resonance half width. We have, from Eq. (7-50),

$$\Delta = \frac{R}{2m} \tag{7-51}$$

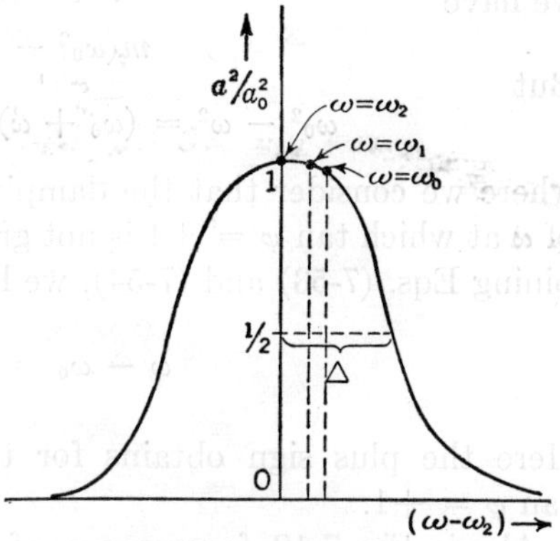

FIG. 7-9

But $R/2m = \delta/\tau_1$, where δ is the logarithmic decrement for the unforced motion and τ_1 is the natural period of the oscillator. Therefore

$$\Delta = \frac{\delta}{\tau_1} = \frac{\omega_1 \delta}{2\pi} \tag{7-52}$$

Clearly the resonance half width is directly proportional to R, the damping constant. Thus a highly dissipative system may not be expected to show sharp resonance qualities. It is instructive to point out that, if we had held ω fixed and instead had allowed ω_0 to vary (but keeping R fixed), we should have found the resonance frequency to be just ω_0. Carrying out the remaining steps in a similar fashion as above, and with similar approximations, a curve of $a_0{}^2/a^2$ as a function of $\omega - \omega_0$ would have been obtained.

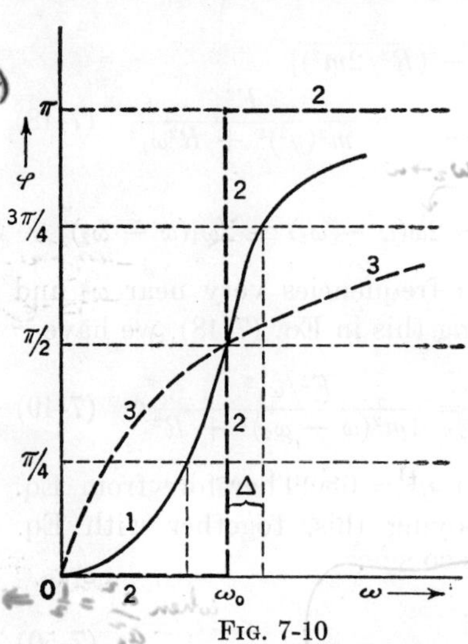

FIG. 7-10

Let us examine the phase of the displacement relative to that of the driving force. To do this, we again consider Eq. (7-41). At $\omega = 0$, we have $\varphi = 0$, at $\omega = \omega_0$, $\varphi = \pi/2$, and finally, at $\omega = \infty$, $\varphi = \pi$. Thus there is a complete phase shift of π on passing through the resonance. In Fig. 7-10, the phase shift φ for an oscillator with small damping ($R^2/4m^2 \ll \omega_0{}^2$) is plotted as a function of ω [solid curve (1)]. The angular frequency interval between the points at which $\varphi = \pi/4$ and $3\pi/4$, or at which $\tan\varphi = 1$ and -1, is just 2Δ. This can readily be shown. From Eq. (7-41), putting $\tan\varphi = \pm 1$, we have

$$m(\omega_0{}^2 - \omega^2) = \pm\omega R \tag{7-53}$$

But

$$\omega_0{}^2 - \omega^2 = (\omega_0 + \omega)(\omega_0 - \omega) \simeq 2\omega(\omega_0 - \omega) \tag{7-54}$$

where we consider that the damping is sufficiently small that the value of ω at which $\tan\varphi = \pm 1$ is not greatly different from ω_0. Hence, combining Eqs. (7-53) and (7-54), we have

$$\omega - \omega_0 = \pm\frac{R}{2m} \equiv \pm\Delta \tag{7-55}$$

Here the plus sign obtains for $\tan\varphi = -1$ and the minus sign for $\tan\varphi = +1$.

Also in Fig. 7-10, for purposes of comparison, the behavior of the phase shift is depicted as dashed curves, (2) and (3), for the cases, respectively, where $R = 0$ and $R^2/4m^2$ is severalfold greater than $\omega_0{}^2$.

We have seen that ω_2 is the driving frequency for which the square of the amplitude is a maximum and thus is the driving frequency for which a maximum average potential energy will be found stored in the oscillating system. It is of interest, also, to examine the corresponding situation with regard to the kinetic energy, that is, to determine the driving frequency for which the average kinetic energy of the oscillator is a maximum. We can write

$$T = \frac{m\dot{x}^2}{2} = \frac{mF^2\omega^2/2}{[m^2(\omega_0{}^2 - \omega^2)^2 + \omega^2R^2]} \sin^2 (\omega t - \varphi)$$
$$= \frac{m}{2} a^2\omega^2 \sin^2 (\omega t - \varphi) \tag{7-56}$$

so that the average kinetic energy is $ma^2\omega^2/4$. The value of ω^2 at which ω^2a^2 is a maximum is determined by putting $d(a^2\omega^2)/d(\omega^2)$ equal to zero. Carrying out this operation, we obtain

$$m^2(\omega_0{}^2 - \omega^2)^2 + \omega^2R^2 + 2m^2\omega^2(\omega_0{}^2 - \omega^2) - \omega^2R^2 = 0$$

the root of which is $\omega = \omega_0$. Accordingly, the oscillator possesses a maximum average kinetic energy when the driving frequency is equal to ω_0. The fact that the average potential energy has a maximum at a different frequency is not strange, since the system is not a conservative one. In passing, it is interesting to notice that, since the velocity $\dot{x}$ has the form $-\sin (\omega t - \varphi)$ and since, at $\omega = \omega_0$, $\varphi = \pi/2$, the velocity is exactly in phase with the force. This is so since

$$-\sin \left[\omega t - \left(\frac{\pi}{2}\right)\right] = \cos \omega t$$

7-9. Rate at Which Work Is Being Done. Let us look at the question of energy in general. Multiplying Eq. (7-37) by $m\dot{x}$, we have

$$m\dot{x}\ddot{x} + R\dot{x}^2 + m\omega_0{}^2x\dot{x} = F\dot{x} \cos \omega t \tag{7-57}$$

This may be rewritten

$$\frac{d}{dt}\left(\frac{m\dot{x}^2}{2} + \frac{m\omega_0{}^2x^2}{2}\right) + R\dot{x}^2 = (F \cos \omega t)\dot{x}$$

or

$$\frac{d}{dt}(T + V) + R\dot{x}^2 = (F \cos \omega t)\dot{x} \tag{7-58}$$

where T and V are the kinetic and potential energies, respectively. The first term on the left is the time rate of change of the sum of the kinetic and potential energies. The second term is the rate at which energy is being dissipated by the damping force, and the right side is the rate at which energy is being supplied by the driving force. Since $\dot{x}^2$ has a maximum average value at $\omega = \omega_0$, we see that energy is being dissipated at a maximum average rate at $\omega = \omega_0$. The right-hand side may be rewritten as

$$\begin{aligned} \dot{x}F \cos \omega t &= -Fa\omega \cos \omega t \sin (\omega t - \varphi) \\ &= Fa\omega(\cos^2 \omega t \sin \varphi - \cos \omega t \sin \omega t \cos \varphi) \end{aligned} \tag{7-59}$$

which is the rate at which the driving force is doing work on the system. The second set of terms within the parentheses may at times outweigh the first set, and at these times the driving force is extracting energy from the system. The upshot of Eq. (7-59) is, therefore, that the driving force is alternately introducing and withdrawing energy to and from the system. But on the average (since the average value of the first set is $\frac{1}{2}$ and that of the second is zero) the driving force is supplying energy to the system. This energy is being dissipated at the rate $R\dot{x}^2$, as has been seen above. The passing of energy back and forth between the system and the driving agency is common to the experience of anyone who has pushed a swing. Now the average rate at which energy is being dissipated is

$$(R\dot{x}^2)_{av} = \frac{Ra^2\omega^2}{2} = \frac{RF^2\omega^2}{2[m^2(\omega_0{}^2 - \omega^2)^2 + \omega^2R^2]} \tag{7-60}$$

The average rate at which the external force is doing work is, from Eq. (7-59) by inspection,

$$(\dot{W})_{av} = (\dot{x}F \cos \omega t)_{av} = \frac{Fa\omega}{2} \sin \varphi$$

but, from Eq. (7-41), $\sin \varphi = \omega R/[m^2(\omega_0{}^2 - \omega^2)^2 + \omega^2R^2]^{\frac{1}{2}}$, and thus

$$\begin{aligned} (\dot{W})_{av} &= \frac{Fa\omega^2R}{2[m^2(\omega_0{}^2 - \omega^2)^2 + \omega^2R^2]^{\frac{1}{2}}} \\ &= \frac{F^2\omega^2R}{2[m^2(\omega_0{}^2 - \omega^2)^2 + \omega^2R^2]} = (R\dot{x}^2)_{av} \end{aligned} \tag{7-61}$$

which is to be expected.

For general interest let us consider the quantity Q of electric-circuit theory, defined as

$$Q = 2\pi \frac{\text{maximum energy stored}}{\text{energy loss in one period}}$$

It becomes here

$$Q = 2\pi \frac{(m\dot{x}^2/2)_{max}}{(R\dot{x}^2)_{av}\tau_1} = \frac{\pi m a^2\omega^2}{(R/2)a^2\omega^2(2\pi/\omega_1)} = \frac{m\omega_1}{R} \tag{7-62}$$

and we have that the resonance breadth is inversely proportional to the Q value.

Example 7-2. A critically damped oscillator has a mass m and a damping coefficient R. At $t = 0$ the displacement has a maximum value a. It is desired to determine the energy dissipated through damping during the time interval $t = 0$ to $t = m/R$.

Employing Eq. (7-30), and since, at $t = 0$, $x = a$ and $\dot{x} = 0$, we have

$$\begin{aligned} x &= ae^{-(R/2m)t}\left(1 + \frac{R}{2m}t\right) \\ \dot{x} &= -\frac{R^2a}{4m^2}te^{-(R/2m)t} \end{aligned} \tag{7-63}$$

Therefore the time rate of dissipation of energy through damping may be written

$$\left(\frac{dW}{dt}\right)_R = R\dot{x}^2 = \frac{R^5a^2}{16m^4}t^2e^{-(R/m)t} \tag{7-64}$$

and the total energy dissipated in the interval $t = 0$ to m/R is

$$\begin{aligned} W_R &= \frac{R^5a^2}{16m^4}\int_0^{m/R} t^2e^{-(R/m)t}\,dt \\ &= \frac{R^5a^2}{16m^4}\left[e^{-(R/m)t}\left(-\frac{t^2}{R/m} - \frac{2t}{R^2/m^2} - \frac{2}{R^3/m^3}\right)\right]_0^{m/R} \\ &= \frac{R^2a^2}{16m}(2 - 5e^{-1}) \simeq 0.01\frac{R^2a^2}{m} \end{aligned} \tag{7-65}$$

7-10. Application to an Elastically Bound Electron. It is of interest to apply the results of the last few sections to the classical absorption of energy by an electron, bound elastically in an atom, when the system is irradiated by a plane electromagnetic wave. So far as the force exerted upon the electron is concerned it is the electric field $E \cos \omega t$ of the incident wave which is important. Here E is the amplitude of the electric field, and ω is the angular frequency of the incident radiation. The equation of motion of the electron, assuming it to be bound to a system of infinite mass, may be written

$$m\ddot{x} + m\gamma\dot{x} + m\omega_0^2x = Ee \cos \omega t \tag{7-66}$$

where m is the mass of the electron, e is its charge, and ω_0 has the same significance as in Sec. 7-1. The quantity γ is the *radiation-damping* term calculated from classical electron theory. The manner of the dissipation of energy through γ is a continual reradiation of some of the absorbed energy. The damping coefficient is given by

$$\gamma = \frac{2}{3}\frac{e^2\omega^2}{mc^3} \tag{7-67}$$

By comparison with Eq. (7-43), the steady-state solution is seen to be

$$x = \frac{eE}{m[(\omega_0^2 - \omega^2)^2 + \omega^2\gamma^2]^{\frac{1}{2}}}\cos(\omega t - \varphi) \tag{7-68}$$

From Eq. (7-61), the average rate of absorption of energy is

$$(\dot{W})_{\text{av}} = \frac{e^2E^2\omega^2\gamma}{2m[(\omega_0{}^2 - \omega^2)^2 + \omega^2\gamma^2]} \tag{7-69}$$

The resonance width at half maximum is, from (7-51),

$$2\Delta = \gamma \tag{7-70}$$

The resonance is a sharp one since, for optical radiation, ω is of the order of 10^{16}, while γ is of the order of 10^{10}. Hence the γ^2 terms in the denominators of (7-68) and (7-69) are small compared with the ω^2 terms.

It is of interest to mention that, for a spectral-emission line, the natural breadth, or breadth due to radiation damping, arises from the presence of the same γ term in the equation of motion.

DEPARTURES FROM HARMONIC OSCILLATIONS

It was pointed out at the beginning of the chapter that the motion of most natural oscillating systems is simple harmonic for only a limited range of conditions. A departure from this state of affairs arises if the restoring force deviates from a linear dependence on the displacement and also if damping forces are present which are nonlinear in the velocity. We may write the equation for the free motion of such a system as

$$m\ddot{x} + g(\dot{x}) + f(x) = 0 \tag{7-71}$$

where $f(x)$ and $g(\dot{x})$ are *nonlinear* in x and $\dot{x}$, respectively.

Complete solutions for Eq. (7-71) for arbitrary functions $f(x)$ and $g(\dot{x})$ are not known. A reason for this, the equation being nonlinear, is that, even if two particular solutions are known, this does not immediately lead to the general solution (cf. Probs. 7-8, 7-17, and 7-18). In general, except for certain cases for which the solutions are known, approximate methods must be employed even to find particular solutions of Eq. (7-71). These procedures are treated in detail in texts on nonlinear mechanics. For our purposes we shall not attempt to study the free motion of such a system except to show, immediately below, that for a nonlinear system the natural period in general depends on the amplitude. We shall also glance briefly at an approximate method of finding a steady-state solution for an undamped nonlinear system (the nonlinear term being small) under the action of a periodic force.

7-11. Natural Period of the Undamped System. In the absence of forcing and damping terms, Eq. (7-71) reduces to

$$m\ddot{x} = -f(x) \tag{7-72}$$

Remembering that $\ddot{x} = \dot{x}\, d\dot{x}/dx$, Eq. (7-72) becomes

$$m\dot{x}\, d\dot{x} = -f(x)\, dx$$

from which

$$\frac{m\dot{x}^2}{2} = \int_{x_0}^{x} [-f(x)]\,dx \tag{7-73}$$

where $\dot{x} = 0$ at $x = x_0$, the maximum displacement. Here $\dot{x}$ is the velocity when the displacement is x. Clearly if, as above, we take the integration over x to be from the maximum displacement x_0 to a still positive value of x closer to the origin, the sign of $\dot{x}$ will be negative. Upon reversing the limits of the integration and extracting the square root, Eq. (7-73) yields

$$\dot{x} \equiv \frac{dx}{dt} = -\left[\frac{2}{m}\int_{x}^{x_0} f(x)\,dx\right]^{\frac{1}{2}} = -\sqrt{\frac{2}{m}}\,[F(x_0) - F(x)]^{\frac{1}{2}} \tag{7-74}$$

whence the time for a quarter period is given by

$$\frac{\tau_0}{4} = -\int_{x_0}^{0} \frac{dx}{\sqrt{2/m\,[F(x_0) - F(x)]}} \tag{7-75}$$

or

$$\tau_0 = 4\sqrt{\frac{m}{2}}\int_{0}^{x_0} \frac{dx}{[F(x_0) - F(x)]^{\frac{1}{2}}} \tag{7-76}$$

Equation (7-76) shows that, for an arbitrary function $f(x)$, the natural period τ_0 and consequently the natural frequency $1/\tau_0$ in general depend upon the amplitude x_0. For the familiar case of the Hooke's law force, in which $f(x) = kx$, where k is a constant, this dependence disappears.

7-12. Forced Motion of an Undamped Nonlinear System. Let us consider the undamped motion of a mass m attached, as shown in Fig. 7-11, to points A and A' by means of two equal light elastic strings of force constant k. We are given that, when the strings are in the configuration AOA' (a distance $2b$), the tension in each string is S_0. We do not necessarily restrict ourselves to small values of the displacement x. We consider, also, that a periodic driving force $F_0 \cos \omega t$ is acting in the horizontal direction and that gravitational forces are absent. For a lateral displacement x, each string increases in length to a quantity L, the tension then being given by

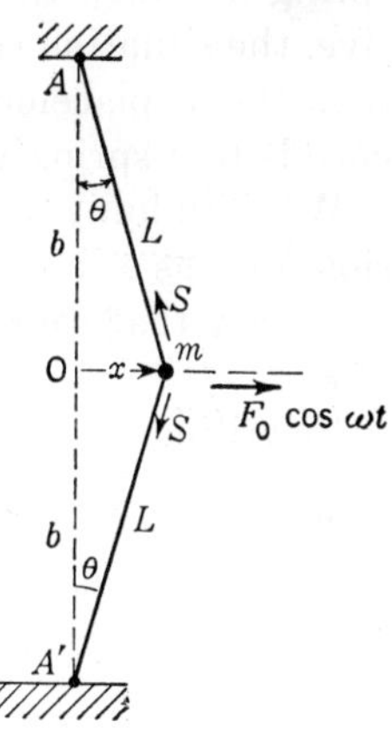

FIG. 7-11

$$S = S_0 + k(L - b) \tag{7-77}$$

Hence the equation of motion is

$$m\ddot{x} = -2S \sin\theta + F_0 \cos\omega t \tag{7-78}$$

$$= -2[S_0 + k(L - b)]\frac{x}{L} + F_0 \cos\omega t \tag{7-79}$$

Now $L = b\sqrt{1 + (x/b)^2}$, whence Eq. (7-79) becomes

$$m\ddot{x} = -2\left(S_0 + kb\left\{\left[1 + \left(\frac{x}{b}\right)^2\right]^{\frac{1}{2}} - 1\right\}\right)\frac{x}{b}\left[1 + \left(\frac{x}{b}\right)^2\right]^{-\frac{1}{2}} + F_0 \cos \omega t$$

$$= -2\left(S_0 + kb\left\{\left[1 + \frac{1}{2}\left(\frac{x}{b}\right)^2 + \cdots\right] - 1\right\}\right)\frac{x}{b}\left[1 - \frac{1}{2}\left(\frac{x}{b}\right)^2 + \cdots\right] + F_0 \cos \omega t \tag{7-80}$$

If we neglect powers of x higher than the third, this reduces to

$$m\ddot{x} = -\frac{2S_0}{b}x - \frac{kb - S_0}{b^3}x^3 + F_0 \cos \omega t \tag{7-81}$$

Equation (7-81) may be rewritten

$$\ddot{x} + \alpha x + \epsilon x^3 = \frac{F_0}{m}\cos \omega t \tag{7-82}$$

where

$$\alpha = \frac{2S_0}{mb} \qquad \epsilon = \frac{kb - S_0}{mb^3} \tag{7-83}$$

The case represented by Eq. (7-82) is physically similar to that of a spring in which the stiffness varies with the displacement. If ϵ is positive, the situation corresponds to a spring for which the stiffness increases with the displacement (*hard* spring), and if ϵ is negative, the correspondence is to a spring with decreasing stiffness (*soft* spring).

We shall limit ourselves to finding an approximate steady-state expression for x as a function of time. In the method to be employed here it is necessary that we restrict ourselves also to small values of the magnitude of ϵ.

Rewriting Eq. (7-82), we have

$$\ddot{x} = -\alpha x - \epsilon x^3 + \frac{F_0}{m}\cos \omega t \tag{7-84}$$

We assume, as a first approximation, that

$$x_1 = A \cos \omega t \tag{7-85}$$

Substituting this in the right side of (7-84), we obtain

$$\ddot{x}_2 = -\alpha A \cos \omega t - \epsilon A^3 \cos^3 \omega t + \frac{F_0}{m}\cos \omega t \tag{7-86}$$

Employing the trigonometric identity

$$\cos^3 \omega t = \tfrac{3}{4}\cos \omega t + \tfrac{1}{4}\cos 3\omega t$$

Eq. (7-86) becomes

$$\ddot{x}_2 = -\left(\alpha A + \frac{3}{4}\epsilon A^3 - \frac{F_0}{m}\right)\cos \omega t - \frac{1}{4}\epsilon A^3 \cos 3\omega t \qquad (7\text{-}87)$$

Integrating (7-87) twice, we have, as a second approximation to the solution of Eq. (7-84),

$$x_2 = \frac{1}{\omega^2}\left(\alpha A + \frac{3}{4}\epsilon A^3 - \frac{F_0}{m}\right)\cos \omega t + \frac{\epsilon A^3}{36\omega^2}\cos 3\omega t \qquad (7\text{-}88)$$

where the integration constants are taken to be zero. One procedure for ϵ small assumes (7-85) to approximately satisfy (7-87). This leads to a value of ω appropriate to a given A.

A further discussion of the approximate solution of (7-84) is above the level of this text but may be pursued further in any detailed treatment of nonlinear vibrations. It is worthwhile, however, in passing, to point out that for a system with nonlinear restoring forces resonance does not exist in quite the same manner as for the case of elastic forces. In the present case, even if no damping is present, the amplitude cannot increase without limit for a driving force of given frequency. This follows since, if the driving frequency is chosen equal to the natural frequency of the system for small amplitudes, it no longer will be equal to the natural frequency for large amplitudes. It can be shown that a plot of the amplitude (and also of the square of the amplitude) as a function of the driving angular frequency ω will display a tilt in the resonance peak if nonlinear terms, such as ϵx^3 in Eq. (7-84), are present in the equation of motion. If ϵ is positive (*hard* spring), the natural frequency increases with increasing amplitude and consequently the tilt of the peak is toward higher values of ω. For ϵ negative (*soft* spring) the tilt is toward lower ω. The tilting of the resonance curves has an interesting consequence. This is that the amplitude and also the square of the amplitude are not single-valued in ω for certain small ranges in ω at the resonance peak. Consequently, if ω is varied over these ranges, sudden changes in the amplitude of the motion are likely to result.

7-13. Thermal Expansion of a Crystal. Let us examine the case of a pair of neighboring K and Cl ions in a crystal of KCl. It has been determined that the approximate expression for the mutual potential energy for such an ion pair in a polar crystal of the type of KCl is

$$V(r) = -\alpha\frac{e^2}{r} + \frac{\beta}{r^p} \qquad (7\text{-}89)$$

where r is the distance of separation of the ions, α and β are positive constants, and p is a constant which is approximately equal to 9. The first term arises from the Coulomb attraction, and the second term is the potential energy corresponding to a repulsive force which is present. The constant α is less than unity because of the presence of neighboring ions. It is approximately 0.29 for this type of crystal. The constant β can be determined in terms of α and the equilibrium separation r_0 of an ion pair, since at the equilibrium distance we must have

$$\left(\frac{dV}{dr}\right)_{r=r_0} = 0 = \frac{\alpha e^2}{r_0{}^2} - \frac{p\beta}{r_0{}^{p+1}} \tag{7-90}$$

from which

$$\beta = \frac{\alpha e^2 r_0{}^{p-1}}{p} \tag{7-91}$$

The equilibrium separation r_0 is known from X-ray data (approximately 3.136×10^{-8} cm for KCl). Consequently a numerical value for β can be obtained readily.

Figure 7-12 shows a plot of the potential energy of an ion pair as a function of the separation r of the constituents. The solid curve is the resultant potential energy. It is the algebraic sum of that owing to the attractive Coulomb force and that owing to the repulsive force, of the form $1/r^{10}$. The minimum of the solid curve occurs at the equilibrium separation r_0.

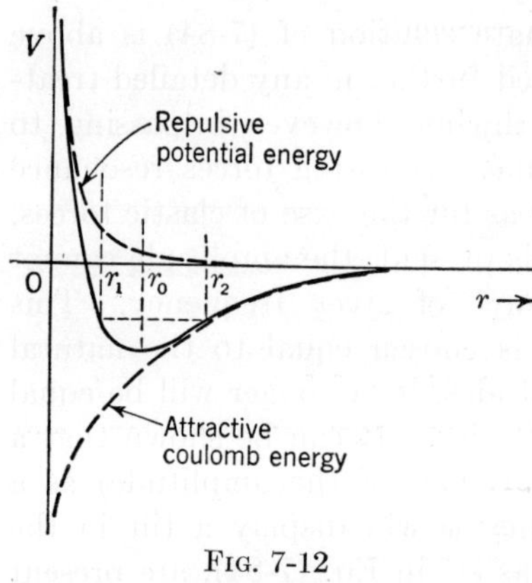

Fig. 7-12

It is of interest to examine small displacements from the equilibrium radius r_0. In order to do this we expand the potential energy in a Taylor series about the point $r = r_0$. We obtain, for r near r_0,

$$V(r) = V(r_0) + x\left(\frac{dV}{dr}\right)_{r_0} + \frac{x^2}{2!}\left(\frac{d^2V}{dr^2}\right)_{r_0} + \frac{x^3}{3!}\left(\frac{d^3V}{dr^3}\right)_{r_0} + \cdots \tag{7-92}$$

where the symbol x denotes $r - r_0$. Now the second term is zero, since the force is zero at $r = r_0$. Equation (7-92) therefore may be rewritten as

$$V(x) = A\frac{x^2}{2!} + B\frac{x^3}{3!} + \cdots \tag{7-93}$$

Here A and B are the differential coefficients evaluated at r_0, and $V(r) - V(r_0) \equiv V(x)$. The first term on the right is the familiar quadratic term corresponding to an elastic force. Hence, for small enough displacements x from r_0, the potential-energy curve approximates the parabolic shape (see Sec. 7-4). This means that, if a small amount of energy is given to the ion pair, the system will execute a small oscillation symmetrically about $r = r_0$. However, if a much larger amount of energy is given to the system, the second term on the right side of Eq. (7-93) will become appreciable and the oscillations will no longer be symmetrical about the point r_0. The amplitude $r_2 - r_0$ in the direction of r increasing will be greater than the amplitude $r_0 - r_1$ in the direction of r decreasing (see the horizontal dotted line between r_1 and r_2 in the figure). Consequently the time average of $r - r_0$ is not zero, as it would be for an elastic force, but is somewhat greater than zero. Hence the time average of r is greater than r_0. Accordingly, if all the ion pairs in the crystal are given this amount of energy (say by heating the crystal), the net effect is that the crystal will expand.

It is useful to write, from (7-93), the expression giving the force between the ions. We have, differentiating with respect to x,

$$F = -\frac{dV}{dx} = -Ax - \frac{B}{2}x^2 - \cdots \tag{7-94}$$

It is important to notice that the quadratic term in Eq. (7-94) is unsymmetric for positive and negative values of x. This follows from the fact that the term contains an even power of x. This is to be contrasted with the cube term occurring in the previous section. An odd-power term, such as was encountered there in the law of force, is symmetric for positive and negative displacements. The lack of symmetry contributed by the quadratic term, in the present instance in Eq. (7-94), is the factor which is responsible for the unequal amplitudes in the two directions.

The equation of motion for an ion pair could easily be written, since the force is given by (7-94). However, as will be evident from the next chapter, the mass, in the inertial reaction term, is not the mass of either ion but rather is an effective mass (reduced mass of the system). It is possible to solve the resulting equation of motion in an approximate manner. A study of the linear term in (7-94) leads to a value of the elastic coefficient of the crystal for a tensile strain (*Young's modulus*), while an investigation of the quadratic term provides information concerning the *coefficient of thermal expansion* of the crystal.

Problems

Problems marked C are reprinted by kind permission of the Cambridge University Press. Those marked J are taken from Jeans, "Theoretical Mechanics" and are used by kind permission of Ginn & Company.

7-1. A point moving with simple harmonic motion is observed to have velocities 3 ft/sec and 4 ft/sec when at distances of 4 ft and 3 ft, respectively, from its equilibrium position. Find the amplitude and period of the motion.

7-2*J*. A train is running smoothly along a curve at the rate of 60 mi/hr, and a pendulum which would ordinarily oscillate with a period of 1 sec is found to oscillate

121 times in 2 min. Show that the radius of the curve described by the train is approximately $\frac{1}{4}$ mile.

7-3. A slingshot is made by tying the two ends of a piece of rubber, of natural length L and modulus λ, to the two prongs of a forked piece of wood at a distance a apart, a being less than L. A stone of mass m is placed at the middle point of the rubber and is drawn back until the rubber is stretched to a length of $2L$. If the stone is then set free, find the velocity with which it will leave the slingshot.

7-4. A plank of mass M rests on four equal massless springs, one at each corner. A mass m is placed symmetrically on the plank, and the whole rests in static equilibrium under gravity. If now the plank is raised a distance b above the equilibrium position and released, what is the force constant of each spring such that m will just not leave the plank during the subsequent motion?

7-5*C*. A horizontal board is made to perform simple harmonic oscillations horizontally, moving to and fro through a distance 30 in. and making 15 complete oscillations per minute. Find the least value of the coefficient of friction in order that a heavy body placed on the board may not slip.

7-6. A mass m is suspended by means of a light elastic string of modulus λ. If the mass is pulled downward a further distance b and then released, show that m will execute a simple harmonic motion of amplitude b provided that the tension in the string is never zero.

7-7*C*. A body is suspended from a fixed point by a light elastic string, of natural length b, whose modulus of elasticity is equal to the weight of the body and makes vertical oscillations of amplitude a. Show that, if as the body rises through its equilibrium position it picks up another body of equal weight, the amplitude of the oscillation becomes $(b^2 + a^2/2)^{\frac{1}{2}}$.

7-8. A damped linear harmonic oscillator is simultaneously subjected to two driving forces $a_1 \cos \omega_1 t$ and $a_2 \cos (\omega_2 t + \varphi)$. Show that the resultant displacement in the steady state is the sum of the displacements due to the impressed forces acting separately. Show also that the rate at which work is being done by the driving forces is the sum of the rates obtaining if the forces were acting separately.

7-9. A straight hole is dug from the north to the south pole of the earth. Taking the radius of the earth to be 6.37×10^8 cm and the gravitational acceleration at the surface to be 980 cm/sec², find the time in minutes required for a body dropped from the north pole to reach the south pole.

7-10. From classical electromagnetic theory an accelerated electron radiates energy at the rate $2e^2\ddot{x}^2/3c^3$, where e is the charge, $\ddot{x}$ the acceleration, and c the velocity of light. Making the approximation that the motion of the electron is always simple harmonic with constant frequency ν_0, determine the energy radiated during one period. The initial amplitude is x_0. Show also that the number of periods required for the energy to decrease to half the initial value is very nearly $mc^3/(4\pi^2e^2\nu_0)$, where m is the mass of the electron.

7-11. In Example 7-1 determine the rate at which the kinetic energy is changing at a time $\pi\sqrt{b/g}$ after the initial instant.

7-12. An oscillator consists of a mass m attached to a massless spring of force constant k, the whole being immersed in a viscous medium which exerts a damping force proportional to the velocity (proportionality factor $2\sqrt{km}$). At $t = 0$ the system is at rest in the equilibrium position, and a force $F \cos \sqrt{k/m}\, t$ is suddenly applied. Find the displacement as a function of time. Find also the rate of change of the potential energy at a time $\pi\sqrt{m/k}$.

7-13. One end of a massless spring of natural length a is attached to a fixed point, while at the other end is attached a mass m, the whole being immersed in a viscous medium which exerts a force on m proportional to the velocity of m (proportionality constant R). The force constant of the spring has the value $R^2/4m$. Initially the spring is stretched to the length s and released. Find the position of m as a function of time, evaluating all arbitrary constants in terms of the initial conditions. Find also the rate at which the total energy is changing at a time $\pi m/R$.

7-14. A damped oscillating system has an effective mass m and a natural undamped angular frequency ω_0 and has a damping coefficient, proportional to the velocity, of magnitude $m\omega_0/\sqrt{2}$. If there is a driving force $F \cos (\omega_0 t/\sqrt{2})$, show that the energy supplied to the system by the driving force during the first quarter period of the force, that is, during the interval $t = 0$ to $t = \pi/(\sqrt{2}\,\omega_0)$, is $F^2(\pi - 2)/4m\omega_0^2$, neglecting all transients.

7-15. Consider a simple pendulum executing small oscillations. The string is shortened an amount Δb very slowly, so that the change in amplitude from one oscillation to the next is negligible. Show that under these conditions the ratio of the energy of the oscillatory motion to the frequency remains constant. (This illustrates Ehrenfest's celebrated *principle of adiabatic invariance.*)

7-16. In the damped linear harmonic oscillator show that, if the damping is small, the average rate at which energy is dissipated is proportional to the energy stored in the system.

7-17. Given the linear differential equation with variable coefficients

$$\ddot{x} + g(t)\dot{x} + f(t)x = 0$$

If $x_1 = \phi_1(t)$ and $x_2 = \phi_2(t)$ are each a particular solution of the equation, show that

$$x = c_1\phi_1(t) + c_2\phi_2(t)$$

where c_1 and c_2 are constants, is also a solution of the differential equation. (This shows that the superposition principle holds for a linear differential equation, whether or not the coefficients are variable.)

7-18. Show that the superposition principle does not hold for a nonlinear differential equation such as

$$\ddot{x} + cx^2 = 0$$

7-19. In Eq. (7-76) show, for the case of the linear spring, where $f(x) = kx$, k being the force constant, that the dependence of the period T upon the amplitude x_0 drops out.

7-20. In the system considered in Sec. 7-12, consider that in the configuration AOA' the strings are just unstretched (that is, the natural length of each is b). Find an approximate steady-state solution of the resulting equation of motion. Employ the method of Sec. 7-12.

7-21. If the amplitude of the motion of a pendulum is not small, the free horizontal motion may be represented approximately by

$$\ddot{x} + \frac{g}{b}x - \frac{g}{2b^3}x^3 = 0$$

where the notation is that of Sec. 7-1. Verify this statement. If a force $F_0 \cos \omega t$ is also acting, find an approximate steady-state expression for $x(t)$, considering $g/2b^3$ to be small.

CHAPTER 8

MOTION OF A SYSTEM OF PARTICLES

In the two preceding chapters, some of the more elementary problems connected with the motion of a single particle were considered. Although many of the problems there treated involved extended bodies, the approximation of treating the bodies as particles was made, rotational motions and the like being neglected. Later, in studying the motion of rigid bodies, we shall consider the added complications which are present when rotational motions are taken into account. In order to prepare for this, it is necessary in the present chapter to develop several general dynamic theorems involving systems of particles and to examine some of their consequences. We shall find that Newton's third law, the law of reaction, plays a very important part when more than one particle is involved. We begin by investigating briefly the analogs of some of these theorems for the much simpler case of a single particle.

8-1. Linear and Angular Momentum for a Single Particle. For the case of a single particle of mass m, if no forces are acting, Newton's equation of motion becomes (for a rest system)

$$\frac{d}{dt}(m\dot{\mathbf{r}}) = 0$$

from which

$$\boxed{m\dot{\mathbf{r}} = \mathbf{p} = \text{const vector}} \tag{8-1}$$

This states that:

I. *If no forces are acting on a particle, the momentum of the particle remains constant in direction and magnitude.*

There is an analogous situation which holds for the *moment of momentum.* If the momentum of a particle is $m\dot{\mathbf{r}}$, then by definition the moment of momentum $\mathbf{J}$ is

$$\boxed{\mathbf{J} = \mathbf{r} \times m\dot{\mathbf{r}}} \tag{8-2}$$

where the moment is taken about O, the fixed origin of $\mathbf{r}$.

In Fig. 8-1, a particle of mass m, moving along the path S_1S_2, is instantaneously at point P. Here $\mathbf{r}$ is the radius vector from the origin to P, $m\dot{\mathbf{r}}$ is the instantaneous value of the momentum of the particle at point P, and the moment of momentum is as given by Eq. (8-2).

In Fig. 8-1, let PR be constructed normal to OP at P and OQ normal to the line of action of $m\dot{\mathbf{r}}$. If we denote the magnitude of $\dot{\mathbf{r}}$ by the symbol v, we see that the magnitude of $\mathbf{J}$ may be written

$$J = |\mathbf{r} \times m\dot{\mathbf{r}}| = rmv \sin \varphi$$
$$= mvr \cos \theta = mr^2\dot{\theta} \qquad (8\text{-}3)$$

where the last equality follows, since $v \cos \theta = r\dot{\theta}$. This gives several alternative forms for the magnitude of the moment of momentum. The last form in (8-3) will become familiar in Chap. 10. The moment of momentum is commonly referred to by another name, the *angular momentum,* and it is the latter term which we shall employ. The importance of the quantity $\mathbf{J}$ for a particle subject to a force $\mathbf{F}$ can be appreciated by taking the moment, about the origin O, of both sides of Newton's equation of motion. Thus

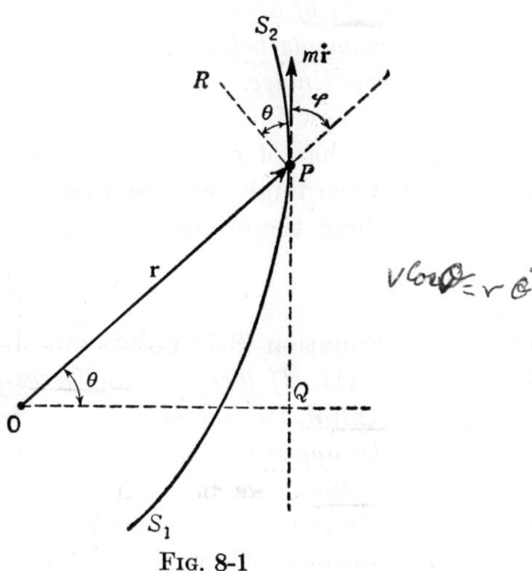

FIG. 8-1

$$\mathbf{r} \times \frac{d}{dt}(m\dot{\mathbf{r}}) = \mathbf{r} \times \mathbf{F}$$

The left side may be rewritten

$$\mathbf{r} \times \frac{d}{dt}(m\dot{\mathbf{r}}) = \frac{d}{dt}(\mathbf{r} \times m\dot{\mathbf{r}}) \qquad (8\text{-}4)$$

The validity of this transformation can easily be established by performing the indicated operation on the right side. Thus

$$\frac{d}{dt}(\mathbf{r} \times m\dot{\mathbf{r}}) = \dot{\mathbf{r}} \times m\dot{\mathbf{r}} + \mathbf{r} \times \frac{d}{dt}(m\dot{\mathbf{r}}) = \mathbf{r} \times \frac{d}{dt}(m\dot{\mathbf{r}})$$

where the last step follows since $\dot{\mathbf{r}} \times m\dot{\mathbf{r}}$ is the cross product of a vector $\dot{\mathbf{r}}$ with itself (or rather with the vector $m\dot{\mathbf{r}}$, which has the same line of action as $\dot{\mathbf{r}}$), a quantity which is identically zero. Hence Eq. (8-4) becomes

$$\frac{d}{dt}(\mathbf{r} \times m\dot{\mathbf{r}}) = \mathbf{r} \times \mathbf{F} \qquad (8\text{-}5)$$

But, in view of Eq. (8-2), the left side is just the time rate of change of the angular momentum $\mathbf{J}$, and since the right side is the moment of the external force, which we may call $\mathbf{L}$, Eq. (8-5) may be written

$$\dot{\mathbf{J}} = \mathbf{L} \qquad (8\text{-}6)$$

This states that:

II. *The time rate of change of the moment of momentum (angular momentum) of a particle is equal to the moment of the applied force (torque), both moments being taken with respect to the origin of a Newtonian system of coordinates.*

Now **L** is zero either if **F** is zero or if the line of action of **F** is parallel to that of **r**. If the latter situation obtains, F is called a *central force*. Accordingly we see that, if the force is central, the moment of force about the origin is zero and Eq. (8-6) integrates to

$$\boxed{\mathbf{J} = \text{const vector}} \tag{8-7}$$

Equation (8-7) points out that:

III. *If there are no forces acting on a particle or if the force acting is central, the line of action of the force passing through the Newtonian origin, the angular momentum, with reference to this origin, is a constant of the motion.*

Furthermore, since **J** is a vector quantity, its direction, if it is constant, will be unchanged and consequently the path of the particle will remain in one plane.

It is worthwhile to make the definition of a central force more general than was stated above. The practical case is that of an interacting pair of particles. If the lines of action of the forces, which each exerts upon the other, lie along the straight line joining the particles, then these forces are termed *central*. In order to study the relative motion of such a pair of particles, the origin may be taken to be any fixed point or, more conveniently, the center of mass of the two (which lies along the line joining the particles). For a system of particles, if the system consists of such interacting pairs, the forces of interaction in each case lying along the straight line joining the two particles, the system is described as being one in which this portion, at least, of the internal forces consists of central forces. For the study of such a system, it is convenient to place the origin either at the center of mass or at a fixed point (see below).

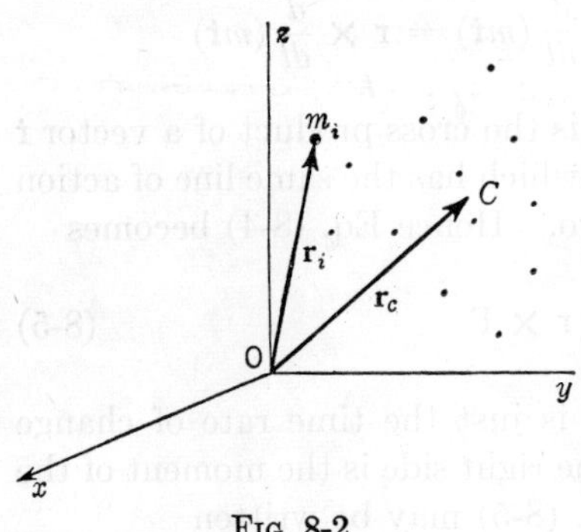

FIG. 8-2

SYSTEM OF PARTICLES

8-2. Linear Momentum of the Center of Mass of a System. Suppose we have a system of n particles having masses $m_1, \ldots, m_n$ to which the radius vectors from a given origin O are, respectively, the quantities $\mathbf{r}_1, \ldots, \mathbf{r}_n$ (see Fig. 8-2). Let us write the equation of motion of the ith particle of mass m_i. If the external force acting on m_i is $\mathbf{F}_i$, and if the internal forces on m_i, arising because of the presence of the other particles, are $\mathbf{F}_{i1}, \ldots, \mathbf{F}_{in}$, we have

$$m_i\ddot{\mathbf{r}}_i = \mathbf{F}_i + \mathbf{F}_{i1} + \mathbf{F}_{i2} + \cdots + \mathbf{F}_{in} = \mathbf{F}_i + \sum_{k=1}^{n} \mathbf{F}_{ik} \tag{8-8}$$

where the internal forces are all collected together under one summation sign. In Eq. (8-8) we are considering that the mass m_i remains constant throughout the motion. We are able to achieve an important result merely by summing over all n particles, that is, by adding together all n equations of motion of the type (8-8). We have

$$\sum_{i=1}^{n} m_i\ddot{\mathbf{r}}_i = \sum_{i=1}^{n} \mathbf{F}_i + \sum_{i=1}^{n}\sum_{k=1}^{n} \mathbf{F}_{ik} \tag{8-9}$$

In Eqs. (8-8) and (8-9), the term $\mathbf{F}_{ik}$ never becomes $\mathbf{F}_{ii}$. Physically, this means that the particle does not exert a force upon itself.

We shall now make use of Newton's third law. It is at this stage in the discussion of the motion of a system that the third law is of importance. The third law says, in effect, that one particle exerts on a second particle a force which is equal and opposite to that which the second exerts on the first. With reference to particles i and k this requires that

$$\mathbf{F}_{ik} = -\mathbf{F}_{ki} \tag{8-10}$$

and the $\mathbf{F}_{ik}$'s need not be central forces. In view of Eq. (8-10), we see that the double sum in Eq. (8-9) consists of pairs of terms which mutually cancel one another, and so we are left with

$$\sum_{i=1}^{n} m_i\ddot{\mathbf{r}}_i = \sum_{i=1}^{n} \mathbf{F}_i \tag{8-11}$$

where it is understood that i takes all values from 1 to n. We can now obtain further insight into this equation by considering the way in which the motion of the center of mass enters. The radius vector $\mathbf{r}_c$ from the arbitrary origin O to the center of mass C of the system is, by definition,

$$M\mathbf{r}_c = \sum_{i=1}^{n} m_i\mathbf{r}_i \tag{8-12}$$

in which $M = \sum_{i=1}^{n} m_i$. By differentiating Eq. (8-12) and requiring the masses to remain constant, we have

$$M\ddot{\mathbf{r}}_c = \sum_{i=1}^{n} m_i\ddot{\mathbf{r}}_i$$

Substituting this result into Eq. (8-11), we obtain

$$M\ddot{\mathbf{r}}_c = \sum_{i=1}^{n} \mathbf{F}_i \tag{8-13}$$

But the summation $\sum_{i=1}^{n} \mathbf{F}_i$ is just the vector sum $\mathbf{F}$ of all the external forces, regardless of their points of application. We therefore have the important result that:

IV. *The center of mass moves as if the entire mass is concentrated there as a single particle and is acted upon by the resultant of all the external forces, irrespective of their points of application.*

It is to be emphasized that we have said only that the motion of the center of mass is independent of the points of application of the external forces. Rotational motions, as we shall see later, will of course be affected by the points of application of the external forces. Finally, it is well to remember that we have shown Eq. (8-13) to be valid only for a system in which the *masses remain constant*, so that the equations of motion are of the type of Eq. (8-8).

It follows from IV, as for a simple particle, that if the resultant of the external forces is zero, the momentum of the center of mass is constant.

8-3. Angular Momentum of a System of Particles. In a manner similar to that for a single particle we may derive an expression involving the angular momentum for a system of particles. We begin, as in Sec. 8-2, by writing the equation of motion for the ith particle, although here we shall be able to write it in more general form. We have

$$\frac{d}{dt}(m_i\dot{\mathbf{r}}_i) = \mathbf{F}_i + \sum_{k=1}^{n} \mathbf{F}_{ik} \tag{8-14}$$

where, as before, $\mathbf{F}_i$ is the external force acting on m_i and $\mathbf{F}_{ik}$ is the force, internal to the system, acting on m_i owing to the kth particle. The index k in the sum is understood to assume all values, save i, from 1 to n. Taking the moment of both sides with respect to the origin, we obtain

$$\mathbf{r}_i \times \frac{d}{dt}(m_i\dot{\mathbf{r}}_i) = \mathbf{r}_i \times \mathbf{F}_i + \sum_{k=1}^{n} \mathbf{r}_i \times \mathbf{F}_{ik} \tag{8-15}$$

As in Sec. 8-1, the left side may be rewritten as the rate of change of the angular momentum. Making this change, and at the same time summing over all particles [that is, adding together all n equations of the type (8-15)], we have

$$\frac{d}{dt}\sum_{i=1}^{n} \mathbf{r}_i \times m_i\dot{\mathbf{r}}_i \equiv \sum_{i=1}^{n} \dot{\mathbf{J}}_i \equiv \dot{\mathbf{J}} = \mathbf{L} + \sum_{i=1}^{n}\sum_{k=1}^{n} \mathbf{r}_i \times \mathbf{F}_{ik} \tag{8-16}$$

in which $\mathbf{J}$ is the total angular momentum of the system and $\mathbf{L}$ is the vector sum of the moments of all the external forces acting on the system. The double sum on the right of Eq. (8-16) is the vector sum of the internal moments. This does not in general vanish but will do so if the lines of action of all the internal forces lie along straight lines joining the particles (that is, if the internal forces are all central forces). To see this, we note that the double sum consists of pairs of terms such as

$$\mathbf{r}_i \times \mathbf{F}_{ik} + \mathbf{r}_k \times \mathbf{F}_{ki} \tag{8-17}$$

where $\mathbf{F}_{ik}$ is the force on m_i exerted by m_k and $\mathbf{F}_{ki}$ is the force on m_k owing to the presence of m_i. By Newton's third law we have $\mathbf{F}_{ik} = -\mathbf{F}_{ki}$, and expression (8-17) assumes the form

$$\mathbf{r}_i \times \mathbf{F}_{ik} - \mathbf{r}_k \times \mathbf{F}_{ik} = (\mathbf{r}_i - \mathbf{r}_k) \times \mathbf{F}_{ik} \tag{8-18}$$

An inspection of the vector triangle in Fig. 8-3 reveals at once that

$$\mathbf{r}_{ki} = \mathbf{r}_i - \mathbf{r}_k$$

in which $\mathbf{r}_{ki}$ is the radius vector from m_k to m_i. Clearly then, in Eq. (8-18), $\mathbf{r}_i - \mathbf{r}_k$ is the radius vector from m_k to m_i, and if the line of action of $\mathbf{F}_{ik}$ is parallel to this, the cross product vanishes identically. Accordingly, for the case in which the internal forces are central forces (and for the case in which they are zero), Eq. (8-16) takes the form

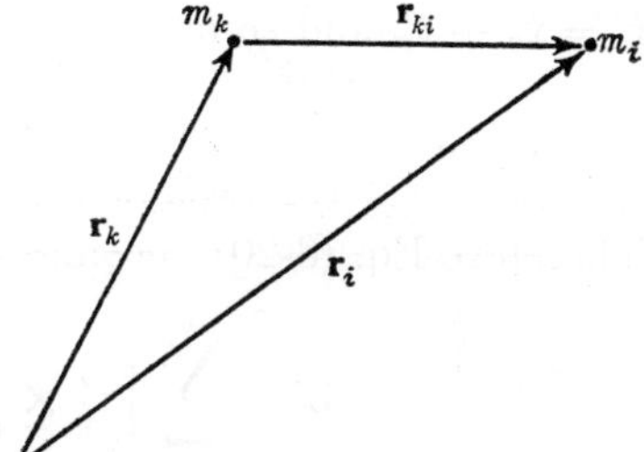

FIG. 8-3

$$\boxed{\dot{\mathbf{J}} = \mathbf{L}} \tag{8-19}$$

This states that:

V. *For a system of particles in which the internal forces all are central forces, the time rate of change of the total angular momentum of the system is equal to the vector sum of the moments of all the external forces.*

In particular, if the system is completely isolated, the external forces all being zero, the total angular momentum is a constant of the motion.

8-4. Torque Equation with Reference to an Arbitrary Origin. It should be emphasized that, since the starting point in the developments leading to Eqs. (8-6) and (8-19) was Newton's equation of motion, it has been demonstrated that Eqs. (8-6) and (8-19) are true in a Newtonian reference system. It is interesting to inquire whether or not an equation similar to (8-19) can be written in a system which is being accelerated in some fashion. In order to do this we again consider a system of particles (see Fig. 8-4). Point O is the origin of a Newtonian system, and O' is an arbitrary origin which may be undergoing an acceleration with reference to O. The vectors $\mathbf{r}_i$ and $\mathbf{r}'_i$ are drawn from O and O', respectively, to the ith particle m_i. Suppose there are external forces $\mathbf{F}_1, \mathbf{F}_2, \ldots, \mathbf{F}_n$ acting upon the corresponding particles and that all internal forces are central forces. We may write the vector sum of the moments of the external forces about O' as

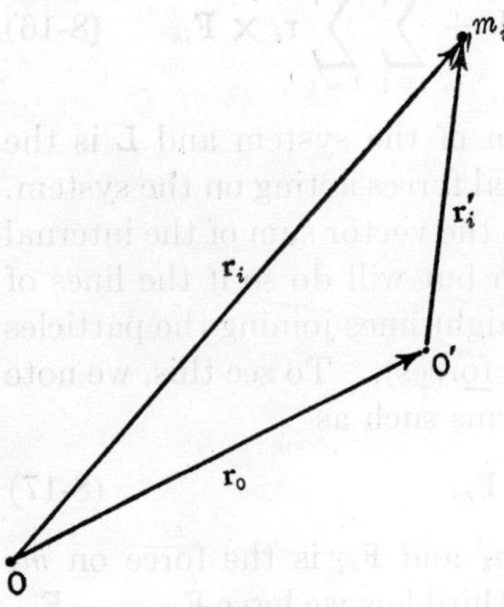

FIG. 8-4

$$\mathbf{L}' = \sum_{i=1}^{n} \mathbf{r}'_i \times \mathbf{F}_i \tag{8-20}$$

Now from Fig. 8-4, $\mathbf{r}_i = \mathbf{r}_0 + \mathbf{r}'_i$, and $\dot{\mathbf{r}}_i = \dot{\mathbf{r}}_0 + \dot{\mathbf{r}}'_i$, and so on. Also in the Newtonian system we have

$$\mathbf{F}_i = \frac{d}{dt}(m_i\dot{\mathbf{r}}_i)$$

Therefore Eq. (8-20) becomes

$$\mathbf{L}' = \sum_{i=1}^{n} \left[\mathbf{r}'_i \times \frac{d}{dt} m_i(\dot{\mathbf{r}}_0 + \dot{\mathbf{r}}'_i) \right] \tag{8-21}$$

$$= -\ddot{\mathbf{r}}_0 \times \sum_{i=1}^{n} m_i\mathbf{r}'_i + \sum_{i=1}^{n} \left[\mathbf{r}'_i \times \frac{d}{dt}(m_i\dot{\mathbf{r}}'_i) \right] \tag{8-22}$$

$$= -\ddot{\mathbf{r}}_0 \times \sum_{i=1}^{n} m_i\mathbf{r}'_i + \frac{d}{dt} \sum_{i=1}^{n} \mathbf{r}'_i \times m_i\dot{\mathbf{r}}'_i \tag{8-23}$$

in which the step from Eq. (8-21) to (8-22) follows since $\ddot{\mathbf{r}}_0$ is not being summed over and may therefore be taken outside of the summation sign, and provided also that m_i is *constant*. The negative sign arises because of the inversion of the order of the cross product. The step from Eq. (8-22) to (8-23) follows in the same manner as that leading to Eq. (8-5). Now the last term on the right of Eq. (8-23) is simply the rate of change of $\mathbf{J}'$, where $\mathbf{J}'$ is the resultant moment of momentum about the arbitrary origin O'. Consequently Eq. (8-23) may be rewritten as

$$-\ddot{\mathbf{r}}_0 \times \sum_{i=1}^{n} m_i\mathbf{r}'_i + \dot{\mathbf{J}}' = \mathbf{L}' \tag{8-24}$$

Hence the torque equation, (8-19), is modified for an arbitrarily moving origin by the presence of the first term on the left. This added term vanishes for any of three situations. The first and most obvious one is when the acceleration $\ddot{\mathbf{r}}_0$, of the origin O' with respect to O, is zero. The second case in which the first term vanishes is when O' is the center of mass of the system of particles, for then $\sum_{i=1}^{n} m_i\mathbf{r}_i' = 0$ by definition. The third and last case is a bit more subtle and requires us to notice that, even if O' is not the center of mass, the quantity $\sum_{i=1}^{n} m_i\mathbf{r}_i'$ is, nevertheless, a vector which is directed along a straight line passing through the center of mass. [See Eq. (8-12). The value of $\sum_{i=1}^{n} m_i\mathbf{r}_i'$ is just $M\mathbf{r}_c'$, in which M is the total mass $\sum_{i=1}^{n} m_i$ and $\mathbf{r}_c'$ is the vector from O' to the center of mass C.] Accordingly if the acceleration $\ddot{\mathbf{r}}_0$, of O', is also a vector directed through the center of mass, the cross product of $\ddot{\mathbf{r}}_0$ with the sum will vanish. This last situation will be found to be a useful one in the next chapter when we deal with rigid bodies which are rolling but not sliding.

It should be noticed that, if the masses are not *constant*, Eq. (8-22) becomes even more complicated.

Example 8-1. Two masses m_1 and m_2, connected by a light inextensible string of length b, are whirled around by hand and subsequently released. The conditions of release are that m_1 is momentarily at rest in the hand and m_2 is traveling with a linear velocity v_0. It is desired to find the tension in the string after release.

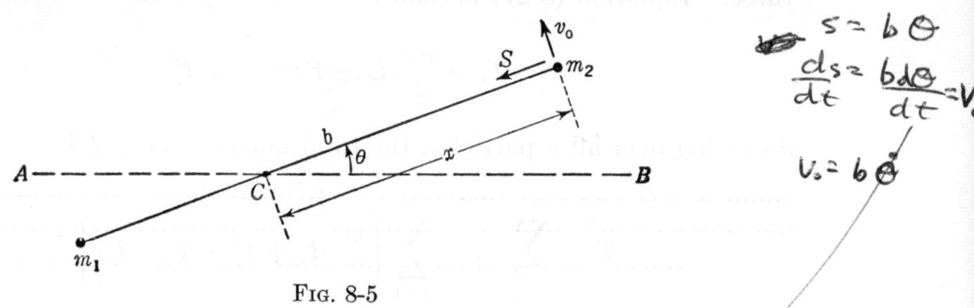

FIG. 8-5

At the instant of release, the situation is as shown in Fig. 8-5. The angle θ is the angle the string makes at any time with a line AB, pointing in a fixed direction in space, passing through the center of mass. Initially we must have

$$b\dot{\theta} = v_0 \tag{8-25}$$

In free flight the only external forces acting are the forces of gravity m_1g and m_2g. The resultant moment of these about the center of mass is zero since the gravitational field is uniform. Thus the angular momentum about the center of mass C, distant x

from m_2, is constant throughout the motion. This requires that the angular velocity of m_1 and m_2 about C be constant. The tension S in the string is determined entirely by the centrifugal reaction of the particles, arising because of this rotation, and therefore is equal in magnitude to $m_2 x \dot{\theta}^2$. But x may be determined from the equation $(m_1 + m_2)x = m_1 b$, and since $\dot{\theta}$ is given in terms of knowns by Eq. (8-25), S becomes

$$S = \frac{m_1 m_2 v_0^2}{(m_1 + m_2)b} \tag{8-26}$$

So far as the motion of the center of mass is concerned, it behaves as if both m_1 and m_2 were one particle situated there. Accordingly, following the results of Sec. 6-5, the center of mass pursues a parabolic path.

8-5. Kinetic Energy of a System of Particles. It is possible to write the total kinetic energy of a system of n particles as the sum of two parts, the first of which is the kinetic energy of translation of the center of mass, and the second the kinetic energy of relative motion of the particles. In order to demonstrate this, we first write the kinetic energy of m_i relative to an inertial system with origin at O (see Fig. 8-6). If v_i is the magnitude of the velocity $\dot{\mathbf{r}}_i$, of m_i relative to O, the kinetic energy T_i is

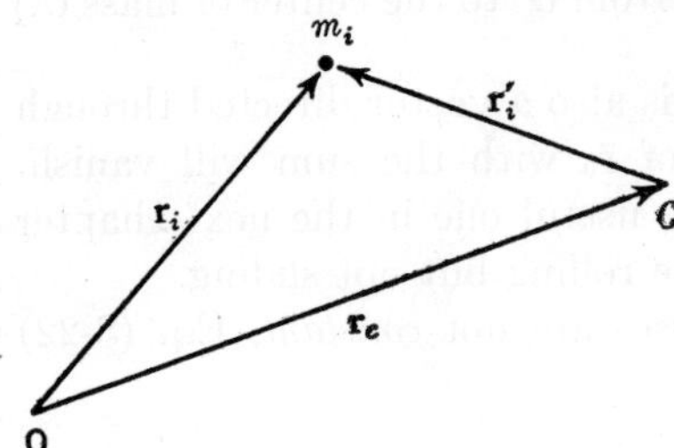

FIG. 8-6

$$T_i = \frac{m_i}{2} v_i^2 = \frac{m_i}{2} (\dot{\mathbf{r}}_i \cdot \dot{\mathbf{r}}_i) \tag{8-27}$$

Defining vectors $\mathbf{r}_c$ from O to another point C and $\mathbf{r}_i'$ from C to m_i, we may put $\dot{\mathbf{r}}_i$, in Eq. (8-27), in terms of these. Equation (8-27) becomes

$$T_i = \frac{m_i}{2} (\dot{\mathbf{r}}_c + \dot{\mathbf{r}}_i') \cdot (\dot{\mathbf{r}}_c + \dot{\mathbf{r}}_i')$$

Summing over all n particles, the total kinetic energy T is

$$T = \sum_{i=1}^{n} T_i = \sum_{i=1}^{n} \left[\frac{m_i}{2} (\dot{\mathbf{r}}_c + \dot{\mathbf{r}}_i') \cdot (\dot{\mathbf{r}}_c + \dot{\mathbf{r}}_i') \right] \tag{8-28}$$

$$= \frac{1}{2} |\dot{\mathbf{r}}_c|^2 \sum_{i=1}^{n} m_i + \dot{\mathbf{r}}_c \cdot \sum_{i=1}^{n} m_i \dot{\mathbf{r}}_i' + \sum_{i=1}^{n} \frac{m_i}{2} |\dot{\mathbf{r}}_i'|^2 \tag{8-29}$$

where Eq. (8-29) follows from (8-28) by carrying out the indicated multiplication and then taking $\dot{\mathbf{r}}_c$ outside the summation sign. If point C is chosen to be coincident with the center of mass for all time, the sum in the middle term on the right side of Eq. (8-29) vanishes. Hence if

C is the center of mass (which in general moves with the velocity $\dot{\mathbf{r}}_c$),

$$T = \frac{M}{2} v_c^2 + \sum_{i=1}^{n} \frac{m_i v_i'^2}{2} \tag{8-30}$$

in which $M = \sum_{i=1}^{n} m_i$, v_c is the magnitude of $\dot{\mathbf{r}}_c$, and v_i' is the magnitude of $\dot{\mathbf{r}}_i'$. Equation (8-30) states that:

VI. *The total kinetic energy of a system of particles may be written as the sum of the kinetic energy of translation of the center of mass, plus the kinetic energy of the motion relative to the center of mass.*

Example 8-2. An amount of energy E is liberated when a gun of mass m_1 discharges a projectile of mass m_2. Assuming that the projectile is discharged horizontally and that the gun carriage is mounted on wheels of negligible mass, it is desired to determine the effect of the recoil of the gun upon the motion of the projectile.

The recoil velocity is v_1, and the velocity of the projectile is v_2. Making use of the principles of conservation of energy and momentum, we have

$$\tfrac{1}{2}m_1v_1^2 + \tfrac{1}{2}m_2v_2^2 = E \qquad m_1v_1 - m_2v_2 = 0 \tag{8-31}$$

from which the velocity v_2 of the projectile is found to be

$$v_2 = \sqrt{\frac{2m_1E}{m_2(m_1 + m_2)}}$$

If the gun carriage is anchored firmly in the surface of the earth, Eqs. (8-31) are replaced by

$$\tfrac{1}{2}Mv_1'^2 + \tfrac{1}{2}m_2v_2'^2 = E \qquad Mv_1' - m_2v_2' = 0$$

where v_1' is the recoil velocity of the earth and gun and the mass of the carriage is included in the mass M of the earth. From this it is clear that the momentum of the projectile is equal and opposite to that of the earth plus carriage. The situation is different in the case of the energy. We may show this by eliminating v_1' from the first by means of the second. We have

$$\frac{1}{2} M \frac{m_2^2}{M^2} v_2'^2 + \frac{1}{2} m_2v_2'^2 = \frac{1}{2} m_2v_2'^2 \left(\frac{m_2}{M} + 1\right) = E$$

Now M is very much greater than m_2, and so the kinetic energy $m_2v_2'^2/2$ of the projectile differs from E by only a very small quantity. Accordingly we see that, although the earth and carriage acquire a momentum equal in magnitude to that of the projectile, they absorb very little of the available energy.

Example 8-3. A shell of mass M is traveling along a parabolic trajectory when an internal explosion, generating an amount of energy E, blows the shell into two portions. One portion, of mass kM, where k is a number less than 1, continues in the original direction, and the remainder is reduced to rest. What is the velocity of the mass kM immediately after the explosion?

Suppose v_0 to be the speed of the projectile just before the explosion, and v_1 to be the speed of kM immediately after. The principles of conservation of energy and momentum provide the equations

$$\tfrac{1}{2}Mv_0^2 + E = \tfrac{1}{2}kMv_1^2 \qquad Mv_0 = kMv_1$$

Eliminating v_0 between these, we obtain

$$v_1 = \sqrt{\frac{2E}{Mk(1-k)}}$$

which is the speed of kM immediately after the explosion. An important point to notice is that the energy E of the explosion is available only for the motion relative to the center of mass. Immediately after the explosion the center of mass of the two bodies continues along the trajectory with the same velocity v_0 that it had originally.

8-6. Angular Momentum of a System of Particles in Terms of the Center of Mass. The angular momentum of a system of particles with reference to a given origin can be split up in a manner similar to that in which the kinetic energy was divided into two parts in the last section. Employing Fig. 8-6 again, the total angular momentum $\mathbf{J}$, with reference to the origin, of the system of n particles is the vector sum of the angular momenta of the individual particles. This may be written

$$\mathbf{J} = \sum_{i=1}^{n} m_i \mathbf{r}_i \times \dot{\mathbf{r}}_i \tag{8-32}$$

In terms of $\mathbf{r}'_i$ and $\mathbf{r}_c$, Eq. (8-32) becomes

$$\mathbf{J} = \sum_{i=1}^{n} m_i[(\mathbf{r}_c + \mathbf{r}'_i) \times (\dot{\mathbf{r}}_c + \dot{\mathbf{r}}'_i)]$$

Expanding the quantity within the brackets, we have

$$\mathbf{J} = M\mathbf{r}_c \times \dot{\mathbf{r}}_c + \mathbf{r}_c \times \sum_{i=1}^{n} m_i\dot{\mathbf{r}}'_i - \dot{\mathbf{r}}_c \times \sum_{i=1}^{n} m_i\mathbf{r}'_i + \sum_{i=1}^{n} m_i\mathbf{r}'_i \times \dot{\mathbf{r}}'_i \tag{8-33}$$

where $M = \sum_{i=1}^{n} m_i$ is the total mass of the system. If C is the center of mass, the summations in the second and third terms on the right of Eq. (8-33) are both zero. Accordingly Eq. (8-33) becomes

$$\mathbf{J} = M\mathbf{r}_c \times \dot{\mathbf{r}}_c + \sum_{i=1}^{n} m_i\mathbf{r}'_i \times \dot{\mathbf{r}}'_i = \mathbf{J}_c + \mathbf{J}' \tag{8-34}$$

In Eq. (8-34), the quantity $\mathbf{J}_c$ (equal to $M\mathbf{r}_c \times \dot{\mathbf{r}}_c$) is the angular momentum, with reference to the origin O, of a particle of mass M situated at the mass center C. The quantity $\mathbf{J}' \left(= \sum_{i=1}^{n} m_i\mathbf{r}'_i \times \dot{\mathbf{r}}'_i\right)$ is the total angular momentum of the particles relative to the mass center. Hence:

VII. *The total angular momentum of a system of particles, referred to a given origin, can be expressed as the sum of two parts: one part being the angular momentum of all the particles relative to the center of mass, and the other being the angular momentum of the total mass, assumed to be at the mass center, relative to the origin.*

It should be pointed out that the origin, in the present section, may be arbitrary. However, the angular momentum, so defined, is *useful* only if the origin is such that Eq. (8-19) is true.

8-7. Relative Motion of Two Bodies. The Reduced Mass. A very useful procedure for treating the relative motion of two interacting particles will now be established. In Fig. 8-7, the positions of the particles of masses m_1 and m_2 are specified, respectively, by the radius vectors $\mathbf{r}_1$ and $\mathbf{r}_2$, drawn from the origin O of a Newtonian system. The only forces acting are $\mathbf{F}_1$ on m_1, owing to m_2, and $\mathbf{F}_2$ on m_2, owing to m_1. The forces need not be central for the results of this section to be valid. Relative to O, the equations of motion of the two particles are

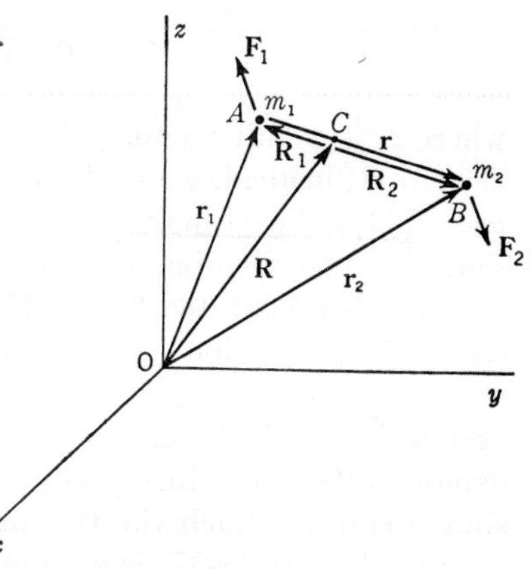

FIG. 8-7

$$m_1\ddot{\mathbf{r}}_1 = \mathbf{F}_1 \qquad m_2\ddot{\mathbf{r}}_2 = \mathbf{F}_2 \tag{8-35}$$

Now in the vector triangle OAB of Fig. 8-7 we may write, where $\mathbf{r}$ is a vector drawn from m_1 to m_2,

$$\mathbf{r}_2 = \mathbf{r}_1 + \mathbf{r}$$
$$\ddot{\mathbf{r}}_2 = \ddot{\mathbf{r}}_1 + \ddot{\mathbf{r}}$$

Multiplying the second of these by m_2, we obtain

$$m_2\ddot{\mathbf{r}}_2 = m_2\ddot{\mathbf{r}}_1 + m_2\ddot{\mathbf{r}}$$

Substituting from Eqs. (8-35), this becomes

$$\mathbf{F}_2 = \frac{m_2}{m_1}\mathbf{F}_1 + m_2\ddot{\mathbf{r}} \tag{8-36}$$

Since $\mathbf{F}_2 = -\mathbf{F}_1$ by Newton's third law, this may be put in the form

$$\frac{m_1 m_2}{m_1 + m_2}\ddot{\mathbf{r}} = \mathbf{F}_2 \tag{8-37}$$

a very useful result indeed. Equation (8-37) states that:

VIII. *The motion of m_2, relative to m_1, takes place precisely as if the particle m_1 were fixed and m_2 were replaced by an effective mass $m_1m_2/(m_1 + m_2)$.*

This latter quantity is often called the *reduced mass* of the system and is frequently denoted by the symbol μ. It is important to notice that the force $\mathbf{F}_2$ is exactly as stated in Eq. (8-35). Thus, if the force on m_2 is a gravitational one, it will have, in both (8-35) and (8-37), the form

$$\mathbf{F}_2 = -\frac{Gm_1m_2}{r^2}\mathbf{r}_0 \tag{8-38}$$

where $\mathbf{r}_0$ is a unit vector pointing in the direction from m_1 to m_2. It is clear that, instead, we could have considered the motion of m_1 relative to m_2. The reduced mass is the same in either case. Proper regard for the sign of r must then be taken, however. Finally it should be noticed that we have demonstrated Eq. (8-37) only for the case in which the masses are *constant.* [Equation (8-36) depends upon the circumstance that the masses do not vary.] Hence, the validity of Eq. (8-37) is limited to a system in which the masses are constant. Further, Eq. (8-37) has been demonstrated only in the *absence* of external forces. However, for the simple case in which the two masses are situated in a uniform gravitational field, Eq. (8-37) is unaltered. The proof of this is left as an exercise for the student.

It is also useful to obtain the equations for motion relative to the center of mass. Let C, in Fig. 8-7, be the center of mass of the two particles, with the vectors $\mathbf{R}$, $\mathbf{R}_1$, and $\mathbf{R}_2$ drawn, respectively, from O to C, from C to m_1 at A, and from C to m_2 at B. Now

$$\mathbf{r}_1 = \mathbf{R} + \mathbf{R}_1 \qquad \mathbf{r}_2 = \mathbf{R} + \mathbf{R}_2$$

If no external forces are acting, $\ddot{\mathbf{R}}$ is zero, and Eqs. (8-35) may be written

$$\begin{aligned} m_1\ddot{\mathbf{R}}_1 &= \mathbf{F}_1 \\ m_2\ddot{\mathbf{R}}_2 &= \mathbf{F}_2 \end{aligned} \tag{8-39}$$

for the motion of each relative to the center of mass C. [It is to be noticed that Eqs. (8-39) require the masses to be *constant.*] Clearly, for Eqs. (8-39) to be solved in terms of $\mathbf{R}_1$ and $\mathbf{R}_2$, the forces $\mathbf{F}_1$ and $\mathbf{F}_2$ must be expressed as functions of $\mathbf{R}_1$ and $\mathbf{R}_2$, respectively. For example, consider the second of these and that $\mathbf{F}_2$ is a gravitational force, given by Eq. (8-38). We have the scalar equations

$$\begin{aligned} m_2R_2 &= m_1R_1 \\ r &= R_1 + R_2 \end{aligned}$$

These yield

$$r = R_2\frac{m_2 + m_1}{m_1}$$

and Eq. (8-38) becomes

$$\mathbf{F}_2 = -\frac{Gm_1{}^3m_2}{(m_1 + m_2)^2R_2{}^2}\mathbf{r}_0$$

a quantity which is expressed in terms of the dependent variable R_2. Consequently the second of Eqs. (8-39) can now be solved, yielding, for example, R_2 as a function of time t.

Example 8-4. We shall use the problem of Example 8-1, employing the results of the present section (see Fig. 8-5).

For this we may employ Eq. (8-37). It is sufficient to equate the absolute values of both sides. Thus, since F_2 is the tension S in the string,

$$\frac{m_1m_2}{m_1 + m_2}|\ddot{\mathbf{r}}| = S$$

But $|\ddot{\mathbf{r}}|$ is just $v_0{}^2/b$, where v_0 is the velocity of m_2 relative to m_1 and b is the length of the string. Therefore the tension becomes

$$S = \frac{m_1m_2}{m_1 + m_2}\frac{v_0{}^2}{b}$$

a result which agrees with Eq. (8-26).

IMPULSIVE FORCES AND IMPACT

8-8. Nature of an Impulse. In certain situations in nature, forces may act upon a system for a very short time, but while they are acting, they may be of a sufficient magnitude to affect significantly the subsequent motion of the system. Suppose that a particle of mass m, moving with an initial velocity $\mathbf{v}_1$, is acted upon by a force $\mathbf{F}$ during a small interval of time τ. The equation of motion is

$$\frac{d}{dt}(m\mathbf{v}) = \mathbf{F}$$

Multiplying by dt and integrating, we have

$$m\mathbf{v}_2 - m\mathbf{v}_1 = \int_0^\tau \mathbf{F}\,dt = \mathbf{P} \qquad (8\text{-}40)$$

in which $\mathbf{v}_2$ is the velocity of the particle at the end of the interval, at which time the force ceases to act. The integral in Eq. (8-40) has a finite limit for all forces which exist in nature. This limit, denoted by the symbol $\mathbf{P}$, is called the *impulse*. It is the momentum imparted to the system (in this case a particle) by the force during the time τ in which the force acts. In practical problems it is usually the rule to assume this time τ to be so small that the system does not move while the force is acting. Thus, in the present case, the particle is assumed to have its velocity instantaneously increased from $\mathbf{v}_1$ to $\mathbf{v}_2$. Since this means an infinite rate of change of momentum, it amounts to regarding **F** as an infinite force which acts for an infinitesimal time such that the limit of the integral is just **P**. Strictly speaking, no such situation exists in nature. However, in many cases such an assumption provides an end result which is a very good approximation to what actually occurs. A force which acts in this manner for a very short time is called an *impulsive force*. The resulting impulse is measured by the change of the momentum which it produces in a system. **P** is a vector which has for its rectangular components the quantities

$$P_x = \int_0^\tau F_x\,dt \qquad P_y = \int_0^\tau F_y\,dt \qquad P_z = \int_0^\tau F_z\,dt \tag{8-41}$$

and the equations of impulsive motion in component form are

$$m(\dot{x}_2 - \dot{x}_1) = P_x \qquad m(\dot{y}_2 - \dot{y}_1) = P_y \qquad m(\dot{z}_2 - \dot{z}_1) = P_z \tag{8-42}$$

It should be noticed that the discontinuous motions characteristic of many impulse problems involve in general the *dissipation* of energy. Hence the application of energy conservation in a manner which does not take into account this dissipation (usually in the form of heat) will not be expected to yield the correct result. Problems 8-33 and 8-34 provide especially good examples of this feature.

Example 8-5. A particle of charge $+e_1$ moves with a very high velocity v_0 along a straight line which passes at a distance b from a particle of mass m and charge $+e_2$. The quantities e_1 and e_2 are positive constants. It is desired to find the energy Q transferred to m during the encounter, assuming a central law of force of magnitude e_1e_2/ρ^2, where ρ is the distance separating e_1 and e_2. (Since e_1 and e_2 both have the same sign, the force is one of repulsion.)

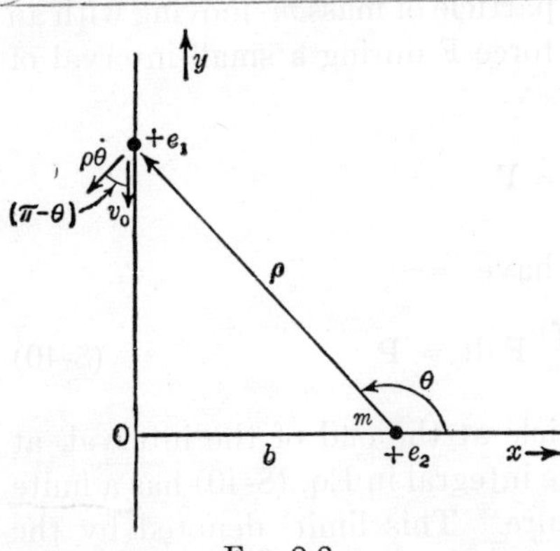

FIG. 8-8

Such a situation is typical of the interaction of a high-speed cosmic-ray particle in traversing matter. We make the assumption that very little momentum and energy are imparted to e_2 during the encounter, so that a fair approximation is to consider the path of e_1 to be a straight line, and also that v_0 is approximately constant. Let us choose this line to be the y axis, and we may place e_2 on the x axis at a distance b from the origin O (see Fig. 8-8). We assume that e_1 goes by so rapidly that e_2 does not have time to move from its position at $x = b$ during the passage. It is convenient to consider the x and y components of the impulse separately. From symmetry we see that the y component is zero, and we have only to consider the x component,

$$dP_x = F_x\,dt = -\frac{e_1e_2}{\rho^2}\cos\theta\,dt = -\frac{e_1e_2}{\rho^2}\cos\theta\,\frac{dt}{d\theta}\,d\theta \tag{8-43}$$

But $\rho\dot{\theta} = -v_0 \cos\theta$, where $\rho = -b/\cos\theta$. [The student should verify this. The minus sign follows here, and also in Eq. (8-43), since $\cos(\pi - \theta) = -\cos\theta$.] Therefore Eq. (8-43) takes the form

$$dP_x = \frac{e_1 e_2}{-bv_0} \cos\theta \, d\theta \tag{8-44}$$

In order to find P_x, we may integrate Eq. (8-44). Since the contribution at y large is very small, it will change the value of the integral very little if we integrate between the limits $y = +\infty$ to $y = -\infty$, or, what is equivalent, between $\theta = \pi/2$ and $3\pi/2$. Accordingly we have

$$P_x = -\frac{e_1 e_2}{bv_0} \int_{\pi/2}^{3\pi/2} \cos\theta \, d\theta = -\frac{e_1 e_2}{bv_0} \sin\theta \Big|_{\pi/2}^{3\pi/2} = \frac{2e_1 e_2}{bv_0} \tag{8-45}$$

This is directed along the positive x axis because all quantities in Eq. (8-45) are positive numbers. We may now compute at once the energy Q transferred from e_1 to e_2, since the impulse (8-45) is just the momentum $m\dot{x}$ imparted to e_2. Thus

$$Q = \frac{1}{2} m\dot{x}^2 = \frac{1}{2m} m^2\dot{x}^2 = \frac{2e_1{}^2 e_2{}^2}{mb^2 v_0{}^2}$$

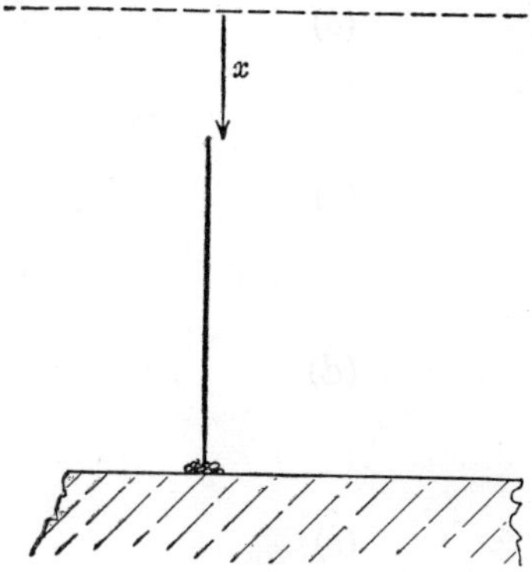

Example 8-6. A uniform heavy chain of mass m per unit length hangs vertically so that the lower end just touches a horizontal table. If it is released at the top, show that, at the time a length x of the chain has fallen, the force on the table is equivalent to the weight of a length $3x$ of the chain.

Consider that the chain has fallen a height x, as shown in Fig. 8-9, and has an instantaneous speed v. The length of chain which falls to the table, during a small interval dt following this instant, is $v\,dt$. Accordingly, the increment dp of momentum communicated by this length in coming to rest is just $m(v\,dt)v$. Thus the rate at which momentum is being communicated to the table is

Fig. 8-9

$$\frac{dp}{dt} = mv^2$$

This is equal to $2mgx$, since the fall takes place with uniform acceleration g and $v^2 = 2gx$. But this is just the force arising because of the arresting of the falling chain. Since a length x, of weight mgx, has already fallen, the total force on the table top is just $3mgx$. Now a length b of the chain has a weight mgb, and we see that the length which has a weight equivalent to the total force on the table top during the fall is just $3x$.

8-9. Impact, Elastic and Inelastic. Coefficient of Restitution. We next turn our attention to the direct impact of two moving elastic bodies such as two smooth spheres, of masses m_1 and m_2, as shown in Fig. 8-10. (In this chapter we limit ourselves to collisions involving particles. In the present example of the spheres, rotational motions are not introduced and the problem is essentially that of the collision of two particles. Collisions involving rotational motions of extended bodies will be considered in the next chapter.) These are traveling with initial velocities v_1 and v_2, both being measured positively to the right. When contact is initiated between the two spheres, deformation results and body forces are brought into play which resist this deformation. The net result of these is a force, which we may call F, acting along the line of centers and tending to push the two bodies apart. It is convenient to consider the impact as a succession of two stages. The first is that which is represented by the passage from (*b*) to (*c*) in Fig. 8-10, that is, the situation which exists

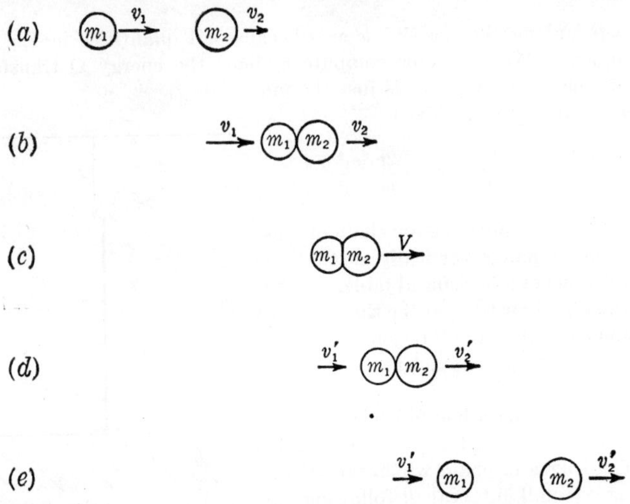

FIG. 8-10

during the interval between initiation of contact and *maximum deformation.* The instant of maximum deformation occurs when both spheres are traveling with the same velocity V. The body forces continue to act as long as deformation exists, and the second stage of the collision may be thought of as being initiated at this instant of maximum compression, ending when the two bodies m_1 and m_2 are ultimately pushed apart, moving with velocities v_1' and v_2' as shown.

It is of interest to calculate the final velocities v_1' and v_2' in terms of the initial velocities v_1 and v_2. We suppose that the first stage of the impact lasts for a time τ and that the second stage lasts from τ until a later time τ'. We consider the two stages separately. Denoting the momenta of the two spheres at any time by the symbols p_1 and p_2, we may write the equations of motion as

$$\frac{dp_1}{dt} = -F \qquad \frac{dp_2}{dt} = +F \tag{8-46}$$

Multiplying through by dt and integrating over the interval τ, we obtain

$$m_1V - m_1v_1 = -\int_0^\tau F\,dt = -P \qquad m_2V - m_2v_2 = P$$

or

$$v_1 = V + \frac{P}{m_1} \qquad v_2 = V - \frac{P}{m_2} \tag{8-47}$$

from which

$$v_1 - v_2 = P\frac{m_1 + m_2}{m_1m_2} \tag{8-48}$$

In a similar manner we may integrate Eqs. (8-46) over the interval $t = \tau$ to $t = \tau'$. Corresponding to Eqs. (8-48), we obtain

$$v_1' = V - \frac{P'}{m_1} \qquad v_2' = V + \frac{P'}{m_2}$$

where $P' = \int_\tau^{\tau'} F'\,dt$. (The possibility is considered that the force F', acting from τ to τ', may be different from F.) In analogy to Eq. (8-48), we obtain

$$v_1' - v_2' = -P'\frac{m_1 + m_2}{m_1m_2} \tag{8-49}$$

The two quantities P and P' are called, respectively, the *impulse of compression* and the *impulse of restitution*. Experiment tells us that they are related by the equation

$$P' = eP \tag{8-50}$$

where e is approximately a constant for a given pair of colliding substances. This constant e is known as the *coefficient of restitution*. Eliminating P' from Eq. (8-49) provides us with the important relation

$$v_1' - v_2' = -e(v_1 - v_2) \tag{8-51}$$

This result, known as *Newton's rule*, states that the relative velocity after collision is equal and opposite to e times the relative velocity before the collision. Since momentum is conserved during the collision, we may write another relationship among the velocities which expresses this conservation. It is

$$m_1v_1' + m_2v_2' = m_1v_1 + m_2v_2 \tag{8-52}$$

Either v_1' or v_2' may be eliminated from (8-51) by means of (8-52). This gives us the two expressions

$$\begin{aligned} v_1' &= \frac{m_1v_1 + m_2v_2 - em_2(v_1 - v_2)}{m_1 + m_2} \\ v_2' &= \frac{m_1v_1 + m_2v_2 + em_1(v_1 - v_2)}{m_1 + m_2} \end{aligned} \tag{8-53}$$

These furnish the final velocities in terms of the initial velocities, the masses, and the coefficient of restitution e. Since, in Fig. 8-10, the convention is chosen that velocities are positive to the right, and if the initial velocities are oppositely directed, appropriate account must be taken of the difference in sign when substituting in Eqs. (8-51) and (8-53). The individual directions of v_1' and v_2' are then determined by the signs of the right sides of (8-53) after the correct substitutions have been made.

The quantity e, as applied to a substance, is a measure of the elasticity of that substance. For this reason it is sometimes called the *coefficient of elasticity*. It is a dimensionless number which assumes values, depending on the substance, from zero to 1. For a perfectly inelastic substance $e = 0$. If the two spheres above were perfectly inelastic, the final stage of the collision would be that the bodies would remain in contact with maximum deformation, traveling with the velocity V. For a perfectly inelastic substance internal frictional forces exist which oppose the deformation, but, once the maximum deformation has been realized, they cease to act. There is no tendency to return to the original shape. An example of the collision of two perfectly inelastic bodies might be approximated by two balls of soft putty. For actual substances which occur in nature the values of e vary over almost the entire range from zero to 1. A highly elastic substance is exemplified by glass. The coefficient of restitution for glass on glass is in the neighborhood of 0.95. A perfectly elastic substance, an idealization, would have $e = 1$.

Example 8-7. A particle of mass m lies at the middle, A, of a hollow tube of length $2b$ and mass M. The tube, which is closed at both ends, lies on a smooth table. The coefficient of restitution between m and M is e. Let m be given an initial velocity v_0 along the tube.

a. Find the velocities of m and M after the first impact.

b. Find the loss in energy during the first impact.

c. Find the time required for m to arrive back at A traveling in the original direction.

If v_0' and V', respectively, denote the velocities of m and M after the first collision, we have from Eqs. (8-53)

$$v_0' = \frac{m - eM}{m + M} v_0 \qquad V' = \frac{m}{m + M}(1 + e)v_0$$

Initially the total kinetic energy is $\frac{1}{2}mv_0^2$. After the first collision the total kinetic energy T' is

$$T' = \frac{1}{2}MV'^2 + \frac{1}{2}mv_0'^2 = \frac{mv_0^2}{2(m + M)^2}[m^2 + mM(1 + e^2) + e^2M^2]$$

whence the change in kinetic energy ΔT is given by

$$\Delta T = T' - \frac{1}{2}mv_0^2 = -\frac{1}{2}\frac{mM}{(m + M)}(1 - e^2)v_0^2 \tag{8-54}$$

Part c is readily determined by use of Eq. (8-51). Accordingly the relative velocities before and after the first collision are v_0 and $-ev_0$. Similarly, after the second collision the relative velocity is $+e^2v_0$. Thus the total time elapsed until m is again at A and traveling in the original direction is

$$t = \frac{b}{v_0} + \frac{2b}{ev_0} + \frac{b}{e^2v_0} = \frac{b}{v_0}\left(1 + \frac{1}{e}\right)^2 \tag{8-55}$$

8-10. Motion Relative to the Center of Mass. Loss of Energy during Impact. It is useful to express the velocities of two colliding bodies, such as the billiard balls in Sec. 8-9, in terms of a reference system with origin at the center of mass. If the center of mass is not being accelerated, the equations of motion relative to the mass center are the same as Eqs. (8-46). Consequently the same subsequent procedure may be carried out, to give us, finally,

$$v_1' - v_2' = -e(v_1 - v_2) \tag{8-56}$$

which is Newton's rule, unchanged save that here the velocities are with respect to the center of mass. Relative to the center of mass the conservation of momentum is expressed by

$$m_1v_1 + m_2v_2 = 0 \qquad m_1v_1' + m_2v_2' = 0 \tag{8-57}$$

Equations (8-57) may be combined with (8-56) to yield

$$v_1' = -ev_1 \qquad v_2' = -ev_2 \tag{8-58}$$

and the final velocities are oppositely directed from the corresponding initial velocities. For the case of perfectly inelastic bodies $e = 0$, and the final situation is one of rest at the center of mass.

It is of interest to compute the loss in energy which results in a collision in which e is different from unity. The initial and final values of the kinetic energy relative to the center of mass are, respectively,

$$T_r = \tfrac{1}{2}m_1v_1^2 + \tfrac{1}{2}m_2v_2^2 \qquad T_r' = \tfrac{1}{2}m_1v_1'^2 + \tfrac{1}{2}m_2v_2'^2$$

In view of Eqs. (8-58), the second of these becomes

$$T_r' = \tfrac{1}{2}m_1e^2v_1^2 + \tfrac{1}{2}m_2e^2v_2^2 = e^2T_r \tag{8-59}$$

from which the change ΔT_r in the kinetic energy becomes

$$\Delta T_r = T_r' - T_r = -T_r(1 - e^2) \tag{8-60}$$

Thus no energy is lost if the bodies are perfectly elastic, and all the kinetic energy of motion relative to the mass center will be lost if the bodies are perfectly inelastic.

Example 8-8. (*a*) In the system of Example 8-7, determine the velocities relative to the center of mass before and after the first collision. (*b*) Perform part *b* of Example 8-7 by means of the present section. (*c*) In the system of Example 8-7, how far has the center of mass traveled during the time given by Eq. (8-55)?

Let v_1 and v_1' be the velocities of m relative to the mass center before and after the first impact and v_2 and v_2' the corresponding quantities for M. If v_c designates the velocity of the center of mass (which remains constant), we must have

$$v_c = \frac{m}{m + M} v_0$$

and

$$v_1 = v_0 - v_c = \frac{M}{m + M} v_0 \qquad v_2 = -\frac{m}{m + M} v_0$$

Thus, from Eqs. (8-58),

$$v_1' = -\frac{eM}{m + M} v_0 \qquad v_2' = +\frac{em}{m + M} v_0$$

The initial kinetic energy T_r relative to the center of mass may be written

$$T_r = \frac{1}{2} m v_1^2 + \frac{1}{2} M v_2^2 = \frac{1}{2} \frac{mM}{m + M} v_0^2 \tag{8-61}$$

a result which already could have been written down on the basis of Sec. 8-7. From Eqs. (8-60) and (8-61) the change in the kinetic energy relative to the mass center can be obtained at once and is identical to Eq. (8-54). When the number of collisions becomes very large, the final result is that the kinetic energy relative to the center of mass approaches zero, leaving only the kinetic energy T_c of translation of the mass center. This is

$$T_c = \tfrac{1}{2}(M + m) v_c^2$$

It is easy to show that this is just $(m v_0^2/2) - T_r$, as is to be expected. The distance through which the center of mass moves is $mb(1 + e)^2/(m + M)e^2$.

8-11. Generalization of Newton's Rule. Oblique Impact of Two Smooth Spheres.

Newton's rule is more general than has been stated in Sec. 8-9. It is found empirically that Eq. (8-51) applies to the components of velocity resolved along the common normal to the surfaces of the colliding bodies at the point of contact. In the case of the two spheres the velocity components involved are the components resolved along the line of centers during the contact. If we take as the x axis the line of centers during the impact, we have

$$\dot{x}_1' - \dot{x}_2' = -e(\dot{x}_1 - \dot{x}_2) \tag{8-62}$$

where $\dot{x}_1$ and $\dot{x}_1'$ are the initial and final components of the velocity of m_1 along the line of centers and $\dot{x}_2$ and $\dot{x}_2'$ are the corresponding quantities for m_2. So far as the y components of the velocities (perpendicular to the line of centers) are concerned, we can easily see that the impulse P_y will be zero provided that the surfaces are smooth, for then we have

$$\frac{d(m\dot{y})}{dt} = 0$$

for both bodies during the impact. Consequently the components of velocity perpendicular to the line of centers will be unchanged during the impact, and we have

$$\dot{y}_1' = \dot{y}_1 \qquad \dot{y}_2' = \dot{y}_2 \tag{8-63}$$

where $\dot{y}_1$ and $\dot{y}_1'$ are the initial and final values of this component of velocity for m_1 and $\dot{y}_2$ and $\dot{y}_2'$ are the corresponding values for m_2.

Example 8-9. A smooth sphere of mass m_2 is tied to a fixed point by a light inextensible string. Another sphere of mass m_1, having a velocity v_1 in a direction making an angle θ with the string, makes a direct impact with m_2. Find the velocity with which m_2 begins to move. The coefficient of restitution is e.

The situation is pictured in Fig. 8-11. Clearly, since the string AB is inextensible and under tension, the sphere m_2 is constrained to move in a circle about point A. [We take the case where e is the same between AB and m_2 as between m_1 and m_2. In this event Newton's rule applied to the impact of m_1 and m_2 reduces to the form of Eq. (8-51). See Prob. 8-41.] Furthermore, since the impulse between the spheres is along the line of centers, the direction of motion of m_1 is unaltered by the collision. It will have a final velocity v_1', say, in the *same* direction as v_1. We may write the expression for the conservation of the component of momentum at right angles to AB as

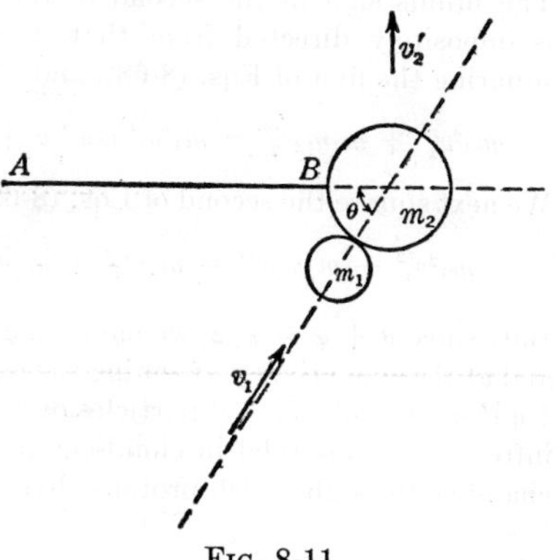

FIG. 8-11

$$m_2 v_2' + m_1 v_1' \sin\theta = m_1 v_1 \sin\theta \tag{8-64}$$

And by Newton's rule we have

$$v_1' - v_2' \sin\theta = -e v_1 \tag{8-65}$$

Combining Eqs. (8-64) and (8-65), we obtain, for the velocity v_2' of m_2 immediately after impact,

$$v_2' = \frac{(e+1)m_1 v_1 \sin\theta}{m_2 + m_1 \sin^2\theta} \tag{8-66}$$

If it is desired, we may also determine v_1', the velocity of m_1 immediately after impact, from (8-65) and (8-66). The quantity v_1' has the same line of action as v_1 and has the magnitude

$$v_1' = \frac{v_1(m_1 \sin^2\theta - e m_2)}{m_2 + m_1 \sin^2\theta}$$

An illustration of oblique impact in which it is not necessary to consider the details during the collision and in which only the initial and final momentum and energy equations need be taken into account is provided by the following example:

Example 8-10. A smooth sphere impinges on another one at rest. After the collision their directions of motion are at right angles. It is desired to show that, if they are perfectly elastic, their masses must be equal.

This problem is readily treated by means of the equations expressing the conservation of energy and momentum. In Fig. 8-12, the mass m_1, traveling at a velocity v_1, strikes the mass m_2, which is at rest. Subsequently m_1 and m_2 come off at right angles.

as shown. Let the final velocities be denoted by v_1' and v_2'. The principle of conservation of energy yields

$$\tfrac{1}{2}m_1v_1^2 = \tfrac{1}{2}m_1v_1'^2 + \tfrac{1}{2}m_2v_2'^2 \tag{8-67}$$

The principle of conservation of momentum in two perpendicular directions gives

$$\begin{aligned} m_1v_1 &= m_1v_1' \cos\theta + m_2v_2' \cos\varphi \\ m_1v_1' \sin\theta &- m_2v_2' \sin\varphi = 0 \end{aligned} \tag{8-68}$$

The minus sign in the second of these arises because the vertical component of v_2' is oppositely directed from that of v_1'. Multiplying Eq. (8-67) through by $2m_1$, squaring the first of Eqs. (8-68), and combining the two, we have

$$m_1^2v_1'^2 + m_1m_2v_2'^2 = m_1^2v_1'^2\cos^2\theta + m_2^2v_2'^2\cos^2\varphi + 2m_1m_2v_1'v_2'\cos\theta\cos\varphi \tag{8-69}$$

We next square the second of Eqs. (8-68) and add it to Eq. (8-69). We obtain

$$m_1^2v_1'^2 + m_1m_2v_2'^2 = m_1^2v_1'^2 + m_2^2v_2'^2 + 2m_1m_2v_1'v_2'(\cos\theta\cos\varphi - \sin\theta\sin\varphi)$$

But, since $\theta + \varphi = \pi/2$, we have $\sin\varphi = \cos\theta$ and $\cos\varphi = \sin\theta$. Making use of this, and at the same time performing the possible cancellations, we have left that $m_1 = m_2$. Right-angle collisions of particles of equal mass, such as those in this example, are not infrequently observed in cloud-chamber photographs, for example, a hydrogen-filled chamber through which protons (hydrogen nuclei) are passing.

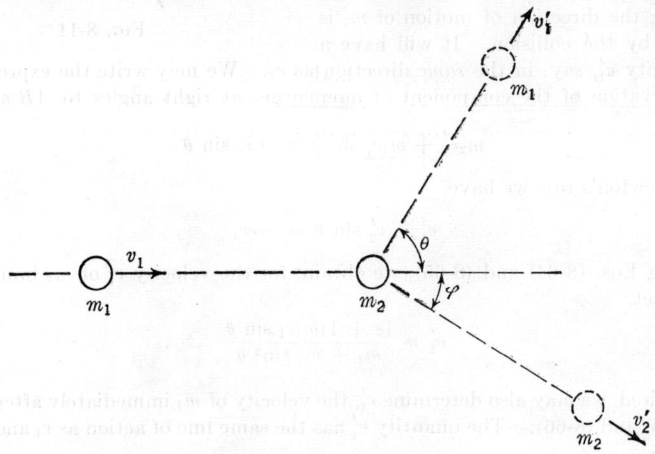

FIG. 8-12

8-12. Comparison of the Rest and Center-of-mass Systems of Coordinates for the Oblique Impact of Two Particles. In laboratory experiments involving the impact of particles, the observer views the process from a rest system (*laboratory system*). Theoretical calculations concerning the collision are usually made in terms of a *center-of-mass system* of coordinates. Accordingly it is of interest to examine the way in which the two points of view are related, especially as to the angle, with the direction of the original incident particle, at which one of the particles leaves the scene of the collision.

Let us consider that a particle of mass m_1 is incident at a velocity v_1 upon a particle of mass m_2 which is initially at rest. The velocity V of the center of mass will be in the direction of motion of the incident particle and will be such that

$$(m_1 + m_2)V = m_1 v_1$$

from which

$$V = \frac{m_1}{m_1 + m_2} v_1 \tag{8-70}$$

Consequently m_1 possesses an initial velocity V_1 relative to the center of mass, given by

$$V_1 = v_1 - V = \frac{m_2}{m_1 + m_2} v_1 \tag{8-71}$$

Likewise the initial velocity V_2 of m_2 relative to the center of mass is found to be

$$V_2 = -V = -\frac{m_1}{m_1 + m_2} v_1 \tag{8-72}$$

This is consistent with the principle of momentum conservation, which requires that, in the center-of-mass system,

$$m_1 V_1 + m_2 V_2 = 0 \tag{8-73}$$

In Fig. 8-13, parts (a) and (b) show the collision as seen, respectively, from a coordinate system the origin of which is at the center of mass and

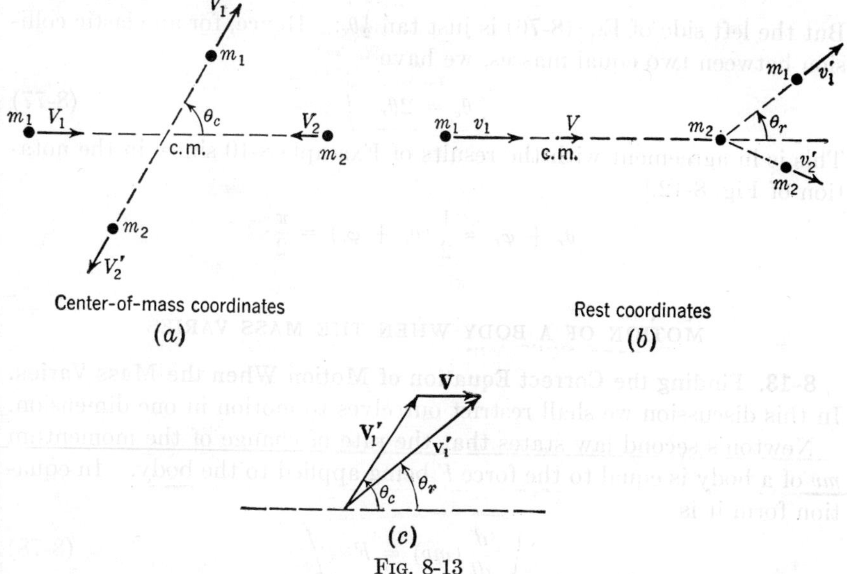

FIG. 8-13

from a coordinate system which is at rest. It is apparent, since the linear momentum relative to the mass center must remain zero after the collision, that, from the point of view of the center-of-mass coordinate system, the two particles must go in opposite directions following the impact. If the collision is an elastic one, the velocities V_1' and V_2' will be identical to V_1 and V_2, respectively. If the collision is inelastic, the energy relative to the center of mass will change (cf. Sec. 8-10).

The magnitude of the angle between the final and original directions of m_1 (θ_c in the center-of-mass system and θ_r in the rest system) is different in the two systems. Figure 8-13c reveals the way in which θ_c is related to θ_r. Clearly

$$V_1' \cos \theta_c + V = v_1' \cos \theta_r \tag{8-74}$$

$$V_1' \sin \theta_c = v_1' \sin \theta_r \tag{8-75}$$

a pair of relations from which θ_r may be expressed in terms of θ_c. (Similar equations are found for the case of m_2.) Equations (8-74) and (8-75) obtain for both elastic and inelastic collisions, the velocities V_1' and v_1' being determined from energy and momentum considerations, due account being taken of the energy loss in the event of an inelastic collision.

The relation between θ_c and θ_r becomes especially simple if the impact is elastic and with $m_1 = m_2$, for then $V_1 = V = V_1'$. Consequently, dividing Eq. (8-75) by Eq. (8-74), we obtain

$$\frac{\sin \theta_c}{\cos \theta_c + 1} = \tan \theta_r \tag{8-76}$$

But the left side of Eq. (8-76) is just $\tan \frac{1}{2}\theta_c$. Hence, for an elastic collision between two equal masses, we have

$$\theta_c = 2\theta_r \tag{8-77}$$

This is in agreement with the results of Example 8-10 since, in the notation of Fig. 8-12,

$$\theta_r + \varphi_r = \frac{1}{2}(\theta_c + \varphi_c) = \frac{\pi}{2}$$

MOTION OF A BODY WHEN THE MASS VARIES

8-13. Finding the Correct Equation of Motion When the Mass Varies. In this discussion we shall restrict ourselves to motion in one dimension.

Newton's second law states that the rate of change of the momentum mv of a body is equal to the force F being applied to the body. In equation form it is

$$\frac{d}{dt}(mv) = F \tag{8-78}$$

If the mass m is changing, caution must be exercised in the use of Eq. (8-78). A good example of the precautions which must be observed is provided by the case of a box which is filled with sand and is sliding down a smooth hill, given that the sand is leaking out at a constant rate. If the inclination of the hill is α, the equation of motion of the box plus all the sand, whether spilled or not, is

$$m \frac{dv}{dt} = mg \sin \alpha \tag{8-79}$$

where v is taken positive down the hill. Here, the mass m is not changing, since the entire system consists of the box plus all the sand, and Eq. (8-79) refers to the motion of the center of mass. The sand that has already spilled out continues down the hill with the same instantaneous velocity as the box and with the same acceleration.

Another way of looking at the situation is to notice that the motion of the box is not changed by the leaking sand since the sand, in spilling out, imparts no momentum to the box. We may study this in detail in the following way: Suppose that we now take m to be the mass, at any time, of the box plus the sand remaining in the box. At a time t the momentum of the box and its contents is mv. During an interval dt following this, the mass changes from m to $m - dm$, and the velocity changes from v to $v + dv$. The initial momentum is mv, and the momentum of the box plus contents at the end of the interval dt is $(m - dm)(v + dv)$. The momentum carried away by the mass dm is $v\,dm$. Accordingly, since the total change in momentum over the interval dt must be equal to the impulse $F\,dt$, where $F = mg \sin \alpha$, we have

$$v\,dm + (m - dm)(v + dv) - mv = mg \sin \alpha \, dt$$

or

$$v\,dm + mv - v\,dm + m\,dv - dm\,dv - mv = mg \sin \alpha \, dt \tag{8-80}$$

Neglecting infinitesimals of the second order, performing the possible cancellations in Eq. (8-80), and dividing by dt, we have

$$m \frac{dv}{dt} = mg \sin \alpha$$

as the equation of motion of the box and its contents. Clearly this is the same result as Eq. (8-79).

A different result is obtained if, for example, we assume that the escaping sand is thrown out in a manner such as always to reduce the sand to rest. This will be accomplished if the element dm is given a velocity $-v$, that is, a velocity in the direction opposite to that of the box and equal in magnitude to that of the box. In this case the element dm of sand will possess zero momentum at the end of the interval dt. Accordingly the principle of the conservation of momentum gives us

$$(m - dm)(v + dv) - mv = mg \sin \alpha \, dt \tag{8-81}$$

which yields

$$m \frac{dv}{dt} - v \frac{dm}{dt} = mg \sin \alpha \tag{8-82}$$

This may be rewritten, for purposes of interpretation, as

$$m \frac{dv}{dt} = v \frac{dm}{dt} + mg \sin \alpha \tag{8-83}$$

and the first term on the right may be regarded as the rate at which momentum is being imparted to the box by the sand being ejected. In Eqs. (8-81) to (8-83), dm and dm/dt are both taken as positive quantities, due account of the decrease in m being taken by means of the minus sign in $m - dm$.

Example 8-11. Work out the case represented by Eq. (8-82), finding the speed of the box as a function of time. Assume that at any time the mass of the box plus contents is a quantity m and that it had an initial value m_0, the sand being ejected at a constant rate b. Assume that the box starts from rest.

The mass m may be expressed as a function of time

$$m = (m_0 - bt)$$

Hence, substituting this in Eq. (8-80), we have [since the minus sign is already taken into account in Eqs. (8-81) and (8-82), $dm/dt = +b$, for purposes of substitution]

$$(m_0 - bt) \frac{dv}{dt} - bv = (m_0 - bt) g \sin \alpha \tag{8-84}$$

Equation (8-84) is a linear equation of the first order, with v as the dependent variable and t as the independent variable. Moreover it is an exact equation and may be rewritten

$$\frac{d}{dt} [(m_0 - bt)v] = (m_0 - bt) g \sin \alpha$$

This may be integrated at once. Since $v = 0$ at $t = 0$, we obtain

$$v = \frac{(m_0 - bt/2) tg \sin \alpha}{m_0 - bt}$$

8-14. Motion of a Rocket-propelled Body. A familiar example of this type of device is the ordinary rocket. The assumption is made that the burned gas issues from the rear of the rocket with a fixed velocity v_0 relative to the rocket. Accordingly we let m denote, at any time t, the mass of the rocket plus that of the unburned fuel, and during the subsequent interval dt a mass dm will be ejected to the rear with a velocity v_0 relative to the rocket. At the end of the interval dt the rocket plus the unburned fuel has the momentum $(m - dm)(v + dv)$, and the momentum of the mass dm is $(v - v_0)\,dm$. The momentum at the beginning of the interval is mv. The change in the momentum must be equated to the impulse contributed by the other forces. Thus we have

$$(v - v_0)\,dm + (m - dm)(v + dv) - mv = F\,dt$$

where the forces acting on the rocket are contained in F. Performing all possible cancellations and dividing through by dt, we have

$$m\frac{dv}{dt} - v_0\frac{dm}{dt} = F \tag{8-85}$$

where we have neglected second-order infinitesimals. Equation (8-85) is the equation of motion of the rocket. Information concerning F may be found in reference texts dealing with rockets. It is found to contain such quantities as the weight mg of the rocket, whatever aerodynamic forces may exist, and the quantity $(P - P_0)A$, where P is the pressure of the exit gases over the area of the exit surface, P_0 is the static atmospheric pressure, and A is the area of the exit nozzle. The factor $(P - P_0)A$, together with the term $v_0\dot{m}$ in Eq. (8-85), may be determined by a static measurement of the thrust on the rocket. The thrust is

$$v_0\dot{m} + (P - P_0)A$$

Problems

Problems marked C are reprinted by kind permission of the Cambridge University Press.

8-1. A projectile is fired at an angle of 60° with the horizontal with an energy E. At the top of its path it explodes with equal energy E into two fragments of equal mass. After the explosion one of the fragments is observed to be moving directly upward. What is the velocity of each fragment? What angle with the horizontal is made by the initial velocity of the other fragment?

8-2. A hammer of 1 ton weight falls from a height of 16 ft on the end of a vertical pile weighing 500 lb and drives it $\frac{1}{2}$ in. deeper into the ground. Assuming the resisting force to be constant, find its amount and the direction of its action.

8-3. An inelastic pile of mass m is driven vertically into the ground a distance s at each blow of a hammer of mass M. Find the average resistance of the ground to penetration by the pile.

8-4C. A gun of mass M fires a shell of mass m horizontally, and the energy of the explosion is such as would be sufficient to project the shell vertically to a height h. Show that the velocity of the recoil of the gun is

$$\left[\frac{2m^2gh}{M(M + m)}\right]^{\frac{1}{2}}$$

Assume the gun barrel is rigidly attached to the carriage.

8-5C. A shell of mass $m_1 + m_2$ is fired with a velocity whose horizontal and vertical components are U, V, and at the highest point in its path the shell explodes into two fragments m_1, m_2. The explosion produces an additional kinetic energy E, and the fragments separate in a horizontal direction. Show that they strike the ground at a distance apart which is equal to

$$\frac{V}{g}\left[2E\left(\frac{1}{m_1}+\frac{1}{m_2}\right)\right]^{\frac{1}{2}}$$

8-6. A bullet of mass m_1 is fired with velocity v_1 at a body of mass m_2 advancing with velocity v_2 and emerges with velocity u_1, the direction of v_1, v_2, and u_1 being along the same straight line. If the length perforated by the bullet is L, find the average resistance of the body to the bullet.

8-7. Two masses m_1 and m_2, connected by a massless spring of force constant k, are at rest in equilibrium on a smooth horizontal table. Suddenly a velocity v is imparted to m_1 in a direction away from m_2. Find (*a*) the velocity of m_2 relative to the table at the instant the spring becomes again unstretched, (*b*) the velocity of the center of mass, and (*c*) the period of the oscillatory motion.

8-8. A block of wood of mass M rests on a smooth horizontal table. A bullet of mass m traveling with a velocity v embeds itself in the block. How much energy is lost? Also, if the bullet experiences a constant retarding force f as it penetrates the block, how deeply will it penetrate?

8-9*C*. A uniform chain of length $2a$ hangs in equilibrium over a smooth peg. If it is started from rest, prove that its velocity when it is leaving the peg is $\sqrt{ga}$.

8-10*C*. Two particles m_1 and m_2 are connected by a string passing over a smooth pulley, as in Atwood's machine. Prove that, if the inertia of the pulley is neglected, the mass center of the particles has a downward acceleration

$$\left(\frac{m_1 - m_2}{m_1 + m_2}\right)^2 g$$

8-11*C*. A man of mass m is standing in a lift of mass M, which is descending with velocity V, the counterweight being of mass $M + m$. Suddenly the man jumps with an energy which would raise him to a height h if he were jumping from the ground. Calculate the velocities of the man and the lift immediately after he jumps. Find also their accelerations. Find the height, relative to the lift, to which the man jumps.

not **8-12.** In the previous problem consider the lift to be initially at rest. Find the height, from the initial level of the lift, to which the man can jump.

8-13*C*. Two equal flat scale pans are suspended by an inextensible string passing over a smooth pulley so that each remains horizontal. An elastic sphere falls vertically, and, when its velocity is u, it strikes one of the scale pans and rebounds vertically. Show that the sphere takes the same time to come to rest on the scale pan as it would if the scale pan were fixed.

8-14. A steel block of mass M initially at rest is fired upon by bullets each having a mass m, a velocity v_0, and a coefficient of restitution e. Assuming that there is no friction and that the bullets strike the surface of the block normally at the rate of N per second, find the average initial force on the block. Find also the velocity of the block at any time t after the firing has begun. What is the maximum velocity that the block can attain? Take $M \gg m$.

8-15. A perfectly elastic ball weighs 4 oz and is discharged in a manner such that it hits a smooth wall horizontally at an angle of incidence of 45° at a point 6 ft above the ground. It rebounds and reaches the ground at a point 17.2 ft measured horizontally from the wall. Find the velocity with which it left the wall.

8-16. A ball of mass m and a ball of unknown mass approach each other from opposite directions and have the same velocity v_0 (but oppositely directed). The ball having the unknown mass is reduced to rest by the impact, while the ball of mass m is not. What is the mass of the unknown, and what is the final velocity of the ball of mass m in terms of v_0? Consider the impact to be perfectly elastic.

8-17*C*. A chain rests across a smooth circular cylinder whose axis is horizontal, its length being equal to half the circumference. Prove that, if it be slightly disturbed, its velocity when a length $a\theta$ has slipped over the cylinder will be

$$\left\{\frac{ga}{\pi}[\theta^2 + 2(1 - \cos\theta)]\right\}^{\frac{1}{2}}$$

where a is the radius of the cylinder.

8-18. A uniform heavy rope is coiled up on a smooth horizontal table. If one end is raised by hand with a uniform velocity v_0, show that when this end is at a distance y above the table the force on the hand is equal to the weight of a length $y + v_0^2/g$ of the rope.

8-19*C*. Two equal spheres each of mass m are in contact on a smooth horizontal table, and a third equal sphere of mass m' impinges symmetrically on them. Prove that this sphere is reduced to rest by the impact if $2m' = 3me$, where e is the coefficient of restitution, and find the loss of kinetic energy by the impact.

8-20. A ball of mass m moving with a speed v strikes a ball of equal mass which is initially at rest. The line of contact of the balls makes an angle α with the initial velocity of the first ball. The balls are smooth, and the coefficient of restitution is e. Find the directions and magnitudes of the velocities after impact.

8-21. A ball of mass M initially at rest is struck by a ball of mass m having an energy E. After the collision, the ball of mass m is observed to come away from the impact in a direction which is at right angles to the original path. What is the energy of the ball of mass M after the impact? Consider the impact to be perfectly elastic and all motion to be in a horizontal plane. M and m are smooth.

8-22. A raindrop falls through a uniform fog, sweeping out mist (that is, collecting mass) from the region it traverses. If it starts from rest with zero radius and remains at all times spherical, deduce that the acceleration with which it falls is $g/7$.

8-23. A particle of mass M rests in the bottom of a hemispherical bowl. Another lighter particle of mass m is permitted to slide from rest down the side of the bowl, being released from the edge (the coefficient of sliding friction is μ). After the two particles collide, the ligher one is observed to rebound halfway up the side of the bowl, that is, at the maximum position the line joining the particle of mass m with the center of curvature of the bowl makes an angle of 45° with the vertical. Calculate the coefficient of restitution between the particle of mass M and the particle of mass m.

8-24. A gun of mass M is mounted on frictionless wheels which rest on a horizontal surface. When the muzzle is pointed directly upward, the gun is able to project a shell to a vertical height h above the ground. If the gun is shifted to an angle of elevation α with the horizontal and a projectile of the same mass m is fired by means of an identical powder charge, calculate the range of the projectile in terms of M, m, h, α. The barrel of the gun is rigidly attached to the carriage, which may recoil freely over the horizontal surface.

8-25*C*. From a gun of mass M which can recoil freely on a horizontal platform a shell of mass m is fired, the angle of elevation of the gun being α. Determine the direction in which the shell is moving when it leaves the gun. Also show that, if the shell strikes at right angles the plane that passes through the point of projection and is inclined to the horizontal at an angle β, then

$$\tan\alpha = \frac{M(\cot\beta + 2\tan\beta)}{M + m}$$

8-26*C*. A smooth wedge of mass M and angle α is free to move on a smooth horizontal plane in a direction perpendicular to its edge. A particle of mass m is projected directly up the face of the wedge with velocity V. Prove that it returns to the point on the wedge from which it was projected after a time

$$\frac{2V(M + m \sin^2 \alpha)}{(m + M)g \sin \alpha}$$

Also find the force between the particle and the wedge at any time.

8-27*C*. A particle of weight 2 lb is placed on the smooth face of an inclined plane of weight 7 lb and slope 30°, which is free to slide on a smooth horizontal plane in a direction perpendicular to its edge. Show that, if the system starts from rest, the particle will slide down a distance of 15 ft along the face of the plane in 1.25 sec.

8-28. A ball of mass m slides with a velocity v across a smooth table, striking a ball of mass M of the same radius and initially at rest. At the instant of impact the line of centers makes an acute angle α with the original direction of motion, and, immediately following the impact, the ball of mass m goes off in a direction perpendicular to the line of centers. Determine the coefficient of restitution. M and m are smooth.

8-29. A particle is projected from a point on a smooth floor such that a horizontal distance R_0 is traversed before it again strikes the floor. The particle then rebounds from the floor, the coefficient of restitution being e. Find, in terms of R_0 and e, the total horizontal distance traversed while the particle is in the air. If we are now given that initially the velocity of the particle was v_0 and the angle of elevation β, what is the final velocity of the particle?

8-30. A uniform flexible cord of length b and weight W hangs in the equilibrium configuration over a smooth peg. If the equilibrium is disturbed, find the force on the peg when the length of rope hanging on one side of the peg is x. Express the result in terms of W, x, b.

8-31. A box of mass M slides from an initial position of rest down a smooth hill of inclination α. The box contains sand which is being ejected at a constant rate σ g/sec in the forward direction with a speed v_0 relative to the box. Initially the mass of the sand was m_0. Find the speed of the box as a function of time.

8-32. A racing car with rocket-type propulsion weighs M lb without fuel and carries an initial load of M_0 lb of fuel. The action of the rocket motor can be described as that of ejecting its mass at a rate of k lb/sec ($k = dm/dt$) with a velocity V relative to the rocket. Find the acceleration as a function of time. Find the velocity of the rocket at the time all the fuel has been ejected.

8-33. A uniform chain of length L and linear density ρ lies coiled in a heap at the edge of a smooth table. The chain slides over the edge in such a manner that each link remains at rest until it falls off. At the instant when the last link falls off, determine (*a*) the velocity of the chain, (*b*) the total time elapsed, and (*c*) the amount of energy dissipated.

8-34. A uniform chain of length L and linear density ρ is lying in a heap on a table when a force equal to its total weight is exerted upward on one end. (*a*) Derive the differential equation of motion which obtains while the chain is being elevated from the table. (*b*) Find the speed as the last link leaves the table. (*c*) Before the last link has left the table, what is the rate at which mechanical energy is being converted into heat? (*d*) When the upper end is at a height $2L$ above the table top, find the potential energy, the kinetic energy, and the work done by the force.

8-35. In order to dissociate a hydrogen molecule into two atoms, a minimum of 4.4 electron volts must be supplied. What is the least energy a bombarding proton must have in order to cause this reaction, the molecule being initially at rest? One electron volt = 1.60×10^{-12} erg, and the mass of a proton ($\simeq$ mass of a hydrogen atom) = 1.67×10^{-24} g.

8-36. Find the equation of motion of a rocket, projected vertically upward in a uniform gravitational field, in terms of its instantaneous mass and the speed of the escaping gases relative to the rocket. Neglect atmospheric resistance. Find the required ratio of the initial mass of fuel to the final mass of the empty rocket if a final speed of 7 mi/sec is desired. Take the relative speed of the escaping gas as 1 mi/sec and the fuel mass loss per second as one-sixtieth of the initial fuel mass.

8-37. A particle of mass m traveling with a speed v strikes a smooth wall at an angle θ with the normal to the surface. The coefficient of restitution is e. Find the recoil speed of the particle and the angle at which it rebounds.

8-38. The same as the previous problem save that the contact is imperfectly rough, with coefficient μ of sliding friction.

8-39. An ion of mass M and charge $+e$ is held at an equilibrium position by a three-dimensional potential which may be considered equivalent to a spring of force constant k. The ion experiences weak damping forces proportional to the velocity, with proportionality constant R. (This is similar to the state of binding of an ion in a crystal lattice.)

A charged particle of mass m and charge $+q$ moves with a velocity V. It has an impact parameter b with the equilibrium position of the ion. Assume that b is much larger than the displacement of the ion from equilibrium at all times and that the incident particle is essentially undeflected by the ion. Find an approximate value for the maximum displacement of the ion from equilibrium under the following conditions: (*a*) V is very large; (*b*) V is very small. Estimate the magnitude of the lowest V and of the highest V, respectively, for which the approximations of (*a*) and (*b*) are still valid.

8-40. In Secs. 8-2 and 8-7 permit the masses to depend explicitly upon the time. Show, as the result, the complications which enter Eqs. (8-13) and (8-37).

8-41. In the system of Example 8-9, and employing a procedure similar to that in Sec. 8-9, show that Newton's rule governing the impact between m_1 and m_2 is as given by Eq. (8-65) provided that the impulses of compression and restitution at point B are related through the same coefficient e as that between m_1 and m_2.

CHAPTER 9

MOTION OF A RIGID BODY IN A PLANE

9-1. Specification of a Rigid Body. The configuration of a system of particles with reference to an arbitrary set of axes in space can be specified by stating the positions of all the particles. This means that, if there are n particles present, $3n$ coordinates are required to locate the system completely. The situation becomes simpler if the particles comprise a *rigid body,* since in that case the distance between any pair of particles remains fixed. Accordingly the position of a rigid body can be completely specified by a much smaller number of coordinates, six, to be exact.

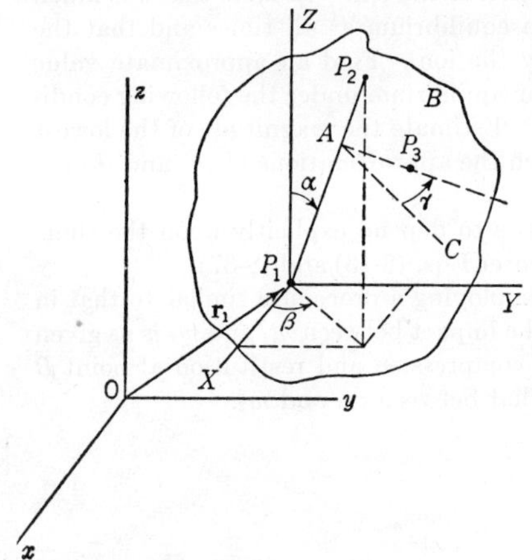

FIG. 9-1

In Fig. 9-1, B is a rigid body the position of which we desire to specify with respect to the system of axes $Oxyz$. In order to accomplish this, it is necessary to select three noncollinear points P_1, P_2, and P_3 of the body. The coordinates of P_1 in the $Oxyz$ system are x_1, y_1, and z_1 and provide three of the six coordinates necessary to describe completely the position of the body. The remaining three, usually angles, are introduced by the process of stating the orientation of the body about P_1. Let P_1X, P_1Y, and P_1Z be drawn through P_1 and parallel, respectively, to Ox, Oy, and Oz. We next draw a straight line P_1P_2. The orientation of P_1P_2 in space is obtained by stating, for example, the angles α and β as shown. The angle α is the angle which P_1P_2 makes with the line P_1Z, and β is the angle which the projection of P_1P_2 in the XY plane makes with P_1X.

The information is not yet sufficient to specify the body B completely, however, since it may rotate about P_1P_2 as an axis without changing x_1, y_1, z_1, α, or β. The final step is accomplished by making use of the third point P_3, which does not lie along P_1P_2. We draw the line P_3A perpendicular to P_1P_2 at A and the line AC perpendicular to P_1P_2 and lying in the ZP_1P_2 plane. The angle γ between AC and AP_3 supplies the final information as to the orientation of the body about the axis P_1P_2. Hence the six coordinates required to specify the position of the rigid body B are x_1, y_1, z_1, α, β, and γ. It will be seen later that the angles α, β, and γ are not the angles usually chosen but are merely a convenience in explanation at this point.

The foregoing discussion was concerned with the specification of the position and orientation of a rigid body in three dimensions. In the present chapter we shall investigate a very much simpler situation, the motion of a rigid body *parallel to a fixed plane,* and moreover we shall be interested only in those cases in which the axis of rotation remains perpendicular to the latter. However, since the rigid body may execute translational motion parallel to the plane, the point of intersection of the axis of rotation with the plane may move. For relatively simple problems of this type, three coordinates are sufficient to specify the body, two for translation and one for rotation.

KINEMATICS OF A RIGID BODY

9-2. General Displacement of a Rigid Body. The most general finite displacement of a rigid body is composed of a pure translation plus a rotation about a suitably chosen point. We are able to verify this statement in a simple two-dimensional case by means of Fig. 9-2. In Fig. 9-2 an L-shaped rod is to be displaced from an initial configuration X to a final one Y. A means of accomplishing this end is given by employing A as a reference point, first performing the translation S_A from A to A', such that the rod is then in the configuration Y', followed by the rotation θ about A' as a center. The rotation carries the rod from the configuration Y' to that of Y. A second possibility is to use B as a reference point, executing the translation S_B from B to B'', followed by the same rotation θ about B'' as the center of rotation. In this procedure the rod is carried from the configuration X to Y'' and then to Y. Clearly it is possible to accomplish the end result in an infinite variety of ways. Also it is worthwhile to notice that, although the translations S_A and S_B in the above two cases are not equal, they both may be followed by the same rotation.

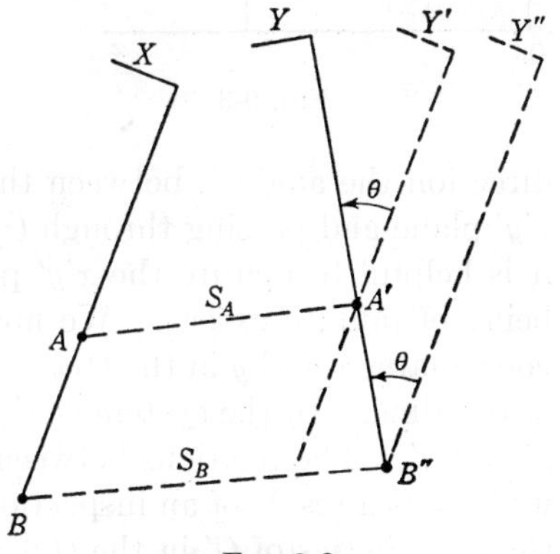

FIG. 9-2

It is evident from these considerations that a two-dimensional change in the location and orientation of a rigid body may be accomplished by a single translation followed by a single rotation (or vice versa), and, moreover, the rotation may be taken about any point of the body. The only quantity which is affected by the choice of the center of rotation is the translation. The fact that the most general finite displacement of a rigid body is composed of a translation plus a rotation may be shown to be true also in three dimensions.

9-3. Space and Body Centrodes for Velocity. Point of Instantaneous Rest. Of frequent interest in connection with a rigid body moving parallel to a fixed plane is the point of the body which is instantaneously fixed in the coordinate system which is at rest. Such a point always exists for this type of motion and readily may be found. For the case of a pure translation the point is at infinity. In Fig. 9-3 let us consider that the body is moving in the xy plane of a system of axes Oxy which are at rest. Selecting any point O' of the body, let it be the origin of a system of coordinates $O'x'y'$, such that the xy and $x'y'$ planes are parallel, and with the $O'x'y'$ system rigidly attached to the body and rotating with it. In Fig. 9-3, the $x'y'$ plane is rotating in the counterclockwise sense, in which direction the angle θ, between the $O'x'$ axis and a horizontal line, in the $x'y'$ plane and passing through O', is positive. For reasons given below, it is helpful to picture the $x'y'$ plane, which is attached to the body, as being of infinite extent. We now select a point P of the body, having coordinates x and y in the Oxy system. In general, x and y are functions of the time. In the system $O'x'y'$ the coordinates of P are the constants x' and y'. The relations between the two systems of coordinates may be written as a result of an inspection of Fig. 9-3. We have, if x_0 and y_0 are the coordinates of O' in the Oxy system, that the coordinates x and y of point P are

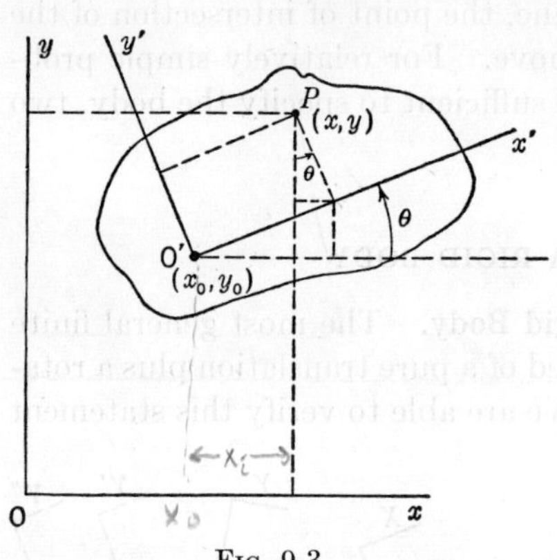

FIG. 9-3

$$\begin{aligned} x &= x_0 + x' \cos\theta - y' \sin\theta \\ y &= y_0 + x' \sin\theta + y' \cos\theta \end{aligned} \tag{9-1}$$

The corresponding inverse relationships are

$$\begin{aligned} x' &= (x - x_0)\cos\theta + (y - y_0)\sin\theta \\ y' &= -(x - x_0)\sin\theta + (y - y_0)\cos\theta \end{aligned} \tag{9-2}$$

The components of the velocity of point P may be computed by means of Eqs. (9-1). We have, considering the x coordinate first,

$$\dot{x} = \dot{x}_0 - x'\dot{\theta} \sin \theta - y'\dot{\theta} \cos \theta = \dot{x}_0 - \dot{\theta}(y - y_0) \tag{9-3}$$

The quantity $\dot{x}_0$ is the x component of the velocity of the moving origin, and $\dot{\theta}$ is the angular velocity of rotation of the body. In a similar manner the y component of the velocity of P is found to be

$$\dot{y} = \dot{y}_0 + \dot{\theta}(x - x_0) \tag{9-4}$$

If point P is to be at rest at the instant being considered, we require that both $\dot{x}$ and $\dot{y}$ in Eqs. (9-3) and (9-4) shall vanish. The values of x and y thus given by these equations are the coordinates of the point of the body which is at rest in the Oxy system at the instant under consideration. From Eqs. (9-3) and (9-4), setting $\dot{x} = \dot{y} = 0$, we have

$$x = x_0 - \frac{\dot{y}_0}{\dot{\theta}} \qquad y = y_0 + \frac{\dot{x}_0}{\dot{\theta}} \tag{9-5}$$

as the coordinates of the point of the body which is instantaneously at rest in the Oxy system. Such a point is often called the *instantaneous center* for the velocity. Furthermore the point is unique, since Eqs. (9-5) determine but one point. Thus there is but one instantaneous center. It should be apparent also that this point need not actually be in the body but may be at some point of the $x'y'$ plane external to the body (see below, the case of the sliding sphere). This latter property applies, as well, to the reference point x_0, y_0; it does not have to be situated on the body. Finally, it is clear that Eqs. (9-5) are true whether or not $\dot{x}_0$, $\dot{y}_0$, and $\dot{\theta}$ are constants.

The locus of the instantaneous center in the Oxy system, the curve traced out by the instantaneous center in space, may be obtained from Eqs. (9-5). This curve is called the *space centrode* (for the velocity). Of equal interest is the locus of the instantaneous center in the $O'x'y'$ system. The latter is called the *body centrode* (for the velocity) and is the curve traced out by the instantaneous center in the $O'x'y'$ system. The equation of the body centrode may be found by substituting x and y, as given by Eqs. (9-5), into Eqs. (9-2) and then suitably combining the latter.

Example 9-1. Find the instantaneous center in the case of a homogeneous sphere of radius a rolling across a perfectly rough horizontal table.

In Fig. 9-4, take the axis Ox to lie in the table top and the sphere to be rolling to the left with a velocity v_0. Since the sphere is rolling without sliding we have that $a\dot{\theta} = v_0$. Let us take the center of the sphere to be the origin of the $O'x'y'$ system. Accordingly, from Eqs. (9-5) the coordinates of the instantaneous center are (since

$\dot{y}_0 = 0$ here)

$$x = x_0$$

$$y = y_0 - \frac{v_0}{v_0/a} = y_0 - a = 0 \tag{9-6}$$

From Eqs. (9-6) the space centrode is the x axis itself. We may obtain the equation of the body centrode by substituting the values of $x - x_0$ and $y - y_0$, given by (9-6), in Eqs. (9-2). We obtain

$$x' = -a \sin \theta \qquad y' = -a \cos \theta$$

from which, after squaring and adding,

$$x'^2 + y'^2 = a^2$$

This is the equation of a circle of radius a with its center at the center of the sphere. Accordingly, for this simple case the body centrode is the periphery of the sphere itself.

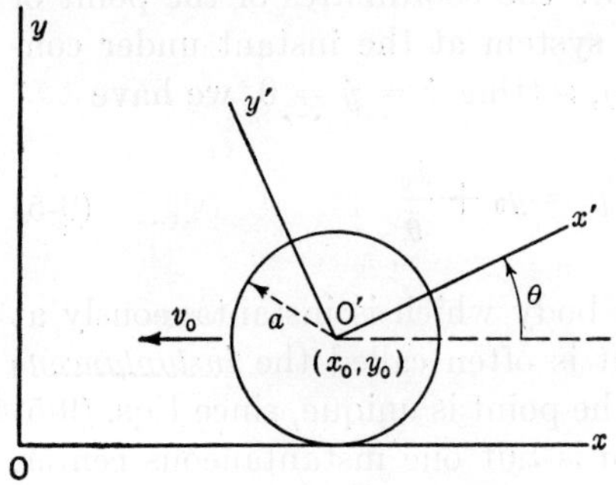

FIG. 9-4

Example 9-2. Take the case of the sphere of Example 9-1, save that the table is no longer perfectly rough and that there is sliding present. Consider that the center of the sphere has a velocity of magnitude v_0 to the left but is rotating with an angular velocity ω_0 in a clockwise sense, or in the direction of θ decreasing. Find the space and body centrodes.

Again employing Fig. 9-4, but remembering that here we no longer have $a\omega_0 = v_0$, we find the coordinates of the instantaneous center to be

$$x = x_0 \qquad y = \frac{-v_0}{-\omega_0} + y_0 = \frac{v_0}{\omega_0} + a \tag{9-7}$$

a point distant $a + (v_0/\omega_0)$ vertically above the point of contact of the sphere with the table. Thus the space centrode is a straight line parallel to, and at a height of $a + (v_0/\omega_0)$ above, the x axis. Substituting these values of x and y in Eqs. (9-2), we obtain

$$x' = \frac{v_0}{\omega_0} \sin \theta \qquad y' = \frac{v_0}{\omega_0} \cos \theta$$

from which the body centrode is

$$x'^2 + y'^2 = \frac{v_0^2}{\omega_0^2}$$

This is a circle of radius v_0/ω_0 about the center of the sphere and is in contact with the space centrode.

The results of Examples 9-1 and 9-2 are characteristic of a general situation. Kinematically speaking, the motion of a rigid body in a plane may be interpreted in terms of the rolling of the body centrode along the space centrode.

9-4. Infinitesimal Rotations and the Angular-velocity Vector. In contrast to the vector representation of finite linear displacements, it is not possible to represent *finite rotations*, such as θ in Fig. 9-2, by means of vectors. Finite rotations cannot be represented by directed segments which may be combined according to the parallelogram law. Consider Fig. 9-5. The unit coordinate vectors $\mathbf{i}$, $\mathbf{j}$, and $\mathbf{k}$ are initially oriented as shown in Fig. 9-5a. The rotations (1) and (2), of $\mathbf{k}$ into $\mathbf{i}$ and $\mathbf{i}$ into $\mathbf{j}$, are

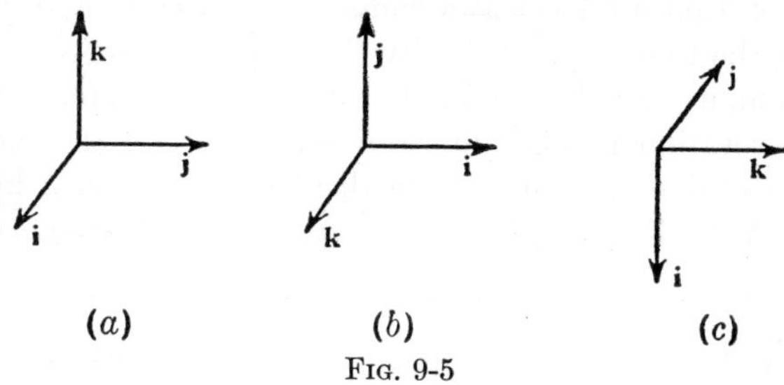

FIG. 9-5

to be taken. Both are rotations through an angle $\pi/2$. If (1) and (2) are taken in that order, the result is as shown in Fig. 9-5b. If, however, the order is reversed, the result is Fig. 9-5c, which is entirely different. Thus the order of the operations affects the result, a property incompatible with the parallelogram law of addition (see Sec. 1-6).

The situation is different for *infinitesimal rotations*. In Fig. 9-6 let OQ be the axis of rotation of a rigid body. At a given instant the position of a volume element dV at point P of the body is specified by the radius vector $\mathbf{r}$ from O as shown. During the subsequent interval of time dt, the body rotates through a small angle $d\theta$, after which dV is located at a new point P', the position of which is specified by the vector $\mathbf{r} + d\mathbf{r}$. The infinitesimal $d\mathbf{r}$ is to be interpreted as a small vector pointing perpendicularly to the plane of OQP. From the geometry of Fig. 9-6 it is evident that the magnitude $|d\mathbf{r}|$ of the infinitesimal vector $d\mathbf{r}$ may be written

$$|d\mathbf{r}| = r \sin \varphi \, d\theta \tag{9-8}$$

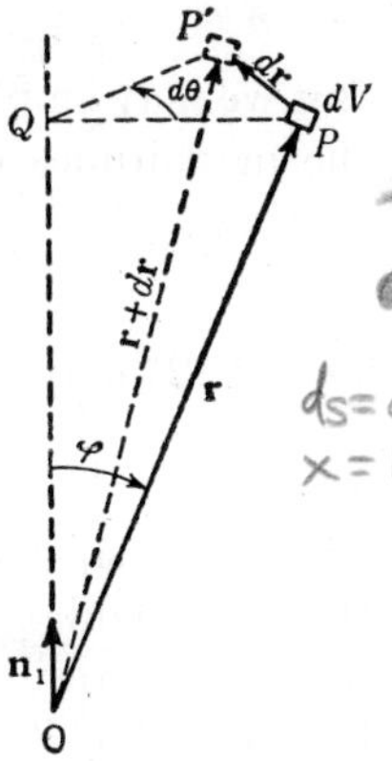

FIG. 9-6

But knowing that the direction of $d\mathbf{r}$ is perpendicular to OQP, and in view of its magnitude being given by (9-8), we see that both the direction and magnitude of $d\mathbf{r}$ may be realized by regarding $d\mathbf{r}$ to be the cross product, with the radius vector $\mathbf{r}$, of a vector of magnitude $d\theta$ pointing along OQ. Thus, if we take $\mathbf{n}_1$ to be a unit vector directed along OQ, we may write

$$d\mathbf{r} = \mathbf{n}_1 \, d\theta \times \mathbf{r} \tag{9-9}$$

an expression which gives $d\mathbf{r}$ correctly, both in direction and in magnitude. Equation (9-9) makes the useful suggestion that the infinitesimal rotation $d\theta$ may be represented by a vector pointing along the axis of rotation in the direction of advance of a *right-hand screw* turning in the sense of $d\theta$. In order to test this supposition further, we need to determine whether or not two such infinitesimal rotations may be added vectorially. Consider two successive infinitesimal rotations $\mathbf{n}_1\, d\theta_1$ and $\mathbf{n}_2\, d\theta_2$, in which the two axes of rotation intersect at a common point O, $\mathbf{n}_1$ and $\mathbf{n}_2$ being unit vectors along the two axes. Initially the radius vector to a point P is $\mathbf{r}$. After a rotation, $\mathbf{n}_1\, d\theta_1$, $\mathbf{r}$ has suffered an infinitesimal change $d\mathbf{r}_1$. Following this, the rotation $\mathbf{n}_2\, d\theta_2$ induces a change $d\mathbf{r}_2$ in the radius vector. However, $d\mathbf{r}_1$ and $d\mathbf{r}_2$, being linear displacements, may be combined vectorially. Accordingly the resultant $d\mathbf{r}$, of $d\mathbf{r}_1$ and $d\mathbf{r}_2$, may be written

$$d\mathbf{r} = d\mathbf{r}_1 + d\mathbf{r}_2 = (\mathbf{n}_1\, d\theta_1 \times \mathbf{r}) + [\mathbf{n}_2\, d\theta_2 \times (\mathbf{r} + d\mathbf{r}_1)] \tag{9-10}$$

On retaining only first-order infinitesimals, Eq. (9-10) becomes

$$d\mathbf{r} = (\mathbf{n}_1\, d\theta_1 \times \mathbf{r}) + (\mathbf{n}_2\, d\theta_2 \times \mathbf{r}) = (\mathbf{n}_1\, d\theta_1 + \mathbf{n}_2\, d\theta_2) \times \mathbf{r} \tag{9-11}$$

Equation (9-11) states that the resultant $d\mathbf{r}$ is the same as if a single rotation $\mathbf{n}_1\, d\theta_1 + \mathbf{n}_2\, d\theta_2$ had been taken, thus establishing the desired result.

If Eq. (9-11) is divided by the infinitesimal dt, we find that the velocity $\dot{\mathbf{r}}$ ($\equiv d\mathbf{r}/dt$) may be expressed in terms of the rotational velocities as

$$\frac{d\mathbf{r}}{dt} = \left(\mathbf{n}_1 \frac{d\theta_1}{dt} + \mathbf{n}_2 \frac{d\theta_2}{dt}\right) \times \mathbf{r} \tag{9-12}$$

But we may employ the symbol ω_1 for $d\theta_1/dt$ and ω_2 for $d\theta_2/dt$, these being the magnitudes of the two angular velocities of rotation. Thus, upon writing

$$\mathbf{n}_1\omega_1 = \boldsymbol{\omega}_1 \qquad \mathbf{n}_2\omega_2 = \boldsymbol{\omega}_2$$

Eq. (9-12) may be expressed as

$$\dot{\mathbf{r}} = (\boldsymbol{\omega}_1 + \boldsymbol{\omega}_2) \times \mathbf{r} = \boldsymbol{\omega} \times \mathbf{r} \tag{9-13}$$

where $\boldsymbol{\omega} = \boldsymbol{\omega}_1 + \boldsymbol{\omega}_2$. Equations (9-12) to (9-13) contain the information that the angular velocity is a vector which has the instantaneous direction of the infinitesimal rotation, has the magnitude of the rotation divided by the scalar dt, and obeys the parallelogram law of vector addition. If referred to a set of rectangular coordinate axes, the angular velocity $\boldsymbol{\omega}$, like any other vector, may be expressed in terms of its components along these axes. We have

$$\boldsymbol{\omega} = \omega_x\mathbf{i} + \omega_y\mathbf{j} + \omega_z\mathbf{k} \tag{9-14}$$

A useful result may be obtained by examining again the case of the rod introduced in Fig. 9-2, but this time considering that the finite translations and rotations are replaced by infinitesimal quantities. Figure 9-7 shows the displacement $d\mathbf{r}_P$ of a given point P, on the rod, resulting from the displacement $d\mathbf{s}_Q$ from Q to Q' of an arbitrary base point, together with a rotation about an axis passing through Q'. Clearly, from the small vector triangle we may write

$$d\mathbf{r}_P = d\mathbf{s}_Q + \mathbf{n}\, d\theta \times \mathbf{r}_{QP} \qquad (9\text{-}15)$$

Here, we make use of the fact that $Q'P'' = QP \equiv r_{QP}$. Supposing the translation and rotation to be executed simultaneously during a time dt, and dividing both sides of Eq. (9-15) by dt, we obtain

$$\frac{d\mathbf{r}_P}{dt} = \frac{d\mathbf{s}_Q}{dt} + \mathbf{n}\frac{d\theta}{dt} \times \mathbf{r}_{QP}$$

or

$$\boxed{\dot{\mathbf{r}}_P = \dot{\mathbf{s}}_Q + \boldsymbol{\omega} \times \mathbf{r}_{QP}} \qquad (9\text{-}16)$$

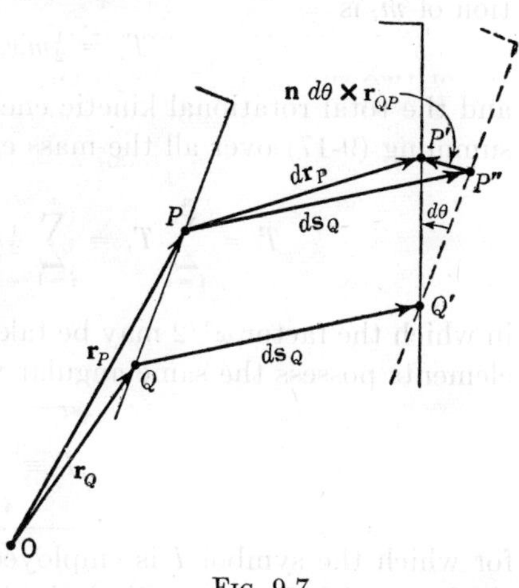

Fig. 9-7

in which $\dot{\mathbf{r}}_P$ is the velocity of point P expressed in terms of the translational velocity $\dot{\mathbf{s}}_Q$ of the arbitrary base point Q, and the angular velocity $\boldsymbol{\omega}$, of rotation about an axis passing through Q. We see that the velocity of a point of a rigid body is equal to the translational velocity of an arbitrary base point plus the cross product of the angular velocity with the radius vector from the base point to the point in question. We see further that, no matter what base point is chosen, the angular velocity is the same.

A word of caution is required at this stage. Equation (9-16) is purely a kinematic relationship. In general it is not desirable to employ a completely arbitrary base point in terms of which to write the dynamic equations of the motion. For only a few such points will the equations be simple. This will be discussed in more detail in Secs. 9-9 and 9-11.

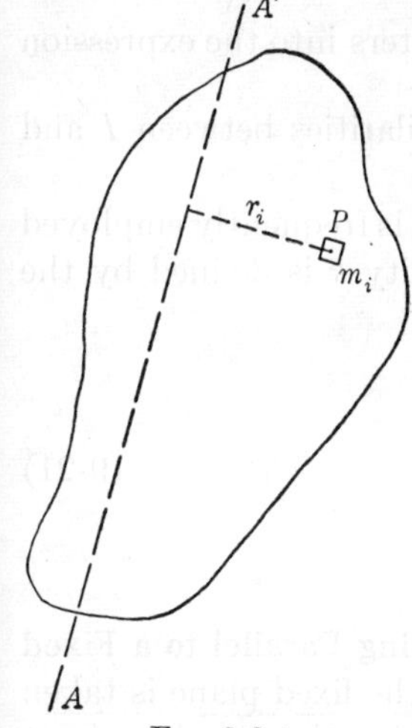

Fig. 9-8

DYNAMICS OF A RIGID BODY IN A PLANE

9-5. Kinetic Energy of Rotation. The Moment of Inertia.

Consider that the rigid body in Fig. 9-8 is rotating about the axis AA'. In particular, let us notice the small element of mass m_i at point P. The element m_i lies at a perpendicular distance r_i from the axis of rotation. Now, if ω is the magnitude of the angular velocity of the body about AA', the magnitude v_i of the linear velocity of m_i is simply ωr_i. Hence the kinetic energy of rotation of m_i is

$$T_i = \tfrac{1}{2}m_i v_i^2 = \tfrac{1}{2}m_i r_i^2 \omega^2 \tag{9-17}$$

and the total rotational kinetic energy of the entire body is obtained by summing (9-17) over all the mass elements. We have

$$T = \sum_{i=1}^{n} T_i = \sum_{i=1}^{n} \tfrac{1}{2}m_i r_i^2 \omega^2 = \tfrac{1}{2}\omega^2 \sum_{i=1}^{n} m_i r_i^2 \tag{9-18}$$

in which the factor $\omega^2/2$ may be taken outside the sum, since all the mass elements possess the same angular velocity. The quantity

$$I \equiv \sum_{i=1}^{n} m_i r_i^2 \tag{9-19}$$

for which the symbol I is employed, is termed the *moment of inertia* of the body with respect to the axis AA'. The moment of inertia is sometimes spoken of as the *rotational mass* of the body (with respect to the axis AA'; in general it is different for different axes of rotation). It occurs in the expression of the rotational kinetic energy

$$T = \tfrac{1}{2}I\omega^2 \tag{9-20}$$

in the same way that the mass $M \left(= \sum_{i=1}^{n} m_i\right)$ enters into the expression for the kinetic energy of translation. Other similarities between I and M will become apparent in Sec. 9-6.

Another quantity k, called the *radius of gyration*, is frequently employed in place of the moment of inertia I. The quantity k is defined by the relation

$$k^2 = \frac{I}{M} = \frac{\sum_{i=1}^{n} m_i r_i^2}{\sum_{i=1}^{n} m_i} \tag{9-21}$$

9-6. Angular Momentum of a Rigid Body Moving Parallel to a Fixed Plane. The Rotational Equation of Motion. If the fixed plane is taken to be the xy plane, the equation of motion describing the rotation about the z axis is just the z component of Eq. (8-6), or

$$\dot{J}_z = L_z \tag{9-22}$$

The quantity J_z is the moment of momentum about the z axis. In Eq. (9-22), the internal forces acting between adjacent volume elements are assumed to be central in nature, and consequently, when the sum of the moment is taken over the entire body, the internal forces vanish from the equation of motion.

If we consider only those cases in which the axis of rotation is the z axis, the angular momentum has a simple form. Taking r_i to be the perpendicular distance of a mass element m_i from the z axis, about which the body is rotating with an angular velocity ω, the angular momentum owing to m_i is

$$J_i = r_i m_i r_i \omega = m_i r_i^2 \omega \tag{9-23}$$

Consequently

$$J_z = \sum_{i=1}^{n} J_i = \sum_{i=1}^{n} m_i r_i^2 \omega = \left(\sum_{i=1}^{n} m_i r_i^2\right) \omega \tag{9-24}$$

[Equations (9-23) and (9-24) also may be obtained from Eq. (12-5), by putting $\omega_x = \omega_y = 0$ there.] The expression within the parentheses is just the moment of inertia I about the z axis. Accordingly

$$J_z = I\omega \tag{9-25}$$

for the simple class of rigid-body motions being considered in the present chapter. Equation (9-22) then becomes

$$I\dot{\omega} = L \tag{9-26}$$

Although in Eq. (9-26) the z subscripts have been dropped, it is to be remembered that the equation holds for rotation only about the z axis, which axis does not change in direction. In the present chapter we confine ourselves to motions only of this type.

In Eq. (9-26), $\dot{\omega}$ is the angular acceleration α about the z axis. Hence Eq. (9-26) may be rewritten as

$$I\alpha = L \tag{9-27}$$

Note that the occurrence of I in Eqs. (9-25) and (9-27) is similar to the occurrence of M in the corresponding translational equations.

9-7. Theorem of Parallel Axes. Consider an arbitrary rigid body, a cross section of which is shown in Fig. 9-9. Let there be two alternative axes, both perpendicular to the plane of the paper, one passing through O, and one passing through an arbitrary point O'. Points O and O' are in the plane of the paper, point O being selected so that the axis

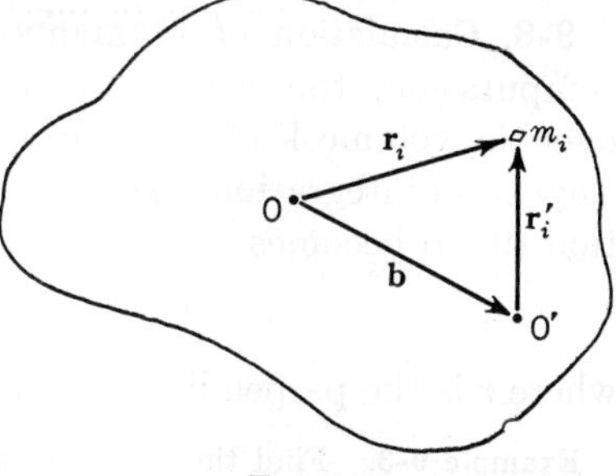

FIG. 9-9

passing through it also passes through the center of mass C (not shown in the figure) of the body. The position of a mass element m_i is specified either by a radius vector $\mathbf{r}_i$ drawn from the axis OC to m_i, perpendicular to OC, or by a radius vector $\mathbf{r}_i'$ to m_i, perpendicular to the parallel axis passing through O'. The vector $\mathbf{b}$ is drawn from O to O'. The moment of inertia I' of the body about the axis passing through O' is, by virtue of Eq. (9-19),

$$I' = \sum_{i=1}^{n} m_i r_i'^2 = \sum_{i=1}^{n} m_i(\mathbf{r}_i' \cdot \mathbf{r}_i') \tag{9-28}$$

but, from the vector triangle in Fig. 9-9, $\mathbf{r}_i' = \mathbf{r}_i - \mathbf{b}$, and thus (9-28) becomes

$$\begin{aligned} I' = \sum_{i=1}^{n} m_i(\mathbf{r}_i - \mathbf{b}) \cdot (\mathbf{r}_i - \mathbf{b}) &= \sum_{i=1}^{n} m_i(b^2 + r_i^2 - 2\mathbf{b} \cdot \mathbf{r}_i) \\ &= Mb^2 + \sum_{i=1}^{n} m_i r_i^2 - 2\mathbf{b} \cdot \sum_{i=1}^{n} m_i \mathbf{r}_i \end{aligned} \tag{9-29}$$

in which $M = \sum_{i=1}^{n} m_i$. Now the second term on the right is just the moment of inertia I_0 about the parallel axis passing through the center of mass O, and the third term on the right vanishes identically from the definition of the center of mass. Accordingly, the moment of inertia about the axis through O' may be written

$$I' = I_0 + Mb^2 \tag{9-30}$$

Consequently we may always relate the moment of inertia about an arbitrary axis to that about a parallel axis passing through the center of mass. The quantity b is the perpendicular distance between the two axes.

9-8. Calculation of Moments of Inertia. For practical purposes of computation, the summation notation is replaced by the integration over the volume V of the body. The mass element m_i is replaced by the element of integration $\rho\, dV$, in which ρ is the density of the body. Equation (9-19) becomes

$$I = \int_V \rho r^2\, dV \tag{9-31}$$

where r is the perpendicular distance from the element dV to the axis.

Example 9-3. Find the moment of inertia of a uniform thin rod of mass m and length $2b$ about an axis perpendicular to the rod and passing through one end.

If the origin is taken at the axis, in terms of the coordinate x the moment of inertia is

$$I_e = \frac{m}{2b} \int_0^{2b} x^2\, dx = \frac{4}{3} mb^2 \tag{9-32}$$

By means of Eqs. (9-30) and (9-32) the moment of inertia I_0 about a parallel axis passing through the center of the rod is found to be $mb^2/3$.

Example 9-4. Find the moment of inertia of a homogeneous circular disk, of radius b and mass m, about an axis perpendicular to the sheet, passing through the center.

The element of integration conveniently may be chosen to be a ring of radius r and thickness dr. The moment of inertia is

$$I_0 = \frac{m}{\pi b^2} \int_0^b 2\pi r^3 \, dr = \frac{1}{2} mb^2 \qquad (9\text{-}33)$$

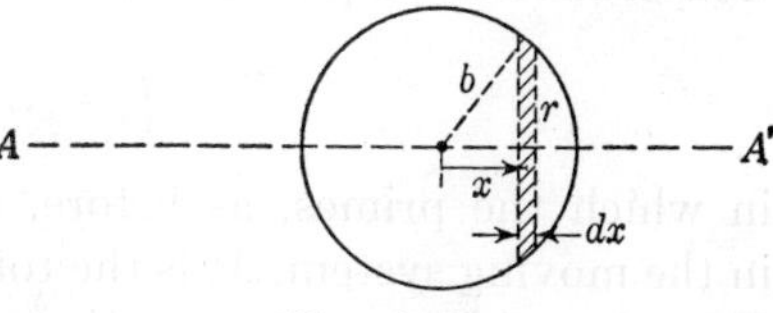

FIG. 9-10

Example 9-5. Find the moment of inertia of a homogeneous sphere of radius b and mass m about a diameter.

We select the coordinate x, measured from the center along the axis AA', in Fig. 9-10. The moment of inertia dI of the disk-shaped element of radius $r = \sqrt{b^2 - x^2}$ and thickness dx about AA' is, by virtue of Eq. (9-33),

$$dI = \frac{1}{2} \pi r^2 \rho r^2 \, dx = \frac{\pi \rho}{2} (b^2 - x^2)^2 \, dx$$

where ρ is the density of the sphere. After integrating we have

$$I = \pi \rho \int_0^b (b^2 - x^2)^2 \, dx = \tfrac{2}{5} mb^2 \qquad (9\text{-}34)$$

9-9. Coordinate Systems for Rigid Bodies. It has been stated that the types of motion of rigid bodies being considered in the present chapter are those in which the translation is parallel to a fixed plane and the rotation is about an axis which is always perpendicular to that plane. In general, these conditions are realized when the fixed plane contains the center of mass of the body and when all the forces acting upon the body can be replaced by forces the lines of action of which are likewise in that plane (cf. Sec. 3-14; evidently the replacement of the forces must not introduce a couple tending to shift the direction of the axis of rotation). For this class of problem it is convenient to restrict ourselves to coordinate axes which are not rotating. Therefore, the direction in which each coordinate axis points does not vary. If the motion of the body is such that a point of the body remains fixed in space, it is usually convenient to choose the origin at this point (see the example of the compound pendulum below). If no point of the body remains fixed in space, the origin is usually placed at the center of mass of the body and therefore adopts whatever translational motion is executed by the center of mass. From Sec. 8-4 and Sec. 9-6, it is apparent, for a coordinate system such as either of those mentioned above, that the rotational equation of motion for the body about the z axis is

$$\dot{J}_z = L_z \qquad (9\text{-}35)$$

in which J_z and L_z have the meanings given them in Sec. 9-6.

It is permissible, although usually not desirable, to select a point other than either a permanently fixed point or the center of mass. It is shown in Sec. 8-4, however, that Eq. (9-35) must be modified for an arbitrary origin. If we take the origin to be moving with an acceleration $\ddot{\mathbf{r}}_0$, but with the axes still fixed in direction, the z component of the equation of motion [Eq. (8-24)] referred to the moving axes is

$$\dot{J}'_z = L'_z + (\ddot{\mathbf{r}}_0 \times M\mathbf{r}'_c)_z \tag{9-36}$$

in which the primes, as before, denote that the quantities are defined in the moving system, M is the total mass of the body, and $\mathbf{r}'_c$ is the vector from the arbitrary origin to the center of mass. The cross-product term on the right is just the z component of the similar term appearing in Eq. (8-24). Equation (9-36) has more academic interest than practical importance, and we shall restrict ourselves to coordinate systems in which the last term vanishes. Aside from the simple cases mentioned above, this will be the situation if either $\ddot{\mathbf{r}}_0$ or $\mathbf{r}'_c$ has a component only in the z direction, since the cross product is a vector perpendicular to both $\ddot{\mathbf{r}}_0$ and $\mathbf{r}'_c$. It will also follow for an instantaneous axis of rotation which is being accelerated toward a parallel axis passing through the center of mass (see also Sec. 8-4).

Much information often can be obtained by use of the principle of energy, without the need of employing the equations of motion. It was shown in Chap. 8 that the kinetic energy of a system of particles may be written as the sum of two parts. The first is the kinetic energy of translation of the center of mass, the entire mass being regarded as concentrated at that point, while the second is the kinetic energy of motion relative to the center of mass. This is true whether or not the particles are rigidly connected to one another as in the present case of a rigid body. Moreover, it was shown in Sec. 9-5 that the kinetic energy of rotation of a rigid body has the form $\frac{1}{2}I\omega^2$. If, in addition, a potential energy V exists and there are no dissipative forces present, we may write the equation expressing the conservation of energy as

$$V + \tfrac{1}{2}Mv^2 + \tfrac{1}{2}I\omega^2 = \text{const} \tag{9-37}$$

where M is the total mass of the body, v the velocity of translation of the center of mass, I the moment of inertia about the axis of rotation passing through the center of mass, and ω the angular velocity of rotation about that axis. It should be emphasized that the third term on the left of Eq. (9-37) has this simple form only when the axis of rotation (that is, the direction of ω) is parallel to the angular-momentum vector. This is the case for the class of problem treated in the present chapter, in which the axis of rotation is at all times perpendicular to the plane of translation.

Equation (9-37) can be made more general by an inspection of Eq. (8-29). There, the total kinetic energy of a system of n particles was written

$$T = \tfrac{1}{2}M|\dot{\mathbf{r}}_c|^2 + \dot{\mathbf{r}}_c \cdot \sum_{i=1}^{n} m_i\dot{\mathbf{r}}_i' + \sum_{i=1}^{n} \tfrac{1}{2}m_i|\dot{\mathbf{r}}_i'|^2 \tag{9-38}$$

in which $\dot{\mathbf{r}}_c$ is the velocity of a point C (here arbitrary) of the body, m_i the mass of the ith particle, M the total mass of the body, and $\dot{\mathbf{r}}_i'$ the velocity of the ith particle relative to C. The first term on the right is the kinetic energy of translation of a mass point at C the mass of which is equal to the total mass of the body, and the third term may be interpreted as the kinetic energy relative to C. In the case of a rigid body the latter is just the kinetic energy of rotation about an axis passing through C. Equation (9-38) is further complicated by the presence of the second term on the right, which, for an arbitrary point C, is not zero. This term vanishes in three cases. In the first instance it vanishes, as does the first term, if point C is at rest, for then $\dot{\mathbf{r}}_c$ is zero. Second it vanishes if point C is the center of mass, since in that instance the summation vanishes by virtue of the definition of the center of mass (see Chap. 3). Lastly, if C is neither at rest nor the center of mass, the dot product can be made to vanish if $\dot{\mathbf{r}}_c$ and $\sum_{i=1}^{n} m_i\dot{\mathbf{r}}_i'$ are mutually perpendicular. It is usually convenient, however, to restrict ourselves to the first two alternatives.

9-10. The Compound Pendulum. A good example of the motion of a rigid body, having one point fixed, is afforded by the compound pendulum shown in Fig. 9-11. We have a rigid body of mass m suspended at point O a distance b from the center of mass C. The body is free to rotate about an axis which is perpendicular to the paper and passing through O. Since gravity is acting downward, there is a force mg acting at point C. The equation of motion may be written in terms of the coordinate θ measured, as shown, from the fixed vertical line OQ. The equation of motion is

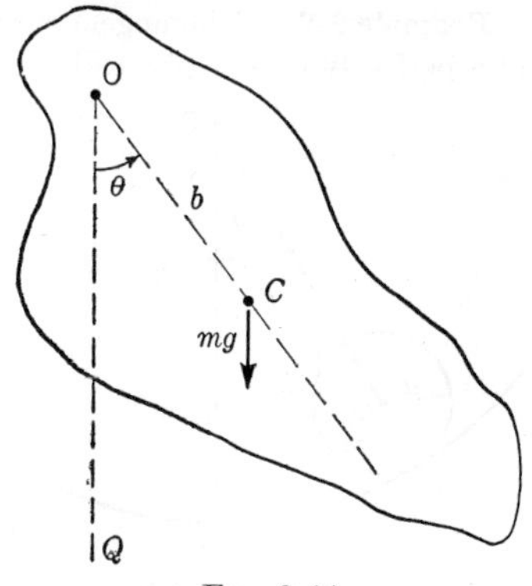

FIG. 9-11

$$I\ddot{\theta} = -mgb \sin\theta \tag{9-39}$$

in which the minus sign occurs because the moment of force tends to produce rotation in the direction of decreasing θ. I is the moment of inertia of the body about the axis of rotation passing through O. If we confine ourselves to small angular displacements, we may write $\sin \theta$ approximately equal to θ and Eq. (9-39) becomes for that case

$$\ddot{\theta} + \frac{mgb}{I}\theta = 0 \tag{9-40}$$

the familiar equation of simple harmonic motion. Since mgb/I is a positive quantity, Eq. (9-40) has the solution

$$\theta = A \sin\left(\sqrt{\frac{mgb}{I}}\,t + B\right) \tag{9-41}$$

in which A and B are arbitrary constants. Equation (9-41) represents an oscillation of frequency ν and period T given by

$$\nu = \frac{\omega}{2\pi} = \frac{1}{2\pi}\sqrt{\frac{mgb}{I}} \qquad T = 2\pi\sqrt{\frac{I}{mgb}} \tag{9-42}$$

in which ω is the angular frequency. The angular frequency of a simple pendulum of length L has the value $\sqrt{g/L}$. Thus the present compound pendulum is equivalent (in frequency and period) to a simple pendulum of length L defined by

$$L = \frac{I}{mb} \tag{9-43}$$

L, defined in this way, is called the *length of the equivalent simple pendulum.*

Example 9-6. A homogeneous sphere of radius a is free to roll on the inside surface of a perfectly rough spherical bowl of radius b. The motion is confined to a vertical plane passing through the lowest point. It is desired to determine the length of the equivalent simple pendulum for the motion.

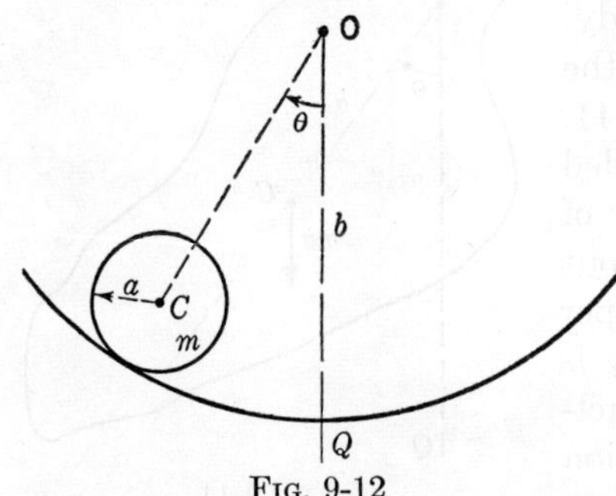

FIG. 9-12

The conditions of the problem may be seen with the aid of Fig. 9-12. A convenient choice of a coordinate with which to describe the translation of the center of mass is the angle θ which OC makes with the vertical line OQ. The point O is the center of curvature of the bowl. The kinetic energy of the sphere is composed of two parts, one involving the translation of the center of mass, and the other involving rotation about the center of mass. Thus, if ω is the angular velocity of rotation (rolling motion) of the sphere, the kinetic energy T may be written

$$T = \tfrac{1}{2}m(b-a)^2\dot{\theta}^2 + \tfrac{1}{2}\cdot\tfrac{2}{5}ma^2\omega^2 \tag{9-44}$$

Now, since we have pure rolling, we must have $a\omega = (b - a)\dot{\theta}$. (It will be a useful exercise for the student to demonstrate this.) Hence Eq. (9-44) becomes

$$T = \tfrac{7}{10}m(b-a)^2\dot{\theta}^2$$

Since the potential energy is $mg(b - a)(1 - \cos\theta)$, the equation of energy takes the form

$$\tfrac{7}{10}m(b - a)^2\dot{\theta}^2 + mg(b - a)(1 - \cos\theta) = \text{const}$$

Differentiating this with respect to time and rearranging, we obtain

$$\ddot{\theta} + \frac{g}{\frac{7}{5}(b - a)} \sin\theta = 0 \tag{9-45}$$

For the case of θ small we may replace $\sin\theta$ by θ, and Eq. (9-45) assumes the familiar form of that of a simple pendulum of length $7(b - a)/5$.

9-11. Use of the Instantaneous Axis. Of special interest is the use of the so-called *instantaneous axis* in the motion of rigid bodies. The instantaneous axis is an axis perpendicular to the plane of motion, passing through that point of the body which is instantaneously at rest, that is, through the instantaneous center for the velocity. A simple problem illustrating this is that of a disk rolling down an inclined plane. In

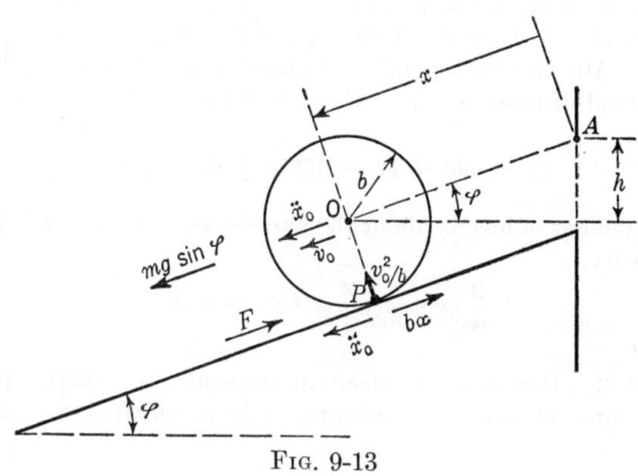

FIG. 9-13

Fig. 9-13, the disk is constrained to remain in the plane of the paper. If no sliding exists, the point of the disk which is momentarily at rest is the point P, which is in contact with the inclined plane. At this instant the motion of the disk may be regarded kinematically as being entirely one of rotation about the instantaneous axis passing through P. According to Eq. (9-38), we may consider the total kinetic energy as being that of rotation about the instantaneous axis, since the first two terms both vanish. Are we justified, however, in taking moments about the instantaneous axis and equating them to the rate of change of the angular momentum $\mathbf{J}'$, where $\mathbf{J}'$ is measured with reference to an origin situated at P? At this point caution is required. We note that, although P is instantaneously at rest, it still is being accelerated. There are three accelerations acting on P. These are shown in Fig. 9-13 to be the acceleration $\ddot{x}_0$ of the center of curvature O of the curved surface, the linear acceleration $b\alpha$ due to the angular acceleration α, and the centripetal acceleration v_0^2/b directed toward the center of curvature. Here, v_0 is the velocity of the center of curvature, and b is the radius of curvature. Now, the quantities $\ddot{x}_0$ and $b\alpha$ are equal and oppositely directed and therefore cancel. Thus the resultant acceleration of P is just v_0^2/b, a quantity which is directed through the center of curvature O. If the disk is homogeneous, point O is also the center of mass. Accordingly, the second set of terms on the right side of Eq. (9-36) vanishes (because the cross product of two parallel vectors is zero), and the equation of motion reduces to $\dot{\mathbf{J}}' = \mathbf{L}'$ for the origin placed at P. Hence, in the very symmetrical case which we have chosen, the instantaneous axis is one with reference to which we may equate the torque to the rate of change of the angular momentum.

Example 9-7. In the example discussed immediately above, determine the linear acceleration $\ddot{x}_0$ of the center of mass by placing the origin both at O and at P. In both cases employ two methods of solution, the energy method and the method of moments.

1. *Origin at P; Energy Method.* Take the coordinate x of the center of mass to be zero at point A. After a time t, point O will have traveled a distance x and will have descended a vertical distance h. The moment of inertia I_P about the instantaneous axis is

$$I_P = mb^2 + I_0 = mb^2 + \tfrac{1}{2}mb^2 = \tfrac{3}{2}mb^2$$

where I_0 is the moment of inertia about the parallel axis through O. We have, if ω is the angular velocity,

$$\frac{3}{4}\,mb^2\omega^2 = \frac{3}{4}\,mb^2\,\frac{v_0^2}{b^2} = mgh$$

whence $v_0^2 = 4gh/3$. Here v_0 is the velocity of the center of mass O. But $x \sin\varphi = h$, and since the center of mass is being uniformly accelerated, we have $v_0^2 = 2\ddot{x}_0 x$. Thus we have

$$\ddot{x}_0 = \tfrac{2}{3}g \sin\varphi \tag{9-46}$$

2. *Origin at O; Energy Method.* Here, part of the kinetic energy will be translational, and we have

$$\tfrac{1}{4}mb^2\omega^2 + \tfrac{1}{2}mv_0^2 = mgh$$

from which Eq. (9-46) follows at once, since $v_0 = b\omega$ and $v_0^2 = 2\ddot{x}_0 x$.

3. *Origin at O; Torque Method.* The forces acting on the body parallel to the plane are $mg \sin\varphi$ at the mass center and a force F at the point of contact (F must be present to prevent sliding). If the origin is at O, we require the equation of rotational motion and the equation of translation of O. We have

$$\frac{1}{2}\,mb^2\,\frac{\ddot{x}_0}{b} = Fb$$

$$m\ddot{x}_0 = mg \sin\varphi - F$$

We eliminate the unknown force F between these, and since $v_0 = b\omega$, we obtain the result (9-46).

4. *Origin at P; Torque Method.* We require only the rotational equation of motion

$$\frac{3}{2} mb^2 \frac{\ddot{x}_0}{b} = mgb \sin \varphi$$

from which Eq. (9-46) follows at once.

9-12. Rolling and Sliding Motion of a Sphere. We consider now the case of a billiard ball of mass m and radius b which is initially given a translational velocity v_0, without rotation, down the line of greatest slope of an imperfectly rough plane inclined at an angle α to the horizontal (see Fig. 9-14). We shall select as coordinates the coordinate x of the center of mass (measured parallel to the plane) and the angle θ which a line OA, fixed in the ball, makes with the normal to the plane. θ is positive in the clockwise direction, and x is positive in the direction down the plane. It is convenient to take $x = 0$ and $\theta = 0$ at $t = 0$. The forces acting on the ball are mg downward at the center of mass and the reactions N and F acting normal to and parallel to the plane, respectively, as shown.

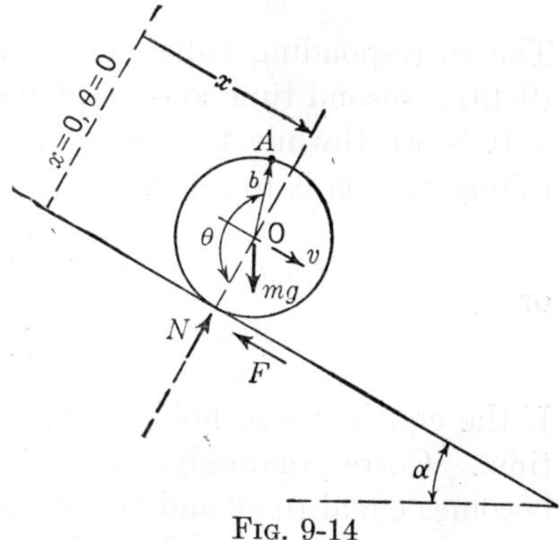

FIG. 9-14

If μ is the coefficient of sliding friction, F has the magnitude μN during the first stage of the motion since the ball slides during this stage. Initially the sphere is not rotating, but the moment of the force F imparts to it an angular acceleration. Therefore the complete motion of the ball is described by the two equations

$$m\ddot{x} = mg \sin \alpha - \mu mg \cos \alpha \qquad I\ddot{\theta} = \mu bmg \cos \alpha \tag{9-47}$$

in which $I = 2mb^2/5$. Upon making this substitution and rearranging, Eqs. (9-47) assume the form

$$\ddot{x} = g(\sin \alpha - \mu \cos \alpha) \qquad \ddot{\theta} = \frac{5\mu g}{2b} \cos \alpha \tag{9-48}$$

These may be integrated to

$$\dot{x} = g(\sin \alpha - \mu \cos \alpha)t + v_0 \qquad \dot{\theta} = \frac{5\mu gt}{2b} \cos \alpha \tag{9-49}$$

If at any time t_1 the linear velocity becomes related to the angular velocity by the expression

$$\dot{x} = b\dot{\theta} \tag{9-50}$$

the subsequent motion is that of pure rolling. The time t_1 at which the situation is represented by Eq. (9-50) may be determined from Eqs. (9-49). We have, employing (9-50),

$$g(\sin \alpha - \mu \cos \alpha)t_1 + v_0 = \frac{5\mu g t_1}{2} \cos \alpha$$

from which

$$t_1 = \frac{v_0}{g(\frac{7}{2}\mu \cos \alpha - \sin \alpha)} \tag{9-51}$$

The corresponding values of x and θ can be found by integrating Eqs. (9-49) a second time and employing Eq. (9-51).

It is worthwhile to inspect Eq. (9-51) briefly. Evidently, if a pure rolling motion is to set in at a positive value of t, we must have

$$\tfrac{7}{2}\mu \cos \alpha - \sin \alpha \geq 0$$

or

$$\mu \geq \tfrac{2}{7} \tan \alpha \tag{9-52}$$

If the equality sign holds, pure rolling commences only after an infinite time. Correspondingly, if $\mu < \frac{2}{7} \tan \alpha$, the linear velocity $\dot{x}$ never becomes equal to $b\dot{\theta}$ and the state of pure rolling motion never obtains.

If conditions are such that Eq. (9-50) is ultimately true, the sphere ceases to slide at a time t_1. Consequently, at this state the frictional force F is no longer equal to $\mu mg \cos \alpha$. Thus, Eqs. (9-47) must be altered for this phase of the motion. For pure rolling we have

$$m\ddot{x} = mg \sin \alpha - F \qquad \tfrac{2}{5}mb^2\ddot{\theta} = bF \tag{9-53}$$

in which F is a force necessary to maintain the rolling motion. Much information can be obtained without actually determining F since it may be eliminated between Eqs. (9-53). Thus, since

$$b\dot{\theta} = \dot{x} \tag{9-54}$$

we have

$$m\ddot{x} = mg \sin \alpha - \tfrac{2}{5}m\ddot{x}$$

or

$$\ddot{x} = \tfrac{5}{7}g \sin \alpha \tag{9-55}$$

If v_1 is the velocity at time t_1, Eq. (9-55) may be integrated to

$$\dot{x} = \tfrac{5}{7}gt \sin \alpha + v_1 - \tfrac{5}{7}gt_1 \sin \alpha$$

It is interesting also to determine F. In this case we may employ Eq. (9-54) and eliminate $\ddot{x}$ between Eqs. (9-53). We obtain

$$F = \tfrac{2}{7}mg \sin \alpha \tag{9-56}$$

The result (9-56) throws additional light on the inequality in Eq. (9-52), for if $\mu = \frac{2}{7} \tan \alpha$, we have the interesting relation

$$\mu mg \cos \alpha = \tfrac{2}{7} \tan \alpha \cdot mg \cos \alpha = \tfrac{2}{7} mg \sin \alpha$$

Thus we see that, in the case of equality in Eq. (9-52), the maximum frictional force ($\mu mg \cos \alpha$) is just that necessary to maintain a pure rolling motion.

IMPULSIVE MOTION OF A RIGID BODY IN A PLANE

9-13. Impulsive Torque. The impulsive motion of particles was considered in Sec. 8-8. The problem of an impulsive force $\mathbf{F}$ acting upon a single particle was to regard $\mathbf{F}$ as acting for a short interval of time τ, the impulse $\mathbf{P}$ transmitted to the particle being

$$\mathbf{P} = \int_0^\tau \mathbf{F}\, dt = m\mathbf{v}_2 - m\mathbf{v}_1 \tag{9-57}$$

In many practical cases it was convenient to regard the impulse as having been transmitted in a sufficiently short interval of time such that the particle did not move appreciably during this interval (cf. Example 8-5).

In the present section the impulsive motion of extended bodies will be considered. The bodies will be regarded as being perfectly rigid except during impulsive impact, at which times deformations will occur. The important feature added, when extended bodies are treated, is the concept of a moment of an impulsive force, or an *impulsive torque*. The angular impulse $\mathbf{H}$, which is transmitted to a body by the moment $\mathbf{L}$ of a force acting during a short time τ, is obtained from the rotational equation of motion

$$\dot{\mathbf{J}} = \mathbf{L} \tag{9-58}$$

in which $\mathbf{J}$ is the angular momentum. Integrating Eq. (9-58) over the time τ, we have

$$\mathbf{H} = \int_0^\tau \mathbf{L}\, dt = \mathbf{J}_2 - \mathbf{J}_1 \tag{9-59}$$

where $\mathbf{J}_1$ is the initial angular momentum of the body and $\mathbf{J}_2$ is the final angular momentum. For the cases to which we confine ourselves in the present chapter, the axis of rotation of the body remains perpendicular to a fixed plane, and $\mathbf{H}$ has only a component normal to the plane. Designating this scalar component as H, we may write

$$H = I\omega_2 - I\omega_1 \tag{9-60}$$

where I is the moment of inertia of the body about the rotational axis and ω_1 and ω_2 are the initial and final magnitudes of the angular velocity.

9-14. Center of Percussion. In Fig. 9-15, a uniform rod of mass M and length $2b$ is given a blow at point A, a distance a from the center of mass C. The blow transmits a horizontal impulse of magnitude P, as shown.

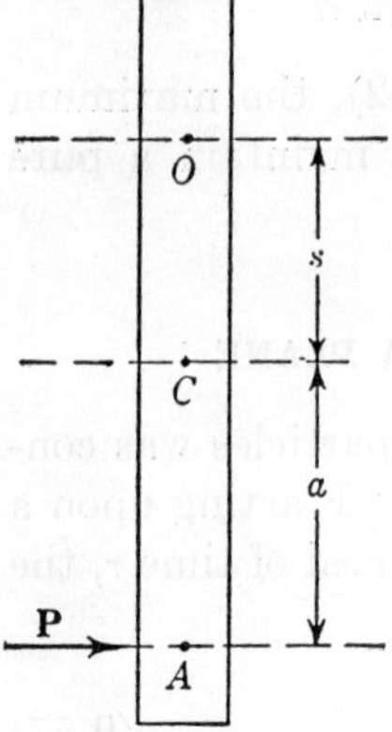

FIG. 9-15

Immediately after the impact, the velocity v of the mass center is given by

$$v = \frac{P}{M}$$

and the initial angular velocity of the rod after the impact is

$$\omega = \frac{aP}{I} = \frac{3aP}{Mb^2}$$

where I is the moment of inertia about an axis normal to the paper and passing through C (equal to $Mb^2/3$). After the impact, the instantaneous velocity v_0 of point O, located a distance s from C, on the opposite side of C from point A, is given by

$$v_0 = v - \omega s = \frac{P}{M} - \frac{3aPs}{Mb^2} = \frac{P}{M}\left(1 - \frac{3as}{b^2}\right) \tag{9-61}$$

This is zero if

$$3as = b^2$$

or (9-62)

$$k^2 = as$$

where k is the radius of gyration.

It is evident, if the rod is suspended from point O by means of a frictionless pivot, that Eq. (9-62) is just the condition for which there is no impulsive reaction between the rod and the pivot during the impact. This can also be seen in the following way: Let it be supposed that the rod is suspended at point O in the manner just described. As is depicted in Fig. 9-16, if a and s are arbitrary, because of the pivot at O there will be an additional impulse, which we may call P_1, transmitted to the rod during the impact. Accordingly, the initial velocity of C after the impact is now given by

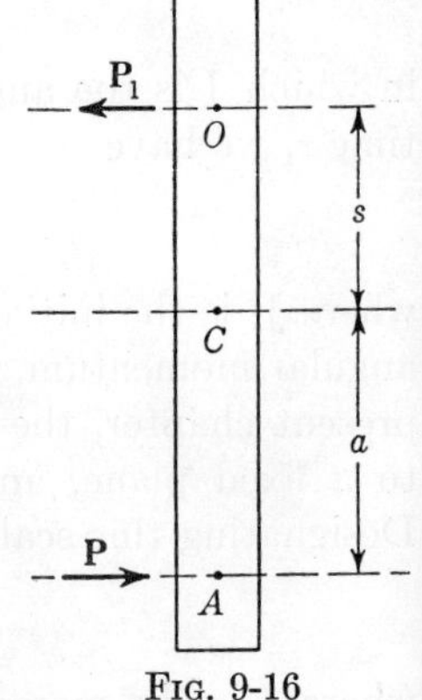

FIG. 9-16

$$v = \frac{P - P_1}{M} = \omega s$$

where v, ω, s are as defined previously. If $I_0\,[= Ms^2 + (Mb^2/3)]$ is the moment of inertia of the rod about the axis through O, we may write the angular velocity ω of the rod as

$$\omega = \frac{P(a+s)}{I_0}$$

Therefore

$$P_1 = P - M\omega s = P\left[1 - \frac{Ms(a+s)}{I_0}\right] = P\,\frac{b^2 - 3as}{3s^2 + b^2} \qquad (9\text{-}63)$$

Clearly, Eq. (9-63) gives the same condition as (9-62) if P_1 is to vanish. The point of application A of the impulse P for which P_1 vanishes is called the *center of percussion* for point O. If point O is at the end of a uniform thin rod of length $2b$, the center of percussion is located at a distance $b/3$ on the other side of C from O, or a distance $4b/3$ from the point of suspension. It should also be noticed that point O is the instantaneous center with respect to the impulse P being applied at A.

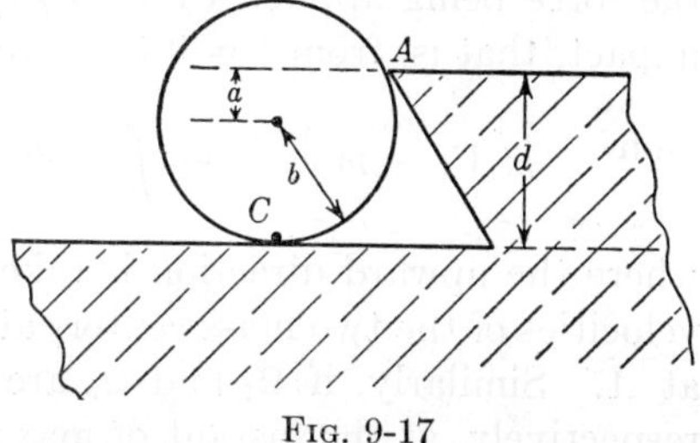

FIG. 9-17

Example 9-8. Figure 9-17 represents the geometry of a billiard ball near the edge of a billiard table. It is desired that, at the instant of impact, the point C of the ball in contact with the table top shall be an instantaneous center for the velocity. This condition is necessary if there is to be no sliding of the ball during impact. Only if sliding does not occur will the angle of incidence be equal to the angle of reflection.

If b is the radius of the ball, $k^2 = \frac{2}{5}b^2$. Employing Eq. (9-62), we have

$$k^2 = ab = \frac{2b^2}{5}$$

$$d = a + b = \frac{2b}{5} + b = \frac{7}{5}b \qquad (9\text{-}64)$$

Thus the edge A of the table must be a distance above the table top equal to $\frac{7}{5}$ of the radius of the billiard ball if the ball is to rebound from the edge at an angle equal to the incident angle.

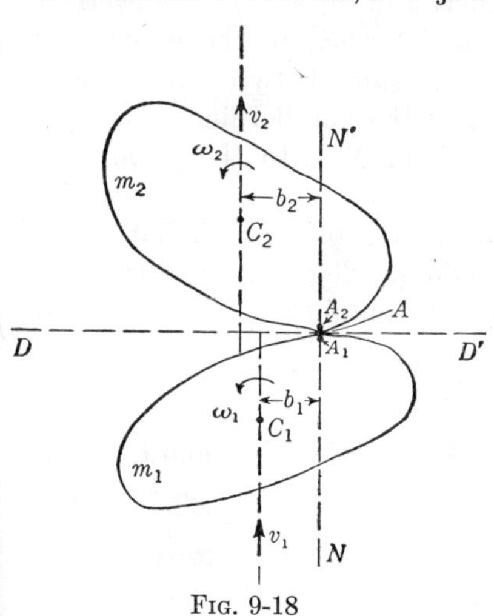

FIG. 9-18

9-15. Newton's Rule for the Smooth Impact of Two Extended Bodies.

Two smooth bodies of masses m_1 and m_2 are shown in the act of colliding at point A in Fig. 9-18. A_1 and A_2 are, respectively, the points of m_1 and m_2 in contact at A. The line DD' is tangent to the two surfaces at the point of contact. Before the collision, the two mass centers C_1 and C_2 are traveling with velocities which have components v_1 and v_2, respectively, normal to DD'. (In the figure, both are directed upward.) Since the impact is smooth, the components u_1 and u_2 of the velocities parallel to DD' will be unaltered, and these need not be considered in this discussion. The angular velocities of the bodies before the collision are ω_1 and ω_2. All motions are taken to lie in the plane of the paper. As in the case of the impact of two smooth spheres treated in Sec. 8-9, the collision may be considered in two stages. The first stage begins with the contact being made at point A and lasts until maximum deformation occurs. This latter situation is realized when point A_1 of m_1 and point A_2 of m_2, in contact at A, are moving with a common velocity. If F is the force being applied by m_2 to m_1 at A_1 during the first stage of the impact, that is, from $t = 0$ to τ, we have

$$m_1V_1 - m_1v_1 = -\int_0^\tau F\,dt = -P \qquad m_2V_2 - m_2v_2 = P \qquad (9\text{-}65)$$

where the upward direction is taken positive. Here, V_1 and V_2 are the velocities of the two mass centers at the instant of maximum deformation at A. Similarly, if Ω_1 and Ω_2 are the angular velocities of m_1 and m_2, respectively, at the instant of maximum deformation, we have

$$I_1(\Omega_1 - \omega_1) = -\int_0^\tau Fb_1\,dt = -\int_0^\tau L_1\,dt = -H_1 = -b_1P \qquad (9\text{-}66)$$

$$I_2(\Omega_2 - \omega_2) = b_2P \qquad (9\text{-}67)$$

where I_1 and I_2 are the moments of inertia of m_1 and m_2, respectively, about axes through the mass centers and normal to the paper. The symbol b_1 represents the perpendicular distance between the common normal NN', at the point of contact, and the parallel line through C_1, while b_2 is the perpendicular distance from NN' to the parallel line through C_2.

The second stage of the collision may be treated in a similar manner and lasts from the time τ at which maximum deformation occurs until the time τ' at which separation occurs. We obtain

$$m_1v_1' - m_1V_1 = -P' \qquad m_2v_2' - m_2V_2 = P' \qquad (9\text{-}68)$$

where v_1' and v_2' are the components of the velocities of C_1 and C_2 normal to DD' immediately after the impact and P' is defined by $P' = \int_\tau^{\tau'} F'\,dt$. From Chap. 8, $P' = eP$ where e is the coefficient of restitution for the colliding pair of bodies. Also, if ω_1' and ω_2' are the angular velocities immediately after impact, we may write

$$I_1(\omega_1' - \Omega_1) = -b_1P' = -b_1eP \qquad I_2(\omega_2' - \Omega_2) = b_2eP \qquad (9\text{-}69)$$

Now Eqs. (9-65) may be combined and Eqs. (9-68) also, to yield

$$\begin{aligned}(v_1 - v_2) - (V_1 - V_2) &= P\,\frac{m_1 + m_2}{m_1 m_2} \\ (v_1' - v_2') - (V_1 - V_2) &= -eP\,\frac{m_1 + m_2}{m_1 m_2}\end{aligned} \tag{9-70}$$

Equations (9-70) may be utilized to eliminate P, m_1, and m_2. We obtain

$$(v_1' - v_2') - (V_1 - V_2) = -e[(v_1 - v_2) - (V_1 - V_2)] \tag{9-71}$$

From Eqs. (9-66) and (9-67), we have

$$b_1 I_1(\Omega_1 - \omega_1) = -b_1^2 P \qquad b_2 I_2(\Omega_2 - \omega_2) = b_2^2 P \tag{9-72}$$

from which

$$b_1(\omega_1 - \Omega_1) + b_2(\Omega_2 - \omega_2) = P\left(\frac{b_1^2}{I_1} + \frac{b_2^2}{I_2}\right)$$

or

$$(b_1\omega_1 - b_2\omega_2) - (b_1\Omega_1 - b_2\Omega_2) = P\left(\frac{b_1^2}{I_1} + \frac{b_2^2}{I_2}\right) \tag{9-73}$$

Similarly, from Eqs. (9-69), we have

$$(b_1\omega_1' - b_2\omega_2') - (b_1\Omega_1 - b_2\Omega_2) = -eP\left(\frac{b_1^2}{I_1} + \frac{b_2^2}{I_2}\right) \tag{9-74}$$

Combining Eqs. (9-73) and (9-74), we obtain

$$(b_1\omega_1' - b_2\omega_2') + e(b_1\omega_1 - b_2\omega_2) = (1 + e)(b_1\Omega_1 - b_2\Omega_2) \tag{9-75}$$

Now, at the instant of maximum deformation, A_1 and A_2 have the common velocity

$$V_A = V_1 + b_1\Omega_1 = V_2 + b_2\Omega_2$$

from which

$$V_1 - V_2 = -(b_1\Omega_1 - b_2\Omega_2) \tag{9-76}$$

Combining Eqs. (9-75) and (9-76) and then eliminating $V_1 - V_2$ from Eq. (9-71), we obtain, after rearranging,

$$(v_1' + b_1\omega_1') - (v_2' + b_2\omega_2') = -e[(v_1 + b_1\omega_1) - (v_2 + b_2\omega_2)] \tag{9-77}$$

But the quantity on the left is just the velocity of separation v_s of points A_1 and A_2, and the quantity within brackets on the right is the velocity of approach v_a of A_1 and A_2. Thus we have Newton's rule in the form

$$v_s = -ev_a \tag{9-78}$$

in which v_a and v_s are, respectively, the components along the common normal NN' at the point of contact of the relative velocity of the points of contact before and after the collision.

9-16. Connected Systems. As an example of these we shall consider a system consisting of two rods, ABD in Fig. 9-19a, smoothly hinged at B as shown. Rod AB, of length $2b_1$, has a mass m_1 and moment of inertia I_1 about the mass center C_1. The corresponding quantities for BD are $2b_2$, m_2, and I_2. An impulse P_0 is applied at D in the manner indicated. It is desired to find, immediately after the impulse is delivered, the components of the velocities of the two mass centers, the angular velocity of each rod, and the impulse at the joint.

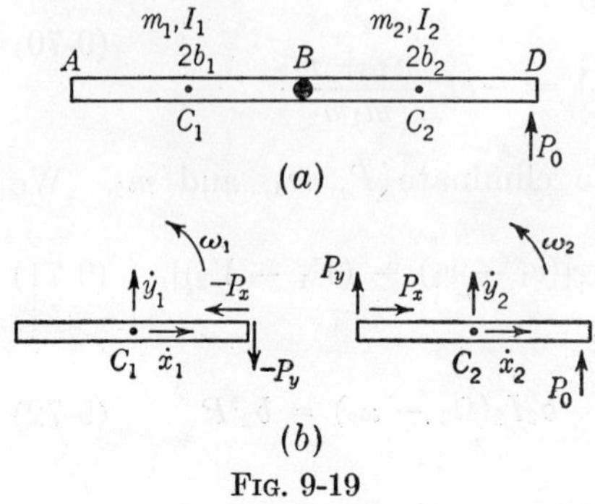

FIG. 9-19

The velocities and the impulses are depicted in part (b) of the figure. The impulse P being delivered to BD at B is expressed as the two components P_x and P_y, chosen, arbitrarily, to point in the positive directions of x and y, respectively. The corresponding components delivered to AB at B, by Newton's third law, must be represented as being equal and opposite to those being communicated to BD at B. We choose, as shown, the components of the translational velocities of the mass centers to be $\dot{x}_1$, $\dot{x}_2$, $\dot{y}_1$, and $\dot{y}_2$. The angular velocities ω_1 and ω_2 are represented in the counterclockwise (positive) sense.

The impulse equations governing translation are

$$m_1\dot{x}_1 = -P_x \qquad m_2\dot{x}_2 = P_x \tag{9-79}$$

$$m_1\dot{y}_1 = -P_y \qquad m_2\dot{y}_2 = P_y + P_0 \tag{9-80}$$

The impulse equations describing the rotational motion are

$$I_1\omega_1 = -P_y b_1 \qquad I_2\omega_2 = -P_y b_2 + P_0 b_2 \tag{9-81}$$

An additional pair of equations follows from the fact that the joined ends of the rods at B must have a common velocity. Hence, we must have

$$\dot{x}_1 = \dot{x}_2 \tag{9-82}$$

$$\dot{y}_1 + b_1\omega_1 = \dot{y}_2 - b_2\omega_2 \tag{9-83}$$

where the form of the equations holds at the instant after the application of P_0.

Equations (9-79) to (9-83) furnish eight equations from which to determine the eight unknowns $\dot{x}_1$, $\dot{x}_2$, $\dot{y}_1$, $\dot{y}_2$, ω_1, ω_2, P_x, and P_y in terms of the given quantities b_1, b_2, m_1, m_2, I_1, I_2, and P_0. It is evident at once, from Eqs. (9-79) and (9-82), without simplifying the problem, that

$$\dot{x}_1 = \dot{x}_2 = 0 \qquad P_x = 0 \tag{9-84}$$

This follows since P_0 is entirely in the y direction.

The remainder of the solution, in the interests of simplicity, will be obtained for the case in which the two rods are identical, with $b_1 = b_2 = b$, $m_1 = m_2 = m$, and $I_1 = I_2 = I = mb^2/3$. Solving Eqs. (9-80), (9-81), and (9-83) simultaneously, we obtain

$$P_y = \frac{P_0}{4} \tag{9-85}$$

$$\dot{y}_1 = -\frac{P_0}{4m} \qquad \dot{y}_2 = \frac{5P_0}{4m} \tag{9-86}$$

$$\omega_1 = -\frac{3P_0}{4mb} \qquad \omega_2 = \frac{9P_0}{4mb} \tag{9-87}$$

Problems

Problems marked C are reprinted by kind permission of the Cambridge University Press.

9-1. A wheel of radius a rolls along a level road with a velocity v_0. Find the velocity and acceleration of an arbitrary point on the rim in terms of its height b above the ground. Finally consider the special cases $b = 0, a, 2a$.

9-2. A ladder of length $2a$ stands on a level floor and leans against a vertical wall. If the ladder slides to the floor but is meanwhile constrained to have its extremities in contact with the wall and the floor, find the space and body velocity centrodes.

9-3. A wheel of radius a rolls around the inside of a hoop of larger radius b. If their angular velocities are ω_1 and ω_2, respectively, and are oppositely directed, find the space and body centrodes of the wheel for the case when the center of the hoop is fixed.

Show that the moments of inertia of the following homogeneous bodies of mass m are:

9-4. $mr^2/4$ for a circular disk of radius r, about a diameter.

9-5. $2mr^2/3$ for a spherical shell of radius r, about a diameter.

9-6. $m(a^2 + b^2)/12$ for a rectangular sheet of sides a and b, about an axis through its center, normal to its plane.

9-7. $m(b^2 + c^2)/5$ for an ellipsoid of principal axes $2a$, $2b$, $2c$, about the major axis $2a$.

9-8. A small uniform solid sphere is initially at rest at point A in Fig. 9-20. It is permitted to roll down the perfectly rough incline, the bottom of which ends in the circular path of large radius b as shown. Show that $h \geq 27b/10$ if the small sphere is to roll completely around the circular path without falling off.

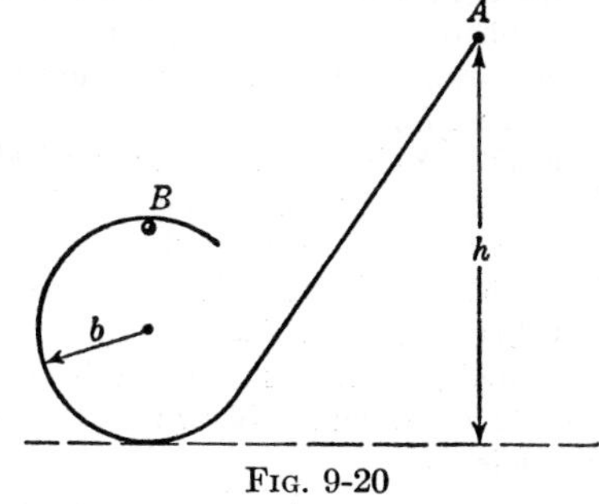

FIG. 9-20

9-9. A homogeneous circular disk has a light string wrapped around its circumference a number of times. One end of the string is attached to a fixed point. If the disk is permitted to fall under gravity, with the string unwinding, find the acceleration of the center of the disk.

9-10. A homogeneous rod of length $2a$ stands vertically upright on a perfectly rough floor. If it is very slightly disturbed, find its angular velocity as it strikes the floor.

9-11. A uniform plank of length L is balanced on a fixed horizontal cylinder of radius a. The length of the plank is at right angles to the axis of the cylinder. If the plank is set rocking, without slipping, show that the motion is simple harmonic for small amplitudes, and find the period.

9-12. A uniform plank is suspended horizontally by two vertical ropes, one at each end. If one is severed, find the tension, immediately afterward, in the remaining rope.

9-13. Show that the ratios of the times of descent, along an inclined plane, for a spherical shell and solid sphere are $5/\sqrt{21}$ if both bodies are homogeneous.

9-14. A smooth wire is bent into the form of a circle of radius a in the vertical plane. Two beads of mass m each are constrained to slide on the wire and are connected by a massless rigid rod of length $2b$, where $b < a$. Find, in terms of a and b, the length of the equivalent simple pendulum.

9-15. A homogeneous sphere rests on top of a larger sphere, the latter being fixed. If the small sphere is slightly disturbed find the angle which the line of centers makes with the vertical when the two spheres separate, assuming the contact between the two spheres to be perfectly rough.

9-16. A uniform rod of mass m is placed like a ladder with one end against a smooth vertical wall and the other end on a smooth horizontal plane. It is released from rest at an inclination α to the vertical. Show that the initial reactions of the wall and floor are, respectively,

$$\tfrac{3}{4}mg \cos \alpha \sin \alpha \qquad mg(1 - \tfrac{3}{4} \sin^2 \alpha)$$

and that the angle of inclination at which the rod will leave the wall is

$$\cos^{-1} (\tfrac{2}{3} \cos \alpha)$$

9-17. A uniform heavy rod is suspended by two equal vertical cords of length b, one at each end of the rod. Find the length of the equivalent simple pendulum for small torsional oscillations in which the center of mass describes only vertical motion.

9-18*C*. A homogeneous solid hemisphere is held with its base against a smooth vertical wall and its lowest point on a smooth floor. The hemisphere is released. Find the initial reactions of the wall and the floor.

9-19*C*. A homogeneous sphere is projected, without rotation, up a rough inclined plane of inclination α and coefficient of friction μ. Show that the time during which the sphere ascends the plane is the same as if the plane were smooth and that the time during which the sphere slides stands to the time during which it rolls in the ratio $(2 \tan \alpha)/7\mu$.

9-20*C*. A circular hoop in a vertical plane is projected down an inclined plane with velocity v_0 and at the same time is given an angular velocity ω_0 tending to make it roll up the plane. Find the relation among v_0, ω_0, the slope of the plane, and the coefficient of friction if the hoop comes to a position of instantaneous rest.

9-21*C*. A sphere of mass m rolls down the rough face of an inclined plane of mass M and angle α which is free to slide on a smooth horizontal plane in a direction perpendicular to its edge. Show that the normal force between the sphere and the inclined plane is

$$\frac{m(2m + 7M)g}{(2 + 5 \sin^2 \alpha)m + 7M} \cos \alpha$$

9-22. A homogeneous circular cylinder of mass M and radius a is free to rotate without friction about its axis, which is fixed. A rough string is wrapped many times around the cylinder and carries at one end a hanging mass m. If the system is permitted to unwind freely, find the tension in the string and the acceleration of m.

9-23. A ball of radius a is set spinning with an angular velocity ω_0 about a horizontal axis. It is then released, just touching the surface of a horizontal table. If the coefficient of sliding friction between the ball and the table is μ, how far will the ball go before pure rolling sets in?

9-24. A homogeneous solid sphere of mass m and radius a is given an angular velocity ω_0 about a horizontal axis passing through the center. It is then placed in contact with an inclined plane of inclination ϵ, where ϵ is also the angle of sliding friction between the sphere and the plane. The initial angular velocity ω_0 of the sphere is such as to tend to make the ball roll up the plane along the line of greatest slope. If the initial point of contact between the sphere and the plane is a distance b from the foot of the plane, find the total time elapsed, from the initial contact, for the sphere to arrive at the foot of the plane.

9-25. A circular hoop of mass m and radius a is projected down a rough inclined plane of inclination α with a velocity v. At the same time the hoop is given an angular velocity such as to tend to make it roll up the plane. Initially the hoop is at a vertical height h above the foot of the plane. If the hoop is constrained to remain in a vertical plane and if it just comes to rest at the foot, find (*a*) the coefficient of friction between the hoop and the inclined plane and (*b*) the initial angular velocity of the hoop.

9-26. A solid homogeneous hemisphere of radius a rests with its flat side up on a perfectly rough horizontal table. Find the length of the equivalent simple pendulum for small oscillations about the equilibrium position. Assume that the radius of gyration about a horizontal axis passing through the center of mass is k. Justify any approximations.

9-27. A semicircular cylindrical shell has a mass M and a radius b. Show that the mass center lies a distance $2b/\pi$ from the axis of the cylinder. Show also that the moment of inertia about the line through the mass center parallel to the axis is

$$mb^2\left(1 - \frac{4}{\pi^2}\right)$$

If the shell is placed open side up on a smooth table, show that, if disturbed slightly from the equilibrium configuration, the cylinder executes simple harmonic motion. Find the angular frequency of this motion, being careful to justify all approximations.

9-28. Find the angular frequency for the case of Prob. 9-27, save that the contact between the shell and the table is now perfectly rough.

9-29. In Fig. 9-21, a rod of mass m and length $2b$ lies on a smooth horizontal table the surface of which is in the plane of the paper. At one end a smooth peg A is fixed into the table. The side of the rod rests against the peg. An impulse P is communicated to the rod at the end B, as shown. Find the magnitude and direction of the impulse given to the peg.

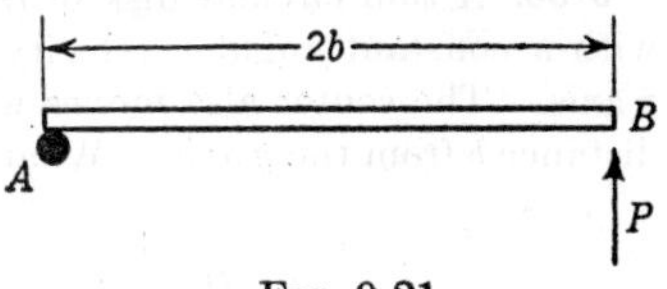

FIG. 9-21

9-30. Two men support by the ends a uniform pole of mass m and length $2b$ in a horizontal position. They wish to change ends without changing their positions on the ground by throwing the pole into the air and catching it. If the pole is to remain horizontal throughout its flight and the magnitude of the impulsive force applied by each man is to be a minimum, find the magnitude and direction of the impulse applied by each man in throwing the pole.

9-31. In Fig. 9-22, a homogeneous sphere of mass m spinning with an angular velocity ω_0 about a horizontal axis strikes vertically with a velocity v_0 on a rough table, the coefficient of friction between the sphere and the table being μ. It rebounds with an angular velocity ω and the horizontal velocity u. The radius of the sphere is b, and the coefficient of restitution is e. What is the largest value of ω_0 the sphere may have and still rebound with an angular velocity ω satisfying $b\omega = u$? Express your answer in terms of b, ω_0, v_0, μ, e.

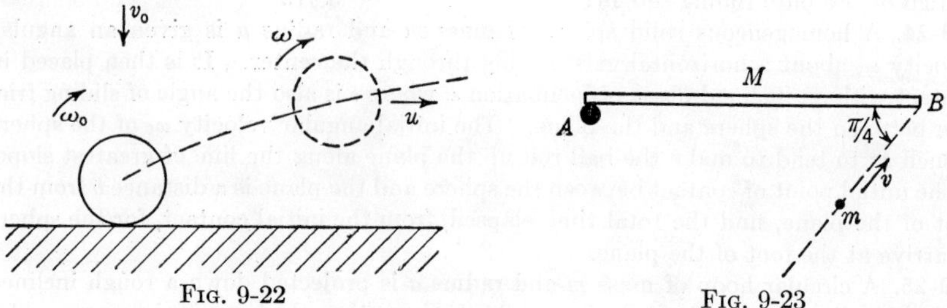

FIG. 9-22 FIG. 9-23

9-32. In Fig. 9-23, a smooth uniform rod of length $2b$ and mass M lies on a smooth table. At one end A the rod rests against a smooth peg which is rigidly attached to the table. A particle of mass m, traveling with a velocity v in a direction as shown, making an angle of $\pi/4$ with the rod, strikes the rod at B. The coefficient of restitution at the points of contact A and B is e. What is the recoil velocity of m? Find also the angular velocity of the rod immediately after the impact, and the impulse delivered to the peg.

9-33C. A horizontal rod of mass m and length $2a$ hangs by two parallel strings of length $2a$ attached to its two ends. If an angular velocity ω is suddenly communicated to it about a vertical axis through its center, show that the initial increase of tension of either string is $ma\omega^2/4$ and that the rod subsequently rises a distance $a^2\omega^2/6g$.

9-34. A circular lamina rests on a smooth horizontal table. Where should the lamina be struck so that it will begin to turn around a point on its circumference?

9-35. A sphere of mass m, radius a, and moment of inertia I rolls with a velocity v on a rough table. The coefficient of sliding friction is μ. The sphere hits a smooth vertical wall at an angle θ with the normal and rebounds. If the coefficient of restitution between the sphere and the wall is e, find the initial linear and angular velocities after the collision.

9-36. A thin circular disk of radius a and mass m lies in the xy plane and rotates with a constant angular velocity ω about an axis through its center parallel to the z axis. The center also moves with a uniform linear velocity v parallel to and at a distance b from the y axis. What is the angular momentum relative to the origin?

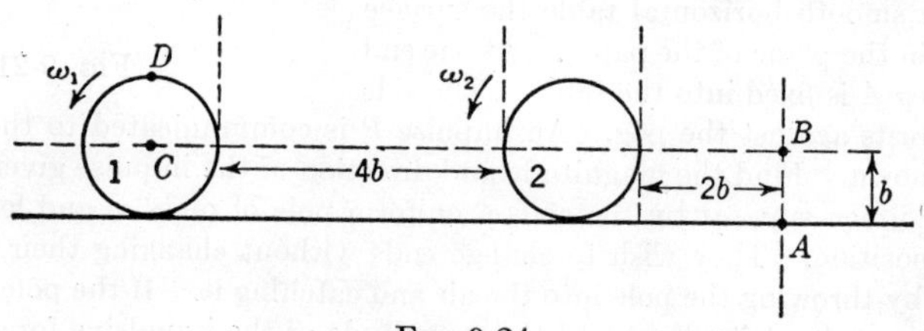

FIG. 9-24

9-37. Two identical disks are rolling on a plane as shown in Fig. 9-24. Each has a mass m and radius b. They are rolling with angular velocities ω_1 and ω_2 as shown. Points A and B are fixed in the paper, and points C and D are fixed in disk 1.

a. What is the total angular momentum of the system about each of points A, B, C, D?

b. If the plane is smooth so that the mass centers move with velocities v_1 and v_2 to the right while still spinning with angular velocities ω_1 and ω_2 as shown, what is the new angular momentum of the system about each of points A, B, C, D?

9-38. The disk in Fig. 9-25 has mass m, moment of inertia I, radius b and is rolling down the plane with an instantaneous angular velocity ω as shown.

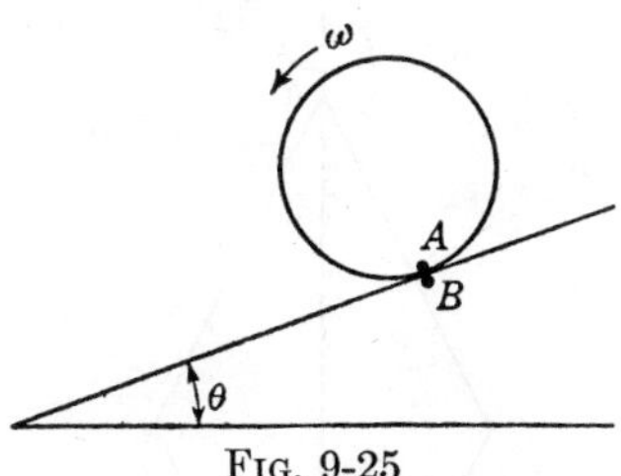

FIG. 9-25

a. What is the angular momentum of the wheel about the point A (the point of the wheel instantaneously in contact with the plane), about the point B (the point, fixed in the plane, that is instantaneously in contact with the wheel), and about the point C (coincident instantaneously with B in the figure) if C is moving down the plane in a manner so as to be always in contact with the wheel?

b. If the wheel is slipping but has a constant angular velocity ω, what is the angular momentum about each of points B and C? The instantaneous value of the translational velocity of the center of the disk is v.

c. In which of the cases above would $\mathbf{J} = \dot{\mathbf{L}}$ with reference to each of the various points taken as the origin of a system of coordinates?

9-39. A cylindrical shell of radius b and mass m rolls on the inclined surface of a wedge which, in turn, rests on a smooth horizontal table. The wedge has a mass M, and the inclined surface of the wedge is at an angle α with the horizontal. If the entire system is released from rest, with the cylinder some distance up the wedge from the foot, find the acceleration of the wedge. Assume the contact between the cylinder and the wedge to be perfectly rough.

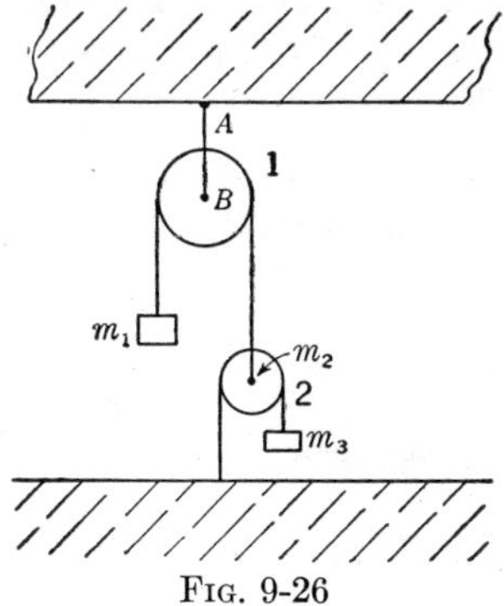

FIG. 9-26

9-40. In Fig. 9-26, pulley 1 is massless, and pulley 2 has a mass m_2. The contacts of the strings passing over each pulley are perfectly rough. Determine the vertical accelerations of each of the three masses, considering that pulley 2 is a uniform disk of radius b. Find also the tension in cord AB.

9-41. In Fig. 9-27, four identical uniform rods, each of mass m and length b, are hinged together to form a perfectly flexible rhombus and are mounted so that they can pivot freely on a fixed vertical shaft. The upper end A is fixed, while the lower end C can slide freely up and down the shaft. Initially C coincides with A, and the system

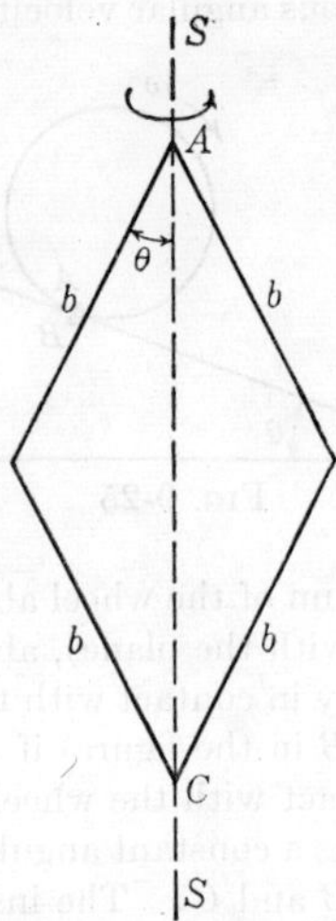

FIG. 9-27

is given an angular velocity ω_0 and released. Find a trigonometric relation involving θ for the condition when C is at its lowest point.

9-42. Two equal rods are smoothly hinged together at one end and lie in a straight line ABC on a smooth horizontal table. A horizontal impulse is delivered at C, normal to ABC. Show that the energy acquired by the rods, smoothly joined at B, is $\frac{7}{4}$ of that which they would acquire if the joint were rigid.

9-43. In the system of Sec. 9-16, consider the two rods to be equal, of length $2b$ and mass m, with AB and BD forming a right angle. AB is horizontal, and BD is vertical, with D toward the top of the paper. The impulse P_0 is applied at D in a direction parallel to BA. Find, immediately after P_0 is applied, the components of the velocities of the mass centers of the rods, their angular velocities, and the impulse at the joint.

9-44. Two gear wheels of radii b_1 and b_2 and axial moments of inertia I_1 and I_2, respectively, can rotate freely about fixed parallel axes. Initially the wheel of radius b_1 is rotating with an angular velocity ω_1, while the other wheel is at rest.

a. Find the total angular momentum of the system relative to the axis of wheel 1; relative to the axis of wheel 2. (Assume that an infinitesimal displacement would engage the wheels.)

b. The wheels are suddenly engaged. Find the angular velocity of each wheel afterward.

c. Find the total angular momentum of the system about the axis of wheel 1 after the gears have been engaged.

d. Explain qualitatively why angular momentum is not conserved.

e. Show that the loss in energy is

$$\frac{1}{2}\, b_1{}^2\omega_1{}^2\, \frac{I_1 I_2}{b_2{}^2 I_1 + b_1{}^2 I_2}$$

f. Discuss the situation for gears which are perfectly elastic.

CHAPTER 10

MOTION OF A PARTICLE UNDER THE ACTION OF A CENTRAL FORCE

In this present chapter we shall be concerned with the motion of a particle under the influence of a *central force.* It will be assumed that the line of action of the force passes always through a point which is fixed in a Newtonian reference system, which point is chosen as the origin. It was shown in Chap. 8 that, if the angular momentum defined with respect to this origin is to be a constant of the motion, a sufficient condition is that the force be central. In such a case the angular momentum is a constant vector. This means not only that the angular momentum is constant in magnitude but also that its direction in space is fixed. The latter fact ensures that the motion will lie entirely in one plane. Consequently it is sufficient to select a system of coordinates lying in this plane in order to treat the problem of central-field motion for a single particle.

10-1. Kinetic Energy in Plane Polar Coordinates. In Appendix 1, the plane polar coordinates r, θ are defined in terms of the rectangular coordinates x and y so that

$$x = r \cos \theta \qquad y = r \sin \theta \tag{10-1}$$

In rectangular coordinates the kinetic energy T of a particle of mass m having a velocity v is

$$T = \frac{m}{2} v^2 = \frac{m}{2} (\dot{x}^2 + \dot{y}^2) \tag{10-2}$$

The corresponding expression can be obtained in polar coordinates by substituting (10-1) in (10-2), performing the necessary differentiations. However, the result also can be obtained from geometrical considerations. In Fig. 10-1, a particle suffers a displacement $d\mathbf{r}$, from P to Q, during a small time interval dt. If $\mathbf{v}$ is its velocity during this time, then

$$d\mathbf{r} = \mathbf{v}\, dt$$

But $d\mathbf{r}$ has components $\mathbf{r}_1\, dr$, along $\mathbf{r}$, and $\boldsymbol{\theta}_1 r\, d\theta$, perpendicular to $\mathbf{r}$, where $\mathbf{r}_1$ and $\boldsymbol{\theta}_1$ are unit vectors, respectively, along r increasing and

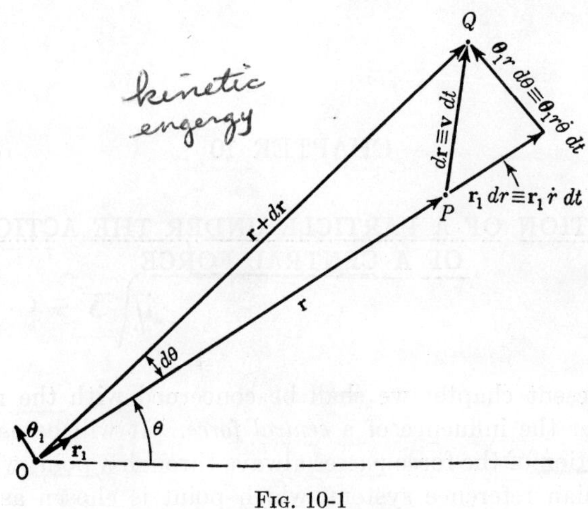

FIG. 10-1

normal to $\mathbf{r}$ in the direction of increasing θ. Now

$$\boldsymbol{\theta}_1 r\, d\theta = r\dot{\theta}\boldsymbol{\theta}_1\, dt$$

Hence

$$\mathbf{v}\, dt = \mathbf{r}_1 \dot{r}\, dt + \boldsymbol{\theta}_1 r\dot{\theta}\, dt$$

Dividing through by dt and squaring both sides (that is, taking the dot product of each side with itself), we obtain

$$v^2 = \dot{r}^2 + r^2\dot{\theta}^2 \qquad (10\text{-}3)$$

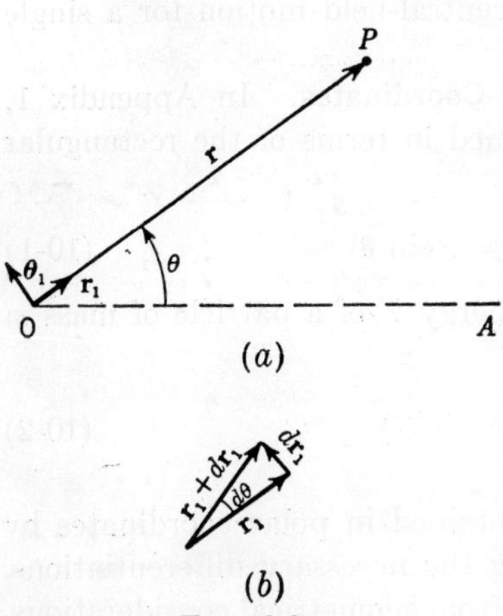

FIG. 10-2

This follows since $\mathbf{r}_1 \cdot \mathbf{r}_1 = 1$, $\boldsymbol{\theta}_1 \cdot \boldsymbol{\theta}_1 = 1$, and $\mathbf{r}_1 \cdot \boldsymbol{\theta}_1 = 0$.

10-2. Acceleration in Plane Polar Coordinates. The component of acceleration along $\mathbf{r}$, and that perpendicular to $\mathbf{r}$, both may be obtained, either from the second time derivatives of Eqs. (10-1) or by a vector method which we now consider. In Fig. 10-2*a*, the origin is at point O, and OA is the polar axis. The polar coordinates r, θ of the point P are as shown. As in the previous section, we define a unit vector $\mathbf{r}_1$ pointing in the sense of increasing r and a unit vector $\boldsymbol{\theta}_1$ pointing in the sense of increasing θ. Both $\mathbf{r}_1$ and $\boldsymbol{\theta}_1$ are dimensionless. Thus the radius vector $\mathbf{r}$ may be written

$$\mathbf{r} = \mathbf{r}_1 r$$

where r is the scalar magnitude of $\mathbf{r}$. From this we are able to calculate the components of the vector acceleration $\ddot{\mathbf{r}}$. We first take the time derivative of $\mathbf{r}$, which is $\dot{\mathbf{r}}$. In doing so, we recall that, even though $\mathbf{r}_1$ is constant in magnitude, it is not fixed in direction and therefore $\dot{\mathbf{r}}_1 \neq 0$. We have

$$\dot{\mathbf{r}} = \dot{r}\mathbf{r}_1 + r\dot{\mathbf{r}}_1 \tag{10-4}$$

From Fig. 10-2*b*, it is apparent that $d\mathbf{r}_1$ is a vector which is perpendicular to $\mathbf{r}_1$ and pointing in the counterclockwise direction. Thus it has the direction of $\boldsymbol{\theta}_1$. Therefore

$$d\mathbf{r}_1 = k\boldsymbol{\theta}_1$$

where k is a scalar factor to be determined. The vertex angle of the triangle in Fig. 10-2 is $d\theta$, and therefore, since $|\mathbf{r}_1| = 1$,

$$|d\mathbf{r}_1| = |\mathbf{r}_1|d\theta = d\theta$$

Clearly the magnitude k of $d\mathbf{r}_1$ is simply $d\theta$, and

$$d\mathbf{r}_1 = d\theta\,\boldsymbol{\theta}_1$$

Dividing through by dt and substituting in Eq. (10-4), we have

$$\dot{\mathbf{r}} = \dot{r}\mathbf{r}_1 + r\dot{\theta}\boldsymbol{\theta}_1$$

Following this, we differentiate a second time, to obtain

$$\ddot{\mathbf{r}} = \ddot{r}\mathbf{r}_1 + 2\dot{\theta}\dot{r}\boldsymbol{\theta}_1 + r\ddot{\theta}\boldsymbol{\theta}_1 + r\dot{\theta}\dot{\boldsymbol{\theta}}_1$$

In a manner similar to that employed in the calculation of $\dot{\mathbf{r}}_1$ we find that $\dot{\boldsymbol{\theta}}_1$ is a vector having the direction of $-\mathbf{r}_1$ and also having the magnitude $\dot{\theta}$. Therefore

$$\ddot{\mathbf{r}} = (\ddot{r} - r\dot{\theta}^2)\mathbf{r}_1 + (r\ddot{\theta} + 2\dot{r}\dot{\theta})\boldsymbol{\theta}_1 \tag{10-5}$$

Accordingly the components of the acceleration in the directions of increasing r and θ are, respectively,

$$\ddot{r} - r\dot{\theta}^2 \qquad r\ddot{\theta} + 2\dot{r}\dot{\theta} \tag{10-6}$$

10-3. Areal Velocity and Angular Momentum of a Particle Moving in a Central Field. It was pointed out in Sec. 8-1 that the quantity $mr^2\dot{\theta}$ is the angular momentum J. It is important to notice a geometrical significance of part of the latter quantity, $r^2\dot{\theta}$, which is very useful in central-field problems. To see this, consider Fig. 10-3, in which a particle of mass m is traversing the trajectory SS' under the influence of a central force the

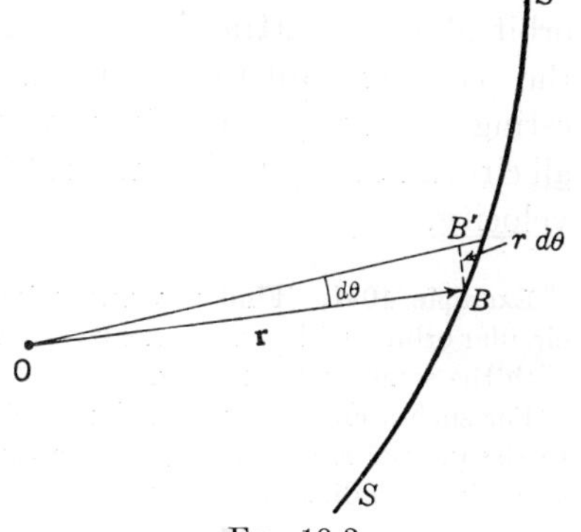

FIG. 10-3

center of which is at O. The area dA of the infinitesimal triangle OBB' is

$$dA = \tfrac{1}{2}rr\,d\theta = \tfrac{1}{2}r^2\,d\theta$$

from which

$$\frac{dA}{dt} = \frac{1}{2}r^2\frac{d\theta}{dt}$$

Thus

$$r^2\dot{\theta} = 2\dot{A} \qquad (10\text{-}7)$$

and the very useful result is obtained that $r^2\dot{\theta}$ (or *moment of velocity*) is equal to twice the rate at which area is being swept out by the radius vector $\mathbf{r}$. It is interesting to note that, if rectangular coordinates had been used, Eq. (10-7) would have become

$$(x\dot{y} - y\dot{x}) = 2\dot{A} \qquad (10\text{-}8)$$

the proof of which is left to the student. The left side is $1/m$ times the component of angular momentum along the z axis, that is,

$$J_z = mr^2\dot{\theta} = m(x\dot{y} - y\dot{x}) = 2m\dot{A} \qquad (10\text{-}9)$$

(Since the motion is confined to the xy plane the only component of J is along the z axis.) We are able to proceed further since we perceive the very simple relation between the areal velocity and the angular momentum. For a central field, since J is a constant of the motion, the *areal velocity* is a constant. The latter half of the statement depends upon the constancy of mass, and it is worthwhile to point out that the areal velocity is constant provided that the translational velocity is not large enough to bring in, appreciably, the variation of mass with velocity. This becomes important, for relativity reasons, at high velocities. This behavior is beyond the scope of the present text.

It is to be emphasized that the areal velocity is a constant for a given orbit about an attracting center, and it is not to be inferred that it is the same constant for all orbits. In general it is different. It is interesting, however, to consider in the following example a force for which all circular orbits, with centers at the force center, do have a common areal velocity.

Example 10-1. Find the attractive central force of the form $f(r)$ in which all circular orbits, with centers of curvature coinciding with the force center, are described with the same areal velocity.

For such a circular orbit to be possible, we need only equate the centripetal force to the inertial reaction mv^2/r, where m and v are the mass and speed of the particle and r is the radius of the orbit. Thus

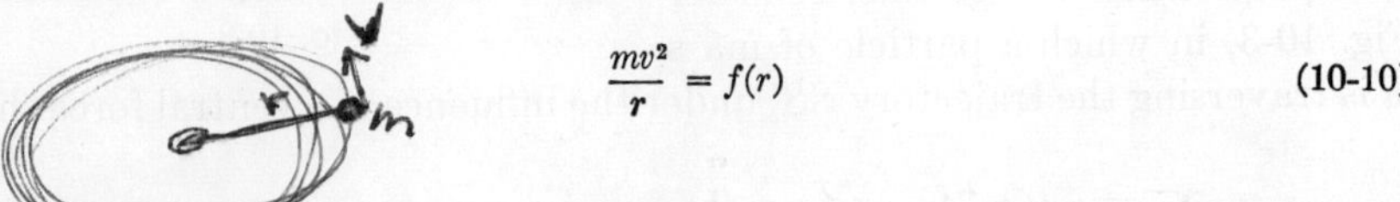

$$\frac{mv^2}{r} = f(r) \qquad (10\text{-}10)$$

But $v = r\dot{\theta}$, and therefore Eq. (10-10) becomes

$$\frac{m\dot{\theta}^2r^2}{r} = \frac{m}{r^3}(r^4\dot{\theta}^2) = f(r) \tag{10-11}$$

In view of Eq. (10-7), the quantity $r^2\dot{\theta}$ is equal to twice the areal velocity, which is a constant. If this constant is $2h$, Eq. (10-11) becomes

$$f(r) = \frac{4mh^2}{r^3}$$

Therefore the attractive force for which all circular orbits are described with the same areal velocity is that of the inverse cube. This means also that all circular orbits for this law of force have the same angular momentum ($J = 2mh$) provided that the centers of curvature are located at the center of force.

INVERSE-SQUARE FIELD: THE ORBIT

10-4. Integration of the Equations of Motion. The determination of the orbit of a particle moving in a central field of force of the form $F = f(r)$ involves the integration of the two equations of motion

$$m(\ddot{r} - r\dot{\theta}^2) = f(r) \tag{10-12}$$
$$m(r\ddot{\theta} + 2\dot{r}\dot{\theta}) = 0 \tag{10-13}$$

where the zero on the right side of Eq. (10-13) follows from the fact that the force is a central one and consequently possesses no θ component. Equations (10-12) and (10-13) demonstrate, also, the complexity of the equations of motion when coordinates other than rectangular ones are employed. In the rectangular case, the inertial reaction has the simple form $m\ddot{x}$, and so on, in which the acceleration is simply the second time derivative in the direction of the coordinate increasing. Equations (10-12) and (10-13) indicate in general that, in an arbitrary set of coordinates, the acceleration may no longer possess such a simple form. Indeed, the term $-r\dot{\theta}^2$ in Eq. (10-12) demonstrates that there exists an acceleration in the **r** direction even if r is not changing in magnitude ($-mr\dot{\theta}^2$ is the centrifugal part of the inertial reaction).

Equations (10-12) and (10-13) are second-order equations, and the solution of Eq. (10-12), at least, introduces complications. However, for the example of the inverse-square force, $f(r) = -k/r^2$, where k is a constant, the procedure is straightforward. Equation (10-13) is equivalent to

$$\frac{m}{r}\frac{d}{dt}(r^2\dot{\theta}) = 0$$

which may be integrated directly to become

$$mr^2\dot{\theta} = \text{const} = J \tag{10-14}$$

the familiar expression for the angular momentum. For the inverse-square case Eq. (10-12) takes the form

$$m(\ddot{r} - r\dot{\theta}^2) = -\frac{k}{r^2} \tag{10-15}$$

from which $\dot{\theta}$ may be eliminated by the use of Eq. (10-14). We obtain

$$\ddot{r} - \frac{J^2}{m^2r^3} = -\frac{k}{mr^2} \tag{10-16}$$

which is an equation in r and t alone. If it were desirable, we could in principle solve both (10-16) and (10-14), obtaining r and θ as functions of t. We could then find the equation of the path by eliminating t between them. However, we shall employ a substitution which is often used in the treatment of central fields. First let us note that

$$\dot{r} = \frac{dr}{d\theta}\dot{\theta} = \frac{dr}{d\theta}\frac{J}{mr^2}$$

from which

$$\ddot{r} = \frac{J^2}{m^2r^4}\left[\frac{d^2r}{d\theta^2} - \frac{2}{r}\left(\frac{dr}{d\theta}\right)^2\right]$$

Making use of the last expression, Eq. (10-16) assumes the form

$$\frac{d^2r}{d\theta^2} - \frac{2}{r}\left(\frac{dr}{d\theta}\right)^2 - r = -\frac{kmr^2}{J^2} \tag{10-17}$$

which is the differential equation of the orbit. Equation (10-17) is nonlinear and very complicated. The procedure mentioned above (useful in many central-field problems) is to make the substitution $r = 1/u$. We have

$$\frac{dr}{d\theta} = -\frac{1}{u^2}\frac{du}{d\theta}$$

and

$$\frac{d^2r}{d\theta^2} = \frac{2}{u^3}\left(\frac{du}{d\theta}\right)^2 - \frac{1}{u^2}\frac{d^2u}{d\theta^2}$$

Employing these in Eq. (10-17), the differential equation takes the simple form

$$\frac{d^2u}{d\theta^2} + u = \frac{km}{J^2} \tag{10-18}$$

which is the equation of a forced oscillator with a constant forcing term km/J^2. It is worthwhile to note that, for the case of a general central force $F = f(r)$, instead of (10-18) we should have found

$$\frac{d^2u}{d\theta^2} + u = -\frac{m}{J^2}r^2f(r) \tag{10-19}$$

The solution of Eq. (10-18) can be written by inspection to be

$$u = A \cos (\theta + \delta) + \frac{km}{J^2} \tag{10-20}$$

In turn, r becomes

$$r = \frac{1}{u} = \frac{1}{(km/J^2) + A \cos (\theta + \delta)} \tag{10-21}$$

In order to interpret Eq. (10-21) let us consult Fig. 10-4. Let SS' depict a portion of a *conic* of which $O'O''$ is the *polar axis*, DD' the *directrix*, O the *focus*, θ the polar angle, and r the radius vector. From the definition of a conic section it easily follows that

$$r = \frac{1}{(1/ep) - (1/p) \cos \theta} \tag{10-22}$$

where e is the *eccentricity* and p is the distance $O'O$. Comparing (10-21) and (10-22), we see that Eq. (10-21) is the equation of a conic. For convenience let us put $\delta = \pi$, which operation renders the major axis coincident with the polar axis. Thus

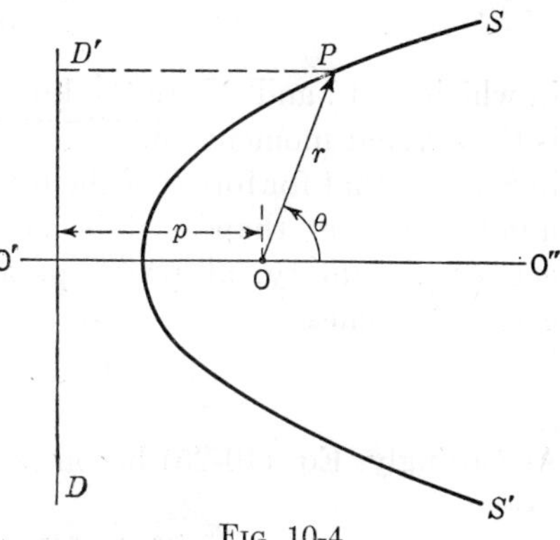

FIG. 10-4

$$r = \frac{1}{(km/J^2) - A \cos \theta} \tag{10-23}$$

The arbitrary constant A may be determined from the boundary conditions in any problem which may be considered.

Example 10-2. A particle of mass m, under the action of a central force, describes an orbit which is a circle of radius a passing through the center of force. It is desired to find the law of force.

This is an excellent example of the use of Eq. (10-19). Now the polar equation of a circle of radius a, with origin on the circumference, is

$$r = 2a \cos \theta \qquad \text{or} \qquad u = \frac{1}{2a \cos \theta} \tag{10-24}$$

From Eq. (10-19) we have

$$f(r) = -\frac{J^2}{m} u^2 \left(\frac{d^2u}{d\theta^2} + u \right)$$

in which $d^2u/d\theta^2$ may be determined by the differentiation of the second of Eqs. (10-24). It follows quite readily that

$$\frac{d^2u}{d\theta^2} + u = \frac{1}{a \cos^3 \theta}$$

and therefore the law of force is

$$f(r) = -\frac{8a^2J^2}{mr^5}$$

having the form of an inverse fifth-power attractive force.

10-5. Equation of the Orbit by the Energy Method. It is of interest to see the simplifications introduced in the determination of the orbit by use of the energy and angular-momentum integrals. We may write

$$T + V = W \tag{10-25}$$
$$mr^2\dot{\theta} = J \tag{10-26}$$

in which T, V, and W are the kinetic, potential, and total energies and J is the angular momentum. Both W and J are constants. We have seen in Sec. 5-7 that for forces of the form k/r^n, where k is a constant and n is a number greater than 1, it is customary to take the zero of potential energy at infinity, at which point the force vanishes. The potential energy becomes

$$V = -\int_\infty^r f(r)\,dr \tag{10-27}$$

Accordingly, Eq. (10-25) becomes

$$\frac{m}{2}(\dot{r}^2 + r^2\dot{\theta}^2) - \int_\infty^r f(r)\,dr = W \tag{10-28}$$

Now

$$\dot{r}^2 = \dot{\theta}^2\left(\frac{dr}{d\theta}\right)^2$$

so that Eq. (10-28) assumes the form

$$\frac{m}{2}\left[\left(\frac{dr}{d\theta}\right)^2 + r^2\right]\left(\frac{d\theta}{dt}\right)^2 - \int_\infty^r f(r)\,dr = W \tag{10-29}$$

But $\dot{\theta}$ may be eliminated by means of Eq. (10-26), and therefore we have

$$\left[\left(\frac{dr}{d\theta}\right)^2 + r^2\right]\frac{J^2}{2mr^4} - \int_\infty^r f(r)\,dr = W \tag{10-30}$$

which is the differential equation of the orbit. In the present case of interest, $f(r) = -k/r^2$, and V becomes $-k/r$. Accordingly, Eq. (10-30) may be written

$$\left[\left(\frac{dr}{d\theta}\right)^2 + r^2\right]\frac{J^2}{2mr^4} - \frac{k}{r} = W \tag{10-31}$$

The variables may be separated in Eq. (10-31), yielding

$$d\theta = \frac{J\,dr}{r(2mWr^2 + 2mkr - J^2)^{\frac{1}{2}}}$$

This may be integrated with the aid of tables. We obtain

$$\theta = \sin^{-1} \frac{mkr - J^2}{r(2J^2mW + m^2k^2)^{\frac{1}{2}}} - \alpha$$

in which α is a constant of integration. Choosing $\alpha = \pi/2$ and solving for r, we obtain

$$r = \frac{J^2/mk}{1 - [(2WJ^2/mk^2) + 1]^{\frac{1}{2}} \cos \theta} \tag{10-32}$$

We again have the familiar equation of a conic section. Comparing with Eq. (10-22), we note that the eccentricity e may be expressed as

$$e = \left(\frac{2WJ^2}{mk^2} + 1\right)^{\frac{1}{2}} \tag{10-33}$$

The various possible values of e, together with the orbits to which they correspond, will be discussed in Sec. 10-6.

The analytical details of integrating the equations of motion to find the orbit, as in Sec. 10-4, or of using the energy principle to determine the orbit, as in the present section, should not blind us to the existence of many simple situations in which the required information may be determined without the necessity of determining the orbit. Such is the case in the following example:

Example 10-3. A particle of mass m moves in the field of a fixed force center O from which it is repelled with a force of magnitude mc/r^3, where c is a constant. At a very large distance from O the particle is moving with a velocity v_0 which, if the particle were not deflected, would carry it along a straight line passing O at a distance b. It is desired to find the closest distance of approach of m to O in the actual motion. (The quantity b, used in this sense, is often referred to as the *impact parameter*, and the distance of closest approach, a, as the so-called *turning radius*.)

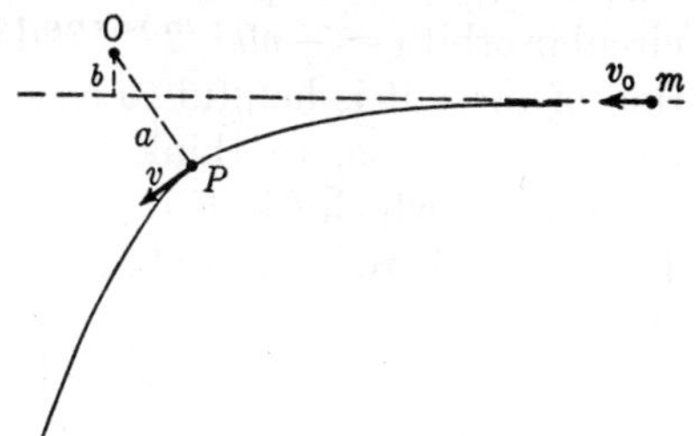

FIG. 10-5

The path is as shown in Fig. 10-5, with point O as the origin. Evidently the closest point of approach is P, at which point the radial velocity $\dot{r}$ is zero and the direction of motion is perpendicular to OP. If we let OP be represented by the symbol a, the potential energy has the value

$$V_P = -\int_{\infty}^{a} \frac{mc}{r^3}\, dr = +\frac{mc}{2a^2}$$

at point P. Accordingly, the equation of energy provides the expression

$$\frac{1}{2} mv_0^2 = \frac{1}{2} mv^2 + \frac{mc}{2a^2} \tag{10-34}$$

where v is the velocity at point P. Still another expression is afforded by the conservation of angular momentum. This is

$$mv_0 b = mva \tag{10-35}$$

Equations (10-34) and (10-35) may be solved simultaneously for a by eliminating v. We find

$$a = \left(b^2 + \frac{c}{v_0^2}\right)^{\frac{1}{2}}$$

10-6. Energy and Classification of the Orbits. It is useful to examine the possible values of the energy of a particle moving under the action of an inverse-square force, since the curve represented by Eqs. (10-23) and (10-32) may be a *circle, ellipse, parabola,* or *hyperbola,* depending upon the value of the eccentricity e. The total energy W of the particle may be written in terms of the angular momentum J and the eccentricity e by rearranging Eq. (10-33). We obtain

$$W = \frac{mk^2}{2J^2}(e^2 - 1) \tag{10-36}$$

Accordingly, the nature of W for the conic sections corresponding to the respective ranges of e may be outlined as follows:

$e = 0$ (circle)	$W = W_c \left(= -\dfrac{mk^2}{2J^2}\right)$
$0 < e < 1$ (ellipse)	$W = W_e,\ W_c < W_e < 0$
$e = 1$ (parabola)	$W = W_p = 0$
$e > 1$ (hyperbola)	$W = W_h > 0$

where W_c, for example, is the total energy of the particle moving in the circular orbit $(= -mk^2/2J^2)$ and is negative. For the further interpretation of these it is helpful to construct a potential-energy diagram. It is convenient, also, to think in terms of the equivalent one-dimensional problem, in which $\dot{\theta}$ in Eq. (10-28) is expressed in terms of J. Thus since $V = -k/r$ here, Eq. (10-28) becomes

$$\frac{m}{2}\dot{r}^2 + \left(\frac{J^2}{2mr^2} - \frac{k}{r}\right) = W \tag{10-37}$$

Hence the $(m/2)r^2\dot{\theta}^2$ part of the kinetic energy of the motion takes the form of a positive potential energy, varying as $1/r^2$, in the equivalent one-dimensional problem. This is sometimes referred to as the potential energy arising from the *centrifugal potential.* Equation (10-37) may be obtained, also, from the r equation of motion, Eq. (10-12), by putting $mr\dot{\theta}^2$ in terms of J, multiplying by $\dot{r}$, and integrating. The centripetal part, $mr\dot{\theta}^2$, of the inertial reaction [see Eq. (10-16)] takes the form J^2/mr^3. Consequently the centripetal term, so far as the equivalent one-dimensional problem is concerned, may be interpreted as an inverse-cube repulsive force. This term, sometimes called the *centrifugal force*, yields, in effect, a positive inverse-square centrifugal potential energy. In Fig. 10-6, the solid lines represent the potential energy $-k/r$, and the apparent potential energy $+J^2/2mr^2$, for the particular values J_0, m_0, k_0. The darker line is the total effective potential energy and is the algebraic sum of the two. The case of a particle traveling in an ellipse with total energy W_e (< 0) is represented in the energy diagram by the horizontal line at the level W_e. The dashed line from B_1 to B_2 is the range for which the values of r are given by $r_1 \leq r \leq r_2$. That this is the range of values of r within which the particle is compelled to remain is easily seen. Consider momentarily that the particle is at a distance r_B from the force center (corresponding to point B in the diagram). At this instant the particle has a potential energy V_B (point A on the potential-energy curve). Furthermore the kinetic energy T, consisting of a radial part T_r and a θ part T_θ, is represented in energy units by the line AB. The radial kinetic energy is represented by the line $A'B$ and the θ part of the kinetic energy (or centrifugal potential energy in the equivalent one-dimensional picture) by AA'. Thus when a particle, having the total energy W_e, is at a distance r_B from the center of force, it has a potential energy V_B, a kinetic energy, represented by $A'B$, for the r motion, and a kinetic energy, represented by AA', for the θ motion. When, on the other hand, the particle is at either point B_1 or point B_2, the radial kinetic energy is zero and the motion is entirely normal to the radius vector. Accordingly, r_1 is the distance of closest approach and r_2 that of farthermost recession. It is to be remembered that Fig. 10-6 is but an energy diagram and that, in the actual motion, B_1 is at a distance r_1 from the force center on the opposite side from B_2.

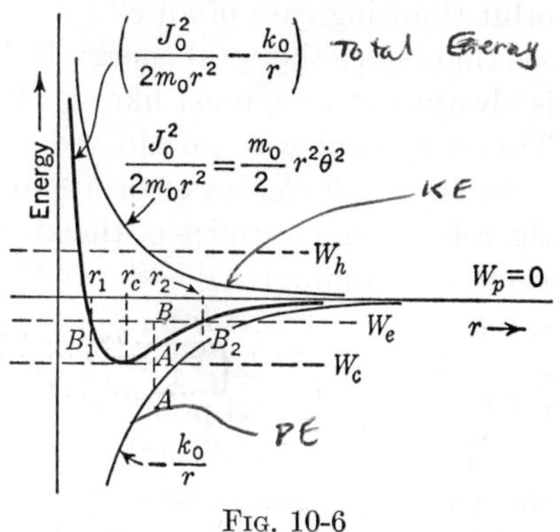

FIG. 10-6

Elliptical motion is possible for all values of W_e between W_c and zero, corresponding to a range of values of e between zero and 1, respectively. When $W = W_c$ (that is, when $e = 0$), there is no radial kinetic energy, the motion is in a circular path of radius r_c, and the centripetal force is balanced by the centrifugal reaction. (It will be shown below that, in such circular orbits in an inverse-square field, the kinetic energy is equal in magnitude to half the potential energy.) When $W = W_p = 0$, the orbit is parabolic ($e = 1$), and when $W = W_h > 0$, the motion follows a hyperbolic path, the closest distance of approach being that magnitude of r at which the appropriate W line intersects the curve of $J_0^2/2m_0r^2 - k_0/r$.

The quantities r_1 and r_2, for an orbit such as the ellipse represented in Fig. 10-6 by the line B_1B_2, may be found from Eq. (10-37) by putting $\dot{r}$ equal to zero. The result is the algebraic equation

$$2mWr^2 + 2kmr - J^2 = 0 \tag{10-38}$$

from which

$$r_1 = -\frac{k}{2W} - \sqrt{\frac{k^2}{4W^2} + \frac{J^2}{2mW}} \qquad r_2 = -\frac{k}{2W} + \sqrt{\frac{k^2}{4W^2} + \frac{J^2}{2mW}} \tag{10-39}$$

In the case of an elliptical orbit, with major axis $2a$, we have the important result

$$2a = r_1 + r_2 = -\frac{k}{W} \qquad \text{or} \qquad W = -\frac{k}{2a} \tag{10-40}$$

Hence, in an elliptical orbit in an inverse-square field, the total energy depends only on the major axis. This fact is useful in atomic theory and is the basis for the known property that the total energy may be expressed in terms of the *principal quantum number*. A further discussion of this point is beyond the scope of this text.

Equation (10-40) provides additional information regarding a circular orbit (limiting case of an ellipse). $V = -k/a$ for such an orbit, and we see that, since the total energy W is half of this, the kinetic energy, which is always positive, must likewise be equal to half of the potential energy. This statement was anticipated earlier.

So far as all classes of orbits in an inverse-square field are concerned, the relative magnitudes of the kinetic and potential energies may be conveniently summarized. Since the total energy $W = T + V$ is a constant, we see that, for an ellipse or circle $|V| > |T|$ at all points of the orbit, for a parabola $|V| = |T|$ at all points of the orbit, and for a hyperbola $|V| < |T|$ at all points of the orbit (for a circle $|V| = 2|T|$ always).

Consider again the total energy W. In the case of the ellipse, W has the value $-k/2a$. The quantity $k/2a$, then, is the amount of energy which would have to be supplied to the system in order to remove the particle, traveling in the elliptical orbit of major axis $2a$, to a state of rest an infinite distance away. In this sense the quantity $-W$ is sometimes called the *binding energy* of the system. It is positive ($k/2a$) for an ellipse and for a circle. The binding energy is zero for a parabolic orbit and signifies that no work is required to remove the particle. In the case of a hyperbolic orbit the binding energy is negative.

KEPLER'S LAWS

10-7. Statement of Kepler's Laws. In 1609 Johannes Kepler stated the first two of his laws of planetary motion:

I. *Every planet describes an ellipse with the sun at one focus.*

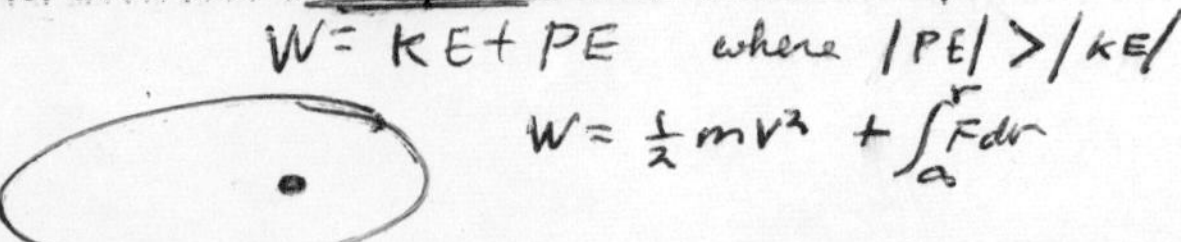

II. *The radius vector drawn from the sun to a planet describes equal areas in equal times.*

We have already deduced theoretically the contents of these two statements. Ten years later, in 1619, Kepler announced the third law:

III. *The squares of the periods of the different planets are proportional to the cubes of their respective major semiaxes.*

These laws were set forth on an empirical basis, and preceded the enunciation of Newton's laws by over half a century. Newton employed these to show that the law of force between the planets and the sun must be that of the inverse square (see below).

It is not difficult to deduce III from the material of the preceding pages. If a and b are the semimajor and semiminor axes, respectively, and if dA/dt is the rate at which the radius vector from the sun to the planet sweeps out area, the period T may be written

$$T = \frac{\pi ab}{dA/dt} = \frac{2\pi abm}{J}$$

where m is the mass of the planet and J is the angular momentum. Comparing Eqs. (10-22) and (10-23), it is possible to replace J in this expression, yielding

$$T = \frac{2\pi abm}{(kmep)^{\frac{1}{2}}} \tag{10-41}$$

Now it may be seen from Eq. (10-22) that the quantity ep is the semilatus rectum, which from a well-known property of an ellipse is just b^2/a. Accordingly, Eq. (10-41) becomes

$$T = \frac{2\pi a^{\frac{3}{2}} m^{\frac{1}{2}}}{k^{\frac{1}{2}}} \tag{10-42}$$

which is the mathematical statement of Kepler's third law. The presence of the mass m of the planet in the numerator is only apparent in this discussion, in which the motion of the attracting center is neglected, since k is proportional to the first power of m. The case in which notice is taken of the motion of the force center will be considered in Sec. 10-9.

Example 10-4. A particle of mass m describes an elliptical orbit about an attracting force center situated at one focus. The force is that of the inverse square. If e is the eccentricity, T the periodic time, and $2a$ the major axis, show that the greatest radial velocity of the particle is $2\pi ae/[T(1 - e^2)^{\frac{1}{2}}]$.

This will occur at the point at which $\ddot{r} = 0$, and therefore, from Eq. (10-16), $J^2 = mkr$. But J also may be expressed as $mr^2\dot{\theta}$. Therefore

$$\dot{r} = \frac{dr}{d\theta}\dot{\theta} = \frac{dr}{d\theta}\frac{mk^2}{J^3} \tag{10-43}$$

The quantity $dr/d\theta$ may be determined with the aid of Eqs. (10-31) and (10-40). We obtain

$$\left(\frac{dr}{d\theta}\right)^2 = \left(-\frac{k}{2a} + \frac{k}{r}\right)\frac{2mr^4}{J^2} - r^2$$

But, since $J = mkr$ when $\ddot{r} = 0$, this takes the form

$$\left(\frac{dr}{d\theta}\right)^2 = \frac{J^4}{m^2k^2}\left(1 - \frac{J^2}{mka}\right) \tag{10-44}$$

Now J can be expressed in terms of the eccentricity. From Eqs. (10-36) and (10-40) we obtain

$$J^2 = kma(1 - e^2)$$

and Eq. (10-44) becomes

$$\frac{dr}{d\theta} = ae(1 - e^2)$$

Inserting this in Eq. (10-43), and making use also of Eq. (10-42), we obtain the desired result.

10-8. Deduction of the Law of Force from Kepler's Laws. The *law of areas*, which is Kepler's second law, gives assurance that the force between the sun and the planets is a central one. Consequently we are free to employ Eq. (10-19) in order to determine the law of force $f(r)$. We have

$$\frac{d^2u}{d\theta^2} + u = -\frac{m}{J^2}r^2f(r) \tag{10-45}$$

in which $r = 1/u$. From Kepler's first law

$$u = \frac{1}{ep} - \frac{1}{p}\cos\theta$$

Accordingly, substituting this in Eq. (10-45) and solving for $f(r)$, we have

$$f(r) = -\frac{J^2}{mep}\frac{1}{r^2} = -\frac{k}{r^2}$$

where k is a constant, as shown. The negative sign indicates that the force is one of attraction. Kepler's third law assures us that the same law of force holds for every planet.

10-9. The Two-body Problem and Kepler's Third Law. In the present chapter we have assumed the attracting center to be at rest. This neglects the fact that it has a finite mass and that therefore the motion is actually a motion of both bodies about their common center of mass. If the finite mass M of the attracting center is taken into account, then, so far as the relative motion is concerned, we must replace the mass m in Eqs. (10-12) and (10-13) by the *reduced mass* $mM/(m + M)$ of the system (see Sec. 8-7). There is a further addition to be made. From the law of gravitation (Chap. 5) the quantity k has the value GMm, where G is the *gravitational constant.* Accordingly, Eqs. (10-14) and (10-15) become

$$\frac{mM}{m+M} r^2\dot{\theta} = J \tag{10-46}$$

$$\frac{mM}{m+M} (\ddot{r} - r\dot{\theta}^2) = -\frac{GMm}{r^2} \tag{10-47}$$

As was pointed out in Sec. 8-7, Eqs. (10-46) and (10-47), for relative motion, are the same no matter which mass is chosen to be at the origin. The solution of (10-47) follows in exactly the same way as did that of Eq. (10-15), and the result, as seen from the sun, is that each planet appears to describe an ellipse about the sun at a focus. Similarly, as seen from one of the planets the sun appears to describe an ellipse with that planet at one focus. The period, however (Kepler's third law), is slightly different from that expressed by Eq. (10-42). It may be found merely by replacing m in Eq. (10-42) by the reduced mass $mM/(m + M)$. Putting in the value $k \equiv mMG$ for the gravitational case, we have that the period of relative motion is

$$T = \frac{2a^{\frac{3}{2}}\pi}{[G(m+M)]^{\frac{1}{2}}} \tag{10-48}$$

Hence we note that the square of the period not only is proportional to the cube of the semimajor axis but also is inversely proportional to the sum of the masses.

DISTURBED CIRCULAR ORBITS

10-10. Stability of Circular Orbits. A circular orbit is always a possible one for any attractive central force if the force center is at the center of curvature. In some instances, circular orbits in which the centers of curvature are not coincident with the centers of force are also possible (cf. Example 10-2). However, in the present section we shall be interested in those circular orbits in which centers of curvature are also centers of force.

It will be convenient now to employ the notation $g(r)$, defined by

$$f(r) = -mg(r) \tag{10-49}$$

in which $f(r)$ is the symbol employed previously for a central force which is a function of r alone. In the new notation $g(r)$ is a positive quantity, and the fact that we are concerned with a force of attraction is brought out by the added minus sign. The quantity m is the mass of the particle describing the orbit. The condition for a circular orbit is obtained by equating the mass times the centripetal acceleration to the centripetal force. We have

$$\frac{mv^2}{a} = mg(a) \tag{10-50}$$

where v is the linear velocity of the particle and a is the radius of the orbit. Although Eq. (10-50) ensures equilibrium, it gives no information as to whether or not the equilibrium is stable. In order to investigate the stability, we must again consider the equation of motion. Equation (10-12) may be rewritten as

$$m(\ddot{r} - r\dot{\theta}^2) = -mg(r) \tag{10-51}$$

We shall not eliminate the time, as we did in Sec. 10-4, but, instead we shall eliminate $\dot{\theta}$. Since $J = mr^2\dot{\theta}$ as before, Eq. (10-51) becomes

$$m\left(\ddot{r} - \frac{J^2}{m^2r^3}\right) = -mg(r) \tag{10-52}$$

Let us consider that the particle is initially traveling in the circular orbit of radius a and that we disturb it slightly in the plane of the orbit and normal to the initial path. This is accomplished analytically by the substitution

$$r = a + x \tag{10-53}$$

where a is a constant (radius of the initial circle) and x is a small quantity. From Eq. (10-53), $\dot{r} = \dot{x}$. Canceling the m's on both sides of (10-52) and making the substitution (10-53) on the left side, we obtain

$$\ddot{x} - \frac{J^2}{m^2(a + x)^3} = -g(r)$$

or

$$\ddot{x} - \frac{J^2}{m^2a^3[1 + (x/a)]^3} = -g(r) \tag{10-54}$$

The quantity x/a is very much less than 1. If we expand the binomial on the left side by means of the binomial theorem and at the same time perform a Taylor series expansion of $g(r)$ about the point $r = a$, we obtain

$$\ddot{x} - \frac{J^2}{m^2a^3}\left(1 - \frac{3x}{a} + \cdots\right) = -\left[g(a) + x\left(\frac{dg}{dr}\right)_{r=a} + \cdots\right] \tag{10-55}$$

Neglecting powers of the small quantity x higher than the first and employing the symbol $g'(a) \equiv (dg/dr)_{r=a}$, we have approximately

$$\ddot{x} - \frac{J^2}{m^2a^3}\left(1 - \frac{3x}{a}\right) = -g(a) - xg'(a) \tag{10-56}$$

But, in view of Eq. (10-50), we have

$$g(a) = \frac{v^2}{a} = \frac{a^2\dot{\theta}^2}{a} = \frac{J^2}{m^2a^3}$$

and thus we see that the first term in the parentheses on the left of Eq. (10-56) will cancel the first term on the right, leaving finally

$$\ddot{x} + \left[\frac{3g(a)}{a} + g'(a)\right] x = 0 \tag{10-57}$$

Equation (10-57) is the familiar equation of the harmonic oscillator and will yield an oscillating solution if the quantity within brackets is positive. In other words, if the particle is disturbed slightly from its initial circular path, there will exist a restoring force, tending to make it return to the initial path, only if the quantity within the brackets of Eq. (10-57) is positive. If there is such a tendency to return to the original path, the initial orbit is said to be a stable one. Hence the criterion of stability is

$$\frac{3}{a} g(a) + g'(a) > 0 \tag{10-58}$$

If this is obeyed, the motion about the initial path is oscillatory and of the form

$$x = x_0 \sin \omega_0 t \tag{10-59}$$

where ω_0 is the angular frequency of the oscillation and is given by

$$\omega_0{}^2 = g'(a) + \frac{3}{a} g(a) \tag{10-60}$$

Now, if we take for the law of force

$$-g(r) = -\frac{c}{r^n} \tag{10-61}$$

where n and c are constants, we find

$$g'(a) = -\left(\frac{nc}{r^{n+1}}\right)_{r=a} = -\frac{n}{a} g(a)$$

Accordingly Eq. (10-60) becomes

$$\omega_0{}^2 = \frac{g(a)}{a}(3 - n) \tag{10-62}$$

and, since this must be positive in order that stability will obtain, we must have that n be less than 3. Hence we conclude that, of the central forces of the form $-c/r^n$, only those for which $n < 3$ will provide stable circular orbits. We notice that the law of the direct distance ($n = -1$) and the law of the inverse square ($n = 2$) both provide stable circular orbits. For the law of the inverse cube ($n = 3$) it is necessary to investigate higher-order terms on both sides of Eq. (10-55) in order to determine the stability. This will not be considered in this text.

APSIDES

10-11. Apsidal Distances and Apsidal Angles. Usually there are certain points in orbits at which the radial component of the velocity is zero. These were mentioned briefly in Sec. 10-6. Such points occur when the radius vector, after having increased for a time, begins to decrease or, after having decreased for a time, begins to increase. The former class of points are called *apocenters* and the latter *pericenters*. Both are quantities to which the general term *apsis* is applied. When the center of force is the sun, the terms *aphelion* and *perihelion* are usually employed. If the apsis is not a singular point of the orbit (for example, a cusp), we must have

$$\frac{dr}{d\theta} = 0 \qquad \text{and} \qquad \frac{du}{d\theta} = 0 \tag{10-63}$$

at the apsis. Here u is again defined by $r = 1/u$. Equation (10-63) states that, at any apsis, the tangent to the orbit at that point is perpendicular to the radius vector.

The distance from the force center to an apsis is called an *apsidal distance,* and the angle between the radii drawn to two consecutive apsides is called an *apsidal angle.* We state that the radius vector to any apsis divides the orbit symmetrically. This may be seen by reasoning as follows: Suppose a point O to be the force center, and take a point P on the orbit to be an apsis. If we imagine two equal particles projected from P with equal velocities in opposite directions perpendicular to OP, it seems clear that, since the force is the same at the same distance from O (that is, the force is single-valued), their subsequent paths must be symmetrical and hence that OP divides the orbit symmetrically.

It may also be seen that, regardless of the number of apsides in an orbit, there are but two apsidal distances and furthermore that all the apsidal angles are equal. Consider Fig. 10-7, in which consecutive apsides of the segment AB of an orbit are P, Q, and R. Now OQ must divide the orbit symmetrically. Consequently OR must be equal to OP, and angle POQ must be equal to angle QOR. Similarly OR must divide the orbit symmetrically, and therefore there must be a subsequent apsis S (not shown) such that OS is equal to OQ and for which angle ROS is equal to angle QOR. It is clear that the ultimate result is that, at most there may be two different magnitudes of apsidal distances in any one orbit and but one magnitude of apsidal angle.

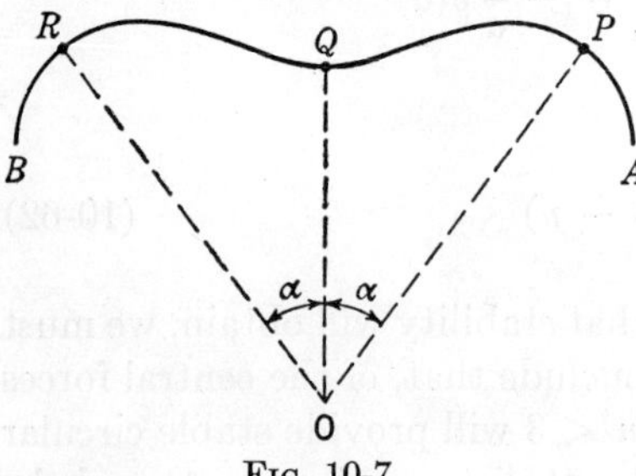

FIG. 10-7

The magnitudes of the apsidal distances readily may be found for a given case by any of several methods. For example, r can be determined from Eq. (10-28) after setting $\dot{r}$ equal to zero and then eliminating $\dot{\theta}$ by means of Eq. (10-14). Similarly, Eq. (10-30) may be used in conjunction with Eq. (10-63).

Suppose we have found the distances to two consecutive apsides by one of these means. We shall call the distances r_1 and r_2. We are then able to find the apsidal angle α (angle between r_1 and r_2) by means of Eq. (10-30), which may be solved for $dr/d\theta$, as

$$\frac{dr}{d\theta} = \left\{\frac{2mr^4}{J^2}\left[W + \int_\infty^r f(r)\,dr\right] - r^2\right\}^{\frac{1}{2}} \tag{10-64}$$

and hence

$$\alpha = \int_{\theta_1}^{\theta_2} d\theta = \int_{r_1}^{r_2} \left\{\frac{2mr^4}{J^2}\left[W + \int_\infty^r f(r)\,dr\right] - r^2\right\}^{-\frac{1}{2}} dr \tag{10-65}$$

Example 10-5. A particle of mass m describes an orbit about a center of force. The law of force is $-km[n^2 + 1 - (2n^2a^2/r^2)]/r^3$, where k, n, a are positive numbers. At $t = 0$, the particle is at an apsis at a distance a from the force center and has an initial velocity equal in magnitude to that it would have acquired if it had dropped from infinity to that point. It is desired to find the equation of the orbit.

For this we may employ Eq. (10-19). Inserting the given expression for $f(r)$, we have

$$\frac{d^2u}{d\theta^2} + u = \frac{m^2k}{J^2}\,[(n^2 + 1)u - 2n^2a^2u^3]$$

and, rewriting,

$$\frac{d^2u}{d\theta^2} + u\left[1 - \frac{m^2k}{J^2}\,(n^2 + 1)\right] = -\frac{2m^2n^2ka^2}{J^2}\,u^3 \tag{10-66}$$

Since the initial velocity, which we may call v_0, is that which would be acquired by free fall from infinity, we have that $mv_0^2/2$ is equal in magnitude to the potential energy V_a at the initial point. But

$$V_a = \int_\infty^a \left[\frac{mk}{r^3}\,(n^2 + 1) - \frac{2mkn^2a^2}{r^5}\right] dr = -\frac{mk}{2a^2}$$

whence

$$J = mv_0a = mk^{\frac{1}{2}} \tag{10-67}$$

By means of (10-67), the angular momentum J may be eliminated from Eq. (10-66), to yield

$$\frac{d^2u}{d\theta^2} - n^2u = -2n^2a^2u^3$$

We make the substitution $p = du/d\theta$, whence $d^2u/d\theta^2 = p\,dp/du$. Therefore we have

$$p\,\frac{dp}{du} = (n^2u - 2n^2a^2u^3)$$

which may easily be integrated. We find

$$p^2 = n^2u^2(1 - a^2u^2) \tag{10-68}$$

in which the constant of integration vanishes, since $p = du/d\theta = 0$ at $t = 0$, the initial point being an apsis of the orbit. Upon substituting for p and extracting the square root, Eq. (10-68) yields, after separating the variables,

$$\frac{du}{u(1 - a^2u^2)^{\frac{1}{2}}} = \pm n\, d\theta$$

Eliminating the variable u in favor of the radius vector r and integrating, we obtain

$$\cosh^{-1}\frac{r}{a} = \pm n\theta$$

the constant of integration being zero if we choose $\theta = 0$ at $r = a$. The equation of the orbit is therefore

$$r = a \cosh n\theta$$

10-12. Apsides in a Nearly Circular Orbit. Advance of the Perihelion. The study of orbits which are very nearly circular is a useful one, since the earth's orbit is closely approximated in this way (the eccentricity is about $\frac{1}{60}$). For this investigation, the result expressed in Eq. (10-60) will be of use, that is, let us view a near-circular orbit as a disturbed circular orbit. Equation (10-60) states that the angular frequency ω_0 of motion perpendicular to the circular path (as contrasted to the orbital angular frequency ω) is

$$\omega_0 = \left[g'(a) + \frac{3}{a}g(a)\right]^{\frac{1}{2}} \tag{10-69}$$

where a is the radius of the circular orbit. Clearly the time required to pass from a maximum value of the radius vector to the succeeding minimum value is just half the period τ_0 corresponding to ω_0, or

$$t_a = \frac{\tau_0}{2} = \frac{1}{2}\frac{2\pi}{\omega_0} = \frac{\pi}{\omega_0}$$

Thus the corresponding apsidal angle, or increment in θ swept through during the time t_a, neglecting the small variation in ω, is

$$\alpha = \omega t_a = \frac{\omega\tau_0}{2} = \frac{\pi\omega}{\omega_0} \tag{10-70}$$

Since $\omega = v/a$ approximately, where v is the linear velocity, we may eliminate ω from Eq. (10-70) by means of Eq. (10-50). Equation (10-70) becomes

$$\alpha = \pi\sqrt{\frac{g(a)}{a}}\frac{1}{\omega_0} = \pi\sqrt{\frac{g(a)}{a}}\frac{1}{\left[g'(a) + 3\dfrac{g(a)}{a}\right]^{\frac{1}{2}}}$$

Rewriting this, we obtain

$$\alpha = \frac{\pi}{\left[3 + \dfrac{ag'(a)}{g(a)}\right]^{\frac{1}{2}}} \tag{10-71}$$

If we choose a law of force of the form

$$g(r) = \frac{k}{r^n} \tag{10-72}$$

where n and k are constants, Eq. (10-71) becomes especially simple. The apsidal angle is

$$\alpha = \frac{\pi}{\sqrt{3 - n}} \tag{10-73}$$

and the dependence on the radius a disappears. This shows that, if the force varies as some power of the distance, for a given value of n the apsidal angle is the same for all the orbits which are nearly circular, regardless of the value of a. From (10-73), for an inverse-square force in which $n = 2$, we obtain $\alpha = \pi$. Hence α is contained in 2π (one revolution) an even number of times. This is the necessary condition that the orbit be reentrant, in other words, that it will repeat itself after one revolution. If n is just slightly different from 2, the orbit is not *reentrant* and a *progressive* motion of the apsis in space, in the plane of the orbit, results. This was brilliantly pointed out by Newton three centuries ago. He stated that even the smallest progressive motion of the perihelia of the solar planets would indicate a deviation from the inverse-square law.

It is of interest to note that the planetary orbits do demonstrate such motions, although they are very slight. Indeed, a portion of these motions can be explained through the medium of *perturbations* arising from the presence of other bodies in the solar system. However, some of the motions cannot be accounted for in that simple manner. For example, the perihelion of Mercury has been observed to advance in space. Part of this, 42.56 seconds of arc per century, cannot be explained by means of effects such as perturbations arising from other planets. The now accepted explanation of this previously unexplained motion is that the equation of motion of a body in a gravitational field is different from that obtained from Eq. (10-19), in which we considered

$$f(r) = -\frac{MmG}{r^2} = mg(r) \tag{10-74}$$

In the simple case for which Eq. (10-74) is employed, Eq. (10-19) becomes

$$\frac{d^2u}{d\theta^2} + u = \frac{m^2GM}{J^2} \tag{10-75}$$

According to general relativity, Eq. (10-75) must be replaced by

$$\frac{d^2u}{d\theta^2} + u = \frac{m^2GM}{J^2} + 3GMu^2 \tag{10-76}$$

the added term behaving effectively as a small additional inverse fourth-power component. To see what effect this has on Eq. (10-71), let us consider

$$g(r) = \frac{k}{r^2} + \frac{b}{r^4} = \frac{k}{r^2}\left(1 + \frac{b}{kr^2}\right)$$

where k and b are constants and $b/ka^2 \ll 1$. We obtain

$$ag'(a) = -\frac{2k}{a^2}\left(1 + \frac{2b}{ka^2}\right)$$

and

$$g(a) = \frac{k}{a^2}\left(1 + \frac{b}{ka^2}\right)$$

from which

$$\begin{aligned}\frac{ag'(a)}{g(a)} &= -2\left(1 + \frac{2b}{ka^2}\right)\left(1 + \frac{b}{ka^2}\right)^{-1} \\ &= -2\left(1 + \frac{2b}{ka^2}\right)\left(1 - \frac{b}{ka^2} + \cdots\right) \simeq -2\left(1 + \frac{b}{ka^2}\right)\end{aligned}$$

where we have neglected powers of b/ka^2 higher than the first. Thus α becomes, from Eq. (10-71),

$$\alpha = \frac{\pi}{\sqrt{3 - 2[1 + (b/ka^2)]}} = \frac{\pi}{[1 - (2b/ka^2)]^{\frac{1}{2}}} \simeq \pi + \frac{b}{ka^2}\pi \qquad (10\text{-}77)$$

From Eq. (10-77) it is apparent that the orbit will not be reentrant after but one revolution. It is also interesting to note that the advance of perihelion

$$\frac{2b}{ka^2}\pi$$

in each revolution varies inversely as the square of the radius of the orbit. Consequently we should expect that the effect would fall off rapidly with the planets more distant from the sun. For this reason Mercury is the only one of the large planets to demonstrate a large effect. Relativity theory predicts 43.03 seconds of arc per century for the planet Mercury. This is considered to be in very good agreement with the above-mentioned observed figure of 42.56 seconds.[1]

Problems

Problems marked *C* are reprinted by kind permission of the Cambridge University Press.

10-1. Work Example 10-2 by differentiating Eq. (10-30).

10-2. A particle of mass m describes the orbit $r = a(1 + \cos\theta)$ under the action of a force which is always directed toward the origin. Find the law of force. Deter-

[1] See G. M. Clemence, *Rev. Mod. Phys.*, **19**: 361 (1947).

mine the total energy of the particle in the orbit. Interpret your result in terms of the amount of work necessary to remove the particle to a state of rest at an infinite distance away for the cases (*a*) from a state of rest at the point $r = 2a$, $\theta = 0$ and (*b*) from an arbitrary point r, θ on the orbit if the particle is moving in the orbit according to the equation of motion.

10-3. Show that, if a particle is describing a hyperbolic orbit about a force center situated at one focus, the total energy may be expressed as $k/2a$, where $2a$ is the distance between the vertices and k is the constant, in the law of force, defined by $F = -k/r^2$.

10-4. A particle of mass m is initially at rest at a distance a from a fixed force center. The law of force is $-k/r^2$. Show that the time of fall is

$$\sqrt{\frac{2m}{k}}\frac{\pi a^{\frac{3}{2}}}{4}$$

and that the ratio of the time average of the velocity over the first half of the path to that over the second half is $(\pi - 2)/(\pi + 2)$.

10-5*C*. The constituents of a double star are observed to describe circles relative to each other with a period T of relative motion. If they are suddenly deprived of velocity, show that they will collide after a time $T/4\sqrt{2}$.

10-6*C*. A comet describes a parabolic orbit the plane of which is the same as that of the earth's orbit, assumed circular. Show by means of the law of areas that the maximum length of time during which the comet is able to remain inside the earth's orbit is $2/3\pi$ of a year.

10-7. A particle of mass m is subject to a repulsive force k/r^5, where k is a constant. Initially it is at a very large distance from the force center O and has a velocity of magnitude such that, if it were directed along a path through the center of force, the closest distance of approach would be a. Actually it is projected, with this same velocity, along a path which would pass the force center at a distance b if it were not deflected. Find the minimum linear velocity (in terms of k, a, b, m) that the particle experiences as it follows its trajectory.

10-8. The closest distance of approach of a comet to the sun is α times the radius of the earth's orbit assumed circular, where α is less than 1. Find the length of time that the comet will remain inside of the earth's orbit. Assume the orbit of the comet to be a parabola.

10-9. A particle is describing a central orbit about a point O, and h is twice its areal velocity. A second particle moves in another orbit such that at any instant its distance from O is the same in magnitude as that of the first particle. The angular velocity of the second particle is always to that of the first particle as $(\sin\alpha)/1$, however. Determine, in terms of h, α, r the difference, at any instant, between the two radial accelerations.

10-10. A comet describes a parabolic orbit about the sun in the plane of the earth's orbit. Assuming the earth's orbit to be circular and that the distance of closest approach of the comet to the sun is one-third of the radius of the earth's orbit, show by means of the law of areas that the time the comet remains within the earth's orbit is approximately 74.5 days.

10-11. If the sun's mass suddenly decreased to half its value, show that the earth's orbit, assumed to be circular originally, would become parabolic.

10-12. A particle of mass m is describing the orbit

$$r = a \sin n\theta$$

where n and a are constants, under the action of a central force. Find the law of force in terms of J, n, a, r, and m, where J is the angular momentum.

10-13. Determine the central law of force for which an orbit may be an ellipse with the attracting center located at the center of the ellipse.

10-14. A particle moves in the field of a force center which attracts with a force inversely proportional to the cube of the distance. Determine the possible orbits.

10-15. A particle is held at rest in a field of force at a distance a from the force center. The force is one of attraction and has the magnitude k/r^n, where k and n are constants, with n greater than 1. The particle is suddenly projected in an arbitrary direction with a velocity which would have carried it to infinity if it had been projected directly away from the force center. Find the greatest values of r for different n.

10-16. A particle of mass m under the action of a central force moves in an orbit

$$r = ab(a^2 \cos^2 \theta + b^2 \sin^2 \theta)^{-\frac{1}{2}}$$

Find the law of force.

10-17. A particle of mass m is projected from infinity with a velocity v in a manner such that it would pass a distance b from a center of inverse-cube repulsive force (magnitude k/r^3, where k is a constant) if it were not deflected. Find the angular deflection which actually occurs.

10-18. Two particles, of masses m_1 and m_2, describe orbits about a center of gravitational force of very large mass M. The particle of mass m_1 describes a circle of radius s, and the particle of mass m_2 describes an ellipse of semiminor axis b. If the period of the ellipse is twice that of the circle, show that the angular momenta of the circle and the ellipse are, respectively, $m_1 \sqrt{GMs}$ and $\dfrac{m_2 b}{2^{\frac{1}{3}}} \sqrt{\dfrac{GM}{s}}$, where G is the gravitational constant. Neglect the interaction between m_1 and m_2.

10-19. A particle of mass m describes an elliptical orbit about a center of inverse-square attracting force situated at one focus. If an inverse-cube attraction of magnitude h/r^3, where h is a very small positive constant, is superimposed upon the original force, show by use of Eq. (10-19) that the resulting orbit may be regarded as an ellipse rotating in the plane of the orbit in which the major axis advances by $mh\pi/J^2$ during each revolution. J is the angular momentum. Consider $mh/J^2 \ll 1$.

10-20. A particle describes an ellipse about a center of attractive force situated at one focus. If the force has the magnitude k/r^2, where k is a constant, show that, if $2a$ is the major axis, the speed v at any point of the orbit is given by

$$v^2 = k\left(\frac{2}{r} - \frac{1}{a}\right)$$

10-21. A string passes through a small hole in a smooth horizontal table and has equal particles attached to its ends, one hanging vertically and the other lying on the table at a distance a from the hole. The latter is projected with a velocity $\sqrt{ga}$ perpendicular to the string. Show that the hanging particle will remain at rest and that, if it be slightly disturbed in a vertical direction, the period of a small oscillation will be $2\pi \sqrt{2a/3g}$.

10-22. A particle moves in a circular orbit of radius a under the action of an attractive force of magnitude $hu^2 + ku^4$, where h and k are positive constants and $u = 1/r$. The origin of force is at the center of the circle. Show that the orbit is stable provided that $a^2h > k$.

10-23. A particle is describing a circle under the influence of an attractive force which varies as $(1/r^2)e^{-r/a}$. Show that the motion is stable or unstable according as the radius of the circle is less or greater than a.

10-24. In a central field of force, a particle of mass m experiences an acceleration

$$k[2(a^2 + b^2)u^5 - 3a^2b^2u^7]$$

where $u = 1/r$ and k, a, b are positive constants. Initially the particle is at a distance a from the origin and has a velocity $k^{\frac{1}{2}}/a$ at right angles to the radius vector. Find the radial velocity of the particle as a function of r.

10-25. A particle of mass m moves in a central field of force

$$-\frac{2mk}{r^3}\left(1 - \frac{a^2}{r^2}\right)$$

where k is a positive constant. Initially the particle is at an apsis a distance a from the force center and has a velocity $k^{\frac{1}{2}}/a$. Find the time, as a function of k, b, a, required for the particle to arrive at a distance b from the force center. Consider the case for which $r < a$.

10-26. A particle of mass m moves in an attractive central field, of magnitude k/r^2, where k is a constant. At a given instant the particle is at an apsis a distance a from the center of force and has a velocity $\sqrt{k/2ma}$. Determine the other apsidal distance.

10-27. The law of force in a central field is $(c/r^2)e^{-kr}$, where k and c are constants, k being small. Show that the apsis line in a nearly circular orbit will advance in each revolution through the angle πka if a is the radius of the orbit.

10-28. In a nearly circular orbit the potential energy is $-(c/r)e^{-kr}$, where c and k are constants. If a is the mean radius, show that the angle of advance is approximately πk^2a^2. Take $ka \ll 1$.

10-29. A particle is describing an ellipse of eccentricity $\frac{1}{2}$ under a force directed to one focus. When it arrives at an apsis, the velocity is suddenly doubled. Show that the new orbit will be a parabola or a hyperbola according as the apsis is the farther or nearer one, respectively.

10-30*C*. If a particle is projected from an apsis at distance a, with the velocity from infinity, under the action of a central force kr^{-2n-3}, prove that the path is $r^n = a^n \cos n\theta$.

10-31*C*. A body is describing an ellipse of eccentricity e under the action of a force tending to a focus, and when at the nearer apsis the center of force is transferred to the other focus. Show that the eccentricity of the new orbit is $e(3 + e)/(1 - e)$.

10-32*C*. A particle moves in a nearly circular orbit with an acceleration $h + k(r - a)$, where h and k are constants and a is the mean radius. Show that the apsidal angle is $\pi n/(3n^2 + k)^{\frac{1}{2}}$, where n is the mean angular velocity.

10-33*C*. A particle is describing a circular orbit of radius a under a force to the center producing an acceleration $f(r)$ at distance r, and a small increment of velocity Δu is given to it along the radial direction. Show that the apsidal distances are approximately

$$a \pm \Delta u \frac{a^{\frac{1}{2}}}{[3f(a) + af'(a)]^{\frac{1}{2}}}$$

10-34. A planet of mass M describes an elliptical orbit of eccentricity $e = 1 - \alpha$, where $\alpha \ll 1$, about a fixed sun. When the planet is at its greatest distance from the sun, it is struck by a comet of mass m, where $m \ll M$, traveling in a tangential direction. The impact is completely inelastic. Find the minimum kinetic energy the comet must have in order that the new orbit be parabolic.

10-35. A particle of mass m_1 is moving in an elliptic orbit relative to a particle of mass m_2. Assume only gravitational forces to be acting. Determine whether or not the orbit is a stable one by finding the equation of motion for small displacements from the initial orbit.

CHAPTER 11

ACCELERATED REFERENCE SYSTEMS AND CONSTRAINED MOTION OF A PARTICLE

In this present chapter we shall consider first the motion of a particle relative to an arbitrarily moving coordinate system. The description of motion relative to moving axes is useful both in problems involving constraining surfaces or curves which are being accelerated and in problems in which it is desirable to describe the free motion of a particle in terms of a coordinate system which is being accelerated. A simple example of the first type is the case of a particle constrained to slide on the surface of a plane, the latter being accelerated in an arbitrary manner. A practical example of the second class is the group of second-order effects which are added to the simple theory of Chap. 6 for the motion of a projectile when the rotation of the earth is taken into account.

Finally, we shall be interested in studying the motion of a particle when it is subjected to various types of constraints. In treating constraints we shall usually employ coordinate systems which are not being accelerated (namely, inertial systems); in certain situations, however, we shall find it expedient to employ reference systems which are being accelerated in some fashion.

MOTION OF A PARTICLE IN AN ACCELERATED REFERENCE SYSTEM

11-1. Nature of the Problem. Frequently it is necessary to write the equations of motion in a coordinate system which is not an inertial one. Let us consider for a moment the motion of a block constrained to slide on the surface of a smooth table, which, in turn, is attached to the rotating earth. In Fig. 11-1, we take the Ox and Oy axes to be fixed in the table top. At $t = 0$ the block is at rest at point A. A constant force F, parallel to the x axis, is then applied (for example, a steady push applied by the hand). On the basis of elementary considerations we might expect the block to move along the

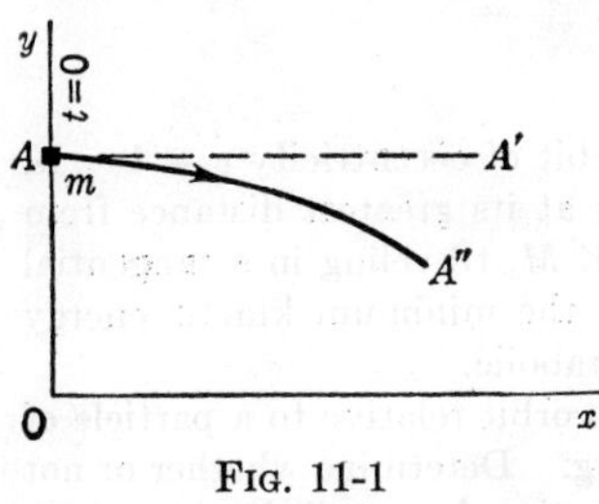

FIG. 11-1

straight line AA', following the equation of motion

$$m\ddot{x} = F \tag{11-1}$$

where m is the mass of the block. This, however, is not the case. The actual path, if the block acquires an appreciable velocity under the action of F, is that of AA''; the block deviates slightly to the right of the path expected on the basis of Eq. (11-1). The difficulty lies in the fact that the system of axes chosen is an accelerated one (see, for example, Sec. 1-20). The path AA'', shown in Fig. 11-1, is that which is experienced in the Northern Hemisphere. In the Southern Hemisphere the deviation is to the left.

We recall, from Sec. 1-20, that the equation of motion of a particle relative to an accelerated reference system contains terms arising from the acceleration of the system. Although in the case discussed in Chap. 1 only a simple translational acceleration, without rotation, was considered, it will be found in the present chapter that effects also arise if the reference system is rotating. Indeed, the most common ones in which we shall be interested are caused by the earth's rotation. As an example, the type of deviation shown in Fig. 11-1 results from a reaction term which in magnitude is proportional to the cross product of the angular velocity of the earth (since the axes are fixed in the surface of the earth) and the velocity of the block. Thus there is effectively a force in the $-y$ direction even though there is no y component of $\mathbf{F}$. Although usually these additional reaction terms are too small to be noticed, they do become of measurable importance in such problems as those involving long-range projectiles fired at high muzzle velocities.

11-2. Calculation of the Inertial Reaction in a Moving Frame. In Fig. 11-2 let the axes $O_0x_0y_0z_0$ denote an inertial system. Sitting at O_0, the origin, we observe a point P, the position of which is determined by the knowledge of the vector $\mathbf{r}_0$. If P is moving, its velocity is $\dot{\mathbf{r}}_0$, where the dot denotes differentiation with respect to time. Now we may choose other vectors $\mathbf{R}$ and $\mathbf{r}$, as shown, bringing in another point O. The position of P is given just as well by the vectors $\mathbf{R}$ and $\mathbf{r}$ as by $\mathbf{r}_0$. The vectors are related as

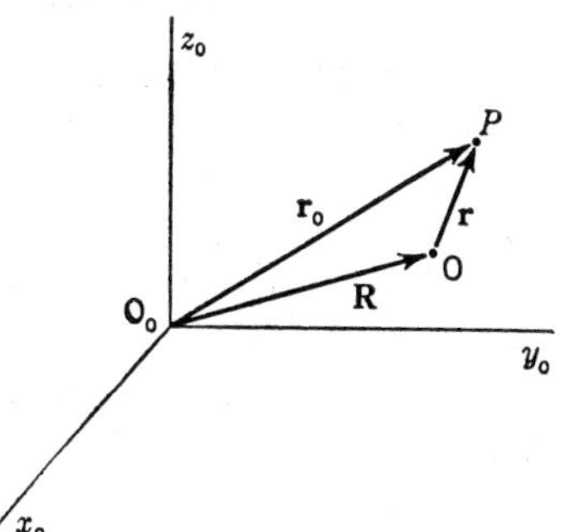

FIG. 11-2

$$\mathbf{r}_0 = \mathbf{R} + \mathbf{r} \tag{11-2}$$

whence, differentiating,

$$\dot{\mathbf{r}}_0 = \dot{\mathbf{R}} + \dot{\mathbf{r}} \tag{11-3}$$

It is not difficult to see that $\dot{\mathbf{R}}$ is the velocity of point O with respect to O_0 and also that $\dot{\mathbf{r}}$ is the velocity of P with respect to point O. Suppose now we let O be the origin of a system of coordinates $Oxyz$, which of course is moving with the velocity $\dot{\mathbf{R}}$ of point O with reference to O_0 (see Fig. 11-3). Writing $\mathbf{r}$ in terms of the unit vectors $\mathbf{i}$, $\mathbf{j}$, $\mathbf{k}$ of the moving system $Oxyz$, we have

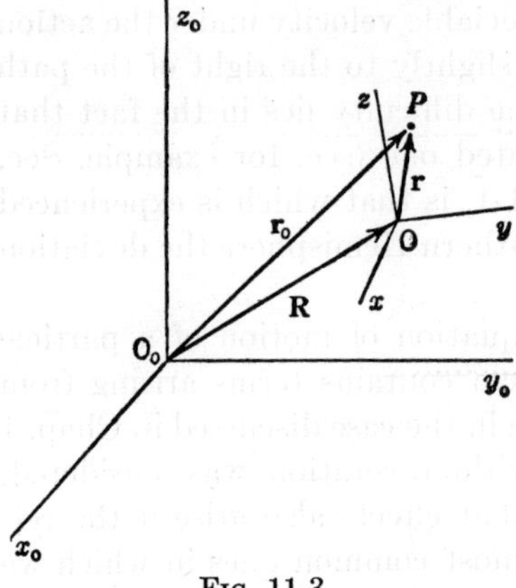

FIG. 11-3

$$\mathbf{r} = x\mathbf{i} + y\mathbf{j} + z\mathbf{k} \tag{11-4}$$

Now the time derivative of $\mathbf{r}$ with respect to the system $Oxyz$ is just

$$\dot{\mathbf{r}} = \dot{x}\mathbf{i} + \dot{y}\mathbf{j} + \dot{z}\mathbf{k} \tag{11-5}$$

However, the time derivative of $\mathbf{r}$ with respect to the inertial system $O_0x_0y_0z_0$ is more complicated since the system $Oxyz$ may be rotating with respect to $O_0x_0y_0z_0$, and consequently, since $\mathbf{r}$ has been expressed in terms of the moving coordinates, such rotation will have to be taken into account. Thus, since the unit vectors $\mathbf{i}$, $\mathbf{j}$, $\mathbf{k}$ are changing in direction, they possess time derivatives which are not zero. Let us employ the notation $[\dot{\mathbf{r}}]_0$ when the time derivative of the quantity $\mathbf{r}$ is taken with respect to the $O_0x_0y_0z_0$ system. Accordingly $[\dot{\mathbf{r}}]_0$ is given by

$$[\dot{\mathbf{r}}]_0 = (\dot{x}\mathbf{i} + \dot{y}\mathbf{j} + \dot{z}\mathbf{k}) + (x\dot{\mathbf{i}} + y\dot{\mathbf{j}} + z\dot{\mathbf{k}}) \tag{11-6}$$

Zero subscripts on the right side of the equation are unnecessary, in the case of the first three terms because x, y, and z are scalars and therefore have the same time derivatives in either system, and in the case of the second three terms because the unit vectors do not change with respect to the moving system. Therefore no confusion is possible. The first quantity within parentheses on the right is what appears in Eq. (11-5), and it is convenient to denote it by the symbol $\mathbf{v}$. Thus we may rewrite Eq. (11-6) as

$$[\dot{\mathbf{r}}]_0 = \mathbf{v} + (x\dot{\mathbf{i}} + y\dot{\mathbf{j}} + z\dot{\mathbf{k}}) \tag{11-7}$$

We remember that the quantity $[\dot{\mathbf{r}}]_0$ is viewed by the observer sitting in the system $O_0x_0y_0z_0$ and that therefore $[\dot{\mathbf{r}}]_0$ is the true velocity of P with respect to point O. It is very important to see this. The quantity $\mathbf{v}$, in Eq. (11-7), is the apparent velocity of P as perceived by an observer seated in $Oxyz$ and rotating with $Oxyz$.

It remains to put the second term (in parentheses) on the right side of Eq. (11-7) into a more familiar form. Let us imagine, for the moment, that P, and consequently $\mathbf{r}$, is fixed in the $Oxyz$ system. In this event $\mathbf{v}$ is zero, and we see that the group of terms within the parentheses in Eq. (11-7) constitutes that part of the rate of change of $\mathbf{r}$ arising from rotation of $\mathbf{r}$ relative to $O_0x_0y_0z_0$. Evidently this is simply $\boldsymbol{\omega} \times \mathbf{r}$ [see Eq. (9-13)], where $\boldsymbol{\omega}$ is the angular velocity of rotation of the $Oxyz$ axes. In more detail, since $\mathbf{i}$, $\mathbf{j}$, and $\mathbf{k}$ are constant in magnitude, the quantities $\dot{\mathbf{i}}$, $\dot{\mathbf{j}}$, and $\dot{\mathbf{k}}$ in Eq. (11-7) may be replaced, respectively, by $\boldsymbol{\omega} \times \mathbf{i}$, $\boldsymbol{\omega} \times \mathbf{j}$, and $\boldsymbol{\omega} \times \mathbf{k}$. The second term on the right side of Eq. (11-7) becomes

$$
\begin{aligned}
x\dot{\mathbf{i}} + y\dot{\mathbf{j}} + z\dot{\mathbf{k}} &= x\boldsymbol{\omega} \times \mathbf{i} + y\boldsymbol{\omega} \times \mathbf{i} + z\boldsymbol{\omega} \times \mathbf{k} \\
&= \boldsymbol{\omega} \times (x\mathbf{i} + y\mathbf{j} + z\mathbf{k}) \\
&= (\boldsymbol{\omega} \times \mathbf{r})
\end{aligned}
\tag{11-8}
$$

the result which had been anticipated immediately above. Accordingly

$$
[\dot{\mathbf{r}}]_0 = \mathbf{v} + \boldsymbol{\omega} \times \mathbf{r} \equiv \dot{\mathbf{r}} + \boldsymbol{\omega} \times \mathbf{r} \tag{11-9}
$$

A little reflection will reveal that this result is even more general than it appears, since any vector could have been employed in place of the radius vector and the form of the result would have been the same. Hence the operation of taking the time derivative of any vector, in the system $O_0x_0y_0z_0$, is equivalent to the operation $d/dt + (\boldsymbol{\omega} \times \;)$ in the $Oxyz$ system. Thus

$$
\left(\frac{d}{dt}\right)_0 = \frac{d}{dt} + (\boldsymbol{\omega} \times \;) \tag{11-10}
$$

We may now rewrite Eq. (11-3) as

$$
\mathbf{v}_0 \equiv [\dot{\mathbf{r}}_0]_0 = [\dot{\mathbf{R}}]_0 + \mathbf{v} + \boldsymbol{\omega} \times \mathbf{r} \tag{11-11}
$$

for the velocity of point P in the $O_0x_0y_0z_0$ system of axes.

Of far greater importance are the terms obtained when the acceleration is calculated, because these are necessary in order to write the equations of motion. Taking the derivative, in the inertial system, of both sides of Eq. (11-11) with respect to time, we have

$$
[\ddot{\mathbf{r}}_0]_0 \equiv [\dot{\mathbf{v}}_0]_0 \equiv \mathbf{a}_0 = [\ddot{\mathbf{R}}]_0 + [\dot{\mathbf{v}}]_0 + [\dot{\boldsymbol{\omega}}]_0 \times \mathbf{r} + \boldsymbol{\omega} \times [\dot{\mathbf{r}}]_0 \tag{11-12}
$$

The usefulness of Eq. (11-10) is now apparent. We have, at once,

$$
[\dot{\mathbf{v}}]_0 = \dot{\mathbf{v}} + \boldsymbol{\omega} \times \mathbf{v} \equiv \mathbf{a} + \boldsymbol{\omega} \times \mathbf{v} \tag{11-13}
$$

$$
[\dot{\boldsymbol{\omega}}]_0 \times \mathbf{r} = \dot{\boldsymbol{\omega}} \times \mathbf{r} + (\boldsymbol{\omega} \times \boldsymbol{\omega}) \times \mathbf{r} = \dot{\boldsymbol{\omega}} \times \mathbf{r} \tag{11-14}
$$

$$
\boldsymbol{\omega} \times [\dot{\mathbf{r}}]_0 = \boldsymbol{\omega} \times \mathbf{v} + \boldsymbol{\omega} \times (\boldsymbol{\omega} \times \mathbf{r}) \tag{11-15}
$$

in which $\mathbf{a} \equiv \dot{\mathbf{v}}$. We are now in a position to rewrite Eq. (11-12) as

$$
\mathbf{a}_0 = [\ddot{\mathbf{R}}]_0 + \mathbf{a} + \dot{\boldsymbol{\omega}} \times \mathbf{r} + 2\boldsymbol{\omega} \times \mathbf{v} + \boldsymbol{\omega} \times (\boldsymbol{\omega} \times \mathbf{r}) \tag{11-16}
$$

The quantity $\mathbf{a}_0$, on the left, is the acceleration of P as viewed by an observer seated in the inertial system. On the right side, $[\ddot{\mathbf{R}}]_0$ is the acceleration of the origin O as seen by an observer in the inertial system. The acceleration $\mathbf{a}$ is the apparent acceleration of P as seen by an observer seated in and rotating with the moving axes; this is what would be written down by this observer if he were unaware of the fact that he was rotating. In terms of its components in the moving system, $\mathbf{a}$ may be written

$$\mathbf{a} = \mathbf{i}\ddot{x} + \mathbf{j}\ddot{y} + \mathbf{k}\ddot{z}$$

The quantity $2\boldsymbol{\omega} \times \mathbf{v}$ is the so-called *Coriolis acceleration*. The term $\dot{\boldsymbol{\omega}} \times \mathbf{r}$ is a linear acceleration due to the angular acceleration (if any) of the axes, and finally $\boldsymbol{\omega} \times (\boldsymbol{\omega} \times \mathbf{r})$ is the centripetal acceleration owing to the rotation of the axes $Oxyz$. It should be noticed in Eq. (11-14) that

$$[\dot{\boldsymbol{\omega}}]_0 = \dot{\boldsymbol{\omega}} + \boldsymbol{\omega} \times \boldsymbol{\omega} = \dot{\boldsymbol{\omega}}$$

and so the angular acceleration is the same in both systems.

11-3. Application of the Principles. If we are observing the motion of a particle from an accelerated coordinate system, we have to be able to write the correct equations of motion in that system. In order to do this, we must know the form of $m\mathbf{a}$ in that system. It is usually convenient to write the equation of motion first in an inertial system. It will be

$$m\mathbf{a}_0 = \mathbf{F}$$

where $\mathbf{a}_0$ is the acceleration of the particle of mass m as it would be perceived from the inertial system. The force which is acting is $\mathbf{F}$. Substituting for $\mathbf{a}_0$ from Eq. (11-16), we have

$$\mathbf{a}_0 = [\ddot{\mathbf{R}}]_0 + \mathbf{a} + \dot{\boldsymbol{\omega}} \times \mathbf{r} + 2\boldsymbol{\omega} \times \mathbf{v} + \boldsymbol{\omega} \times (\boldsymbol{\omega} \times \mathbf{r}) = \frac{\mathbf{F}}{m} \qquad (11\text{-}17)$$

and we now obtain the equation of motion in the accelerated system merely by rearranging Eq. (11-17), as

$$\mathbf{a} = \frac{\mathbf{F}}{m} - [\ddot{\mathbf{R}}]_0 - 2\boldsymbol{\omega} \times \mathbf{v} - \dot{\boldsymbol{\omega}} \times \mathbf{r} - \boldsymbol{\omega} \times (\boldsymbol{\omega} \times \mathbf{r}) \qquad (11\text{-}18)$$

and the interpretation is that the real force $\mathbf{F}$, defined in the inertial system $O_0x_0y_0z_0$, is modified by the presence of reaction terms which arise because we are observing from an accelerated system.

Of special interest is the description of the motion of a particle relative to a point on the surface of the rotating earth. We shall neglect the acceleration of the earth's center. This amounts to regarding a set of nonrotating axes, with origin fixed at the center of the earth, as an inertial system. Let this be the system $O_0x_0y_0z_0$ in Fig. 11-4. The coordinate system of interest is the system $Oxyz$, with origin at point O in the figure and with axes rigidly attached to the earth's surface. Now if the only force acting on a particle of mass m is the gravitational attraction of the earth, $\mathbf{F}$ has the value, for this problem,

$$\mathbf{F} = -\frac{mMG}{r_0^2}\mathbf{r}_1$$

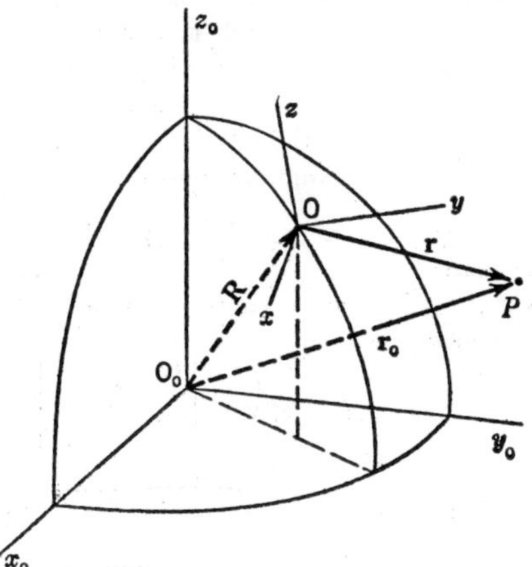

FIG. 11-4

where $\mathbf{r}_1$ is a unit vector pointing in the direction of r_0 increasing, M is the mass of the earth, and G is the gravitational constant. Considering the earth to be rotating at a constant angular velocity $\boldsymbol{\omega}$, Eq. (11-18) becomes, for the system $Oxyz$,

$$\mathbf{a} = -\frac{MG}{r_0^2}\mathbf{r}_1 - [\ddot{\mathbf{R}}]_0 - 2\boldsymbol{\omega} \times \mathbf{v} - \boldsymbol{\omega} \times (\boldsymbol{\omega} \times \mathbf{r}) \qquad (11\text{-}19)$$

where, as before, $\mathbf{v}$ is the velocity of m relative to $Oxyz$. Now

$$[\ddot{\mathbf{R}}]_0 = \boldsymbol{\omega} \times (\boldsymbol{\omega} \times \mathbf{R})$$

and the effective acceleration of gravity $\mathbf{g}$ at the earth's surface is just

$$\mathbf{g} = -\frac{MG}{r_0^2}\mathbf{r}_1 - \boldsymbol{\omega} \times (\boldsymbol{\omega} \times \mathbf{R}) \qquad (11\text{-}20)$$

(The approximation will usually be made that $\mathbf{g}$ is a constant.) If the motion of m is such that it remains close enough to O to neglect the term $\boldsymbol{\omega} \times (\boldsymbol{\omega} \times \mathbf{r})$, Eq. (11-19) reduces approximately to

$$\mathbf{a} = \mathbf{g} - 2\boldsymbol{\omega} \times \mathbf{v} \qquad (11\text{-}21)$$

If we were to consider the motion of a projectile in a case where it would reach a great altitude, the term $\boldsymbol{\omega} \times (\boldsymbol{\omega} \times \mathbf{r})$ would have to be considered, as well as the variation of $\mathbf{g}$ arising from the presence of r_0 in $\mathbf{F}$.

Example 11-1. Let us take the example of a falling body as we would observe it from our position on the earth's surface at a latitude λ. (Although g is really a function of λ, its small variation, as λ is varied, will be neglected.) We must be careful to select a right-hand system of coordinates (see Fig. 11-5) since Eqs. (11-1) to (11-20) all have the forms as stated only in a right-hand system. Let us choose the direction of the z axis to be that of the plumb line (vertical) which may be assumed to pass approximately through the center of the earth. We select the x axis to be positive to the south and the y axis positive to the east. We note also that we have chosen O to be in the Northern Hemisphere. Suppose we drop an object of mass m from a height h above the earth's surface. It is necessary to write the components of Eq. (11-21) in rectangular coordinates. We have

$$\mathbf{v} = \mathbf{i}\dot{x} + \mathbf{j}\dot{y} + \mathbf{k}\dot{z} \qquad \text{and} \qquad \boldsymbol{\omega} = \mathbf{i}\omega_x + \mathbf{j}\omega_y + \mathbf{k}\omega_z$$

where

$$\omega_x = -\,\omega \cos \lambda \qquad \omega_y = 0 \qquad \omega_z = \omega \sin \lambda \tag{11-22}$$

whence

$$\boldsymbol{\omega} \times \mathbf{v} = \begin{vmatrix} \mathbf{i} & \mathbf{j} & \mathbf{k} \\ \omega_x & 0 & \omega_z \\ \dot{x} & \dot{y} & \dot{z} \end{vmatrix} = -\mathbf{i}\omega_z\dot{y} - \mathbf{j}(\omega_x\dot{z} - \omega_z\dot{x}) + \mathbf{k}\omega_x\dot{y}$$

It is not difficult to see from physical grounds that, compared with $\dot{z}$, we may safely neglect terms in $\dot{x}$ and $\dot{y}$ here. Therefore we have approximately

$$2\boldsymbol{\omega} \times \mathbf{v} \simeq -\,2\mathbf{j}\omega_x\dot{z}$$

and the equations of motion, from Eq. (11-21), become approximately

$$\ddot{x} \simeq 0 \tag{11-23}$$

$$\ddot{y} \simeq 2\dot{z}\omega_x = -2\dot{z}\omega \cos \lambda \tag{11-24}$$

$$\ddot{z} \simeq -g \tag{11-25}$$

From Eq. (11-25), if the body is initially at rest, we have

$$\dot{z} = -gt$$

whence Eq. (11-24) becomes

$$\ddot{y} = 2gt\omega \cos \lambda \tag{11-26}$$

and it is evident that, for our choice of λ (in the Northern Hemisphere), the acceleration is directed toward the east. From Eq. (11-26) we obtain

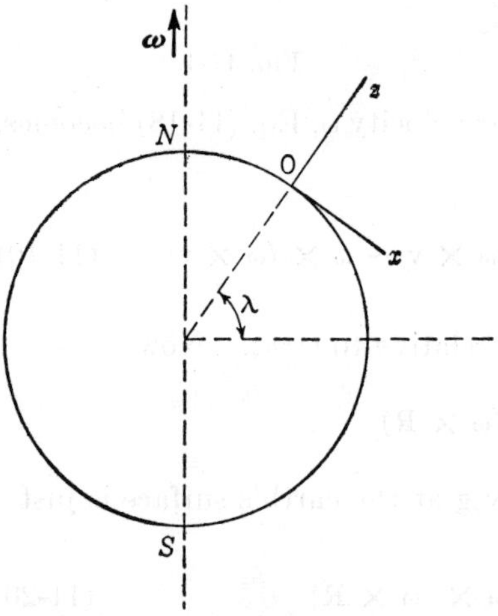

Fig. 11-5

$$\dot{y} = gt^2\omega \cos \lambda$$

and

$$y = \tfrac{1}{3}gt^3\omega \cos \lambda \tag{11-27}$$

taking $\dot{y}$ and y to be zero initially. But we know that Eq. (11-25), when integrated, yields the familiar expression $h = gt^2/2$, from which $t = \sqrt{2h/g}$. Thus, eliminating t from Eq. (11-27), we find that the body drifts to the east a distance

$$y = \frac{1}{3}\,\omega \cos \lambda \sqrt{\frac{8h^3}{g}}$$

In most practical cases this distance is very small, since the angular velocity ω of the earth is a small quantity of the order of 10^{-5} (see Prob. 11-1). A similar type of analysis applied to the case of a particle moving along the earth's surface (in a tangent plane) reveals that the general effect of the Coriolis acceleration, $-2\boldsymbol{\omega} \times \mathbf{v}$, on such a motion is to cause the particle to drift to the right in the Northern Hemisphere and to the left in the Southern Hemisphere. This acceleration influences the sense of the vortex rotation of storms of the cyclone type.

11-4. The Foucault Pendulum. The well-known precession of a pendulum is produced by the effect of the Coriolis acceleration. In looking at this problem let us select axes with origin at the equilibrium position of the pendulum bob and such that z is parallel to the plumb line, x is positive to the south, and y is positive to the east. Now, as is seen in Fig. 11-6, the forces acting on the bob of mass m are the tension $\mathbf{T}$, in the cord, and $m\mathbf{g}$, the weight of the bob. The equation of motion is

$$m\mathbf{a} = m\mathbf{g} + \mathbf{T} - 2m\boldsymbol{\omega} \times \mathbf{v} \tag{11-28}$$

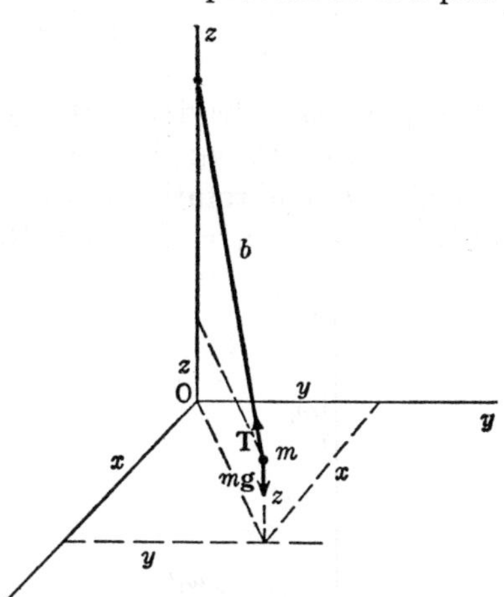

Fig. 11-6

We wish to write the cartesian components of Eq. (11-28). It is clear from Fig. 11-6 that, if the length of the pendulum is b, the components of $\mathbf{T}$ are

$$T_x = -T\frac{x}{b} \qquad T_y = -T\frac{y}{b} \qquad T_z = T\frac{b-z}{b}$$

and, for our choice of axes, the components of $\boldsymbol{\omega}$ are again given by Eq. (11-22). Thus

$$-2m\boldsymbol{\omega} \times \mathbf{v} = 2m\omega_z\dot{y}\mathbf{i} - 2m(\omega_z\dot{x} - \omega_x\dot{z})\mathbf{j} - 2m\omega_x\dot{y}\mathbf{k}$$

and the equations of motion are

$$m\ddot{x} = 2m\omega\dot{y}\sin\lambda - T\frac{x}{b} \tag{11-29}$$

$$m\ddot{y} = -2m\omega(\dot{x}\sin\lambda + \dot{z}\cos\lambda) - T\frac{y}{b} \tag{11-30}$$

$$m\ddot{z} = 2m\omega\dot{y}\cos\lambda + T\frac{b-z}{b} - mg \tag{11-31}$$

Clearly we are interested in the case where the displacement from the equilibrium position is small compared with b, and for this situation $\dot{z}$ can be neglected compared with $\dot{x}$ and $\dot{y}$ (the motion is practically horizontal). Also, for small displacements T is approximately equal to *mg*. Taking these factors into account and writing $\omega_1 \equiv \omega \sin \lambda$, Eqs. (11-29) and (11-30) may be rewritten as

$$\ddot{x} - 2\omega_1 \dot{y} + g\frac{x}{b} = 0 \tag{11-32}$$

$$\ddot{y} + 2\omega_1 \dot{x} + g\frac{y}{b} = 0 \tag{11-33}$$

The physical behavior of the system may be seen by transforming to a new set of axes $Ox_1y_1z_1$ having the same origin O, with the z_1 axis coincident with Oz, and rotating in the negative sense about Oz with a uniform angular velocity $\omega_1 = \omega \sin \lambda$. Referring to a point P, in the xy plane in Fig. 11-7, the transformation equations to the rotating axes are

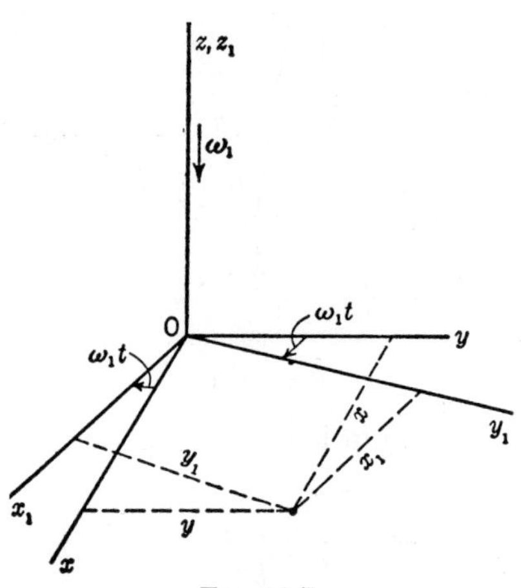

FIG. 11-7

$$x = x_1 \cos \omega_1 t + y_1 \sin \omega_1 t$$
$$y = -x_1 \sin \omega_1 t + y_1 \cos \omega_1 t$$

Differentiating these, substituting for x, $\dot{x}$, $\ddot{x}$, y, $\dot{y}$, $\ddot{y}$ in Eq. (11-32) [or Eq. (11-33)], and making obvious cancellations, we have left

$$\ddot{x}_1 \cos \omega_1 t + \frac{g}{b} x_1 \cos \omega_1 t + \ddot{y}_1 \sin \omega_1 t + \frac{g}{b} y_1 \sin \omega_1 t = 0$$

where we have neglected terms in $\omega_1{}^2$. If this is to be true for all values of $\omega_1 t$, we must be able to equate the coefficients of $\cos \omega_1 t$ and $\sin \omega_1 t$ separately to zero. Carrying out this operation, we obtain the two equations in the rotating system

$$\ddot{x}_1 + \frac{g}{b} x_1 = 0 \tag{11-34}$$

$$\ddot{y}_1 + \frac{g}{b} y_1 = 0 \tag{11-35}$$

These are the familiar equations of the simple pendulum. No cross terms are present, indicating that, in the rotating system $Ox_1y_1z_1$, the appearance of the motion is that of a simple plane pendulum. This shows further that, in the system $Oxyz$ fixed in the surface of the earth in the Northern Hemisphere, the motion is that of a pendulum rotating, in the sense of increasing $\omega_1 t$ in the figure, with an angular velocity of precession $\omega_1 = \omega \sin \lambda$. Thus the period of the precession depends on the latitude λ.

MOTION OF A PARTICLE ALONG A SURFACE OR A CURVE

11-5. Introductory Examples. The majority of dynamical problems involve motion which is subject to constraints of one form or another. A great many of the problems which already have been treated in the present text are of this variety. Let us select two simple illustrations of constrained motion. The first is the case of a particle which is constrained to move over the surface of a smooth plane, and the second is a particle which is free to slide along a smooth rod which rotates in a horizontal plane about a vertical axis passing through one end. In discussing the motion in problems such as these, we have always chosen coordinates which do not violate the constraint. For instance, in the present situation of the particle confined to a plane we would select the plane to be coincident with the xy plane, and therefore the coordinates appropriate to the problem would be x and y. The z coordinate would be ignored since there is no motion in this direction. Similarly, in the example of the particle sliding on the rod the origin would be chosen to be at the axis, and the radius vector r from this origin employed in order to describe the motion. In both cases the constraining forces would be omitted since, for a smooth surface, the force of constraint is always normal to the surface and consequently is normal to the direction of the motion.

It is now desirable to investigate the forces which these constraints are exerting on the particle. In general this information is supplied by merely including that scalar component of the equation of motion which had been omitted previously. The inclusion of the constraining forces, in effect, reduces the problem again to that of the free particle. In all the material below we shall limit ourselves to constraints of the integrable type (see Chap. 1) such as is the case when the motion is confined to a surface or curve.

Example 11-2. A particle slides over a smooth plane which is tangent to the earth's surface at a latitude λ. It is desired to determine the reaction, due to the earth's rotation, of the plane on the particle.

We shall neglect the acceleration of the earth's center in space and also shall restrict the motion to the neighborhood of the point of contact of the plane with the earth's surface. If $\boldsymbol{\omega}$ is the constant angular velocity of the earth, $\mathbf{v}$ the velocity of the particle at any time, and $\mathbf{R}$ the reaction of the plane on the particle, the equation of motion is [see (Eq. (11-21)]

$$m\ddot{\mathbf{r}} = m\mathbf{g} - 2m\boldsymbol{\omega} \times \mathbf{v} + \mathbf{R} \tag{11-36}$$

Equation (11-36) is written with respect to a system of axes rigidly attached to the earth. Since z does not appear explicitly, we may take the origin to be at the point of the plane which is tangent to the surface of the earth. If we select x positive to the south, y to the east, and z vertically upward, the components of $\boldsymbol{\omega}$ in this system are as given in Eq. (11-22). Accordingly, the component equations of motion are

$$m\ddot{x} = 2m\omega\dot{y}\sin\lambda \tag{11-37}$$
$$m\ddot{y} = -2m\omega\dot{x}\sin\lambda \tag{11-38}$$
$$m\ddot{z} = -mg + 2m\omega\dot{y}\cos\lambda + R \tag{11-39}$$

since **g** and **R** have components only along the z axis. Now the left side of Eq. (11-39) is zero, since z is zero for all values of the time. We therefore have left

$$R = mg - 2m\omega\dot{y}\cos\lambda \tag{11-40}$$

in which the quantity $\dot{y}$ may be obtained from the solutions of Eqs. (11-37) and (11-38) (see Prob. 11-2) and may be expressed as a function either of time or of x and y. Thus R also may be written either as a function of time or as a function of x and y.

Example 11-3. A particle slides on a smooth rod, the rod being constrained to rotate uniformly at an angular velocity ω in a plane about an axis passing through one end perpendicular to the rod. (Assume the axis is attached to an inertial system.)

Let us choose a system of coordinates rotating with the rod, with the origin at the intersection of the rod with the axis of rotation. The vector equation of motion is

$$m\ddot{\mathbf{r}} = -m\boldsymbol{\omega} \times (\boldsymbol{\omega} \times \mathbf{r}) - 2m\boldsymbol{\omega} \times \mathbf{v} + \mathbf{R}$$

in which **r** is the radius vector, in the moving system, from the origin to the particle, **v** the velocity of the particle relative to the rod, and **R** the reaction of the rod on the particle. If we choose x to be positive along the rod and the y and z axes perpendicular to it, with the z axis as that of rotation, the component equations of motion are

$$\begin{aligned} m\ddot{x} &= m\omega^2 x \\ m\ddot{y} &= -2m\omega\dot{x} + R_y \\ m\ddot{z} &= R_z \end{aligned} \tag{11-41}$$

We note that **R** has no x component since the rod is smooth and that $\boldsymbol{\omega}$ is along the z axis. Now both y and z are zero for all time, and therefore both $\ddot{z}$ and $\ddot{y}$ must be put equal to zero. This gives

$$\begin{aligned} R_y &= 2m\omega\dot{x} \\ R_z &= 0 \end{aligned}$$

Thus we see that the reaction of the rod is in the positive y direction if ω and $\dot{x}$ are both positive. The quantity $\dot{x}$ may be determined from the solution of the first of Eqs. (11-41).

Example 11-4. This example will be the same as Example 11-3, but employing a coordinate system which is not rotating.

Let us again choose the origin to be at the intersection of the rod with the axis of rotation, with now the coordinates to be plane polar coordinates, θ being measured positively in the usual sense. Since there is no applied force in the r direction, the equations of motion are

$$m(\ddot{r} - r\dot{\theta}^2) = 0 \tag{11-42}$$
$$m(r\ddot{\theta} + 2\dot{r}\dot{\theta}) = R \tag{11-43}$$

But $\dot{\theta}$ is the constant angular velocity ω here. Thus, from Eq. (11-43), R becomes

$$R = 2m\dot{r}\omega$$

in which $\dot{r}$ may be found by carrying out the solution of Eq. (11-42).

11-6. Motion along a Smooth Plane Curve. Normal and Tangential Accelerations. When a particle is moving along a smooth plane curve, it is sometimes desirable to express the reaction of the constraint in terms of the radius of curvature. In Fig. 11-8, a particle is moving along a plane curve S_1S_2 and, at the instant under consideration, is at point P. The radius vector $\mathbf{r}$ is drawn from a fixed origin O. If the only force acting is a force $\mathbf{R}$ of constraint, the equation of motion may be written

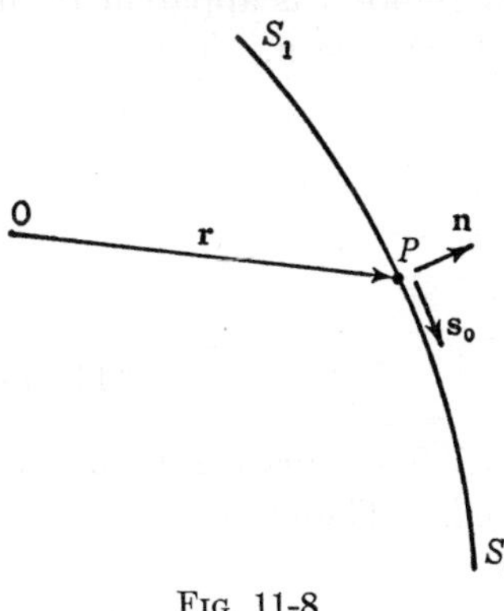

FIG. 11-8

$$m\ddot{\mathbf{r}} = \mathbf{R} \tag{11-44}$$

In Fig. 11-8, $\mathbf{s}_0$ is a unit vector which is always tangent to the path, and $\mathbf{n}$ is a unit vector in the plane of the curve and directed along the normal in the sense shown. In terms of these, the velocity $\dot{\mathbf{r}}$ of the particle at any point of the path may be expressed as

$$\dot{\mathbf{r}} = v\mathbf{s}_0 \tag{11-45}$$

in which v is the magnitude of $\dot{\mathbf{r}}$. Accordingly we may find $\ddot{\mathbf{r}}$ by differentiating Eq. (11-45). In doing this, it is necessary to notice that the direction of $\mathbf{s}_0$ in space is continually changing as the particle moves along the curve and that we must take into account the fact that $\dot{\mathbf{s}}_0$ is therefore not zero. Thus we have

$$\ddot{\mathbf{r}} = \dot{v}\mathbf{s}_0 + v\dot{\mathbf{s}}_0 \tag{11-46}$$

The first term on the right is the usual tangential acceleration. The second term represents the part of the acceleration arising from the change in the direction of motion and requires further clarification. In Fig. 11-9, the particle advances a distance ds from Q to Q' along the path during the infinitesimal interval dt. An examination of Fig. 11-9 reveals that, during this same interval, the unit vector $\mathbf{s}_0$ changes by a small amount $d\mathbf{s}_0$. If C is the center of curvature for the small element of path QQ', we have that $\mathbf{s}_0$ is perpendicular to CQ and $\mathbf{s}_0 + d\mathbf{s}_0$ is perpendicular to CQ'. Thus the angle $d\theta$ between CQ and CQ' is also the angle between $\mathbf{s}_0$ and $\mathbf{s}_0 + d\mathbf{s}_0$.

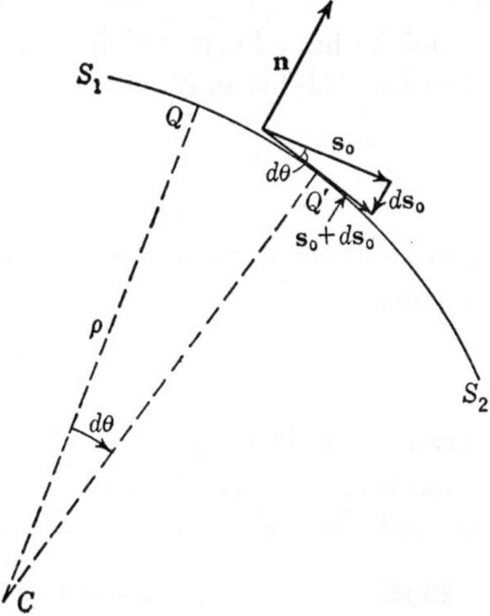

FIG. 11-9

Accordingly the magnitude of $d\mathbf{s}_0$ is

$$|d\mathbf{s}_0| = |\mathbf{s}_0|\,d\theta = d\theta$$

neglecting second-order infinitesimals. Furthermore it is apparent from Fig. 11-9 that the direction of $d\mathbf{s}_0$ is opposite to that of $\mathbf{n}$. Thus

$$d\mathbf{s}_0 = -\mathbf{n}\,d\theta$$

from which

$$\dot{\mathbf{s}}_0 = -\mathbf{n}\dot{\theta}$$

In view of this result, Eq. (11-46) becomes

$$\ddot{\mathbf{r}} = \dot{v}\mathbf{s}_0 - v\dot{\theta}\mathbf{n} \tag{11-47}$$

Now CQ and CQ' are both identical to the radius of curvature ρ of the element ds, provided that $d\theta$ is sufficiently small. Therefore

$$\rho\,d\theta = ds$$

or

$$\rho\dot{\theta} = \frac{ds}{dt} = v \tag{11-48}$$

By means of Eq. (11-48) the quantity $\dot{\theta}$ may be eliminated from Eq. (11-47). We have

$$\ddot{\mathbf{r}} = \dot{v}\mathbf{s}_0 - \frac{v^2}{\rho}\mathbf{n} \tag{11-49}$$

as might have been anticipated from elementary considerations. Therefore Eq. (11-44) may be written

$$m\dot{v}\mathbf{s}_0 - \frac{mv^2}{\rho}\mathbf{n} = \mathbf{R} \tag{11-50}$$

But since the curve is smooth, the reaction $\mathbf{R}$ is normal to the curve, and we have

$$m\dot{v}\mathbf{s}_0 = 0 \qquad \mathbf{R} = -\frac{mv^2}{\rho}\mathbf{n} \tag{11-51}$$

Accordingly the reaction is directed along the inward normal and has the magnitude mv^2/ρ. We notice, from the first of Eqs. (11-51), that the magnitude of the velocity does not change during the motion.

Example 11-5. A particle of mass m is constrained to slide along a smooth wire which is bent into the form of a parabola the axis of which is along a fixed vertical line. Gravity is acting vertically downward. If the particle starts from rest at an end of the latus rectum, find the reaction of the wire on the particle when the latter is at the lowest point. The equation of the parabola is $x^2 = 2ay$, in which the origin is at the vertex and a is a constant.

From the geometrical properties of the parabola the distance from the origin to the focus F is just $a/2$ and the semilatus rectum is a (see Fig. 11-10). The equation of motion of the particle, from Eq. (11-50), may be written

$$m\dot{v}\mathbf{s}_0 - \frac{mv^2}{\rho}\mathbf{n} = m\mathbf{g} + \mathbf{R}$$

Now $\mathbf{R}$ is normal to the wire (since the wire is smooth). At point O we have

$$\mathbf{R}_0 = -\left(mg + \frac{mv_0^2}{\rho_0}\right)\mathbf{n} \qquad \text{(11-52)}$$

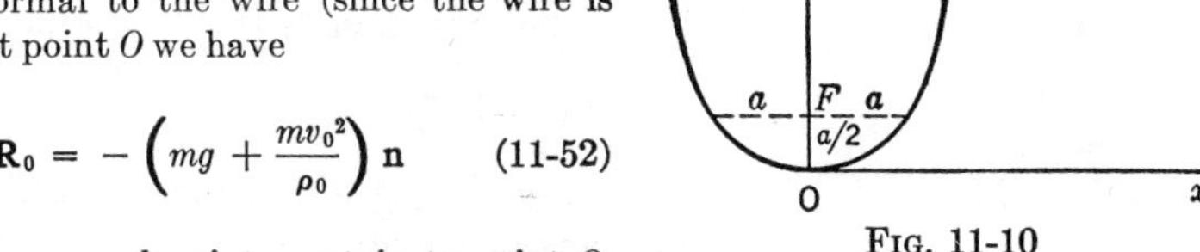

FIG. 11-10

in which the zero subscripts pertain to point O. The speed v_0 may be found at this point by use of the energy equation. We have

$$\frac{1}{2}mv_0^2 = mg\frac{a}{2}$$

and

$$v_0^2 = ga \qquad \text{(11-53)}$$

The radius of curvature ρ_0, at point O, may be found from the equation

$$\frac{1}{\rho} = \frac{d^2y/dx^2}{[1 + (dy/dx)^2]^{\frac{3}{2}}}$$

from which, making use of the equation of the curve, we obtain $\rho_0 = a$. Hence, eliminating ρ_0 and v_0 from Eq. (11-52), we find the reaction of the wire at O to be

$$R_0 = mg + mga\frac{1}{a} = 2mg$$

11-7. More General Treatment of Integrable Constraints. Motion Confined to a Smooth Surface of Arbitrary Form. We have seen that we may replace the geometrical condition of a constraint by a *constraining force*, thus reducing the mechanics of a constrained particle to that of a free particle. Also we have seen that an essential characteristic of all smooth constraints is that the constraining forces are normal to the surfaces to which they confine the motion. This means that they do no work (if the surfaces are independent of time) since they are always normal to the motion. If this were not so, the mere existence of the surface could set the particle in motion.

In the present section we shall consider the motion of a particle which is confined to an arbitrary smooth surface. The equation of motion is

$$m\ddot{\mathbf{r}} = \mathbf{F} + R\mathbf{n} \qquad \text{(11-54)}$$

where $\mathbf{F}$ contains all the acting forces save those of constraint. Equation (11-54) is written with respect to an inertial system. It is convenient to write the constraining forces separately as $R\mathbf{n}$, where we take $\mathbf{n}$

to be a dimensionless unit vector directed along the outward normal to the surface.

Now the equation of a surface may always be written as

$$f(x,y,z,t) = 0 \tag{11-55}$$

where t is present if the form of the surface depends explicitly upon the time. Furthermore, since the reaction $R\mathbf{n}$ has the direction of the normal to the surface, it has also the direction of the gradient of f. Thus

$$R\mathbf{n} = \lambda\,\nabla f$$

where $\lambda = \lambda(x,y,z,t)$, in general, and is a scalar function which may be determined if necessary. In terms of the gradient of f, Eq. (11-54) may be rewritten

$$m\ddot{\mathbf{r}} = \mathbf{F} + \lambda\,\nabla f \tag{11-56}$$

The problem of a particle constrained to move along a given line is a particular case of this, since a line can always be represented as the intersection of two surfaces $f_1 = 0$ and $f_2 = 0$, for example. Thus, in this latter case, Eq. (11-56) becomes

$$m\ddot{\mathbf{r}} = \mathbf{F} + \lambda_1\,\nabla f_1 + \lambda_2\,\nabla f_2$$

In general we know $\mathbf{F}$, f_1, f_2 as functions of the coordinates and time, and, since we must also satisfy $f_1 = 0$ and $f_2 = 0$ for all time, this provides sufficient information to determine λ_1 and λ_2. In so doing, it is useful to remember that, since $f(x,y,z,t) = 0$ for all time,

$$df = 0 \qquad \frac{df}{dt} = 0 \qquad \frac{d^2f}{dt^2} = 0$$

All total derivatives are zero. However, we are not permitted to infer, because of this, that $\partial f/\partial t$ is zero or that the gradient ∇f is zero. This may easily be seen in the following manner: We have, since $f(x,y,z,t) = 0$,

$$df = 0 = \frac{\partial f}{\partial x}\,dx + \frac{\partial f}{\partial y}\,dy + \frac{\partial f}{\partial z}\,dz + \frac{\partial f}{\partial t}\,dt = \nabla f \cdot d\mathbf{r} + \frac{\partial f}{\partial t}\,dt \tag{11-57}$$

We see that, even if $df/dt = 0$, ∇f and $\partial f/\partial t$ are not necessarily zero.

11-8. Equation of Energy. We shall now consider, from the point of view of the energy equation, the problem of a particle moving over a smooth surface. To arrive at the energy integral, we multiply Eq. (11-56) by the factor $\dot{\mathbf{r}}\cdot$ as follows:

$$m\dot{\mathbf{r}} \cdot \ddot{\mathbf{r}} = \mathbf{F} \cdot \dot{\mathbf{r}} + \lambda\,\nabla f \cdot \dot{\mathbf{r}}$$

from which, rewriting the left side, we have at once

$$\frac{m}{2}\frac{d}{dt}(\dot{\mathbf{r}} \cdot \dot{\mathbf{r}}) = \frac{m}{2}\frac{d}{dt}|\dot{\mathbf{r}}|^2 = \mathbf{F} \cdot \frac{d\mathbf{r}}{dt} + \lambda\,\nabla f \cdot \frac{d\mathbf{r}}{dt}$$

Multiplying through by dt, we have

$$\frac{m}{2}\, d|\dot{\mathbf{r}}|^2 = mv\, dv = \mathbf{F}\cdot d\mathbf{r} + \lambda\, \nabla f \cdot d\mathbf{r} \tag{11-58}$$

where v is the magnitude of $\dot{\mathbf{r}}$. In view of Eq. (11-57), the last term in Eq. (11-58) may be written

$$\lambda\, \nabla f \cdot d\mathbf{r} = \lambda\, df - \lambda \frac{\partial f}{\partial t}\, dt = -\lambda \frac{\partial f}{\partial t}\, dt$$

where the last step follows because df is zero. Equation (11-58) now becomes

$$mv\, dv = \mathbf{F}\cdot d\mathbf{r} - \lambda \frac{\partial f}{\partial t}\, dt$$

Integrating between appropriate limits, we have

$$m \int_{v_1}^{v_2} v\, dv = \int_{r_1}^{r_2} \mathbf{F}\cdot d\mathbf{r} - \int_{t_1}^{t_2} \lambda \frac{\partial f}{\partial t}\, dt$$

or

$$\frac{m}{2}(v_2^2 - v_1^2) - \int_{r_1}^{r_2} \mathbf{F}\cdot d\mathbf{r} = -\int_{t_1}^{t_2} \lambda \frac{\partial f}{\partial t}\, dt \tag{11-59}$$

and, if the forces $\mathbf{F}$ are all derivable from a potential energy V, Eq. (11-59) may be further rewritten

$$\frac{m}{2}(v_2^2 - v_1^2) + (V_2 - V_1) = -\int_{t_1}^{t_2} \lambda \frac{\partial f}{\partial t}\, dt \tag{11-60}$$

The first term is the change in kinetic energy of the particle, and the second is the change in potential energy. The term on the right side is the work done by the constraining forces and is different from zero only if the constraint depends explicitly on the time. If the constraining surface does not depend upon the time, Eq. (11-60) then reduces simply to

$$\frac{m}{2} v_2^2 + V_2 = \frac{m}{2} v_1^2 + V_1 = \text{const}$$

which states the familiar form of the conservation of energy.

11-9. The Angular-momentum Integral. It is of interest also to examine the conditions under which the angular momentum is a constant for problems involving constraints. To investigate this, we multiply Eq. (11-56) by the factor $\mathbf{r}\times$. This yields

$$m\mathbf{r} \times \ddot{\mathbf{r}} = \mathbf{r} \times \mathbf{F} + \mathbf{r} \times \lambda\, \nabla f \tag{11-61}$$

We note that the left side may be rewritten as $(d/dt)(\mathbf{r} \times m\dot{\mathbf{r}})$, so that

Eq. (11-61) becomes

$$\frac{d}{dt}(\mathbf{r} \times m\dot{\mathbf{r}}) \equiv \frac{d\mathbf{J}}{dt} = \mathbf{r} \times \mathbf{F} + \lambda\mathbf{r} \times \nabla f \tag{11-62}$$

and $\mathbf{J}$ is a constant only if the right side of Eq. (11-62) vanishes. However, since Eq. (11-62) is equivalent to three scalar equations, it is possible for one or more components of $\mathbf{J}$ to be constant, though the remaining components vary. Such is the case in Example 11-7.

Example 11-6. A particle of mass m is sliding on a smooth wire bent into the form of a vertical circle. Gravity is acting vertically downward. It is desired to determine the reaction of the wire as a function of the position of the particle.

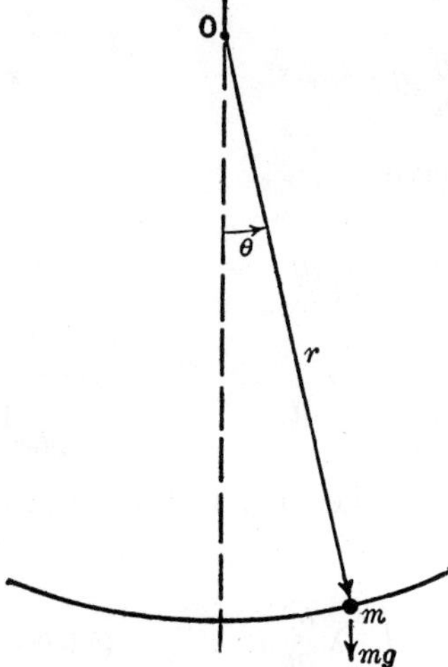

FIG. 11-11

The situation is shown in Fig. 11-11. We choose plane polar coordinates with the origin at the center of the circle as shown. The equations of motion are

$$m(\ddot{r} - r\dot{\theta}^2) = mg\cos\theta + \lambda(\nabla f)_r \tag{11-63}$$

$$m(r\ddot{\theta} + 2\dot{\theta}\dot{r}) = -mg\sin\theta \tag{11-64}$$

Now the equation of the constraint is

$$f(r) = r - a = 0 \tag{11-65}$$

from which we see that

$$\frac{df}{dt} = 0 \qquad \frac{d^2f}{dt^2} = 0 \qquad (\nabla f)_r = 1 \tag{11-66}$$

Equations (11-65) and (11-66) tell us that $\dot{r} = 0$, $\ddot{r} = 0$, and so on. Consequently Eqs. (11-63) and (11-64) become

$$-ma\dot{\theta}^2 = mg\cos\theta + \lambda = mg\cos\theta + R \tag{11-67}$$

$$ma\ddot{\theta} = -mg\sin\theta \tag{11-68}$$

From (11-67) we have

$$R = \lambda = -ma\dot{\theta}^2 - mg\cos\theta \tag{11-69}$$

Equation (11-68) is that of the simple pendulum and may be solved in a manner to yield $\dot{\theta}$ either as a function of time or as a function of θ, depending on what is sought.

Alternatively we could have employed the energy equation for this example. In such a case we have

$$\frac{m}{2}a^2\dot{\theta}^2 - \frac{1}{2}mv_0^2 + amg(1 - \cos\theta) = 0$$

where v_0 is the velocity of the particle at $\theta = 0$. This gives us $\dot{\theta}$ directly in terms of knowns. This may be substituted in Eq. (11-69) to yield

$$R = -m\left(\frac{v_0^2}{a} - 2g + 2g\cos\theta + g\cos\theta\right) = -m\left[\frac{v_0^2}{a} - g(2 - 3\cos\theta)\right]$$

which is the reaction of the wire on the particle. (In the problem of the pendulum, this expression for R is the tension in the string.)

Example 11-7. A particle of mass m is constrained to move on the inner surface of a smooth paraboloid of revolution the axis of which is vertical. The equation of the surface is $x^2 + y^2 = 2az$, where a is a constant. Initially the particle is located at a height z_0, with a horizontal velocity of magnitude $\sqrt{2gz_0}$ and with no vertical velocity. It is desired to find the reaction of the surface upon the particle (see Fig. 11-12).

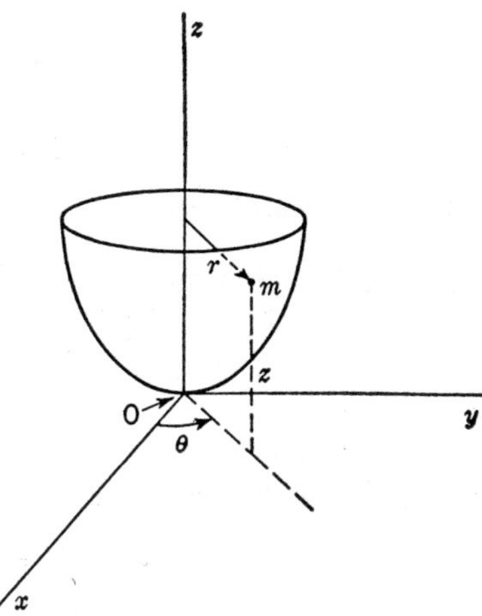

Fig. 11-12

It is desirable to employ cylindrical polar coordinates in this case (that is, plane polar coordinates in the xy plane, plus the addition of the coordinate z). The reaction of the surface will depend upon the nature of the motion of the particle. Let us therefore investigate the character of the motion. Clearly the total energy will be a constant, and we have

$$\frac{m}{2}(\dot{r}^2 + r^2\dot{\theta}^2 + \dot{z}^2) + mgz = \frac{m}{2}v_0^2 + mgz_0 \tag{11-70}$$

in which v_0 is the initial value, $\sqrt{2gz_0}$, of the velocity and $r^2 = x^2 + y^2$. Now the forces acting are $m\mathbf{g}$ vertically downward and the reaction $\mathbf{R}$, which is normal to the surface. Consequently the line of action of $\mathbf{R}$ intersects the z axis. Thus there are no moments present which have components along the z axis, and therefore the z component of the angular momentum $\mathbf{J}$ remains constant. Accordingly we have

$$J_z = mr^2\dot{\theta} = mv_0r_0 = \text{const} \tag{11-71}$$

Now the equation of the surface may be written

$$r^2 = 2az$$

from which

$$\dot{r} = \frac{a\dot{z}}{r} \tag{11-72}$$

By making use of Eqs. (11-71) and (11-72), Eq. (11-70) becomes

$$\frac{a^2\dot{z}^2}{r^2} + \frac{v_0^2r_0^2}{r^2} + \dot{z}^2 = 2g(z_0 - z) + v_0^2$$

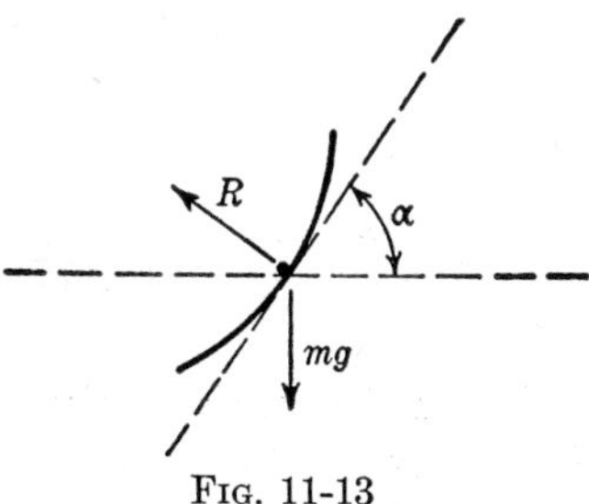

Fig. 11-13

Noting that $r_0^2 = 2az_0$, where r_0 is the value of r at $z = z_0$, we solve for $\dot{z}^2$. We obtain

$$\dot{z}^2 = \frac{2z(z - z_0)[(v_0^2/z) - 2g]}{2z + a} \tag{11-73}$$

In view of the boundary condition $v_0^2 = 2gz_0$, we note that $\dot{z}$ is zero initially. Moreover, if we take the derivative of (11-73) with respect to time, we find that $\ddot{z}$ is zero also. Thus the motion will be a circle in a horizontal plane at a height $z = z_0$, and the tangential velocity of motion in this circle is v_0. Accordingly, the reaction R will remain fixed in magnitude. In Fig. 11-13 we see that R and mg

must be such as to produce the centripetal acceleration v_0^2/r_0. We have therefore

$$R \sin \alpha = \frac{mv_0^2}{r_0} \qquad R \cos \alpha = mg$$

Now $\tan \alpha = dz/dr = r/a$, evaluated at $r = r_0$. Hence $\sin \alpha = r_0/(r_0^2 + a^2)^{\frac{1}{2}}$. Thus since $v_0^2 = 2gz_0$ and $r_0^2 = 2az_0$, we have

$$R = mg\left(\frac{2z_0 + a}{a}\right)^{\frac{1}{2}}$$

which is the desired result.

11-10. Rough Constraints. Particle Sliding on a Rough Wire. Let us again consider the problem of the particle of mass m sliding on a circular wire (Example 11-6). We shall impose the condition, however, that the wire is no longer smooth and that the coefficient of sliding friction between the particle and the wire is μ. Equations (11-67) and (11-68) become $(r - a = 0)$

$$-ma\dot{\theta}^2 = mg \cos \theta + R \tag{11-74}$$

$$ma\ddot{\theta} = -mg \sin \theta - \mu|R| \tag{11-75}$$

where the sign of the second term on the right of (11-75) is appropriate for motion in the sense of θ increasing.

To solve these, it is convenient to eliminate R between the two, yielding

$$\ddot{\theta} + \mu\dot{\theta}^2 + \mu \frac{g}{a} \cos \theta + \frac{g}{a} \sin \theta = 0 \tag{11-76}$$

If it were desired to find R, the procedure would be to solve Eq. (11-76) for $\dot{\theta}$ and then substitute the result in Eq. (11-74) to get R as a function of θ. Equation (11-76) cannot be solved completely, as it stands, in terms of simple functions (such as polynomials, sines, cosines, and so on) without making certain approximations. However, a first integral, that is, $\dot{\theta}$ as a function of θ, can be obtained without making approximations. To do this, we reduce the order of Eq. (11-76) by the substitution $p = d\theta/dt$. At the same time dividing by p, we obtain

$$\frac{dp}{d\theta} + \mu p = -\left(\mu \frac{g}{a} \cos \theta + \frac{g}{a} \sin \theta\right)\frac{1}{p} \tag{11-77}$$

which is now a first-order equation in p and θ. It is not linear, however, but has the form of Bernoulli's equation, in which n here has the value -1 (see Appendix 2). This can always be made linear by a suitable substitution. In the case of Eq. (11-77) the type substitution is simply

$$p = z^{\frac{1}{2}} \tag{11-78}$$

Equation (11-77) becomes

$$\frac{dz}{d\theta} + 2\mu z = -2\frac{g}{a}(\mu \cos \theta + \sin \theta) \tag{11-79}$$

which is a linear differential equation of the first order. This type always admits of solution by means of an exponential integrating factor. In our case the factor is $e^{+2\mu\theta}$. Multiplying Eq. (11-79) by this factor and combining terms on the left side, we obtain

$$\frac{d}{d\theta}(ze^{+2\mu\theta}) = -2\frac{g}{a}(\mu \cos \theta + \sin \theta)e^{+2\mu\theta}$$

Next, multiplying through by $d\theta$ and integrating, we have

$$ze^{+2\mu\theta} = \frac{2g}{a}\int(\mu \cos \theta + \sin \theta)e^{+2\mu\theta}\,d\theta + c \tag{11-80}$$

in which c is a constant of integration. The remainder of the problem is left to the student. The right side of Eq. (11-80) may be integrated readily by parts, or a good set of integral tables can be consulted. Either procedure will furnish z as a function of θ. Then, from Eq. (11-78) it will be possible to obtain $\dot{\theta}$ as a function of θ. Following this, the substitution into Eq. (11-74) can be made, to yield finally R as a function of θ.

11-11. The Pendulum of Arbitrary Amplitude. It is appropriate at this point to consider briefly the motion of a particle moving under gravity on a vertical circle of radius a, in which the amplitude of displacement from the equilibrium position may be large. If the amplitude is small, the period is independent of the amplitude. However, if the amplitude of the displacement is large, the period of the motion *does* depend on the amplitude (see Sec. 7-11). We desire to find the time elapsing between two successive configurations of the system when the amplitude of the motion is large. To set up the problem, we can employ the equation of energy conservation. [Alternatively, the equation of motion, Eq. (11-68), could be used.] We have

$$\tfrac{1}{2}ma^2\dot{\theta}^2 - mga \cos \theta = c \tag{11-81}$$

where θ, varying between 0 and θ_0, is measured from the line between the center of curvature and the lowest point of the circle and c is a constant. In Eq. (11-81) the center of the circle is taken to be the level of zero potential energy. Equation (11-81) may be rearranged as

$$\dot{\theta}^2 = \frac{2g}{a}\cos \theta + c' \tag{11-82}$$

If we impose the boundary condition that, at $\theta = \theta_0$, $\dot{\theta} = 0$, we are able to determine c', from (11-82), to be $-(2g/a)\cos\theta_0$. Equation (11-82) then becomes, after extracting the square root,

$$\frac{d\theta}{dt} = \left[\frac{2g}{a}(\cos\theta - \cos\theta_0)\right]^{\frac{1}{2}} \tag{11-83}$$

We now make the substitution

$$\cos\theta = 1 - 2\sin^2\frac{\theta}{2} \tag{11-84}$$

and since $\frac{1}{2}\,d\theta = d(\theta/2)$, Eq. (11-83) becomes

$$\frac{d\theta/2}{[\sin^2(\theta_0/2) - \sin^2(\theta/2)]^{\frac{1}{2}}} = \sqrt{\frac{g}{a}}\,dt \tag{11-85}$$

We now introduce a new angle φ defined by

$$\sin\frac{\theta}{2} = \sin\frac{\theta_0}{2}\sin\varphi = k\sin\varphi \tag{11-86}$$

in which $k = \sin(\theta_0/2)$. From Eq. (11-86), $\theta/2 = \sin^{-1}(k\sin\varphi)$, and we have

$$d\frac{\theta}{2} = \frac{k\cos\varphi\,d\varphi}{\sqrt{1 - k^2\sin^2\varphi}}$$

Equation (11-85) then assumes the form

$$\frac{d\varphi}{\sqrt{1 - k^2\sin^2\varphi}} = \sqrt{\frac{g}{a}}\,dt \tag{11-87}$$

The time t of passage from φ_1 to φ_2 may be found from (11-87) as

$$t = \sqrt{\frac{a}{g}}\int_{\varphi_1}^{\varphi_2}\frac{d\varphi}{\sqrt{1 - k^2\sin^2\varphi}} \tag{11-88}$$

Now what are the limits of the variable φ? From (11-86) we see that, as θ goes from 0 to θ_0, φ goes from 0 to $\pi/2$, regardless of the value of θ_0. Thus the time required to pass from 0 to θ_0 may be obtained from Eq. (11-88) by putting in the appropriate limits. We find

$$t = \sqrt{\frac{a}{g}}\int_0^{\pi/2}\frac{d\varphi}{\sqrt{1 - k^2\sin^2\varphi}} \tag{11-89}$$

The value of θ_0 has a bearing on the quantity k, since $k = \sin(\theta_0/2)$, by definition. For example, if we choose θ_0 to be 30°, then $\theta_0/2 = 15°$, and we may write $15° = \sin^{-1}k$. In the literature, the quantity K, where

$$K = \int_0^{\pi/2} \frac{d\varphi}{\sqrt{1 - k^2 \sin^2 \varphi}} \tag{11-90}$$

is called the complete elliptic integral of the first kind. Tables of K, as a function of k, are given in most integral tables (for example, Peirce, Dwight, and so on). In these we find that, if $\sin^{-1} k = 15°$, $K = 1.5981$. Thus the quarter period of the motion is

$$t = 1.5981 \sqrt{\frac{a}{g}}$$

Problems

Problems marked C are reprinted by kind permission of the Cambridge University Press.

11-1. A body is dropped from a height of 100 ft above the ground at a latitude of 41°N. If the angular velocity of the earth is 7.29×10^{-5} rad/sec, find the deviation (numerical value) of the particle from a point vertically beneath the initial position.

11-2. A particle slides on a perfectly smooth plane which is located at a latitude λ on the earth and is oriented so as to be perpendicular to the plumb line at that latitude. If the angular velocity of rotation of the earth has the magnitude ω, find the equation of the path of the particle as it appears to an observer sitting on the plane. Choose the z axis upward parallel to the plumb line, x to the south, and y to the east. The initial conditions are that, at $t = 0$, $x = 0$, $y = y_1$, $\dot{x} = \dot{x}_1$, and $\dot{y} = 0$.

11-3. A body is thrown vertically upward to a height h (parallel to plumb line) at a latitude λ in the Northern Hemisphere. Where will it strike the ground?

11-4. Two particles, each of mass m, are connected by a light inextensible string of length b and are free to slide along a smooth wire. The wire is constrained to rotate with a uniform angular velocity ω in a horizontal plane about a vertical axis through one end. If the center of mass is initially at rest at a distance $a - b/2$ from the axis, with the string initially taut, find the radial velocity of the particles as a function of time. Consider that they are initially at rest relative to the wire.

11-5*C*. A tidal current is running northward along a channel of breadth b, in latitude λ. Show that the height of the water on the east coast exceeds that on the west coast by $(2bv\omega \sin \lambda)/g$, where v is the velocity of the water and ω the earth's angular velocity.

11-6. A projectile is fired eastward at a northern latitude λ with a velocity V and angle of elevation α. Neglecting terms in ω^2, the square of the angular velocity of the earth, show that the lateral deviation because of the earth's rotation is

$$\frac{4\omega V^3 \sin^2 \alpha \cos \alpha \sin \lambda}{g^2}$$

11-7. In Prob. 11-6 if, in the absence of the earth's rotation, the range of the projectile is R, show that, neglecting ω^2, the change in the range arising because of the earth's rotation is

$$\sqrt{\frac{2R^3}{g}}\,\omega \cos \lambda \left(\cot^{\frac{1}{2}} \alpha - \frac{1}{3} \tan^{\frac{3}{2}} \alpha\right)$$

11-8. A system of axes $Oxyz$ is moving in a straight line parallel to its own xy plane with a velocity bt, where b is a constant and t is the time. At $t = 0$ the x axis of the moving system is along the direction of motion of the moving origin O. The moving system is rotating about its z axis with a uniform angular velocity ω. Write the component equations of motion in the moving system of a particle of mass m constrained to move in the xy plane of the moving system. There are no forces acting. Integrate the equations of motion, neglecting terms in ω^2, and find x and y as functions of time. Consider motion near the origin and that initially the particle is at the origin of the moving system and at rest relative to it.

11-9. A system of axes $Oxyz$ is moving through space with a uniform velocity. The system is rotating about the Oy axis with an angular velocity which is zero at $t = 0$ but is increasing at a constant rate b. Write the component equations of motion of a particle of mass m in this system, in terms of x, y, z, their time derivatives, b, and t. No forces are acting.

11-10. A mass m is permitted to drop from a position of rest a small distance above the ground at a latitude λ. The air provides a small resistance proportional to the velocity (proportionality constant k). The time of fall of the mass is t_0. Show that, neglecting the translational acceleration of the earth, and neglecting terms of the order of k^2, the deviation of the point on the ground at which it will strike from that if the earth were not rotating is

$$\frac{g\omega \cos \lambda}{3} t_0^3 \left(1 - \frac{k}{4m} t_0\right)$$

where g is the acceleration of gravity and ω is the angular velocity of the earth.

11-11. A particle is constrained to move in a smooth plane which is rotating with a constant angular velocity ω about a horizontal axis in the plane. Find r (the distance of the particle from the axis of rotation) as a function of the time. Find also the reaction of the plane on the particle as a function of θ, where θ is the angle through which the plane has turned after a time t. A constant acceleration g is acting vertically downward.

11-12. A particle of mass m is constrained to slide freely along a smooth wire bent in the form of an ellipse, of major axis $2a$ and minor axis $2b$. The particle is initially given a velocity v. Find the reaction of the wire at the instant when the particle is at an end of the major axis. Assume that gravity is not acting.

11-13. A particle of mass m is constrained to slide on a smooth straight wire, which in turn is forced to rotate in a horizontal plane about a point O on the wire at a constant angular velocity ω. Gravity g is acting vertically downward. Initially the particle is at rest relative to the wire, at a distance b from O. There is a force of magnitude mk/r^2 acting on the particle and directed toward O. (r is the distance from O to the particle, and k is a positive constant.) Show that the magnitude of the reaction of the wire upon the particle is

$$\left\{m^2g^2 + 4m^2\omega^2\left[\omega^2(r^2 - b^2) + 2k\frac{b - r}{rb}\right]\right\}^{\frac{1}{2}}$$

11-14. The radius of a soap bubble is a function of time as follows: $r = \alpha t + r_0$, where α and r_0 are constants. A particle of unit mass is constrained to remain in the surface. At $t = 0$ the particle is given a tangential velocity v. Find the path and the reaction of the surface.

$$\theta = \frac{v}{\alpha}\left(1 - \frac{r_0}{r}\right) \qquad R = -\frac{v^2 r_0^2}{r^3}$$

Show, by integrating the rate at which the constraint is doing work, that the total work done by the bubble during a time t, beginning at the initial instant, is equal to the difference in the kinetic energies at the two times.

11-15. A particle is constrained to move under gravity in a smooth vertical circle of radius a. If it starts from rest where the tangent to the circle is vertical, determine the length of time required for it to reach the lowest point.

11-16. A particle is projected horizontally along the smooth inner surface of a right circular cone with vertex downward. The initial velocity is $\sqrt{gh/2}$, where h is the initial height above the vertex and g is the constant acceleration of gravity, acting vertically downward. Find the height above the vertex of the lowest point of the path.

11-17. A heavy particle is constrained to move on the inside surface of a smooth spherical shell of inner radius a. Gravity is acting vertically downward. Initially the particle is projected with a horizontal velocity v_0 from a point which is at a depth b below the center of the sphere. Deduce that, when the particle is again moving horizontally, it will be at a depth x below the center of the sphere determined by

$$x^2 + \frac{v_0^2}{2g}x + (v_0^2 b - 2ga^2) = 0$$

11-18. A bead moves on a smooth wire which is bent in the form of a vertical circle of radius a. The wire rotates about a fixed vertical diameter with a uniform angular velocity ω. θ is the angular distance from the lowest point, and initially the particle is at rest relative to the tube at $\theta = \alpha$ where $\omega \cos(\alpha/2) = (g/a)^{\frac{1}{2}}$. Show that

$$\ddot{\theta} - \omega^2 \sin\theta \cos\theta + \frac{g}{a}\sin\theta = 0$$

$$\dot{\theta}^2 = \omega^2(1 - \cos\theta)(\cos\theta - \cos\alpha)$$

11-19. A particle of mass m is constrained to move along a smooth wire bent into the form of a horizontal circle of radius a. Initially the particle has a velocity v_0. The motion is subject to an air resistance proportional to the square of the velocity (proportionality constant mk). Find the angular position of the particle as a function of time. Find also the resultant reaction of the wire on the particle. Assume that gravity is not acting.

11-20. Find the position of the particle in Prob. 11-19 if the wire is rough, with coefficient μ of sliding friction. Neglect gravity and air resistance. Find also the resultant reaction of the wire on the particle.

11-21. The same as Prob. 11-20, but with air resistance included.

11-22. A particle of mass m is attached to the end of a string the other end of which is attached to the top of a smooth fixed sphere of radius b. The mass m is in contact with the sphere the string subtending an acute angle α at the center of the sphere. If m is sliding around in a horizontal circle with an angular velocity which is one-half of that necessary for m just to leave the surface, find the reaction of the surface and the tension in the string. Gravity g is acting vertically downward.

11-23. A bead of mass m is free to slide along a rough rod (coefficient of friction μ) which makes an angle θ with a vertical axis passing through the rod. The rod rotates about this axis with an angular velocity αt, where α is a positive constant and t is the time. Gravity g acts vertically downward. Find the equation of motion of the particle along the rod.

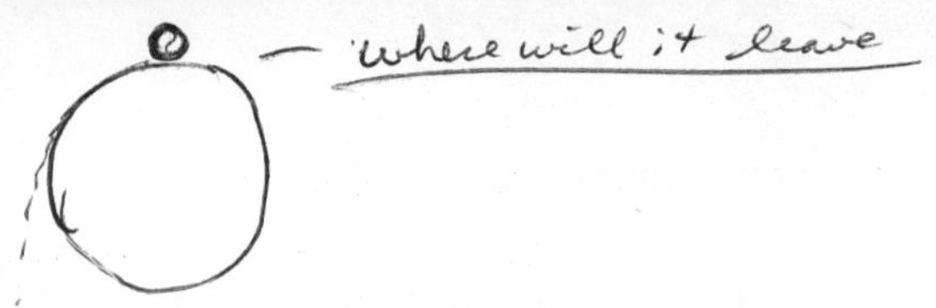

CHAPTER 12

MOTION OF A RIGID BODY IN THREE DIMENSIONS

In Chap. 9 the motion of a rigid body was treated in those simple cases in which the axis of rotation remains perpendicular to a fixed plane. Furthermore, the coordinate systems employed did not rotate relative to this plane. In the present chapter the concepts are generalized to include situations in which the direction of the axis of rotation varies with time. In addition, many problems are treated from the point of view of coordinate systems which are rotating in some manner. In certain cases this rotation is that executed by the rigid body, and the axes may be regarded as being rigidly attached to, and rotating with, the body. In other situations it is convenient to select axes which are describing only a portion of the rotational motions of the body. In such cases only the origin or at most one of the axes may be fixed in the body.

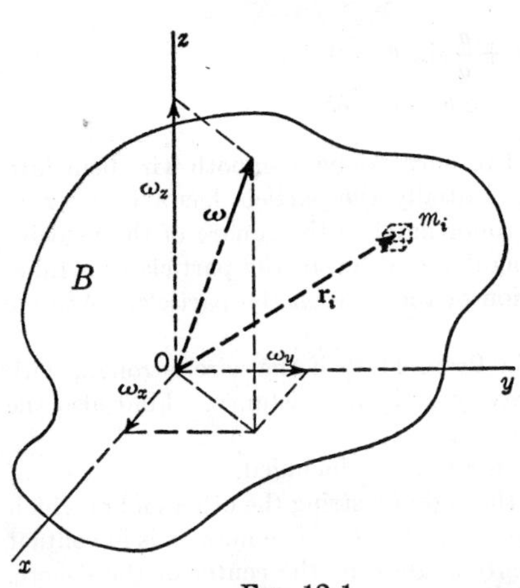

Fig. 12-1

12-1. The Instantaneous Axis. Consider a rigid body B (Fig. 12-1) which at a certain instant may be regarded as rotating with an angular velocity $\boldsymbol{\omega}$ about an axis passing through a given reference point O which is attached to the body (it need not lie within the body). This axis, called the *instantaneous axis*, in general may be undergoing a change, with time, of its direction in space. In terms of instantaneous components along coordinate axes with origin at O, $\boldsymbol{\omega}$ may be written

$$\boldsymbol{\omega} = \mathbf{i}\omega_x + \mathbf{j}\omega_y + \mathbf{k}\omega_z \tag{12-1}$$

where **i**, **j**, and **k** are the customary unit vectors. The quantities ω_x, ω_y, and ω_z will be functions of time if, relative to $Oxyz$, $\boldsymbol{\omega}$ is changing in magnitude or direction, or both.

It is usually useful to define $\boldsymbol{\omega}$ as being the instantaneous angular velocity of the body with respect to a nonrotating coordinate system.

Unless otherwise mentioned this will always be understood to be the case. Of course it is always possible to express such a vector $\boldsymbol{\omega}$, *already defined,* in terms of its components in an arbitrary coordinate system. For example, in Fig. 12-1, B is rotating with the velocity $\boldsymbol{\omega}$. Suppose, for the moment that the axes Ox, Oy, and Oz are all rigidly attached to the body; hence they also rotate with the angular velocity $\boldsymbol{\omega}$. However, $\boldsymbol{\omega}$ still has components ω_x, ω_y, and ω_z in the attached system $Oxyz$. These are just the instantaneous projections of $\boldsymbol{\omega}$ along Ox, Oy, and Oz, respectively.

12-2. Angular Momentum in Terms of Its Components. Moments and Products of Inertia. It is helpful now to restrict the axes $Oxyz$ of Fig. 12-1 to be nonrotating and the origin such that the rotational equation of motion $\dot{\mathbf{J}} = \mathbf{L}$ is valid in this system (see Secs. 8-3 and 8-4), where $\mathbf{J}$ and $\mathbf{L}$ are both defined with reference to O. The instantaneous translational velocity $\dot{\mathbf{r}}_i$, of a mass element m_i relative to O, can be written, following Sec. 8-4, in terms of $\boldsymbol{\omega}$ and $\mathbf{r}_i$ as

$$\dot{\mathbf{r}}_i = \boldsymbol{\omega} \times \mathbf{r}_i \tag{12-2}$$

The angular momentum $\mathbf{J}$, relative to the origin, for a system of masses m_i $(i = 1, 2, \ldots, n)$ has been defined in Chap. 8 to be

$$\mathbf{J} = \sum_{i=1}^{n} m_i\mathbf{r}_i \times \dot{\mathbf{r}}_i \tag{12-3}$$

where the sum is taken over all the n mass elements. In view of Eq. (12-2) this becomes

$$\mathbf{J} = \sum_{i=1}^{n} [m_i\mathbf{r}_i \times (\boldsymbol{\omega} \times \mathbf{r}_i)] \tag{12-4}$$

Now, as may easily be verified,

$$\begin{aligned}\mathbf{r}_i \times (\boldsymbol{\omega} \times \mathbf{r}_i) = {} & \mathbf{i}[y_i(\omega_x y_i - \omega_y x_i) - z_i(\omega_z x_i - \omega_x z_i)] \\ & + \mathbf{j}[z_i(\omega_y z_i - \omega_z y_i) - x_i(\omega_x y_i - \omega_y x_i)] + \mathbf{k}[x_i(\omega_z x_i - \omega_x z_i) \\ & - y_i(\omega_y z_i - \omega_z y_i)]\end{aligned} \tag{12-5}$$

Carrying out the indicated multiplications, and substituting in Eq. (12-4), we obtain

$$\begin{aligned}\mathbf{J} = {} & \mathbf{i}\left[\omega_x \sum_{i=1}^{n} m_i(y_i^2 + z_i^2) - \omega_y \sum_{i=1}^{n} m_i x_i y_i - \omega_z \sum_{i=1}^{n} m_i x_i z_i\right] \\ & + \mathbf{j}\left[-\omega_x \sum_{i=1}^{n} m_i x_i y_i + \omega_y \sum_{i=1}^{n} m_i(x_i^2 + z_i^2) - \omega_z \sum_{i=1}^{n} m_i y_i z_i\right] \\ & + \mathbf{k}\left[-\omega_x \sum_{i=1}^{n} m_i x_i z_i - \omega_y \sum_{i=1}^{n} m_i y_i z_i + \omega_z \sum_{i=1}^{n} m_i(x_i^2 + y_i^2)\right]\end{aligned} \tag{12-6}$$

The summations containing the squared terms are the familiar *moments of inertia* about the coordinate axes. Explicitly

$$I_x = \sum_{i=1}^{n} m_i(y_i^2 + z_i^2)$$
$$I_y = \sum_{i=1}^{n} m_i(x_i^2 + z_i^2) \qquad (12\text{-}7)$$
$$I_z = \sum_{i=1}^{n} m_i(x_i^2 + y_i^2)$$

The remaining summations, that is, those containing the cross terms, are called *products of inertia*. It is convenient to represent these symbolically as

$$P_{xy} = \sum_{i=1}^{n} m_i x_i y_i \qquad P_{yz} = \sum_{i=1}^{n} m_i y_i z_i \qquad P_{xz} = \sum_{i=1}^{n} m_i x_i z_i \qquad (12\text{-}8)$$

Clearly $P_{xy} = P_{yx}$, and so on. In terms of the moments and products of inertia, **J** may be written

$$\mathbf{J} = \mathbf{i}(I_x\omega_x - P_{xy}\omega_y - P_{xz}\omega_z) + \mathbf{j}(-P_{xy}\omega_x + I_y\omega_y - P_{yz}\omega_z) + \mathbf{k}(-P_{xz}\omega_x - P_{yz}\omega_y + I_z\omega_z) \qquad (12\text{-}9)$$

It is apparent that, if the coordinate axes are not rigidly attached to the body and rotating with it, the moments and products of inertia in general will be functions of time. For this reason it will be convenient a little later to employ coordinate systems so attached to the body that the moments and products of inertia are constants when expressed in terms of those coordinates. This means that the axes must be executing some or all of the rotational motions being described by the body, the extent to which the axes follow the motions of the body being dependent upon the degree of symmetry possessed by the body.

In practical cases of computation for a solid body, the moments and products of inertia take the form of integrals over the volume of the body. Thus, if ρ is the density (which is not necessarily constant), Eqs. (12-7) and (12-8) are replaced by

$$I_x = \int_V \rho(y^2 + z^2)\,dx\,dy\,dz \qquad I_y = \int_V \rho(x^2 + z^2)\,dx\,dy\,dz$$
$$I_z = \int_V \rho(x^2 + y^2)\,dx\,dy\,dz \qquad (12\text{-}10)$$

$$P_{xy} = \int_V \rho xy\,dx\,dy\,dz \qquad P_{yz} = \int_V \rho yz\,dx\,dy\,dz$$
$$P_{xz} = \int_V \rho xz\,dx\,dy\,dz \qquad (12\text{-}11)$$

Example 12-1. Two equal masses are connected by a massless rigid rod (a dumbbell) constrained to rotate about an axle which is at an angle α to the rod. It is desired to find the directions and magnitudes of the angular momentum and of the torque being applied to the system. The angular velocity $\boldsymbol{\omega}$ is constant.

In Fig. 12-2 the dumbbell is in the plane of the paper and is of length $2a$. We take the origin fixed at O, and the axes are fixed in the inertial system. From Eq. (12-4)

$$\mathbf{J} = \mathbf{J}_1 + \mathbf{J}_2$$
$$= m\mathbf{r}_1 \times (\boldsymbol{\omega} \times \mathbf{r}_1) + m\mathbf{r}_2 \times (\boldsymbol{\omega} \times \mathbf{r}_2)$$

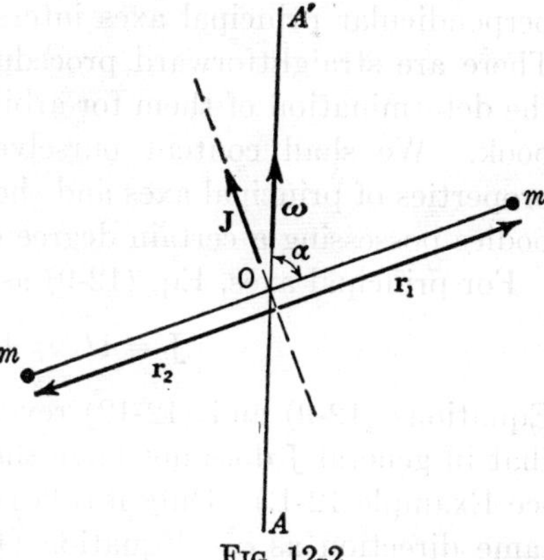

FIG. 12-2

and we note that $\mathbf{J}_1$ and $\mathbf{J}_2$ both point in the same direction, that of $\mathbf{J}$, indicated in Fig. 12-4. (We perceive here a simple example in which $\mathbf{J}$ is not parallel to $\boldsymbol{\omega}$.) It is easy to see that the magnitude of the angular momentum is

$$J = ma^2\omega \sin \alpha + ma^2\omega \sin \alpha = 2ma^2\omega \sin \alpha$$

However, since the direction of $\mathbf{J}$ is continually changing ($\mathbf{J}$ rotates about $\boldsymbol{\omega}$), a torque is necessary to maintain the motion. This is determined by the equation $\dot{\mathbf{J}} = \mathbf{L}$. $\dot{\mathbf{J}}$ is a vector denoting the direction in which the end of the $\mathbf{J}$ vector is moving, and, in analogy to Eq. (12-2), may be written $\boldsymbol{\omega} \times \mathbf{J}$. Thus the torque $\mathbf{L}$ applied to the rotating masses is directed perpendicularly out of the paper for the instantaneous configuration pictured by Fig. 12-2 and is of magnitude

$$L = 2\omega^2 a^2 m \sin \alpha \cos \alpha$$

Furthermore, by Newton's third law, the torque which the rotating dumbbell exerts on the bearings (say at A and A') is equal and opposite to this. Thus it points into the paper at the instant under consideration and is of the same magnitude as that given by the expression immediately above. The direction of the torque on the bearings is readily seen also from elementary considerations since the masses will tend to spring out to a horizontal position in the direction of increasing α.

12-3. Principal Axes. As was pointed out immediately above, a large class of rigid-body problems is best treated from the point of view of coordinates, with origin O, which are so fixed in the body that the moments and products of inertia, referred to those axes, are independent of time. Furthermore, a particular set of such axes is especially useful. This set is such that all the products of inertia are zero. A set of axes possessing this property is called a set of *principal axes* at the point O (which is the origin). Thus:

I. *Three mutually orthogonal straight lines meeting at a point O, taken as coordinate axes, and such that all the products of inertia P_{xy}, P_{xz}, P_{yz} of a rigid body expressed in terms of these axes are zero, are said to be principal axes at point O.*

In the same vein, point O is called a *principal point* for these axes. Furthermore the coordinate planes, that is, the three planes each of which passes through two principal axes, are called *principal planes* at point O. The moments of inertia about these principal axes are called *principal moments of inertia* at point O.

We shall show later that there are always at least three mutually perpendicular principal axes intersecting at any given point of a body. There are straightforward procedures for finding such axes. However, the determination of them for arbitrary cases is beyond the scope of this book. We shall content ourselves here with discussing some of the properties of principal axes and showing how to find them in the cases of bodies possessing a certain degree of symmetry.

For principal axes, Eq. (12-9) assumes the very simple form

$$\mathbf{J} = \mathbf{i}I_x\omega_x + \mathbf{j}I_y\omega_y + \mathbf{k}I_z\omega_z \tag{12-12}$$

Equations (12-9) and (12-12) reveal a very important feature. This is that in general $\mathbf{J}$ does not have the direction of $\boldsymbol{\omega}$. (In this connection, see Example 12-1.) Only if $\boldsymbol{\omega}$ lies along a principal axis does $\mathbf{J}$ have the same direction as $\boldsymbol{\omega}$. Equation (12-9) shows that, if $\boldsymbol{\omega}$ points in some direction which is not parallel to a principal axis, $\mathbf{J}$ possesses components which are perpendicular to that direction: Let this arbitrary direction of $\boldsymbol{\omega}$ coincide with the x axis of a set of axes which are not principal axes; since $\boldsymbol{\omega} = \mathbf{i}w_x$, for such a case $\mathbf{J}$ possesses both a y and a z component in addition to an x component [see Eq. (12-9)].

The material immediately above suggests the usefulness of stating the condition that at least one of the coordinate axes, say the Oz axis, is a principal axis at the origin O. Clearly this will be the case if the two products of inertia associated with that axis both vanish. In the case of the Oz axis, the products of inertia in question are P_{xz} and P_{yz}. A reexamination of Eq. (12-9) will reveal that, whether or not Ox and Oy are principal axes, if $\boldsymbol{\omega} = \mathbf{k}\omega_z$, $\mathbf{J}$ will have only a component, J_z, along Oz.

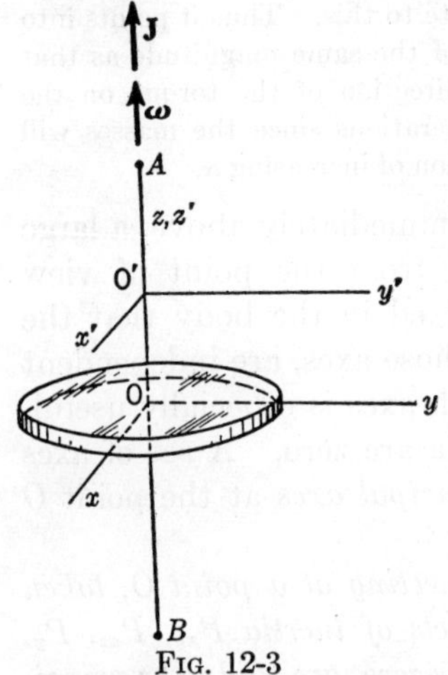

Fig. 12-3

Let us consider a simple case of a wheel rotating about a fixed principal axis (that is, the principal axis is the axle about which the rotation occurs). Such a situation exists in Fig. 12-3, where the axle AB is mounted in bearings at A and B and is attached to the center of the wheel. AB, as will be understood immediately below, is a principal axis of the wheel, and the angular momentum $\mathbf{J}$ lies along the axle.

In the present chapter, for reasons of simplicity we shall be concerned mainly with bodies possessing a certain degree of symmetry. For these it is easy to develop skill in selecting principal axes by inspection. For example, suppose a body possesses a plane of symmetry. Calling this the xy plane, we perceive that, for every element m_i with coordinates x_i, y_i, z_i on one side of the plane, there is another element m_j with coordinates x_j, y_j, z_j located on the other side of the plane such that $z_j = -z_i$, $x_i = x_j$, and $y_i = y_j$. Accordingly, for such a symmetrical body the quantities $\sum_i m_i x_i z_i$ and $\sum_i m_i y_i z_i$ are both zero. Thus the z axis is a principal axis at O. Further, it is important to notice that any line perpendicular to this plane is a principal axis at its point of intersection with the plane since in *any* such case the summations P_{xz} and P_{yz}, just mentioned, are both zero. [Note also that, by designating any line perpendicular to the xy plane as the z axis, and if $\boldsymbol{\omega}$ has only a z component, Eq. (12-6) reveals that there will likewise be only a z component of **J**.]

From what has just been stated it is clear that the axle in Fig. 12-3 (which we identify with the z axis) is a principal axis, at least if the origin is at O, the center of the wheel. However, the wheel is a body of revolution about the z axis, and in such a situation the z axis is a principal axis at any point O' on its length. This is evident since, if we select other axes $O'x'$, and $O'y'$, and $O'z'$ as shown in Fig. 12-3 for every element m_i at x'_i, y'_i, z'_i, there exists another element m_j at x'_j, y'_j, z'_j such that $x'_j = -x'_i$, $y'_j = -y'_i$, and $z'_j = z'_i$. Consequently the quantities $P_{x'z'}$ and $P_{y'z'}$ both vanish.

Summarizing:

II. *If a body possesses a plane of symmetry, any axis perpendicular to this plane is a principal axis at its point of intersection with the plane.*

III. *If a body is one of revolution about a given axis, the axis is a principal one at all points along its length.*

Statement II has an obvious corollary. This is:

IV. *For a body which is a plane lamina, any axis which is perpendicular to the lamina is a principal axis at its point of intersection with the lamina.* Clearly this is so since, upon taking the axis again as the z axis, z will be zero for all points of the lamina. Consequently P_{xz} and P_{yz} are both zero.

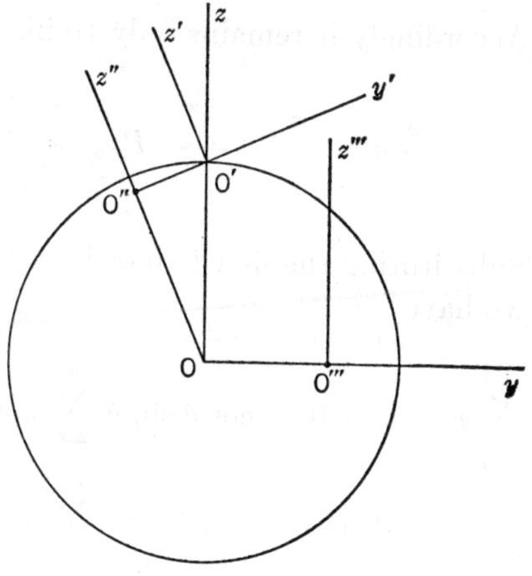

FIG. 12-4

Example 12-2. It is left as an exercise for the student to determine by inspection, for the homogeneous solid sphere depicted in Fig. 12-4, which of the axes shown are principal axes, and at what points in each case. (The figure shown is plane and is a vertical section containing the center O of the sphere. Consider, also, that at each point O, O', and so on, there is an additional axis Ox, $O'x'$, and the like, directed normally out from the page toward the reader.)

12-4. Determination of the Other Two Principal Axes When One Is Given.

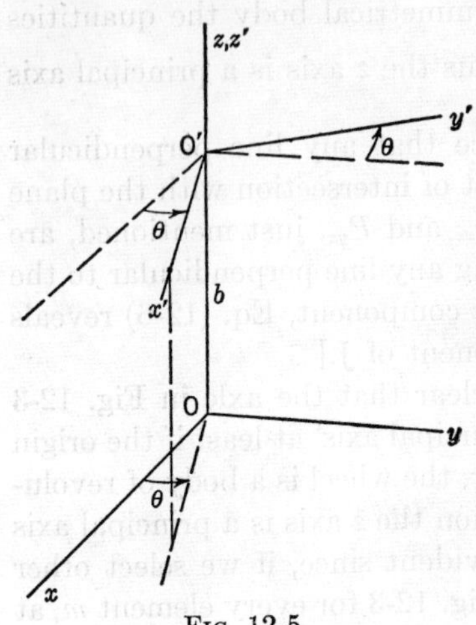

Fig. 12-5

In Fig. 12-5 we are given the axes $Oxyz$, in which the z axis is a principal axis at point O' and where b is the distance OO'. Furthermore the quantities I_x, I_y, I_z, P_{xy}, P_{yz}, P_{xz} all have been found. We desire to determine the axes $O'x'$ and $O'y'$, in a plane through O' parallel to the xy plane, such that they are principal axes at O'. If $O'x'$ and $O'y'$ are rotated through an angle θ from Ox and Oy, as shown, the transformation equations are (see Fig. 11-7)

$$\begin{aligned} x' &= x\cos\theta + y\sin\theta \\ y' &= -x\sin\theta + y\cos\theta \\ z' &= z - b \end{aligned} \tag{12-13}$$

Since $O'z'$ is given to be a principal axis at O', we must have

$$P_{x'z'} = \sum_{i=1}^{n} m_i x_i' z_i' = 0 \qquad P_{y'z'} = \sum_{i=1}^{n} m_i y_i' z_i' = 0 \tag{12-14}$$

Accordingly it remains only to impose the condition

$$P_{x'y'} = \sum_{i=1}^{n} m_i x_i' y_i' = 0 \tag{12-15}$$

Substituting the first two of Eqs. (12-13) in Eq. (12-15) and rearranging, we have

$$\sum_{i=1}^{n} m_i x_i' y_i' = 0 = \cos\theta\sin\theta \sum_{i=1}^{n} m_i(y_i^2 - x_i^2)$$
$$+ (\cos^2\theta - \sin^2\theta)\sum_{i=1}^{n} m_i x_i y_i = \tfrac{1}{2}\sin 2\theta \sum_{i=1}^{n} m_i(y_i^2 - x_i^2)$$
$$+ \cos 2\theta \sum_{i=1}^{n} m_i x_i y_i$$

from which

$$\tan 2\theta = \frac{2\sum_{i=1}^{n} m_i x_i y_i}{\sum_{i=1}^{n} m_i(x_i^2 - y_i^2)}$$

$$= \frac{2\sum_{i=1}^{n} m_i x_i y_i}{\sum_{i=1}^{n} m_i(x_i^2 + z_i^2) - \sum_{i=1}^{n} m_i(y_i^2 + z_i^2)}$$

$$= \frac{2P_{xy}}{I_y - I_x} \tag{12-16}$$

Equation (12-16) yields the angle θ through which $O'x'$ and $O'y'$ must be rotated from Ox and Oy, so that $O'x'$ and $O'y'$ are principal axes at O' provided that $O'z'$ is itself a principal axis at O'. It is worth noting that expression (12-16) is independent of b, and thus if the z axis is a principal axis at more than one point, for example, O' and O'' (O'' not shown), the principal axes $O'x'$ and $O'y'$ are parallel, respectively, to the principal axes $O''x''$ and $O''y''$.

If it is not known whether or not the axis Oz is a principal axis at a point on its length, it is necessary to investigate this possibility. Suppose we examine the situation in Fig. 12-5 in order to see whether or not the Oz axis is a principal axis at O'. To do this, let O' be the origin of coordinates x', y', z', as above. The condition that $O'z'$ be a principal axis is obtained by putting $P_{x'z'}$ and $P_{y'z'}$ equal to zero. Employing Eqs. (12-13), we have

$$\sum_{i=1}^{n} m_i x_i' z_i' = \cos\theta \sum_{i=1}^{n} m_i x_i z_i + \sin\theta \sum_{i=1}^{n} m_i y_i z_i$$
$$- b\left(\cos\theta \sum_{i=1}^{n} m_i x_i + \sin\theta \sum_{i=1}^{n} m_i y_i\right) = 0 \tag{12-17}$$

$$\sum_{i=1}^{n} m_i y_i' z_i' = -\sin\theta \sum_{i=1}^{n} m_i x_i z_i + \cos\theta \sum_{i=1}^{n} m_i y_i z_i$$
$$+ b\left(\sin\theta \sum_{i=1}^{n} m_i x_i - \cos\theta \sum_{i=1}^{n} m_i y_i\right) = 0 \tag{12-18}$$

Eliminating b between (12-17) and (12-18), we obtain

$$\sum_{i=1}^{n} m_i x_i \sum_{i=1}^{n} m_i y_i z_i = \sum_{i=1}^{n} m_i y_i \sum_{i=1}^{n} m_i x_i z_i \tag{12-19}$$

a relation which must be satisfied if the z axis is to be a principal axis at some point O' on its length. The distance b of O' from O is also found from (12-17) and (12-18). We obtain

$$b = \frac{\sum_{i=1}^{n} m_i y_i z_i}{\sum_{i=1}^{n} m_i y_i} = \frac{\sum_{i=1}^{n} m_i x_i z_i}{\sum_{i=1}^{n} m_i x_i} \tag{12-20}$$

The quantities in the numerators are the products of inertia P_{yz} and P_{xz}, and the two denominators are the mass of the body times the y and z coordinates of the mass center. Thus, in more familiar form,

$$b = \frac{P_{yz}}{M\bar{y}} = \frac{P_{xz}}{M\bar{x}} \tag{12-21}$$

where $M = \sum_i m_i$. The particular case in which b is zero, that is, in which the z axis is a principal axis at the initial origin O, requires the already obvious conditions that P_{yz} and P_{xz} both be zero.

Referring to the previously mentioned possibility of the z axis being a principal axis for more than one point on its length, Eq. (12-21) provides interesting information. In such an event b has more than one value, a fact which requires the two ratios in Eq. (12-21) to be indeterminate. Accordingly if b is to have more than one value, we must have

$$P_{yz} = 0 \qquad P_{xz} = 0 \qquad \bar{y} = 0 \qquad \bar{x} = 0 \tag{12-22}$$

from which we see first that the Oz axis must already have been a principal axis at O and second that the Oz axis must pass through the center of mass. However, since the origin was arbitrary, the Oz axis must be a principal axis for every point on its length. Conversely, if (12-22) is true, Eqs. (12-17) and (12-18) are satisfied by all possible values of b. Accordingly we may write:

V. *If a straight line is a principal axis at the center of mass, it is a principal axis at all points on its length.*

12-5. Centrifugal Reactions. Dynamically Balanced Body. Consider a body to be rotating under no forces or torques about an axis which is fixed, for example by means of bearings at each end of the axis. We wish to inquire into the restrictions upon the choice of the axis such that, if the bearings are suddenly removed, the body continues to rotate freely about the same axis, unchanged in position and orientation. This situation results if the mass distribution of the body relative to the chosen axis is such that the body exerts no centrifugal reactions or moments of these reactions upon the axis. If such a state of affairs obtains, the body is said to be dynamically balanced.

The resultant of the centrifugal reactions due to the centripetal accelerations of all the mass elements toward the axis may be written

$$\mathbf{R}_c = -\sum_{i=1}^{n} m_i\boldsymbol{\omega} \times (\boldsymbol{\omega} \times r_i) \tag{12-23}$$

where the sum is understood to be taken over all the mass elements. Here the origin is taken to lie on the axis, $\mathbf{r}_i$ being the radius vector from the origin to the mass element m_i, and $\boldsymbol{\omega}$ is the angular velocity. In terms of axes Ox, Oy, and Oz fixed in the body, Eq. (12-23) becomes

$$\begin{aligned}\mathbf{R}_c = \sum_{i=1}^{n} m_i\{\mathbf{i}[x_i(\omega_y^2 + \omega_z^2) - y_i\omega_x\omega_y - z_i\omega_x\omega_z] \\ + \mathbf{j}[-x_i\omega_x\omega_y + y(\omega_x^2 + \omega_z^2) - z_i\omega_y\omega_z] \\ + \mathbf{k}[-x_i\omega_x\omega_z - y_i\omega_y\omega_z + z_i(\omega_x^2 + \omega_y^2)]\}\end{aligned} \tag{12-24}$$

Now, if the Oz axis is taken to be the axis of rotation, $\boldsymbol{\omega} = \mathbf{k}\omega_z$, and Eq. (12-24) reduces to

$$\begin{aligned}\mathbf{R}_c &= \sum_{i=1}^{n} m_i(\mathbf{i}x_i\omega_z^2 + \mathbf{j}y_i\omega_z^2) \\ &= \omega_z^2\left(\mathbf{i}\sum_{i=1}^{n} m_ix_i + \mathbf{j}\sum_{i=1}^{n} m_iy_i\right) \\ &= \omega_z^2(\mathbf{i}M\bar{x} + \mathbf{j}M\bar{y})\end{aligned} \tag{12-25}$$

where $M = \sum_i m_i$ is the total mass of the body and $\bar{x}$ and $\bar{y}$ are the x and y coordinates of the mass center. Clearly then, if $\mathbf{R}_c$ is to be zero, $\bar{x}$ and $\bar{y}$ each must be zero (save for the trivial case in which $\omega_z = 0$). Hence the center of mass must lie on the z axis, which is the axis of rotation.

The resultant of the moments about the origin O of the centrifugal reactions arising from all the mass elements may be written

$$\mathbf{L}_c = -\sum_{i=1}^{n} m_i\{\mathbf{r}_i \times [\boldsymbol{\omega} \times (\boldsymbol{\omega} \times \mathbf{r}_i)]\} \tag{12-26}$$

If $\boldsymbol{\omega} = \mathbf{k}\omega_z$, Eq. (12-26) reduces at once to

$$\begin{aligned}\mathbf{L}_c &= \sum_{i=1}^{n} \{m_i[-\mathbf{i}y_iz_i\omega_z^2 + \mathbf{j}x_iz_i\omega_z^2 + \mathbf{k}(x_iy_i\omega_z^2 - x_iy_i\omega_z^2)]\} \\ &= \omega_z^2\left(-\mathbf{i}\sum_{i=1}^{n} m_iy_iz_i + \mathbf{j}\sum_{i=1}^{n} m_ix_iz_i\right) \\ &= \omega_z^2(-\mathbf{i}P_{yz} + \mathbf{j}P_{xz})\end{aligned} \tag{12-27}$$

If $\mathbf{L}_e$ is to be zero, P_{yz} and P_{xz} both must be zero. Thus the z axis must be a principal axis at O (and at all other points on its length since the mass center lies on this axis).

To summarize: For a body to be dynamically balanced, the center of mass must lie on the axis of rotation, and the axis must be a principal axis. Hence it can be seen that for the free rotation of a body, if the axis of rotation is coincident with a principal axis passing through the center of mass of the body, the axis of rotation will remain unchanged. External forces and/or external torques are required to shift this axis.

The wheel of Fig. 12-3 is seen to be a dynamically balanced body. The importance of this in engineering practice arises from the fact that a body which is dynamically balanced exerts no forces or torques on the bearings, and consequently the wear on the bearings is reduced to a minimum.

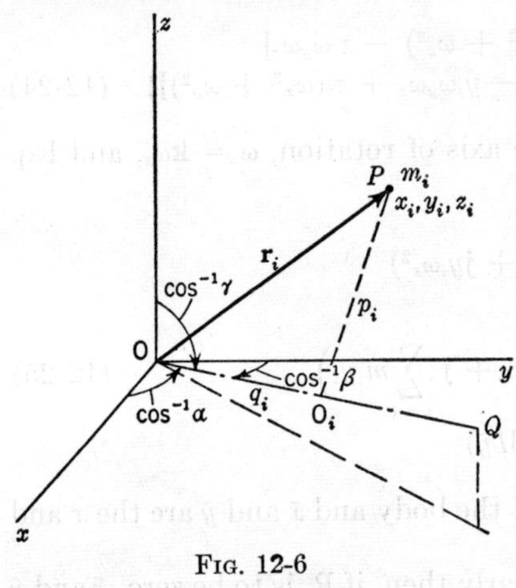

FIG. 12-6

12-6. Moment of Inertia about an Arbitrary Axis. Ellipsoid of Inertia. Given a set of axes $Oxyz$ fixed in a body, together with the moments and products of inertia with respect to these axes, we require the moment of inertia I about an axis OQ passing through O and having direction cosines α, β, γ, respectively, with reference to the x, y, z axes. In Fig. 12-6 the mass element m_i is located at point P, a perpendicular distance p_i from OQ. Let r_i be the radius vector to P and q_i the distance OO_i. The moment of inertia I, about OQ, is

$$I = \sum_{i=1}^{n} m_i p_i^2 = \sum_{i=1}^{n} m_i(r_i^2 - q_i^2) \tag{12-28}$$

but

$$r_i^2 = x_i^2 + y_i^2 + z_i^2$$

and

$$q_i = \alpha x_i + \beta y_i + \gamma z_i \tag{12-29}$$

where Eq. (12-29) follows since q_i is the projection of r_i on OQ (the result of Prob. 1-20 is taken into account). Thus

$$\begin{aligned} I &= \sum_{i=1}^{n} m_i[(x_i^2 + y_i^2 + z_i^2) - (\alpha x_i + \beta y_i + \gamma z_i)^2] \\ &= \sum_{i=1}^{n} m_i[(x_i^2 + y_i^2 + z_i^2)(\alpha^2 + \beta^2 + \gamma^2) - (\alpha x_i + \beta y_i + \gamma z_i)^2] \end{aligned} \tag{12-30}$$

in which the last step has not altered the equality since, from a well-known theorem of analytic geometry, $\alpha^2 + \beta^2 + \gamma^2 = 1$. Expanding (12-30) and recombining, we have

$$I = \sum_{i=1}^{n} m_i[\alpha^2(y_i^2 + z_i^2) + \beta^2(x_i^2 + z_i^2) + \gamma^2(x_i^2 + y_i^2) - 2\alpha\beta x_i y_i - 2\beta\gamma y_i z_i - 2\alpha\gamma x_i z_i]$$

$$= \alpha^2 I_x + \beta^2 I_y + \gamma^2 I_z - 2\alpha\beta P_{xy} - 2\beta\gamma P_{yz} - 2\alpha\gamma P_{xz} \qquad (12\text{-}31)$$

If the original axes $Oxyz$ are principal axes, the products of inertia are zero and Eq. (12-31) reduces to

$$I = \alpha^2 I_x + \beta^2 I_y + \gamma^2 I_z \qquad (12\text{-}32)$$

From (12-31) or (12-32), the moment of inertia I about an arbitrary axis through the origin may be determined. This, in turn, may be related, with the aid of the *parallel-axis theorem* (Sec. 9-7), to the moment of inertia about a parallel axis passing through the center of mass.

It is useful to give a geometrical interpretation of Eqs. (12-31) and (12-32). In Fig. 12-6 let OQ be a radius vector $\mathbf{R}$ which moves about O in any manner, and let its length R be variable so that for any instantaneous orientation of OQ the moment of inertia I about OQ is inversely proportional to R^2. Let the constant of proportionality be K. Thus, from (12-31),

$$K = R^2(\alpha^2 I_x + \beta^2 I_y + \gamma^2 I_z - 2\alpha\beta P_{xy} - 2\beta\gamma P_{yz} - 2\alpha\gamma P_{xz}) \qquad (12\text{-}33)$$

Now R is a real positive quantity, and hence Eq. (12-33) is the equation of a family of ellipsoids, one for each value of K, with center at the origin O. The surface of such an ellipsoid, for a given K, is generated by the tip of the vector $\mathbf{R}$. It is convenient to limit ourselves henceforth to one particular value of K, the value $K = 1$. If, in addition, we write

$$\alpha R = X \qquad \beta R = Y \qquad \gamma R = Z \qquad (12\text{-}34)$$

Eq. (12-33) becomes

$$1 = I_x X^2 + I_y Y^2 + I_z Z^2 - 2P_{xy}XY - 2P_{yz}YZ - 2P_{xz}XZ \qquad (12\text{-}35)$$

The ellipsoid described by Eq. (12-35) is referred to as Poinsot's *ellipsoid of inertia* of the body at point O. It is fixed in the body, with center at the origin O. In general, there is a different one for each point of the body.

It should be noticed that, for $K = 1$, the moment of inertia I about OQ is just $1/R^2$. Hence:

VI. *The moment of inertia about any axis passing through O is just the reciprocal of the square of the radius vector* $\mathbf{R}$ *from O to the point on the surface of the ellipsoid through which the axis passes.*

If $P_{xy} = P_{yz} = P_{xz} = 0$, that is, if the axes Ox, Oy, and Oz are principal axes at O, Eq. (12-35) takes the form

$$1 = I_x X^2 + I_y Y^2 + I_z Z^2 \tag{12-36}$$

which is the equation of an ellipsoid referred to its principal diameters as axes. The quantities I_x, I_y, and I_z are the principal moments of inertia of the body with respect to point O. These considerations lead to an important conclusion:

VII. *At every point of a body there are always at least three principal axes.*

This follows since, if the ellipsoid of inertia is referred to its principal diameters as axes, the products of inertia vanish. Since every ellipsoid has at least three principal diameters, and since it is always possible to rotate the axes until they coincide with these diameters, the statement VII follows immediately. (The procedure of carrying out such a set of rotations to principal axes will not be considered in this text.)

It is helpful to think of the **R** space, or space of the coordinates X, Y, and Z, in terms of which the inertia ellipsoid is described as being superimposed upon the ordinary space coordinates $Oxyz$. They have the common origin O and have OX parallel to Ox, OY parallel to Oy, and OZ parallel to Oz. From Eq. (12-36) it is evident that the length of each of the semiaxes of the ellipsoid in the X, Y, Z space is just the inverse of the square root of the moment of inertia (which is a principal moment) of the body about that axis. For example, the length of the semiaxis of the ellipsoid along OX is $I_x^{-\frac{1}{2}}$, where I_x is the principal moment of inertia of the body about the principal axis Ox.

Of special interest is the case in which the principal moments are all equal, for then the semiaxes likewise are equal, and the ellipsoid with center at O degenerates into a sphere with O as the center of curvature. For this state of affairs all axes passing through O are principal axes, and all the moments are equal. Hence:

VIII. *When the inertia ellipsoid with respect to a point O of a body is a sphere, all axes passing through O are principal axes and have identical moments of inertia which are equal to the reciprocal of the square of the radius R of the inertial sphere.*

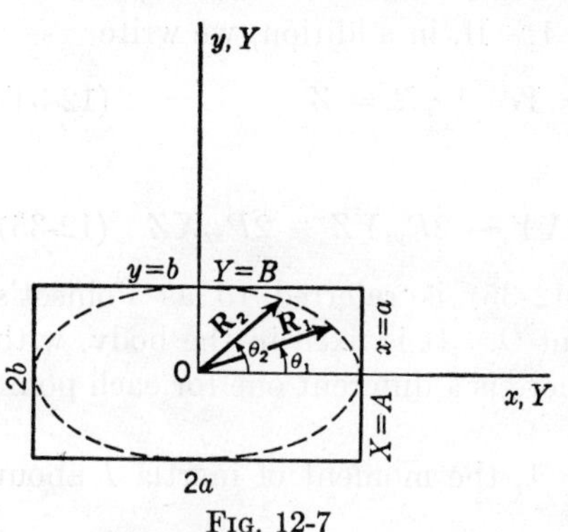

Fig. 12-7

Example 12-3. Determine the inertia ellipsoid for a rectangle of sides $2a$ and $2b$ and mass m, as shown in Fig. 12-7.

Clearly, with O the center of mass of the body, the axes Ox and Oy, in the plane of the rectangle, are principal axes at O (and of course at all points on their lengths, since O is the mass center), and by virtue of IV the Oz axis (not shown) also is a principal axis at O. Hence the plane of the rectangle contains a *principal section*, with center at O, of the inertia ellipsoid. We shall term this principal section as the *inertia ellipse* of the body at O.

It is readily shown that the principal moments of the body are

$$I_x = \frac{mb^2}{3} \qquad I_y = \frac{ma^2}{3} \qquad I_z = \frac{m(a^2 + b^2)}{3}$$

Hence the semiaxes of the inertia ellipsoid are

$$A = I_x^{-\frac{1}{2}} = \frac{1}{b}\sqrt{\frac{3}{m}} \qquad B = I_y^{-\frac{1}{2}} = \frac{1}{a}\sqrt{\frac{3}{m}} \qquad C = I_z^{-\frac{1}{2}} = \sqrt{\frac{3}{m(a^2 + b^2)}} \tag{12-37}$$

From the first and second of Eqs. (12-37)

$$\frac{A}{B} = \frac{a}{b} \tag{12-38}$$

For convenience we may let the point $(X,Y) = (A,0)$ coincide with the point $(x,y) = (a,0)$. Then, in view of (12-38), the point $(X,Y) = (0,B)$ will coincide with $(x,y) = (0,b)$, and the inertia ellipse will have the appearance of being inscribed in the rectangle. This is shown as the dotted line in the figure.

The moments of inertia I_1 and I_2 about the two axes such as $\mathbf{R}_1$ and $\mathbf{R}_2$ in the plane of the rectangle are equal to $1/R_1^2$ and $1/R_2^2$, respectively. From (12-32) these are

$$I_1 = I_x \cos^2\theta_1 + I_y \sin^2\theta_1 = \frac{m}{3}(b^2\cos^2\theta_1 + a^2\sin^2\theta_1)$$

$$I_2 = I_x \cos^2\theta_2 + I_y \sin^2\theta_2 = \frac{m}{3}(b^2\cos^2\theta_2 + a^2\sin^2\theta_2)$$

In the event that $b = a$, the rectangle becomes a square, and the inscribed figure a circle. Then all axes in the plane of the figure and passing through O are principal axes. The moments of inertia about all these axes are the same and are equal to $\frac{1}{3}ma^2$.

12-7. Rotational Kinetic Energy of a Rigid Body. It is desirable to obtain a more general expression for the rotational kinetic energy of a rigid body than that derived for the simple case in Sec. 9-5. It is to be remembered that the only rotations treated in Chap. 9 were those in which the axis of rotation remained always normal to a fixed plane. Equation (9-20) need not be so restricted in its validity. The form of the kinetic energy about any instantaneous axis has precisely the same appearance.

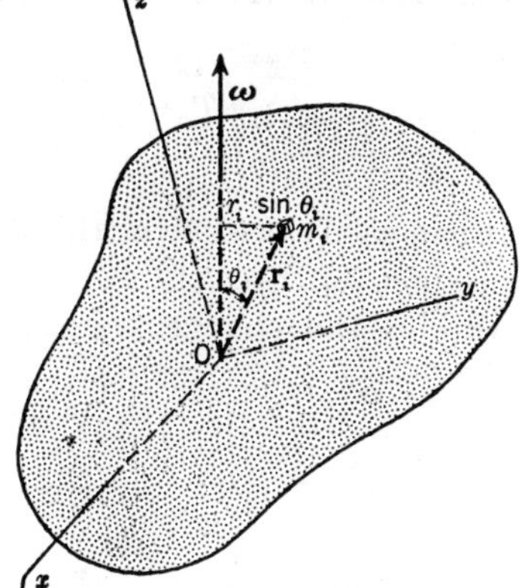

FIG. 12-8

In Fig. 12-8, let point O of the rigid body either be instantaneously fixed in a Newtonian reference system or, alternatively, be the center of mass of the body, which may or may not be fixed. It will usually be convenient, although not necessary, to require the axes Ox, Oy, and Oz to be fixed in the body since then the moments and products of inertia with respect to these axes are constants. The body is rotating with the instantaneous angular velocity $\boldsymbol{\omega}$ as shown (hence the instantaneous axis is the line along which $\boldsymbol{\omega}$ lies and passes through O). The kinetic energy T of rotation about the instantaneous axis can be expressed in terms of a sum over the n mass elements of the type m_i shown in the figure. We have

$$T = \sum_{i=1}^{n} T_i = \tfrac{1}{2} \sum_{i=1}^{n} m_i |\dot{\mathbf{r}}_i|^2 \tag{12-39}$$

$$= \tfrac{1}{2} \sum_{i=1}^{n} m_i |\boldsymbol{\omega} \times \mathbf{r}_i|^2$$

$$= \tfrac{1}{2} \left(\sum_{i=1}^{n} m_i r_i^2 \sin^2 \theta_i \right) \omega^2 \tag{12-40}$$

But the sum inside the parentheses is just the moment of inertia I_ω about the instantaneous axis of $\boldsymbol{\omega}$. Hence the rotational kinetic energy about an instantaneous axis may be written

$$T = \tfrac{1}{2} I_\omega \omega^2 \tag{12-41}$$

Equation (12-41) has but limited usefulness in the present consideration of three-dimensional rotational motions of rigid bodies since the axis may be changing and thus producing an accompanying change in I_ω. We desire to write an expression for T involving the moments and products of inertia with respect to the axes shown. To do this, we note that Eq. (12-39) may be rewritten as

$$T = \tfrac{1}{2} \sum_{i=1}^{n} m_i |\dot{\mathbf{r}}_i|^2 = \tfrac{1}{2} \sum_{i=1}^{n} m_i (\boldsymbol{\omega} \times \mathbf{r}_i) \cdot \dot{\mathbf{r}}_i \tag{12-42}$$

It can easily be shown (this is left to the student) that, in the triple scalar product of Eq. (12-42), the dot and the cross are interchangeable. Making this change, we have

$$T = \tfrac{1}{2} \boldsymbol{\omega} \cdot \left(\sum_{i=1}^{n} \mathbf{r}_i \times m_i \dot{\mathbf{r}}_i \right) \tag{12-43}$$

$$= \tfrac{1}{2} \boldsymbol{\omega} \cdot \mathbf{J} \tag{12-44}$$

where (12-44) follows from (12-43) since the summation is just the angular momentum $\mathbf{J}$ of the body.

It is of interest to express Eq. (12-44) in terms of the components of $\boldsymbol{\omega}$ and $\mathbf{J}$. Upon making use of Eqs. (12-1) and (12-9), Eq. (12-44) becomes

$$T = \tfrac{1}{2}(I_x\omega_x^2 + I_y\omega_y^2 + I_z\omega_z^2 - 2P_{xy}\omega_x\omega_y - 2P_{yz}\omega_y\omega_z - 2P_{xz}\omega_x\omega_z) \quad (12\text{-}45)$$

which, for principal axes, reduces to

$$T = \tfrac{1}{2}(I_x\omega_x^2 + I_y\omega_y^2 + I_z\omega_z^2) \quad (12\text{-}46)$$

12-8. Description of the Free Rotation of a Rigid Body in Terms of the Ellipsoid of Inertia. The concept of the inertia ellipsoid is an exceedingly useful one for picturing the rotational motions of a rigid body when no external torques are acting. In Fig. 12-9 is shown the section of a rigid body in the xy plane, which is also the plane of the paper. The point O of the body is subject to the same restrictions as in the last section. The axes Ox, Oy, and Oz (not shown), fixed in the body, are chosen to be principal axes at O. A principal section of the inertia ellipsoid at point O, drawn to a convenient scale, is also shown in the figure.

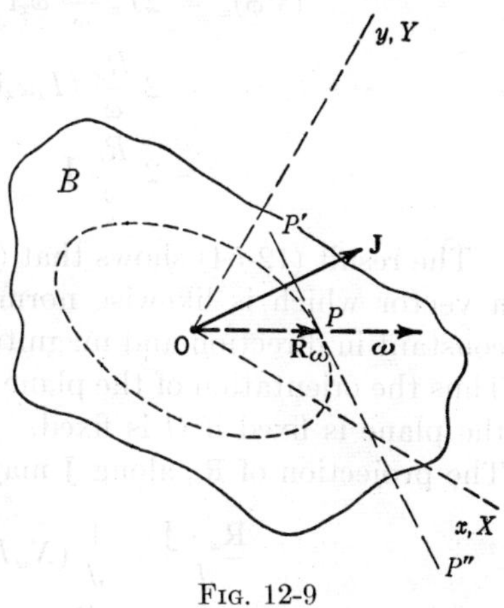

FIG. 12-9

Let us suppose that B is rotating with an angular velocity $\boldsymbol{\omega}$ about an instantaneous axis (along which $\boldsymbol{\omega}$ lies) passing through O and intersecting the surface of the inertia ellipsoid at point P. P is called the *pole* of the instantaneous axis. The radius vector from O to P, in the moment space X, Y, Z, as defined in Sec. 12-6, is designated as $\mathbf{R}_\omega$ in the figure. In terms of its components, $\mathbf{R}_\omega$ may be written

$$\mathbf{R}_\omega = \mathbf{i}X_\omega + \mathbf{j}Y_\omega + \mathbf{k}Z_\omega \quad (12\text{-}47)$$

The line $P'P''$ is the intersection, in the xy plane, of the plane which is tangent to the inertia ellipsoid at P. It is possible to show that the angular momentum $\mathbf{J}$ is normal to this plane. To do this, we shall show that $\mathbf{J}$ is parallel to the gradient, at P, of the surface of the inertia ellipsoid. The equation of the ellipsoid is given by (12-36). The right side is seen to be a function ϕ (which has the particular value 1) of X, Y, and Z. Accordingly, the gradient of ϕ may be written

$$\nabla\phi = \left(\mathbf{i}\frac{\partial}{\partial X} + \mathbf{j}\frac{\partial}{\partial Y} + \mathbf{k}\frac{\partial}{\partial Z}\right)(I_xX^2 + I_yY^2 + I_zZ^2) \quad (12\text{-}48)$$

This becomes, when evaluated at X_ω, Y_ω, Z_ω,

$$(\nabla\phi)_\omega = 2I_xX_\omega\mathbf{i} + 2I_yY_\omega\mathbf{j} + 2I_zZ_\omega\mathbf{k} \quad (12\text{-}49)$$

Now $(\nabla\phi)_\omega$ is a vector which is normal to the plane $P'P''$ at point P. It is useful to investigate how this vector is related to the vector $\mathbf{J}$. To do this we note that $\boldsymbol{\omega}$ and $\mathbf{R}_\omega$ have a common direction, that is,

$$\frac{\boldsymbol{\omega}}{\omega} = \frac{\mathbf{R}_\omega}{R_\omega} \tag{12-50}$$

and hence

$$\omega_x = \frac{\omega}{R_\omega} X_\omega \qquad \omega_y = \frac{\omega}{R_\omega} Y_\omega \qquad \omega_z = \frac{\omega}{R_\omega} Z_\omega \tag{12-51}$$

Substituting Eqs. (12-51) in Eq. (12-49), we obtain

$$(\nabla\phi)_\omega = 2I_x \frac{R_\omega}{\omega} \omega_x\mathbf{i} + 2I_y \frac{R_\omega}{\omega} \omega_y\mathbf{j} + 2I_z \frac{R_\omega}{\omega} \omega_z\mathbf{k} \tag{12-52}$$

$$= 2\frac{R_\omega}{\omega}(I_x\omega_x\mathbf{i} + I_y\omega_y\mathbf{j} + I_z\omega_z\mathbf{k}) \tag{12-53}$$

$$= 2\frac{R_\omega}{\omega}\mathbf{J} \tag{12-54}$$

The result (12-54) shows that $(\nabla\phi)_\omega$ is parallel to $\mathbf{J}$, and hence that $\mathbf{J}$ is a vector which is likewise normal to the plane $P'P''$. Moreover, $\mathbf{J}$ is constant in direction and magnitude since no external torques are acting. Thus the orientation of the plane in space must not change. In addition, the plane is fixed if O is fixed. This can be seen in the following way. The projection of $\mathbf{R}_\omega$ along $\mathbf{J}$ may be written

$$\frac{\mathbf{R}_\omega \cdot \mathbf{J}}{J} = \frac{1}{J}(X_\omega I_x\omega_x + Y_\omega I_y\omega_y + Z_\omega I_z\omega_z)$$

$$= \frac{R_\omega}{J\omega}(I_x\omega_x^2 + I_y\omega_y^2 + I_z\omega_z^2) \tag{12-55}$$

in which Eqs. (12-51) have been employed. Now R_ω is just $I_\omega^{-\frac{1}{2}}$, where I_ω is the moment of inertia about the instantaneous axis. Hence Eq. (12-55) may be rewritten as

$$\frac{\mathbf{R}_\omega \cdot \mathbf{J}}{J} = \frac{1}{JI_\omega^{\frac{1}{2}}\omega}(I_x\omega_x^2 + I_y\omega_y^2 + I_z\omega_z^2) \tag{12-56}$$

Employing Eqs. (12-41) and (12-46), this becomes

$$\frac{\mathbf{R}_\omega \cdot \mathbf{J}}{J} = \frac{2T}{J\sqrt{2T}} = \frac{\sqrt{2T}}{J} \tag{12-57}$$

where T is the kinetic energy of rotation. Now since there are no external torques acting, both T and J are constant. Thus the projection of $\mathbf{R}_\omega$ along $\mathbf{J}$ is constant. This means that the perpendicular drawn from O to $P'P''$ is fixed in length. Hence, if O is fixed, $P'P''$ also is fixed. The plane $P'P''$ is referred to as the *invariable plane* (with a line through O parallel to $\mathbf{J}$ similarly being called the *invariable line*).

The rotational motions of the body B can be thought of in terms of the motion of the inertia ellipsoid relative to the invariable line and to the invariable plane. The ellipsoid moves so that O remains fixed in the invariable line at a constant distance from the plane $P'P''$. Thus $\mathbf{R}_\omega$ (and consequently $\boldsymbol{\omega}$) may change, subject to the restriction that the projection of $\mathbf{R}_\omega$ along $\mathbf{J}$ is constant. Since the pole P is on the axis of rotation, the instantaneous velocity of P is zero and the ellipsoid rolls, without sliding, on the invariable plane. The locus of P (since P is the pole) in the ellipsoid is termed the *polhode*, and the locus of P in the invariable plane the *herpolhode*. Hence the motion may be described as a rolling of the polhode on the herpolhode.

The above description of the free rotation of a rigid body is due to Poinsot.

12-9. Classes of Problems to Be Considered in Rigid Dynamics. It has been pointed out that in the present chapter we are interested in rotational motions of rigid bodies in which the axis of rotation is not necessarily fixed in orientation. It is convenient to divide such problems into two categories, both of which will permit of investigation with the present material. The first is the determination of the rotational motions of a rigid body with respect to a system of axes fixed in the body (that is, rigidly attached to it). This class would include any wandering of the axis of rotation with respect to the system fixed in the body, such as was mentioned in the preceding section. The polhode, for example, would be viewed from such a system fixed in the body. The second class includes the rotational motions of the rigid body with reference to a set of axes the origin of which coincides with that of the above set but which is not rotating, its orientation in space having been fixed. The herpolhode would be fixed in such a system.

The free rotational motions of the earth provide examples, easily treated, of both these classes. (We shall omit the part of the rotational phenomena of the earth arising from the existence of a torque about the center of mass.) In the first instance, the axis of rotation of the earth does not coincide with the axis of symmetry. It is very slightly tilted, although it passes through the center of mass of the earth. Hence the instantaneous axis does not lie along a principal axis and thus precesses slowly about the axis of symmetry (in the same sense as the rotation itself). The period of this precession is about 440 days. Viewed from a set of axes fixed in space, the motion is seen to involve a precession of the axis of symmetry about a fixed direction in space (the invariable line).

Finally we shall consider some of the aspects, from both the above sets of axes, of the motion of a rigid body when a torque is acting. The principal example of this will be a symmetrical top with one point fixed and with gravity present.

12-10. Motion of a Rigid Body Referred to Rotating Axes. Euler's Dynamical Equations. It is convenient to begin with the first class of problems. To do this, we make use of the torque equation $\dot{\mathbf{J}} = \mathbf{L}$, where $\mathbf{J}$ and $\mathbf{L}$ are both defined in a system of axes which has a fixed orientation in space. It is clear, also, that the origin of this system must be either at a fixed point (fixed in space) of the body, if one exists, or at the center of mass. (Still a third possibility is the case where the origin may be at neither of these points but is at a point which has an acceleration the line of action of which passes through the center of mass of the body. The third possibility is usually merely of academic interest.)

It is very important to realize, as stated above, that $\mathbf{J}$ and $\mathbf{L}$ are defined in the system of axes which is not rotating. For this set of axes the equation

$$\dot{\mathbf{J}} = \mathbf{L} \tag{12-58}$$

is valid. Now for the purpose at hand we go to a system of axes fixed in the body, and with the origin coincident with that of the set of axes just mentioned, for which Eq. (12-58) holds. The next step is to express the $\mathbf{J}$ and $\mathbf{L}$, defined in the system of fixed orientation, in terms of the rotating axes. To do this, let us first express $\mathbf{J}$, defined in the nonrotating system, in terms of its instantaneous components in the moving system. We have

$$\mathbf{J} = \mathbf{i}J_x + \mathbf{j}J_y + \mathbf{k}J_z \tag{12-59}$$

[Note that the time derivative in Eq. (12-58) is with respect to the nonrotating system.] Now, although the time derivative is taken with respect to the nonrotating system, $\mathbf{J}$ in Eq. (12-59) is expressed in terms of its components in the rotating system, and the time derivative of (12-59) must therefore take into account the fact that the unit vectors $\mathbf{i}$, $\mathbf{j}$, and $\mathbf{k}$ are changing direction in space. Accordingly the rate at which $\mathbf{J}$ is changing will be

$$\mathbf{i}\dot{J}_x + \mathbf{j}\dot{J}_y + \mathbf{k}\dot{J}_z + J_x\dot{\mathbf{i}} + J_y\dot{\mathbf{j}} + J_z\dot{\mathbf{k}} \tag{12-60}$$

Recalling the notation of Chap. 11, we may rewrite Eqs. (12-58) and (12-60), respectively, as

$$[\dot{\mathbf{J}}]_0 = \mathbf{L} \tag{12-61}$$

and

$$[\dot{\mathbf{J}}]_0 = \dot{\mathbf{J}} + \boldsymbol{\omega} \times \mathbf{J} \tag{12-62}$$

where the second term on the right side of Eq. (12-62) is equivalent to the last three terms on the right of Eq. (12-60). In Eq. (12-62) the time derivative of $\mathbf{J}$, on the left side, is taken with respect to the system of fixed orientation in space, and that on the right side is taken with respect to the second, or rotating, system. In the case of **L**, we merely take the components, along the moving axes, of the **L** defined in the system of fixed orientation. Therefore, in the system rigidly attached to the rotating body, we have

$$\begin{aligned}\mathbf{L} &= \mathbf{i}L_x + \mathbf{j}L_y + \mathbf{k}L_z \\ &= \mathbf{i}\dot{J}_x + \mathbf{j}\dot{J}_y + \mathbf{k}\dot{J}_z + \boldsymbol{\omega} \times \mathbf{J} \\ &= \mathbf{i}(\dot{J}_x + \omega_y J_z - \omega_z J_y) + \mathbf{j}(\dot{J}_y + \omega_z J_x - \omega_x J_z) \\ &\qquad + \mathbf{k}(\dot{J}_z + \omega_x J_y - \omega_y J_x) \end{aligned} \tag{12-63}$$

which holds for axes rigidly attached to the rotating body, and with the origin either at a fixed point or at the center of mass (or at the third class of points of the type mentioned above). It is important to realize again that, for the axes rigidly attached to the body, the moments and products of inertia are constants and independent of the time. We shall also make the simplification of selecting a set of principal axes (all products of inertia are zero). For principal axes, Eq. (12-59) has the form

$$\mathbf{J} = \mathbf{i}I_x\omega_x + \mathbf{j}I_y\omega_y + \mathbf{k}I_z\omega_z \tag{12-64}$$

The equation of motion $\mathbf{L} = \dot{\mathbf{J}}$ becomes for principal axes [putting $P_{xy} = P_{yz} = P_{xz} = 0$ in (12-63)]

$$\begin{aligned}\mathbf{L} = \mathbf{i}[I_x\dot{\omega}_x + (I_z - I_y)\omega_y\omega_z] &+ \mathbf{j}[I_y\dot{\omega}_y + (I_x - I_z)\omega_x\omega_z] \\ &+ \mathbf{k}[I_z\dot{\omega}_z + (I_y - I_x)\omega_x\omega_y]\end{aligned} \tag{12-65}$$

and the components of this may be written

$$\begin{aligned} L_x &= I_x\dot{\omega}_x + (I_z - I_y)\omega_y\omega_z \\ L_y &= I_y\dot{\omega}_y + (I_x - I_z)\omega_z\omega_x \\ L_z &= I_z\dot{\omega}_z + (I_y - I_x)\omega_x\omega_y \end{aligned} \tag{12-66}$$

These are called *Euler's dynamical equations*. It is to be emphasized that they are merely the components of $\dot{\mathbf{J}} = \mathbf{L}$ referred to rotating principal axes.

12-11. Constancy of Energy and Angular Momentum by Means of Euler's Equations. There are some interesting operations that can be performed with Euler's equations. Taking the case where $\mathbf{L} = 0$, Eqs. (12-66) become

$$I_x\dot{\omega}_x + (I_z - I_y)\omega_y\omega_z = 0 \tag{12-67}$$

$$I_y\dot{\omega}_y + (I_x - I_z)\omega_z\omega_x = 0 \tag{12-68}$$

$$I_z\dot{\omega}_z + (I_y - I_x)\omega_x\omega_y = 0 \tag{12-69}$$

Multiplying (12-67) by ω_x, (12-68) by ω_y, and (12-69) by ω_z and adding, we obtain

$$I_x\dot{\omega}_x\omega_x + I_y\dot{\omega}_y\omega_y + I_z\dot{\omega}_z\omega_z = \frac{1}{2}\frac{d}{dt}(I_x\omega_x^2 + I_y\omega_y^2 + I_z\omega_z^2) = 0 \tag{12-70}$$

But, from Eq. (12-46), the quantity within parentheses on the right side of Eq. (12-70) is just twice the kinetic energy of rotation, referred to principal axes. Hence Eq. (12-70) expresses the conservation of energy. Equation (12-70) may be rewritten

$$\frac{dT}{dt} = 0 \tag{12-71}$$

If instead, we had multiplied Eqs. (12-67), (12-68), and (12-69) by $I_x\omega_x$, $I_y\omega_y$, and $I_z\omega_z$, respectively, and had added the results, we should have obtained

$$\begin{aligned} I_x{}^2\dot{\omega}_x\omega_x + I_y{}^2\dot{\omega}_y\omega_y + I_z{}^2\dot{\omega}_z\omega_z &= \frac{1}{2}\frac{d}{dt}(I_x{}^2\omega_x{}^2 + I_y{}^2\omega_y{}^2 + I_z{}^2\omega_z{}^2) \\ &= \frac{1}{2}\frac{d(J)^2}{dt} = J\dot{J} = 0 \end{aligned} \tag{12-72}$$

The second line follows because for principal axes

$$\mathbf{J} = I_x\omega_x\mathbf{i} + I_y\omega_y\mathbf{j} + I_z\omega_z\mathbf{k}$$

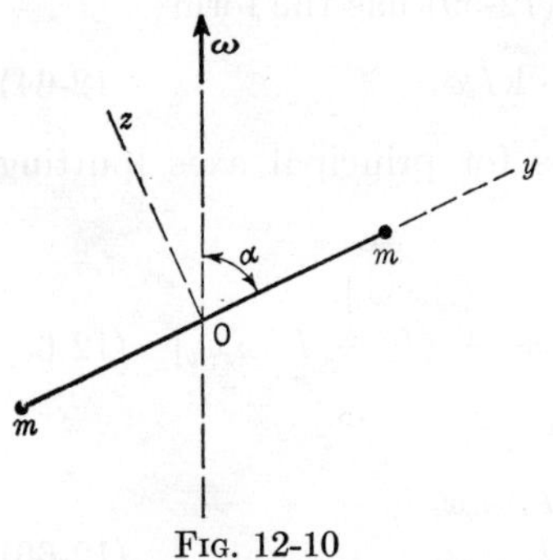

FIG. 12-10

Equation (12-72) states that either $\dot{J}$ or J is zero. In short, if there is any motion at all (and with $\mathbf{L} = 0$), $\mathbf{J}$ remains constant in magnitude. Note that no information concerning the direction of $\mathbf{J}$ is obtained from (12-72).

Example 12-4. Treat the dumbbell problem (Example 12-1) by means of Euler's equations.

We choose axes which are fixed in the rotating system and which furthermore are principal axes. These are shown in Fig. 12-10, where the yz plane is at the moment that of the paper. The Ox axis at this same instant is understood to be normal to the page and positive toward the reader. We note that

$$\begin{gathered} I_x = I_z = 2ma^2 \qquad I_y = 0 \\ \omega_x = 0 \qquad \omega_y = \omega\cos\alpha \qquad \omega_z = \omega\sin\alpha \\ \dot{\omega}_x = \dot{\omega}_y = \dot{\omega}_z = 0 \end{gathered} \tag{12-73}$$

From Eq. (12-64), together with (12-73), we have

$$\mathbf{J} = \mathbf{i}I_x\omega_x + \mathbf{j}I_y\omega_y + \mathbf{k}I_z\omega_z = \mathbf{k}I_z\omega_z = (2ma^2\omega\sin\alpha)\mathbf{k}$$

We obtain the torque required to maintain the motion by use of Eqs. (12-66). Here these take the form

$$L_x = I_z\omega_y\omega_z = 2ma^2\omega^2\sin\alpha\cos\alpha \qquad L_y = 0 \qquad L_z = 0$$

These results agree with those found in Example 12-1.

12-12. Free Rotation of the Earth. We shall now consider part of the problem mentioned briefly before, that is, the case of the free motion of the earth with reference to a set of axes fixed in the earth. The earth closely resembles a spheroid, slightly flattened at the poles. Figure 12-11 depicts a vertical plane section, including the poles and the center of mass, of such a body. The x axis is taken to be normal to the page and positive toward the reader. The origin is taken at the center of mass, and the z axis is the axis of symmetry. Consider the axes to be fixed in the body. (This will permit the use of Euler's equations.) It is clear, since the figure is one of revolution about the z axis, that $I_x = I_y$; we represent both these by the symbol I_1, and I_z by the symbol I. It is evident that $I > I_1$, and also, because of the symmetry of the figure with respect to the axes chosen, that the axes are principal axes. We shall further assume that the only forces acting are through the center of mass (in reality but a good approximation).

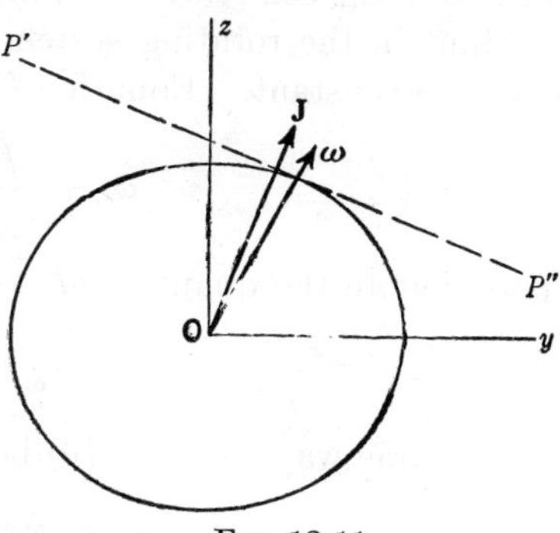

FIG. 12-11

The axis of $\boldsymbol{\omega}$ is taken not to be coincident with the z axis. The figure represents, in a rather exaggerated fashion, the aspect presented by the earth.

Clearly we have here a situation such as is described in Sec. 12-8, in which $\boldsymbol{\omega}$ does not lie along a principal axis. Consequently we should expect $\boldsymbol{\omega}$ to be shifting with respect to the $Oxyz$ axes fixed in the body. The object of the present investigation is to ascertain the manner in which $\boldsymbol{\omega}$ is changing.

In terms of I, I_1, and the coordinates X, Y, and Z defined in Sec. 12-6, the equation of the inertia ellipsoid is, from Eq. (12-36),

$$1 = I_1X^2 + I_1Y^2 + IZ^2 \tag{12-74}$$

Equation (12-74) describes an ellipsoid of revolution about OZ and consequently about Oz. As will be pointed out below I_1 is known to be only slightly less than I. Hence the spheroidal earth may be considered to approximate closely the inertia ellipsoid of Eq. (12-74) provided that a suitable scale for x/X, y/Y, z/Z is chosen. In view of this, the line $P'P''$, shown in the figure to be tangent to the surface at the point of its intersection with the $\boldsymbol{\omega}$ vector, is the intersection of the invariable plane with the paper. The angular momentum $\mathbf{J}$ is shown normal to this line.

Since all the forces are acting through the center of mass at O, Euler's dynamical equations become, since $\mathbf{L} = 0$ and $I_x = I_y = I_1$,

$$I_1\dot{\omega}_x + (I - I_1)\omega_y\omega_z = 0 \tag{12-75}$$
$$I_1\dot{\omega}_y + (I_1 - I)\omega_z\omega_x = 0 \tag{12-76}$$
$$I\dot{\omega}_z + (I_1 - I_1)\omega_x\omega_y = 0 \tag{12-77}$$

Equation (12-77) becomes

$$I\dot{\omega}_z = 0$$

whence

$$I\omega_z = J_z = \text{const} \tag{12-78}$$

which says that, as a result of the earth being a figure of revolution about the z axis, the component of $\mathbf{J}$ along this axis is constant (note that this is constant in the rotating system). Since I is constant, this means that ω_z is also constant. Equation (12-75) becomes, in view of (12-78),

$$\dot{\omega}_x + \frac{(I - I_1)J_z}{II_1}\,\omega_y = 0 \tag{12-79}$$

If we denote the quantity $(I - I_1)J_z/II_1$ by the symbol k, Eq. (12-79) becomes

$$\dot{\omega}_x + k\omega_y = 0 \tag{12-80}$$

In the same way, Eq. (12-76) becomes

$$\dot{\omega}_y - k\omega_x = 0 \tag{12-81}$$

Differentiating (12-81) with respect to time, we obtain

$$\ddot{\omega}_y - k\dot{\omega}_x = 0 \tag{12-82}$$

and, substituting for $\dot{\omega}_x$ in (12-80), we have

$$\ddot{\omega}_y + k^2\omega_y = 0 \tag{12-83}$$

Since k^2 is a positive quantity, Eq. (12-83) is just the equation of harmonic motion. The solution is

$$\omega_y = \omega_0 \sin (kt + \alpha) \tag{12-84}$$

where ω_0, a constant, is the amplitude of ω_y, and α is a phase constant. Also, from (12-81),

$$\omega_x = \frac{1}{k}\,\dot{\omega}_y = \omega_0 \cos (kt + \alpha) \tag{12-85}$$

It is worthwhile to note that

$$\omega_x^2 + \omega_y^2 = \omega_0^2 = \text{const} \tag{12-86}$$

Since ω_z, from (12-78), has already been found to be constant, and since ω_0, from Eq. (12-84), is seen to be constant as well, the magnitude of $\boldsymbol{\omega}$ also does not change. Figure 12-12 shows the geometrical relationship of ω_x to ω_y at a particular instant. Equations (12-84) and (12-85) show that, if α is chosen zero, at $t = 0$, $\omega_y = 0$ and $\omega_x = \omega_0$; as t increases, ω_x becomes smaller, while ω_y increases. This shows that $\boldsymbol{\omega}$ precesses about the z axis in the same sense as the angular velocity $\boldsymbol{\omega}$, that is, counterclockwise as seen along the negative z axis from O. Since ω_x and ω_y have the same amplitude, ω_0, the polhode, or locus of the intersection of $\boldsymbol{\omega}$ with the inertia ellipsoid, is a circle with center on the symmetry axis Oz. Hence the vector $\boldsymbol{\omega}$ describes a right circular cone (the *body cone*) about Oz. Furthermore the angular velocity of this precession is k, a constant. The frequency ν_p and the period T_p of this precession are given by

$$\nu_p = \frac{k}{2\pi} \qquad T_p = \frac{2\pi}{k} = \frac{2\pi I_1}{\omega_z(I - I_1)} \tag{12-87}$$

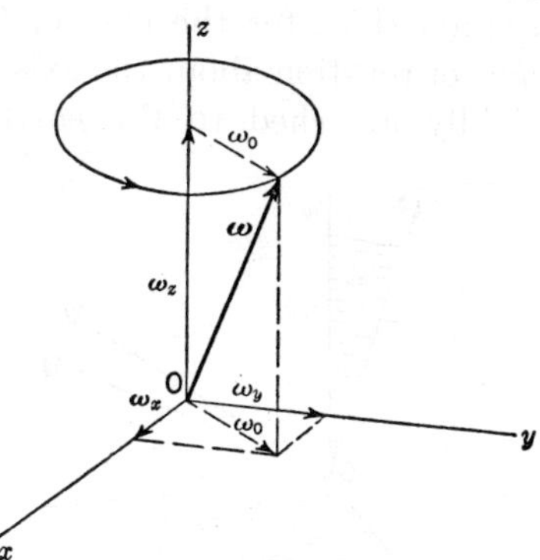

FIG. 12-12

We have seen that the angular velocity of the earth is 7.29×10^{-5} rad/sec. The quantity $(I - I_1)/I_1$ has been determined astronomically, from the independent phenomenon of the *precession of the equinoxes*, to be 0.00327. Substituting in (12-87), we find

$$T_p = 305 \text{ days}$$

The Swiss mathematician Euler derived this result in the latter part of the eighteenth century. The first crude measurements were made in 1888, but no accurate knowledge was obtained until comparatively recently. The difficulty of the measurement is apparent when it is realized that the diameter of the circle formed by the intersection of the cone with the earth's surface is only a few feet! The observed period is 440 days. The difference from the above theoretical value has been ascribed to the deviation of the earth from perfect rigidity.

It should be kept in mind that we are so far discussing a precessional motion of $\boldsymbol{\omega}$ not about an axis fixed in space but about an axis fixed in the body. Thus to an observer situated on the earth it would appear as a *variation of latitudes*. This is true since latitudes are measured from the equator, which is fixed with respect to the poles. Consequently, if the poles wander we should expect a variation in the latitude of a given point on the earth's surface. (The actual motion is complicated by the presence of an annual precession. The latter has been ascribed to meteorological causes such as the depositing of rain, snow, and ice on one hemisphere one part of the year and on the other hemisphere during another part of the year. Neither of these motions should be confused with the so-called precession of the equinoxes mentioned above, this latter motion arising because the forces on the earth due to other heavenly bodies, principally the sun and the moon, do not all pass through the center of mass, and so the torque is, in reality, different from zero. The effect is small, however, since the torque is only slightly different from zero. The period of this precession is about 26,000 years.)

12-13. Free Motion of a Rigid Body Referred to Axes Having a Fixed Direction in Space. Motion of the Earth. It was shown in the previous section that, for the case of the earth, there exists a precession of the axis of rotation about the axis of symmetry with reference to a set of axes rigidly attached to the earth. It is also interesting to consider the rotational motions of such a body relative to a set of axes the orientation of which is fixed in space.

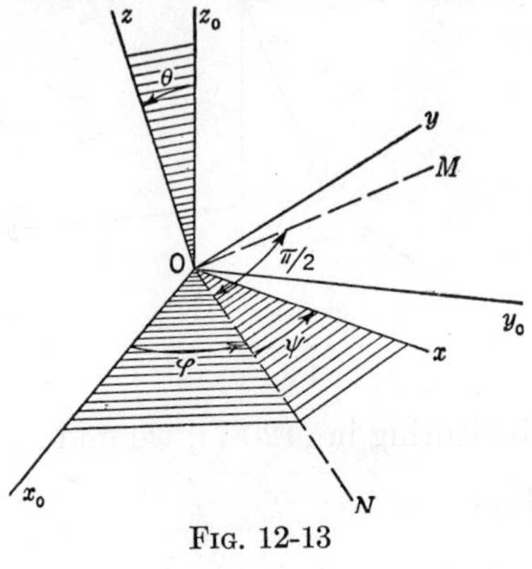

Fig. 12-13

In Fig. 12-13 the $Ox_0y_0z_0$ system of axes has its orientation fixed in space. The $Oxyz$ system is fixed in the body, which may be rotating. (The origin must also be such a point that $\dot{\mathbf{J}} = \mathbf{L}$ is valid in the $Ox_0y_0z_0$ system.) In the present problem of the rotating earth it is convenient to take the origin to lie at the center of mass. The line ON is the intersection of the xy plane with the x_0y_0 plane and is often called the *line of nodes.* θ is the angle between Oz and Oz_0, being positive from z_0 to z. φ is the angle between Ox_0 and ON (in the x_0y_0 plane) and increases in a counterclockwise direction about Oz_0. ψ, the angle between ON and Ox, is measured in the xy plane and increases in a counterclockwise direction about Oz. The line Oz_0 is usually selected to be the line of $\mathbf{J}$ (which is fixed in space if the body is free, the invariable line). The angles θ, φ, and ψ, as defined above, are known as *Euler's angles.*

We choose the $Oxyz$ system to be fixed in the earth, with Oz again coinciding with the axis of symmetry of the earth. The components of $\boldsymbol{\omega}$ along x, y, and z are just those in Euler's dynamical equations considered previously. The next step is to express these components of $\boldsymbol{\omega}$ in terms of the angles θ, φ, ψ, and their time derivatives.

In Sec. 9-4, it was shown that an angular velocity may be regarded as a vector pointing along the axis of rotation in the direction of the advance of a right-hand screw turning in the sense of the angular velocity. Accordingly, in Fig. 12-13, $\dot{\boldsymbol{\varphi}}$ may be regarded as a vector pointing along Oz_0, $\dot{\boldsymbol{\psi}}$ a vector along Oz, and $\dot{\boldsymbol{\theta}}$ a vector along ON. If $\mathbf{k}_0$, $\mathbf{k}$, and $\mathbf{n}$ are unit vectors pointing, respectively, in these directions, we may write

$$\dot{\boldsymbol{\varphi}} = \dot{\varphi}\mathbf{k}_0 \qquad \dot{\boldsymbol{\psi}} = \dot{\psi}\mathbf{k} \qquad \dot{\boldsymbol{\theta}} = \dot{\theta}\mathbf{n} \tag{12-88}$$

It remains to find the components of these along the x, y, and z axes, the algebraic sum in each case being the components ω_x, ω_y, ω_z of $\boldsymbol{\omega}$.

Construct line OM perpendicular to ON. (Thus angle MOy equals ψ in magnitude.) It is clear from the figure that since $\dot{\boldsymbol{\psi}}$ lies along Oz it will have no component along Ox and thus will not contribute to ω_x. However, since $\dot{\boldsymbol{\theta}}$ lies along ON, its component along Ox is $\dot{\theta} \cos \psi$. Similarly $\dot{\boldsymbol{\varphi}}$ has a component along OM which is $\dot{\varphi} \sin \theta$. This follows since ON is perpendicular to plane zOz_0, and consequently OM lies in plane zOz_0. Hence the component of $\dot{\boldsymbol{\varphi}}$ along Ox is just $\dot{\varphi} \sin \theta \sin \psi$.

In a similar fashion $\dot{\boldsymbol{\theta}}$ and $\dot{\boldsymbol{\varphi}}$ have components along Oy which are, respectively, $-\dot{\theta} \sin \psi$ and $\dot{\varphi} \sin \theta \cos \psi$. ($\dot{\boldsymbol{\psi}}$ has no component along Oy.) Also, $\dot{\boldsymbol{\psi}}$ and $\dot{\boldsymbol{\varphi}}$ alone have components along Oz. These are, respectively, $\dot{\psi}$ and $\dot{\varphi} \cos \theta$. Accordingly we may write

$$\omega_x = \dot{\theta} \cos \psi + \dot{\varphi} \sin \theta \sin \psi \tag{12-89}$$
$$\omega_y = -\dot{\theta} \sin \psi + \dot{\varphi} \sin \theta \cos \psi \tag{12-90}$$
$$\omega_z = \dot{\psi} + \dot{\varphi} \cos \theta \tag{12-91}$$

These are referred to as *Euler's geometrical equations.*

It is also useful to write the components of $\mathbf{J}$ in the moving system. Since $\mathbf{J}$ lies along Oz_0, these are

$$\begin{aligned} J_x &= J \sin \theta \sin \psi \\ J_y &= J \sin \theta \cos \psi \\ J_z &= J \cos \theta \end{aligned} \tag{12-92}$$

which may be combined as

$$\frac{J_x}{J_y} = \tan \psi \qquad \frac{J_x}{J} = \sin \theta \sin \psi \qquad \frac{J_y}{J} = \sin \theta \cos \psi \tag{12-93}$$

Employing Eqs. (12-89) to (12-91) and (12-93), it is possible to express $\dot{\theta}$, $\dot{\varphi}$, and $\dot{\psi}$ in terms of the coefficients of inertia. First we multiply Eq. (12-89) by $\cos \psi$, and Eq. (12-90) by $\sin \psi$. Subtracting the second from the first, we obtain

$$\dot{\theta} = \omega_x \cos \psi - \omega_y \sin \psi$$

Dividing this by $\sin \theta$, we have

$$\begin{aligned} \frac{\dot{\theta}}{\sin \theta} &= \omega_x \frac{\cos \psi}{\sin \theta} - \omega_y \frac{\sin \psi}{\sin \theta} \\ &= \omega_x \frac{\cos \psi \sin \psi}{\sin \theta \sin \psi} - \omega_y \frac{\sin \psi \cos \psi}{\sin \theta \cos \psi} \end{aligned} \tag{12-94}$$

Making use of Eqs. (12-93), this takes the form

$$\frac{\dot{\theta}}{\sin \theta} = \left(\omega_x \frac{J}{J_x} - \omega_y \frac{J}{J_y}\right) \sin \psi \cos \psi \tag{12-95}$$

If we choose $Oxyz$ to be principal axes, then

$$J_x = I_x\omega_x \qquad J_y = I_y\omega_y \qquad J_z = I_z\omega_z$$

Hence, for principal axes, Eq. (12-95) may be rewritten as

$$\frac{\dot\theta}{\sin\theta} = \left(\frac{1}{I_x} - \frac{1}{I_y}\right) J \sin\psi \cos\psi = \frac{I_y - I_x}{I_x I_y} J \sin\psi\cos\psi \tag{12-96}$$

It is important to notice that, for bodies which have $I_x = I_y$, $\dot\theta$ is zero. (Note that, for $I_x = I_y$, the body need not necessarily be one of revolution about the z axis. See Prob. 12-14, for example.)

$\dot\varphi$ can be found from Eqs. (12-89) and (12-90) in a manner similar to that leading to Eq. (12-95). We multiply (12-89) by $\sin\psi$ and (12-90) by $\cos\psi$ and add. We obtain

$$\dot\varphi \sin\theta = \omega_x \sin\psi + \omega_y \cos\psi$$

and

$$\begin{aligned}\dot\varphi &= \omega_x \frac{\sin\psi}{\sin\theta} + \omega_y \frac{\cos\psi}{\sin\theta} \\ &= \omega_x \frac{J}{J_x}\sin^2\psi + \omega_y \frac{J}{J_y}\cos^2\psi \\ &= \frac{J}{I_x I_y}(I_y \sin^2\psi + I_x \cos^2\psi)\end{aligned} \tag{12-97}$$

Equation (12-97) shows that $\dot\varphi$ is a constant if $I_x = I_y$.

Referring in particular to the problem of the motion of the earth, we note that Eq. (12-96) gives us the rate of change of θ with time, that is, the rate at which the axis of symmetry of the earth is changing in direction with respect to the line of **J** (invariable line) in space. Since, for the earth, we have $I_z = I$, and $I_x = I_y = I_1$, we have a simple example of the above-mentioned special case in which θ is a constant. Likewise $\dot\varphi$ is a constant, given by

$$\dot\varphi = \frac{J}{I_1} \tag{12-98}$$

This is the rate at which the axis of symmetry, Oz, precesses in space about the Oz_0 axis, or axis of **J**.

So far as $\dot\psi$ is concerned, an examination of Eq. (12-91) shows that it contributes to ω_z. The part of ω_z which is $\dot\varphi\cos\theta$ is the contribution, to ω_z, of the precession of the axis of symmetry about the invariable line. In the present problem we can throw light on $\dot\psi$ by employing the first of Eqs. (12-93), as

$$\tan\psi = \frac{J_x}{J_y} = \frac{I_x\omega_x}{I_y\omega_y} = \frac{I_1\omega_x}{I_1\omega_y} = \frac{\omega_x}{\omega_y}$$

Making use of (12-84) and (12-85), we have

$$\tan \psi = \frac{\cos (kt + \alpha)}{\sin (kt + \alpha)} = \cot (kt + \alpha)$$

Thus

$$\psi = \frac{\pi}{2} - (kt + \alpha) \tag{12-99}$$

whence

$$\dot{\psi} = -k = -\frac{I - I_1}{I_1}\omega_z \tag{12-100}$$

An inspection of Fig. 12-13 shows that the sense of $\dot{\psi}$ is negative (clockwise when viewed from above at the north pole). This motion has the period $2\pi/k$, which is the same as that of the precession, in the body, of $\boldsymbol{\omega}$ about the axis of symmetry.

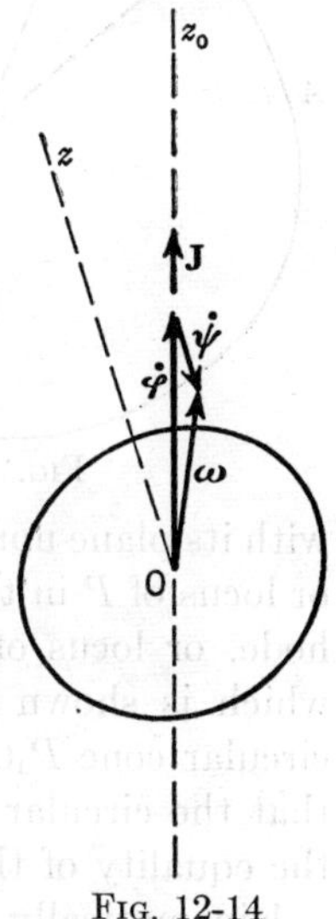

Fig. 12-14

The square of the magnitude of the angular velocity vector $\boldsymbol{\omega}$ may be written

$$\omega^2 = \omega_x^2 + \omega_y^2 + \omega_z^2$$

and, substituting from Eqs. (12-89) to (12-91), we obtain

$$\omega^2 = \dot{\theta}^2 + \dot{\varphi}^2 + \dot{\psi}^2 + 2\dot{\varphi}\dot{\psi} \cos \theta \tag{12-101}$$

For the present problem, since $I_x = I_y$, $\dot{\theta} = 0$. Therefore Eq. (12-101) here reduces to

$$\omega^2 = \dot{\varphi}^2 + \dot{\psi}^2 + 2\dot{\varphi}\dot{\psi} \cos \theta \tag{12-102}$$

which indicates that $\boldsymbol{\omega}$ may be regarded as the sum of the two vectors $\dot{\boldsymbol{\varphi}}$ and $\dot{\boldsymbol{\psi}}$ and is thus coplanar with them (see Fig. 12-14, which is exaggerated in the interests of clarity). In view of this, $\dot{\boldsymbol{\varphi}}$ is the angular velocity of precession of the $OzOz_0\boldsymbol{\omega}$ plane. It is apparent that $\dot{\boldsymbol{\varphi}}$ is larger in magnitude than $\boldsymbol{\omega}$. We can get an idea of its numerical magnitude for the case of the earth by substituting the expression for ω_z from (12-100) into (12-91). This yields (taking $\theta \simeq 0$, in which case $\cos \theta \simeq 1$)

$$\dot{\varphi} = -\dot{\psi}\frac{I}{I - I_1} \tag{12-103}$$

from which the period of precession corresponding to $\dot{\varphi}$ is

$$\frac{2\pi}{\dot{\varphi}} = -\frac{I - I_1}{I}\frac{2\pi}{\dot{\psi}} = \frac{I - I_1}{I}\frac{2\pi}{k} = 0.997 \text{ day} \tag{12-104}$$

In summary, the free rotation of the earth (see Fig. 12-14) consists, first in a rapid precession $\dot{\varphi}$ (slightly greater than ω) of the $Ozz_0\boldsymbol{\omega}$ plane in space in the sense of increasing φ. Accompanying this is a much slower precession of $\boldsymbol{\omega}$, in the body, in the sense of increasing ψ about the Oz axis. Finally there is a slow rotation $\dot{\psi}$ of the earth in space, in the sense of decreasing ψ about its own axis of symmetry Oz. It has the same magnitude as the precession of $\boldsymbol{\omega}$ about Oz, given in Sec. 12-12.

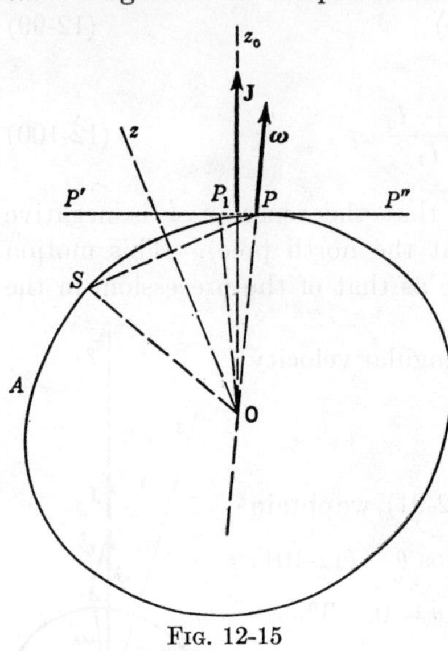

Fig. 12-15

Figure 12-15 shows, in terms of the inertia ellipsoid, the geometry of the free motions of the earth. As before, since the earth is almost spherical, its periphery A is assumed to coincide approximately with the chosen inertia ellipsoid. The line $P'P''$ is the intersection of the invariable plane with the page and is tangent to the earth at the point P of the instantaneous axis of rotation. The invariable line is the axis Oz_0, along which lies the angular momentum **J**. The *body cone* (mentioned above), or cone which the precessing $\boldsymbol{\omega}$ vector describes in the body, is shown in section as SOP. The base of the cone is a circle of diameter SP and with its plane normal to the symmetry axis Oz of the earth. The polhode, or locus of P in the body, is just this circle of diameter SP. The herpolhode, or locus of P in the invariable plane, is a circle the diameter of which is shown as P_1P. This circle comprises the base of the right circular cone P_1OP, the so-called *space cone*. (It should be pointed out that the circular nature of the polhode and the herpolhode is a result of the equality of the moments I_x and I_y.)

Kinematically the motion may be described in terms of either a rolling of the polhode on the fixed herpolhode or a rolling of the body cone SOP on the fixed space cone P_1OP. The period $2\pi/\dot{\varphi}$ is the period associated with this rolling motion, being the time required for the body cone to roll completely around the fixed space cone. This may be seen more easily with the aid of Fig. 12-16, which shows the projections of the polhode and the herpolhode upon a horizontal plane (the effect of the tilt of the body cone with respect to the space cone is neglected in this figure). Let the projections of the xy and x_0y_0 axes be initially oriented as shown in part (a) of the figure, the initial point of contact (through which $\boldsymbol{\omega}$ passes) coinciding with points A of the polhode and P of the herpolhode. By the time the polhode has rolled once around the herpolhode, the point B, where arc AB is equal to the circumference of the herpolhode, becomes the point of the polhode tangent to the herpolhode at P. During the time $2\pi/\dot{\varphi}$ required for this to take place, the xy axes (fixed in the body) have experienced a net rotation through the angle α in the clockwise direction. The period of this net rotation of the xy axes is just that associated with the angular velocity $\dot{\psi}$ mentioned above, that is, $2\pi/\dot{\psi}$.

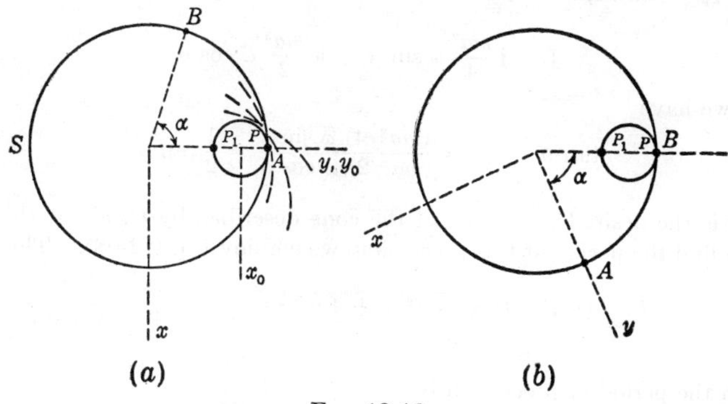

FIG. 12-16

Also, as shown before, the $\boldsymbol{\omega}$ vector (which passes always through the point of tangency of the two circles) appears to move in the counterclockwise direction, relative to the xy axes, with the same period $2\pi/\dot{\psi}$.

Figures 12-11, 12-12, and 12-14 to 12-16 are greatly exaggerated in the interests of clarity. The diameter d_p of the polhode is approximately only 30 ft. Since the motion of the polhode on the herpolhode is one of pure rolling, the diameter d_h of the herpolhode may easily be found. It is given by

$$d_h = \left|\frac{\dot{\psi}}{\dot{\varphi}}\right| d_p = \frac{I - I_1}{I} d_p = 0.00327 d_p \tag{12-105}$$

For $d_p \simeq 30$ ft, the value of d_h is approximately 1 in. These quantitative considerations show strikingly that the earth differs very little from a spherical body and that the axis of rotation deviates only slightly from a principal axis.

Example 12-5. A uniform thin circular disk is constrained to spin with an angular velocity $\boldsymbol{\omega}$ about an axis passing through the center but making an angle α with the axis of symmetry (normal to the disk). It is suddenly released. It is desired to find the half angle of the cone in space described by the axis of symmetry and also the time in which this cone is described.

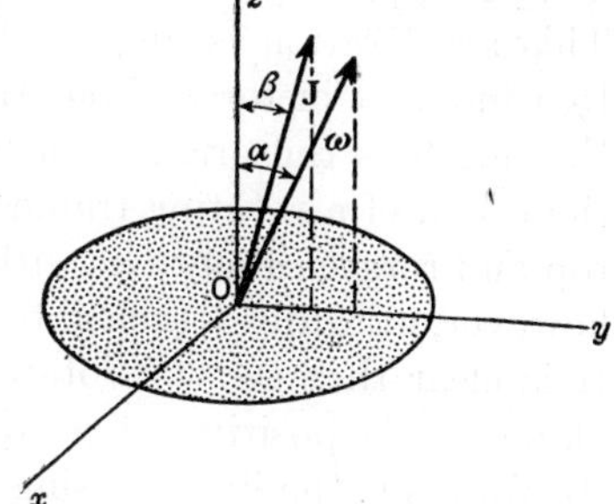

FIG. 12-17

The situation is shown in Fig. 12-17. Let us choose axes rigidly attached to the disk, with the instantaneous axis of $\boldsymbol{\omega}$ in the yz plane and making an angle α with Oz. Since there are no external moments acting, the direction in space of the angular momentum $\mathbf{J}$ will not vary. Let it make an angle β

with Oz, where β is to be determined. Clearly the axes chosen are principal axes, and since $\boldsymbol{\omega}$ has no x component, neither, then, has $\mathbf{J}$. (Since $I_x = I_y$, the axis Oz and the vectors $\boldsymbol{\omega}$ and $\mathbf{J}$ are all coplanar.) We have

$$\mathbf{J} = \mathbf{j}J_y + \mathbf{k}J_z = \mathbf{j}I_y\omega_y + \mathbf{k}I_z\omega_z$$

Now $I_x = I_y = ma^2/4$, and $I_z = ma^2/2$. Accordingly $\mathbf{J}$ becomes

$$\mathbf{J} = \mathbf{j}\,\frac{ma^2}{4}\,\omega \sin \alpha + \mathbf{k}\,\frac{ma^2}{2}\,\omega \cos \alpha$$

Therefore we have

$$\tan \beta = \frac{J_y}{J_z} = \frac{(ma^2/4)\,\omega \sin \alpha}{(ma^2/2)\,\omega \cos \alpha} = \frac{1}{2} \tan \alpha$$

in which β is the desired half angle of the cone described by Oz about the $\mathbf{J}$ vector. In order to find the period of this precession, we employ Eq. (12-98). This yields

$$\dot{\varphi} = \frac{J}{I_y} = \frac{(ma^2\omega/4)\,\sqrt{\sin^2 \alpha + 4 \cos^2 \alpha}}{ma^2/4} = \omega\,\sqrt{1 + 3 \cos^2 \alpha}$$

from which the period of precession is

$$T_\varphi = \frac{2\pi}{\dot{\varphi}} = \frac{2\pi}{\omega(1 + 3 \cos^2 \alpha)^{\frac{1}{2}}}$$

With reference to the space and body cones mentioned above in this section, the half angles of these are, respectively, $\alpha - \beta$ and α.

MOTION OF A TOP

12-14. Choice of Coordinates. Equations of Motion. The methods described on the preceding pages are appropriate for treating the problem of a top. We assume that the top is a solid of revolution, rotating about an axis passing through a point fixed in an inertial system. This point is the point of contact of its *peg* with a rough horizontal plane. (In reality the area of contact between the peg and the plane is of finite size, and consequently, owing to frictional forces, point O suffers acceleration. This acceleration is neglected here.) The weight of the top gives rise to a moment of force about the fixed point on the plane. In Fig. 12-18 the peg O is the origin of a fixed frame of reference $Ox_0y_0z_0$, as well as the origin of a rotating frame $Oxyz$. The z axis is the axis of *spin* of the top and rotates about Oz_0 with the precessional velocity $\dot{\varphi}$. The axis Oz, however, is rigidly attached to the top and is, in fact, its axis of symmetry. It is clear that Ox will rotate with the angular frequency $\dot{\varphi}$ in the x_0y_0 plane. φ is positive in a counterclockwise direction from Ox_0. Also, Ox always is the line (passing through Oz) of intersection of the xy plane with the fixed x_0y_0 plane. The line OR can be regarded as attached to the top. The angular velocity $\dot{\psi}$ is just the so-called *spin* of the top.

The angular velocity $\boldsymbol{\omega}$ of the rotating axes consists of the component $\dot{\varphi}$

along Oz_0 and a component $\dot{\theta}$ along Ox. The components of $\boldsymbol{\omega}$ along the $Oxyz$ axes are, therefore,

$$\omega_x = \dot{\theta} \tag{12-106}$$

$$\omega_y = \dot{\varphi} \sin \theta \tag{12-107}$$

$$\omega_z = \dot{\varphi} \cos \theta \tag{12-108}$$

We remember that we deduced Euler's equations for principal axes which were fixed in the body. The important point to remember is that, for axes fixed in the body, the coefficients of inertia are constants and Eqs. (12-66) possessed a simple form because of this. For the present set of axes in Fig. 12-18, even though the x and y axes are not

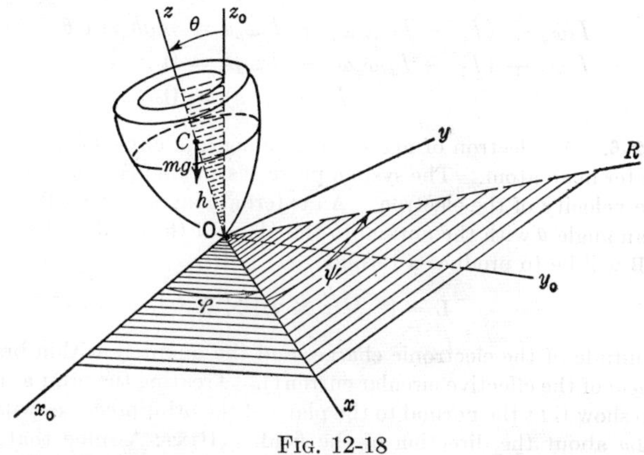

FIG. 12-18

rigidly attached to the top, the coefficients I_x and I_y remain constant since the Oz axis is an axis of revolution. Both I_x and I_y have the same value no matter in what direction perpendicular to Oz the axes point. Let this constant value be I_1. The products of inertia are zero for the present set of axes, and they are therefore principal axes. Accordingly we are able to deduce a simple form of Euler's equations for this problem, remembering, however, that ω_z is the component, along Oz, of the rotation only of the axes. There are two angular velocities which contribute to the component of angular momentum J_z along the Oz axis. These are ω_z, the z component of the rotation of the axis of symmetry about Oz_0, and $\dot{\psi}$, the angular velocity of spin of the top about its own axis of symmetry. It is convenient to employ the symbol s for $\dot{\psi}$. Thus the total angular momentum $\mathbf{J}$ of the system is

$$\mathbf{J} = \mathbf{i}I_1\omega_x + \mathbf{j}I_1\omega_y + \mathbf{k}I_z(\omega_z + s) \tag{12-109}$$

and

$$[\dot{\mathbf{J}}]_0 = \mathbf{i}I_1\dot{\omega}_x + \mathbf{j}I_1\dot{\omega}_y + \mathbf{k}I_z(\dot{\omega}_z + \dot{s}) + I_1\omega_x\dot{\mathbf{i}} + I_1\omega_y\dot{\mathbf{j}} + I_z(\omega_z + s)\dot{\mathbf{k}} \tag{12-110}$$

But since $\dot{\mathbf{i}} = \boldsymbol{\omega} \times \mathbf{i}$, and so on, Eq. (12-110) becomes

$$[\dot{\mathbf{J}}]_0 = \mathbf{i}I_1\dot{\omega}_x + \mathbf{j}I_1\dot{\omega}_y + \mathbf{k}I_z(\dot{\omega}_z + \dot{s}) + I_1\omega_x(\mathbf{j}\omega_z - \mathbf{k}\omega_y) + I_1\omega_y(\mathbf{k}\omega_x - \mathbf{i}\omega_z) + I_z(\omega_z + s)(\mathbf{i}\omega_y - \mathbf{j}\omega_x) \qquad (12\text{-}111)$$

Rearranging Eq. (12-111), we obtain

$$[\dot{\mathbf{J}}]_0 = \mathbf{i}[I_1\dot{\omega}_x + (I_z - I_1)\omega_z\omega_y + I_z\omega_y s] + \mathbf{j}[I_1\dot{\omega}_y + (I_1 - I_z)\omega_x\omega_z - I_z\omega_x s] + \mathbf{k}[I_z(\dot{\omega}_z + \dot{s})] = \mathbf{L} = \mathbf{i}mgh \sin\theta \qquad (12\text{-}112)$$

where h is the distance OC in Fig. 12-13. Thus the component equations (Euler's equations) become

$$I_1\dot{\omega}_x + (I_z - I_1)\omega_z\omega_y + I_z\omega_y s = mgh \sin\theta \qquad (12\text{-}113)$$

$$I_1\dot{\omega}_y + (I_1 - I_z)\omega_x\omega_z - I_z\omega_x s = 0 \qquad (12\text{-}114)$$

$$I_z(\dot{\omega}_z + \dot{s}) = 0 \qquad (12\text{-}115)$$

Example 12-6. An electron of mass m is moving in a circle of radius r around an attracting center in an atom. The system possesses an angular momentum $m\mathbf{r} \times \mathbf{v}$ in which $\mathbf{v}$ is the velocity of the electron. An external magnetic field $\mathbf{B}$ acts in a direction making an angle θ with the normal to the plane of the orbit. We are given that the effect of $\mathbf{B}$ will be to produce a torque

$$\mathbf{L} = [-\tfrac{1}{2}e(\mathbf{r} \times \mathbf{v})] \times \mathbf{B} \qquad (12\text{-}116)$$

(e is the magnitude of the electronic charge, and the quantity within brackets is the *magnetic moment* of the effective circular current). Treating the orbit as a rigid circle, it is desired to show that the normal to the plane of the orbit precesses with an angular velocity $e\mathbf{B}/2m$ about the direction of the field. (HINT: Assume that the angular velocity $\Omega_l = v/r$ of the electron in the orbit is very much greater than the precessional velocity.)

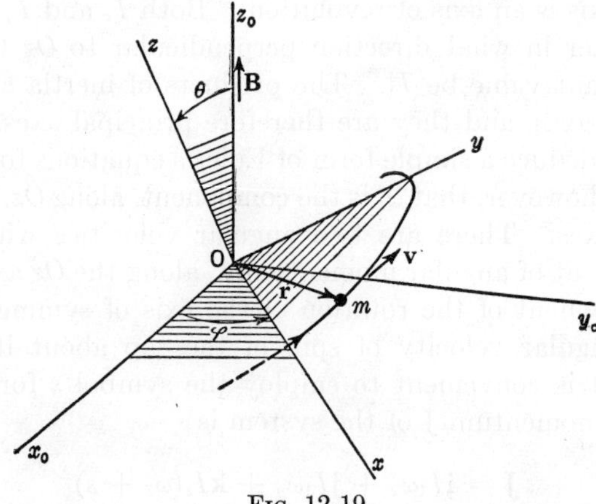

FIG. 12-19

The situation is shown in Fig. 12-19. The axis Oz_0 is the direction of $\mathbf{B}$. The axes $Oxyz$ are moving with the precessional velocity of the orbit, which is in the xy plane, the Ox axis being constrained to remain in the x_0y_0 plane. Since the angular momentum due to the orbital velocity is

$$\mathbf{J}_l = m\mathbf{r} \times \mathbf{v} = \mathbf{k}mrv$$

the total angular momentum is

$$\mathbf{J} = I_x\omega_x\mathbf{i} + I_y\omega_y\mathbf{j} + (I_z\omega_z + J_l)\mathbf{k}$$

in which $\boldsymbol{\omega}$ is the angular velocity of the moving axes. Thus the first of Euler's equations, from Eq. (12-113), becomes

$$I_x\dot{\omega}_x + (I_z - I_y)\omega_z\omega_y + I_z\omega_y\Omega_l = \tfrac{1}{2}ervB \sin\theta$$

The assumption is made that the angular velocity Ω_l of motion in the orbit is very much larger than any of the components of $\boldsymbol{\omega}$. Accordingly we shall neglect terms of the type $\omega_z\omega_y$ and moreover shall neglect $\dot{\omega}_x$. Thus, to a sufficient approximation, we have

$$I_z\omega_y\Omega_l = \tfrac{1}{2}ervB \sin\theta$$

from which

$$\omega_y = \frac{\tfrac{1}{2}ervB \sin\theta}{I_z\Omega_l} \qquad (12\text{-}117)$$

But $I_z\Omega_l = mrv$, and therefore Eq. (12-117) becomes

$$\omega_y = \frac{eB \sin\theta}{2m}$$

We now obtain the angular velocity of precession $\dot{\varphi}$ by use of Eq. (12-107). We have

$$\dot{\varphi} = \frac{\omega_y}{\sin\theta} = \frac{eB}{2m}$$

This gives the rate at which the normal to the plane of the orbit at O ($\mathbf{J}_l$ vector) precesses about the direction of the field $\mathbf{B}$. The frequency is the so-called *Larmor precession* frequency.

12-15.[1] Energy and Angular-momentum Integrals.

Equation (12-115) may be integrated at once. We have

$$(\omega_z + s) = \text{const} = S_1 \qquad (12\text{-}118)$$

Thus the component of the angular momentum along the Oz axis is a constant, equal to I_zS_1. It is possible to obtain an energy integral from Eqs. (12-113) and (12-114) in a manner similar to that leading to Eq. (12-71). We multiply (12-113) by ω_x and (12-114) by ω_y and add the results, obtaining

$$I_1(\dot{\omega}_x\omega_x + \dot{\omega}_y\omega_y) = mgh\omega_x \sin\theta = mgh\dot{\theta} \sin\theta$$

After transposing, this integrates directly to

$$\tfrac{1}{2}I_1(\omega_x^2 + \omega_y^2) + mgh\cos\theta = c_1 \qquad (12\text{-}119)$$

where c_1 is a constant of integration. An inspection of Eq. (12-119)

[1] May be omitted in a first reading.

reveals that c_1 is related to the total energy W by the equation

$$c_1 + \tfrac{1}{2}I_zS_1^2 = W \tag{12-120}$$

that is, it is equal to the total energy minus the kinetic energy of rotation about the Oz axis.

Now the important information is the character of the motion relative to the $Ox_0y_0z_0$ axes fixed in space. In other words, we wish to know how θ and φ vary. To investigate this, we eliminate ω_x and ω_y in Eq. (12-119) by means of Eqs. (12-106) and (12-107). We obtain

$$\dot{\theta}^2 + \dot{\varphi}^2 \sin^2\theta = \frac{2c_1 - 2mgh\cos\theta}{I_1}$$

or, in simpler form,

$$\dot{\theta}^2 + \dot{\varphi}^2 \sin^2\theta = A - B\cos\theta \tag{12-121}$$

in which A and B are constants, defined by

$$A = \frac{2c_1}{I_1} \qquad B = \frac{2mgh}{I_1} \tag{12-122}$$

Now Eq. (12-121) contains both $\dot{\theta}$ and $\dot{\varphi}$, the angular velocities of the axes. It is necessary to find another equation by means of which one of these may be eliminated. Such an equation is furnished by the component J_{z0} of the total angular momentum along the Oz_0 axis. J_{z0} must be a constant since there is no z_0 component of the torque. We may write J_{z0} in terms of the projections along Oz_0 of all the components of $\mathbf{J}$ along the moving axes. Thus we have

$$I_1\omega_y \sin\theta + I_z(\omega_z + s)\cos\theta = J_{z0} = \text{const} \tag{12-123}$$

from which, by employing Eq. (12-118),

$$I_1\omega_y \sin\theta + I_zS_1\cos\theta = J_{z0} \tag{12-124}$$

in which S_1 is constant. We may employ Eq. (12-107) in order to eliminate ω_y from Eq. (12-124). We obtain

$$I_1\dot{\varphi}\sin^2\theta + I_zS_1\cos\theta = J_{z0} \tag{12-125}$$

This may be rewritten as

$$\dot{\varphi} = \frac{(J_{z0}/I_1) - (I_zS_1/I_1)\cos\theta}{\sin^2\theta} = \frac{C - D\cos\theta}{\sin^2\theta} \tag{12-126}$$

in which C and D are constants, defined by

$$C = \frac{J_{z0}}{I_1} \qquad D = \frac{I_zS_1}{I_1} \tag{12-127}$$

12-16.[1] **Limits of the θ Motion.** Equation (12-126) furnishes the precession $\dot{\varphi}$ if θ is known. In order to investigate θ, we make use of Eq. (12-121). $\dot{\varphi}$ may be eliminated from Eq. (12-121) by means of (12-126). We obtain, finally,

$$\dot{\theta}^2 = A - B \cos \theta - \frac{(C - D \cos \theta)^2}{\sin^2 \theta} \qquad (12\text{-}128)$$

Let us make the substitution $\cos \theta \equiv u$. The geometrical significance of u is that it is the vertical height above the x_0y_0 plane of a point which is situated at unit distance along Oz from the origin. This point is called the *apex* of the top. We may express $\dot{\theta}$ in terms of $\dot{u}$ by taking the derivative, as

$$-\dot{\theta} \sin \theta = \dot{u}$$

from which

$$\dot{\theta} = -\frac{\dot{u}}{\sin \theta} = -\frac{\dot{u}}{(1 - u^2)^{\frac{1}{2}}} \qquad (12\text{-}129)$$

Accordingly Eq. (12-128) becomes

$$\dot{u}^2 = (A - Bu)(1 - u^2) - (C - Du)^2 \qquad (12\text{-}130)$$

The quantity $\dot{u}$ is the vertical component of the velocity of the apex and is a real quantity. In principle, Eq. (12-130) may be integrated. Taking the square roots of both sides and separating variables, we have

$$t = \int \frac{du}{\sqrt{(A - Bu)(1 - u^2) - (C - Du)^2}} \qquad (12\text{-}131)$$

The polynomial inside the radical is of the third degree in u and is therefore not integrable in terms of simple functions. In spite of this, however, the behavior of u is periodic with time. A complete discussion of this will be omitted, since it involves the properties of the class of periodic functions called *elliptic functions*. Many of the features of the motion of a top, though, can be understood without considering such complications.

In Eq. (12-130) the velocity $\dot{u}$ is seen to depend on the four constants A, B, C, and D. A, C, and D depend both upon properties of the top (such as moments of inertia) and also upon its state of motion. For example, A depends upon the energy, the spin, and also the moment of inertia I_1. C depends upon the z_0 component of the angular momentum, as well as upon I_1, and, finally, D is related to both the total angular velocity along Oz, including spin, and the moments of inertia I_1 and I_z. The remaining constant B, however, is independent of the state of motion of the top and depends only upon the physical properties of the top. Also, from Eq. (12-122), B is positive.

[1] May be omitted in a first reading.

The constants A, C, and D are determined by the initial conditions of the motion. Thus we have

$$A = \dot{\theta}_0^2 + \dot{\varphi}_0^2 \sin^2 \theta_0 + B \cos \theta_0 \tag{12-132}$$

$$C = \dot{\varphi}_0\left(\sin^2 \theta_0 + \frac{I_z}{I_1} \cos^2 \theta_0\right) + \frac{I_z}{I_1} s \cos \theta_0 \tag{12-133}$$

$$D = \frac{I_z}{I_1} (\dot{\varphi}_0 \cos \theta_0 + s) \tag{12-134}$$

Equation (12-132) follows from Eq. (12-121). Equation (12-133) follows from Eqs. (12-107), (12-108), (12-123), and (12-127). Finally, Eq. (12-134) results from combining Eqs. (12-108), (12-118), and (12-127).

Now $\dot{u}$ is proportional to $\dot{\theta}$. Hence, if one asks the question, at what values of u ($= \cos \theta$) will $\dot{u}$ vanish, this amounts to asking at what values of θ will $\dot{\theta}$ vanish, that is, at what inclinations of the axis of symmetry to the vertical will the motion of this axis reverse itself. Putting $\dot{u}^2 = 0$ in Eq. (12-130), we have the cubic equation

$$f(u) = (A - Bu)(1 - u^2) - (C - Du)^2 = 0 \tag{12-135}$$

From the theory of such algebraic equations this possesses three roots u_1, u_2, and u_3. Now, from Eq. (12-135),

$$\begin{aligned} f(+1) &= -(C - D)^2 & \qquad f(-1) &= -(C + D)^2 \\ f(+\infty) &= +\infty & f(-\infty) &= -\infty \end{aligned} \tag{12-136}$$

[where the lower two follow, since, as u becomes large, $f(u)$ approaches $+Bu^3$]. Thus $f(+1)$ and $f(-1)$ are both negative, whatever the values of C and D. However, $\dot{u}$ is a real quantity, and thus $\dot{u}^2$ is positive except where it is zero. The physical limits of the motion must be somewhere in the range $\theta = 0$ to $\pi/2$, and consequently, somewhere in the range $u = 0$ to $+1$, $f(u)$ must be greater than zero. In view of this, together with (12-136), it is evident that the three roots of (12-135) are real, and $f(u)$ is of the form shown in Fig. 12-20. The roots u_1 and u_2 lie in the interval $u = 0$ to $u = +1$, while u_3 in general must lie to the right of $+1$. Thus, unless u_3 is a coincident root at $+1$, it does not correspond to a real angle. In terms of the angle θ, this means that the symmetry axis Oz of the top will be bobbing back and forth between the two right circular cones of half angles θ_1 and θ_2 (corresponding to u_1 and u_2), meanwhile precessing with the angular velocity

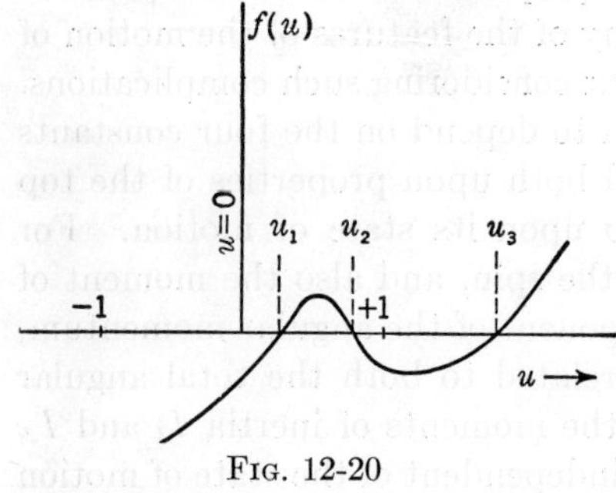

FIG. 12-20

$\dot{\varphi}$ about Oz_0. These cones are shown in Fig. 12-21. Such a bobbing motion is termed a *nutation*.

12-17.[1] **Precession with Nutation.** We may determine the angular velocity $\dot{\varphi}$ from Eq. (12-126). In terms of u this may be written

$$\dot{\varphi} = \frac{C - Du}{1 - u^2} \qquad (12\text{-}137)$$

The sign and magnitude of $\dot{\varphi}$ depend largely on the relative magnitudes of C and D. (In order to simplify the discussion, we restrict ourselves to positive values of C and D. Also we shall limit ourselves to the case when, near $\theta = \theta_1$, that is, near $u = u_1$, $\dot{\varphi}$ is positive.) For increasing u (that is, decreasing θ), $\dot{\varphi}$ decreases in magnitude, meanwhile maintaining the same sign unless, at some value of u less than or equal to u_2 (θ greater than or equal to θ_2), u attains the value C/D. At this point $\dot{\varphi}$ vanishes. If this value of θ is greater than θ_2, $\dot{\varphi}$ is subsequently negative until, in the return motion, u again passes through the value C/D. In Fig. 12-22a, b, and c are shown the three cases of motion of the symmetry axis for positive $\dot{\varphi}$ at $\theta = \theta_1$. The curves depicted are the loci of the intersection of the symmetry axis with a sphere of unit radius with center at the origin (or peg of the top). The horizontal circles are the intersections of the cones of half angles θ_1 and θ_2 with the same sphere. Case (a) is that when $\dot{\varphi}$ never vanishes, case (b) when $\dot{\varphi}$ vanishes at $\theta = \theta_2$, and case (c) when $\dot{\varphi}$ changes sign at $\theta > \theta_2$. In terms of initial conditions applied to $\dot{\varphi}$ at $\theta = \theta_2$, these may be understood to be, respectively, (a) $\dot{\varphi}_0 > 0$, (b) $\dot{\varphi}_0 = 0$, and (c) $\dot{\varphi}_0 < 0$.

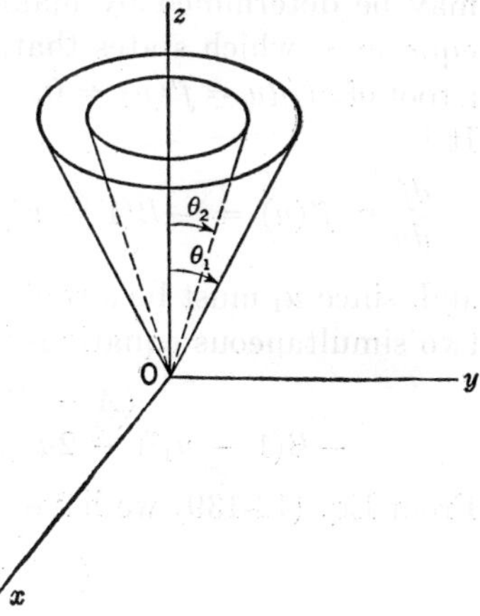

Fig. 12-21

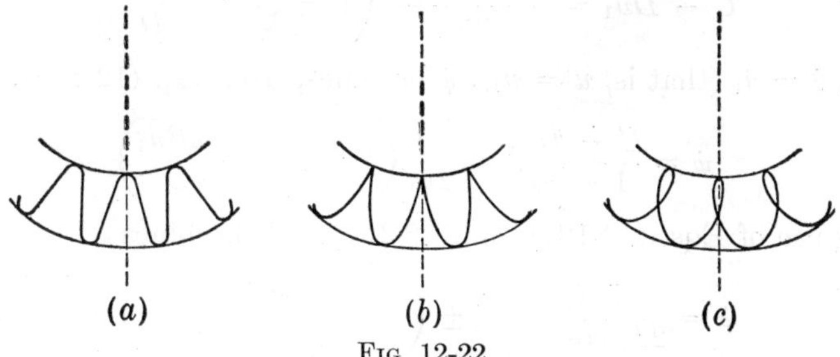

Fig. 12-22

12-18.[1] **Precession without Nutation.** If there is no nutation, the limiting cones in Fig. 12-21 coincide, say at θ_1, and u_1 is a double root of

[1] May be omitted in a first reading.

Eq. (12-135), that is, of $f(u) = 0$. The quantity $C - Du$ in Eq. (12-137) may be determined by making use of the theorem, from the theory of equations, which states that, if u_1 is a double root of $f(u) = 0$, it is also a root of $df/du = f'(u) = 0$. Now df/du may be found from Eq. (12-130). It is

$$\frac{df}{du} \equiv f'(u) = -B(1 - u^2) - 2u(A - Bu) + 2D(C - Du) \quad (12\text{-}138)$$

and, since u_1 must be a root of both $f(u) = 0$ and $f'(u) = 0$, we have the two simultaneous equations in u_1,

$$(A - Bu_1)(1 - u_1^2) - (C - Du_1)^2 = 0 \quad (12\text{-}139)$$

$$-B(1 - u_1^2) - 2u_1(A - Bu_1) + 2D(C - Du_1) = 0 \quad (12\text{-}140)$$

From Eq. (12-139) we have

$$(A - Bu_1) = \frac{(C - Du_1)^2}{1 - u_1^2}$$

This may be substituted into Eq. (12-140) to give, ultimately,

$$(C - Du_1)^2 - \frac{D(1 - u_1^2)}{u_1}(C - Du_1) + \frac{B}{2u_1}(1 - u_1^2)^2 = 0 \quad (12\text{-}141)$$

Equation (12-141) is a quadratic equation in the quantity $C - Du_1$. Accordingly, employing the quadratic formula, we obtain, finally,

$$C - Du_1 = \frac{D(1 - u_1^2)}{2u_1}\left(1 \pm \sqrt{1 - \frac{2Bu_1}{D^2}}\right) \quad (12\text{-}142)$$

Now at $\theta = \theta_1$ (that is, $u = u_1$), $\dot{\varphi}$ becomes, from Eq. (12-137),

$$\dot{\varphi} = \frac{C - Du_1}{1 - u_1^2} = \frac{D}{2u_1}\left(1 \pm \sqrt{1 - \frac{2Bu_1}{D^2}}\right)$$

Making use of Eqs. (12-122) and (12-127), this becomes

$$\dot{\varphi} = \frac{I_z S_1}{2I_1 \cos\theta_1}\left[1 \pm \left(1 - \frac{4mghI_1}{I_z^2 S_1^2}\cos\theta_1\right)^{\frac{1}{2}}\right] \quad (12\text{-}143)$$

which is the angular velocity of precession when no nutation occurs. We shall take the case in which the spin s is large, which means that S_1 $(= \omega_z + s)$ is also large. We write

$$\dot{\varphi} = \frac{I_z S_1}{2I_1 \cos\theta_1}(1 \pm \sqrt{1 - \epsilon})$$

where $\epsilon \equiv (4mghI_1 \cos\theta_1)/I_z^2 S_1^2 \ll 1$. Therefore, employing the binomial theorem, we obtain, neglecting ϵ^2,

$$\dot{\varphi} \simeq \frac{I_z S_1}{2I_1 \cos\theta_1}\left[1 \pm \left(1 - \frac{\epsilon}{2}\right)\right] \quad (12\text{-}144)$$

From Eq. (12-144) we obtain two values of $\dot{\varphi}$ depending on whether we choose the plus or the minus sign. Employing first the latter, we obtain

$$\dot{\varphi}_- \simeq \frac{I_z S_1}{2I_1 \cos \theta_1} \frac{2mghI_1 \cos \theta_1}{I_z^2 S_1^2} = \frac{mgh}{I_z S_1} \tag{12-145}$$

which is a small quantity since S_1 is large. This is the precession which is usually observed. Selecting next the plus sign, we obtain, neglecting ϵ compared with 1,

$$\dot{\varphi}_+ \simeq \frac{I_z S_1}{I_1 \cos \theta_1} \tag{12-146}$$

which is a very large quantity, being of the order of S_1. This is usually not observed.

It should be emphasized that motion with $\dot{\theta} = 0$ and $\dot{\varphi}$ constant (*regular precession*), as described above, is actually the exception rather than the rule, since it can occur only for a very particular set of initial conditions. The apparent absence of the θ motion in practical cases is due to the fact that the spin s is usually very large. In this event it is easy to show that the range of variation of θ will be very small during the motion and thus is often overlooked.

Consider a typical situation in which a top is given a spin s, meanwhile holding the axis of symmetry at a fixed orientation $\theta = \theta_0$ such that $\dot{\theta}_0$ and $\dot{\varphi}_0$ are both zero [an example of case (*b*) in Sec. 12-17]. It is then released, and the symmetry axis starts to fall. The question of interest is: At what value of θ (θ_1 of Sec. 12-17) will the downward motion cease? The value of s is chosen such that the kinetic energy of spin $\frac{1}{2}I_z s^2$ is very large compared with the potential energy mgh. Employing Eqs. (12-132) to (12-134) it is apparent that, for the initial conditions just stated, the constants A, B, C, and D are related as

$$A = B \cos \theta_0 = Bu_0 \qquad C = D \cos \theta_0 = Du_0 \tag{12-147}$$

where u_0 ($= \cos \theta_0$) is the initial value of u. Employing (12-147) and putting $f(u)$ ($= \dot{u}^2$) equal to zero in (12-135), we have

$$f(u) = 0 = (u_0 - u)[B(1 - u^2) - D^2(u_0 - u)] \tag{12-148}$$

One root is, of course, the initial value u_0 (and is to be identified with u_2, mentioned previously), while the remaining roots are given by putting the quadratic quantity within the brackets equal to zero. We obtain the equation

$$u^2 - \frac{D^2}{B} u + \left(\frac{D^2}{B} u_0 - 1\right) = 0 \tag{12-149}$$

for the determination of the remaining roots u_1 and u_3. Employing the quadratic formula,

$$u_{1,3} = \frac{D^2}{2B} \pm \frac{1}{2}\left[\frac{D^4}{B^2} - 4\left(\frac{D^2}{B} u_0 - 1\right)\right]^{\frac{1}{2}} \tag{12-150}$$

Now from the initial conditions, together with Eqs. (12-122) and (12-134), D^2/B is just $I_z{}^2s^2/2I_1mgh$, a quantity which is very much greater than 1. Accordingly we apply the binomial theorem to (12-150). We get

$$u_{1,3} = \frac{D^2}{2B} \pm \frac{D^2}{2B}\left[1 - 2\left(\frac{B}{D^2}u_0 - \frac{B^2}{D^4}\right) - 2\left(\frac{B}{D^2}u_0 - \frac{B^2}{D^4}\right)^2 - \cdots\right]$$

Employment of the plus sign gives us a root much larger than 1, which is thus not of physical interest since it does not represent a real angle. This root is designated u_3. We have approximately

$$u_3 = \frac{D^2}{B} - u_0 = \frac{I_z{}^2s^2}{2I_1mgh} - \cos\theta_0 \tag{12-151}$$

Employing the negative sign and retaining only the first power of the ratio B/D^2, we obtain the remaining root of physical interest. It is

$$u_1 = u_0 - \frac{B}{D^2} + \frac{B}{D^2}u_0{}^2 = u_0 - \frac{B}{D^2}(1 - u_0{}^2) = u_0 - \frac{B}{D^2}\sin^2\theta_0$$

and therefore

$$u_0 - u_1 = \frac{2mghI_1}{I_z{}^2s^2}\sin^2\theta_0 \tag{12-152}$$

Thus the change in $\cos\theta$ is very small indeed, the range of the variation being inversely proportional to s^2.

12-19.[1] **The Sleeping Top.** An interesting case is provided by starting the top with a large spin s, and with its symmetry axis vertical. Thus, initially $u = u_0 = 1$, and Eq. (12-148) becomes

$$f(u) = \dot{u}^2 = (1 - u)^2[B(1 + u) - D^2] \tag{12-153}$$

We are interested in the condition that $\dot{u}$ (and consequently $\dot{\theta}$) will remain zero. The first parentheses on the right provide two coincident roots at $u = +1$. Let these be u_1 and u_2. Accordingly, if $\dot{u}^2$ is to remain zero, the third root u_3, furnished by the quantity within the brackets, must remain greater than 1. However, in that case u_3 will not correspond to a physically possible angle. Hence the only physically possible motion is that provided by the roots u_1 and u_2, with the axis vertical. Thus since

$$u_3 = \frac{D^2}{B} - 1 = \frac{I_z{}^2s^2}{2I_1mgh} - 1$$

we must have

$$s^2 > \frac{4I_1mgh}{I_z{}^2} \tag{12-154}$$

As soon as friction at the peg and air resistance have slowed s down to the point where (12-154) no longer holds, nutation will set in, together

[1] May be omitted in a first reading.

with a precession. As s decreases still further, the nutation will become so pronounced that the top will finally fall over.

12-20. Gyroscopic Action. The Rising Top. Qualitative information concerning the effect of torques applied to a rapidly spinning body can be obtained by neglecting the time derivatives of the ω's and also neglecting the quadratic terms in the $\boldsymbol{\omega}$ components in Eqs. (12-113) to (12-115). This is legitimate since, in Fig. 12-18, $\boldsymbol{\omega}$ is the angular velocity only of the axes, and in general the components of $\boldsymbol{\omega}$ are slowly varying and also are very small compared with the spin s. A convenient example to choose is the case of a top rotating on an imperfectly rough table, and in which the peg is not regarded as a mathematical point but rather has a finite radius of curvature. Such a situation is depicted in an exaggerated fashion in Fig. 12-23. The forces acting are $m\mathbf{g}$ vertically downward through the center of mass O and the normal and frictional reactions $\mathbf{N}$ and $\mathbf{F}$ of the surface upon the peg at point C. Owing to the existence of $\mathbf{F}$, point C is being accelerated and therefore is not a suitable origin of coordinates. Consequently the center of mass is selected as origin, with the axes otherwise identical to the $Oxyz$ axes of Fig. 12-18. Now $\mathbf{F}$, which is parallel to Ox and passes through C, exerts both a y and a z component of torque. $\mathbf{N}$, on the other hand, produces only an x component since it is in the yz plane.

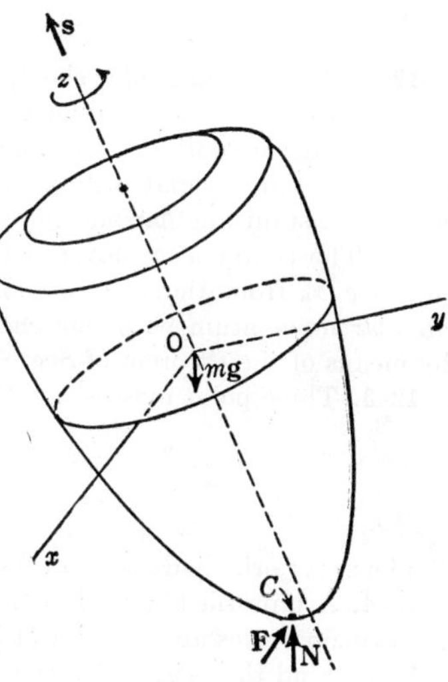

Fig. 12-23

Equations (12-113) and (12-114) become, neglecting such terms as $\dot{\omega}_x$, $\omega_x\omega_z$, and so on,

$$I_z\omega_y s = L_x \qquad -I_z\omega_x s = L_y \tag{12-155}$$

There is no necessity for considering Eq. (12-115) since the effect of L_z is simply to change $\omega_z + s$ and it is therefore of little interest. From (12-155) we have

$$\omega_y = \frac{L_x}{I_z s} = \frac{L_x s}{I_z s^2} \qquad \omega_x = -\frac{L_y}{I_z s} = -\frac{L_y s}{I_z s^2} \tag{12-155a}$$

Accordingly, so far as the x and y components of $\boldsymbol{\omega}$ are concerned, a vector relationship between $\boldsymbol{\omega}$, $\mathbf{L}$, and $\mathbf{s}$ may be written,

$$\boldsymbol{\omega} = \frac{1}{I_z s^2}(\mathbf{s} \times \mathbf{L}) \tag{12-156}$$

Equation (12-156) provides a useful rule of thumb for ascertaining the qualitative behavior of a rotating body under the action of applied torques. Thus, in Fig. 12-23, **F** produces a torque in the $+y$ direction. Application of the right-hand rule of vector products to Eq. (12-156) shows that this torque is such as to cause the symmetry axis of the top to *rise*, a familiar phenomenon indeed. (**F** also provides a torque in the $-z$ direction which tends to slow the spinning motion.) The force **N** produces a $+x$ component of torque, which, following Eq. (12-156), gives rise to a component of $\boldsymbol{\omega}$ in the y direction. In view of Eq. (12-107) this is evident as a precession $\dot{\varphi}$.

The above considerations are applicable, as well, to the case of a gyroscope. In that instrument, the rotor is so mounted in gimbal rings that the center of mass is fixed in space.

Problems

12-1. A thin rectangular sheet, of length b and width a, is rotating about one of its diagonals with a uniform angular velocity ω, the axis being fixed in space. Find the direction and magnitude of the angular momentum.

12-2. A thin circular disk of radius a and mass m lies in the xy plane and rotates with a constant angular velocity ω about an axis through its center parallel to the z axis. The center also moves with a uniform linear velocity v parallel to and at a distance x_0 from the y axis. Find, by integration, the three components of the angular momentum with reference to the origin. Use Eq. (12-3). Solve this also by means of the theorem of Sec. 8-6.

12-3. Three point masses, rigidly connected together, are situated as follows:

$$4m \text{ at } (a,-a,a)$$
$$3m \text{ at } (-a,a,a)$$
$$2m \text{ at } (a,a,a)$$

Find a set of principal axes and the moments of inertia about these axes.

12-4. A uniform lamina has the shape of a right triangle of sides a and b. Find the principal axes at the point of intersection of a and b.

12-5. Find the principal axes at the center of a quadrant of an ellipse of semiaxes a and b (that is, at the intersection of a and b).

12-6. Show that the principal axes at any point P of a homogeneous cube are the straight line joining P to the center of mass O and any two mutually orthogonal straight lines meeting at P, both of which are perpendicular to OP.

12-7. A uniform elliptic lamina of mass M has semiaxes a and b. Find the moment of inertia about a diameter making an angle θ with the major axis $2a$.

12-8. Find the moment of inertia of a rectangular lamina of mass M and sides a and b about a diagonal.

12-9. A table rotates with an angular velocity ω about the polar axis. A billiard ball of radius a and mass m is set rolling on the table. If the table is perfectly rough find the equation of the path of the ball relative to a set of axes at rest. (Hint: Use nonrotating axes.)

12-10. A homogeneous sphere of mass m moves upon a perfectly rough horizontal plane under the action of a center of force situated at a point on the plane. The force is directly proportional to the distance from the force center, the constant of proportionality being k. Initially the sphere is at a distance b from the attracting center and is given a velocity $(20kb/7m)^{\frac{1}{2}}$ in a direction normal to the line between the force center and the center of mass of the sphere. Find the equation, with time eliminated, of the path of the sphere. (Hint: Use nonrotating axes.)

12-11. Show by the use of Euler's equations that, for any free motion of a lamina, the component of the angular velocity in the plane of the lamina is a constant in magnitude.

12-12. A thin uniform disk of radius a and mass m is rotating with a uniform angular velocity ω about a fixed axis passing through its center but inclined at an angle α to the axis of symmetry. Find the direction and magnitude of the torque which is exerted on the bearings.

12-13. A homogeneous circular disk of mass m and radius a is set rotating with an angular velocity ω about an axis passing through the center of the disk and making an angle α with the normal. The disk is suddenly released.

a. Deduce the period of the subsequent motion of the axis of rotation relative to the disk.

b. Find the period of precession of the axis of rotation about the invariable line.

c. During the time required for the axis of revolution to precess through an angle 2π in space, how much has the body rotated about this axis?

12-14. A square sheet is constrained to rotate with an angular velocity ω about an axis passing through its center of mass and making an angle α with the axis through the center of mass and normal to the sheet. At the instant the axis of rotation lies in the plane determined by this axis of symmetry and a diagonal, the body is released.

a. Find the rate at which the axis of symmetry precesses about the invariable line.

b. During the time required for the instantaneous axis of rotation to rotate through an angle $\pi/2$ about the invariable line, find the angle through which the body has rotated about the axis of symmetry.

12-15. A rigid body rotates freely about a principal axis. From Euler's equations show that this motion is stable when it is about an axis of largest or smallest moment of inertia but unstable about the axis of intermediate moment. Find the periods of oscillation for the cases of stability.

12-16. A sphere is rotating in a manner given by the vector sum of two variable angular velocities $\boldsymbol{\omega}$ and $\mathbf{s}$, both axes of which pass through the center of mass. The angular momentum $\mathbf{J}$ and the torque $\mathbf{L}$ are defined with respect to a nonrotating system whose origin is at the center of mass. Transform the equation $\dot{\mathbf{J}} = \mathbf{L}$ to a system of axes rotating with the velocity $\boldsymbol{\omega}$ and having the same origin. If ω_x, ω_y, ω_z, s_x, s_y, s_z, are the components of $\boldsymbol{\omega}$ and $\mathbf{s}$ in this rotating system, write the components of the transformed equation.

12-17. A rigid body (of revolution) is rotating freely in space. If α is the angle which the principal axis Oz makes with the invariable line Oz_0, show that the angular velocity with which the zOz_0 plane is turning about Oz_0 is

$$\frac{1}{\sin^2 \alpha}\left(\frac{2T}{J} - \frac{J}{I_z}\cos^2 \alpha\right)$$

in which T is the kinetic energy of rotation, I_z is the moment of inertia about the z axis, and J is the angular momentum. Body need not be one of revolution.

12-18. A circular disk of radius a and mass m is spinning with an angular velocity s about its axis of symmetry, which is tilted at an angle α to the vertical, the lowest point of the disk being in contact with a smooth horizontal table. Gravity is acting vertically downward. Show that if the disk is executing a steady precession n, with no nutation, n may be determined from the equation

$$an^2 \sin \alpha \cos \alpha + 2asn \sin \alpha + 4g \cos \alpha = 0$$

12-19. A sphere of radius a rolls without sliding on a perfectly rough vertical plane. Gravity is acting vertically downward. The plane rotates with a uniform angular velocity s about a vertical axis which lies in the plane itself. If the ball is initially at rest with respect to the plane, find the vertical distance of the ball from its starting point, expressed as a function of time. Show that the maximum distance the ball is able to descend is $5g/s^2$. Assume that the ball is constrained to remain in contact with the plane.

12-20. In the previous problem find the horizontal distance of the ball from the axis of rotation as a function of time.

12-21. Show that, for the fast-spinning top with its symmetry axis initially released from rest at a small angle with the vertical, the approximate angular frequency of nutation is $I_z s/I_1$, where s is the spin angular velocity and I_z and I_1 are as defined in Sec. 12-14.

12-22. Examination of the motion pictures of a football game shows that in a certain punt the football wobbled in its flight. The long axis of symmetry described a cone of semiangle 10° twice per second. Find the component of the total angular velocity (with respect to a nonrotating coordinate system) along the long axis of symmetry.

12-23. An S shaped crankshaft is mounted on bearings at the points B and B'. It is constructed of two semicircles, each of radius a and mass $m/2$. (*a*) Compute the moments and products of inertia relative to the axes x, y, z shown (y into the paper). (*b*) The crankshaft is accelerated at a constant rate from rest to angular velocity ω_1,

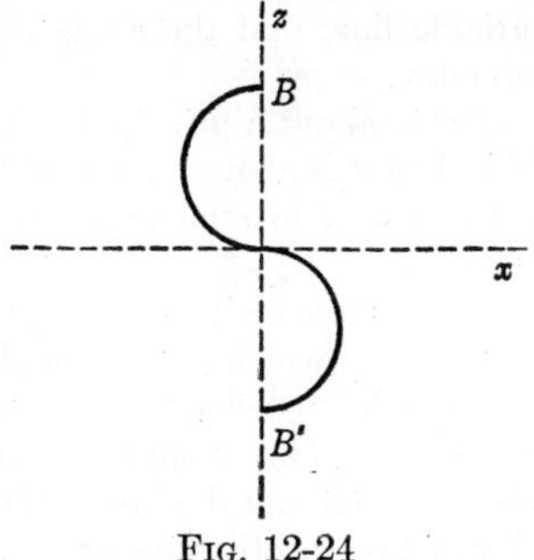

FIG. 12-24

about the axis through the points B and B', in a time T. What torque (including that of the bearings) must be applied to the crankshaft to produce this acceleration? (*c*) How much work must the external torques do in bringing the system to its final angular velocity? (See Fig. 12-24.)

12-24. A homogeneous solid of mass m is in the shape of a rectangular cube of sides $2a$, $2a$, and $4a$.

a. Find a set of principal axes at the mass center, and compute the principal moments of inertia.

b. Draw a picture of the inertia ellipsoid. If the body has an angular velocity $\boldsymbol{\omega}$ along a diagonal, show on your picture the direction of **J**. What is the magnitude of **J**?

12-25. In the previous problem consider the body to be supported at its fixed mass center but free to turn about it in any manner.

a. Write an expression for the kinetic and potential energies in terms of the Euler angles. Find also the equations of motion in terms of these angles.

b. Determine the motion, assuming that $\boldsymbol{\omega}$ is initially directed along a diagonal. Find the angular velocity of the long axis of symmetry (through the mass center and parallel to the edges of length $4a$) about the invariable line.

CHAPTER 13

GENERALIZED COORDINATES

In this chapter we shall be concerned with writing the equations of motion in an arbitrary coordinate system. We have seen that when other than rectangular coordinates are employed the acceleration may be complicated. It is possible, by employing Newton's equations as a basis, to develop a generalized equation of motion. This equation, called *Lagrange's equation*, provides a quick method of computing the inertial reaction when arbitrary coordinates are used. Much of this chapter will be devoted to solving problems by the use of Lagrange's equation and other generalized coordinate methods. In most of the theoretical development we shall employ the symbol q to represent an arbitrary coordinate x, θ, and so on.

13-1. Holonomic and Nonholonomic Constraints. Degrees of Freedom. It was pointed out in Chap. 1 that, in order to describe completely the motion of a given system, a minimum number of distinct coordinates is required. This minimum number will depend upon the nature of the system. In certain cases not all these coordinates are independent in the sense that it may be that one or more of the coordinates cannot be varied in an arbitrary manner and still leave the others unaltered. Such a situation is represented by the simple example of a ball or a disk rolling over a perfectly rough plane, no slipping being permitted. Take, for ease of discussion, the case of a circular disk of radius a rolling over a horizontal plane in a manner such that the plane of the disk is at all times vertical. At a given instant the configuration is as shown in Fig. 13-1, with the instantaneous direction of motion at an angle θ with the x axis. Let the angle φ describe the rotation of the disk about its center of mass. The translational motion of the disk may be described by the coordinates x and y of the center of the disk.

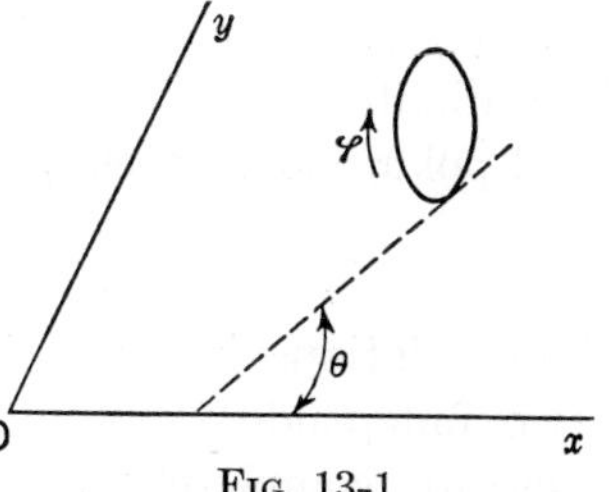

FIG. 13-1

(The constraints preventing translation in the z direction and holding the plane of the disk vertical are, respectively, $z = a$ and $\alpha = \pi/2$, where α is the angle between the plane of the disk and the xy plane.) The coordinates x, y, and φ are not all independent but are related by the conditions

$$a\,d\varphi \cos\theta = dx \qquad a\,d\varphi \sin\theta = dy \tag{13-1}$$

The conditions (13-1) are *not integrable* since there are no geometrical conditions relating θ to x and y. Such a constraint, in which the differential form is not integrable, is called a *nonholonomic constraint.* Any system which is subject to one or more nonholonomic constraints is called a *nonholonomic system.* If the plane were imperfectly rough, slipping being therefore permitted, Eqs. (13-1) would no longer be necessary and x, y, and φ all could be independently varied. In this instance the constraint becomes *holonomic,* and the system is then said to be holonomic. The rolling constraint becomes holonomic, also, if the disk is constrained to roll along a prescribed curve. For such a situation let the running coordinate measuring the length traversed along the path be s. Conditions (13-1) are then replaced by a single equation

$$a\,d\varphi = ds \tag{13-2}$$

which integrates readily to

$$a\varphi = s + \text{const} \tag{13-3}$$

Simple examples of holonomic constraints involving a single particle are those where the motion is confined to a single curve or surface. Suppose the motion of the particle is confined to a spherical surface of radius a with center at the origin. In rectangular coordinates x, y, z, the equation of constraint is

$$x^2 + y^2 + z^2 = a^2 \tag{13-4}$$

Thus, although the coordinates cannot be independently varied, it is clear that since the displacements are related by

$$x\,dx + y\,dy + z\,dz = 0 \tag{13-5}$$

the differential relationship (13-5) is integrable [it integrates to (13-4)]. It was indicated in Sec. 1-2 that the imposition of such a simple constraint as that expressed by (13-4) and (13-5) reduces the number of distinct coordinates required to describe the system completely. Hence, if spherical polar coordinates r, θ, and φ are employed, Eq. (13-4) reduces to

$$r = a = \text{const} \tag{13-6}$$

This relationship serves to eliminate the radial coordinate r, leaving the two independent angle coordinates θ and φ. These are sufficient to describe completely the motion of the particle if it is confined to the surface.

It will now be perceived that in previous chapters a few problems were considered involving nonholonomic constraints (see, for example, Probs.

12-9, 12-10, and 12-19). However, in the present chapter only holonomic systems will be considered, since the Lagrangian treatment of nonholonomic constraints involves a procedure somewhat above the level of this text.

It is useful at this point to define a term which is widely used in connection with a dynamical system. This is its number of *degrees of freedom*. Let n distinct coordinates $q_1, \ldots, q_n$ be required to specify a given system. If the system is holonomic, the arbitrary infinitesimal increments $dq_1, \ldots, dq_n$ will define a possible displacement of the system. However, if the system is nonholonomic, a certain number of nonintegrable relations must be satisfied among the increments in order that they may describe a possible displacement. Let these relations be m in number. In the latter case the system is said to possess $n - m$ degrees of freedom, while in the former case the number of degrees of freedom is n. It is apparent that, for a holonomic system, the distinct coordinates which are required to describe the system are all independent (in the sense of the first paragraph of this section), whereas for a nonholonomic system this is not the case.

In the example of the disk cited above, the system possesses two degrees of freedom if no sliding is permitted. However, if the surface is imperfectly rough, sliding may occur and the system then possesses four degrees of freedom. For the case given by Eq. (13-2), in which the motion is confined to rolling along a given fixed curve, there is but one degree of freedom.

13-2. Kinetic Energy in Curvilinear Coordinates. In rectangular coordinates the element of arc ds (*line element*) is given by

$$ds^2 = dx^2 + dy^2 + dz^2 \tag{13-7}$$

In the case of an element of arc lying in a spherical surface of radius r, this becomes (see Appendix 1)

$$ds^2 = r^2 \sin^2 \theta \, d\varphi^2 + r^2 \, d\theta^2 \tag{13-8}$$

From (13-8) we may form the kinetic energy T of a particle of mass m constrained to move on the surface by dividing by the square of the time element dt and multiplying by $m/2$, as

$$T = \frac{m}{2} \dot{s}^2 = \frac{m}{2} (r^2 \sin^2 \theta \, \dot{\varphi}^2 + r^2 \dot{\theta}^2) \tag{13-9}$$

In Eq. (13-9), T is a homogeneous quadratic function of the velocities $\dot{\theta}$ and $\dot{\varphi}$. However, T is not only a function of the velocities; it is also a function of the coordinates. This usually is true of the kinetic energy when expressed in terms of coordinate systems other than rectangular ones. We now show that T will always have this general form for a particle constrained to move on any surface, provided that surface does not vary with time (that is, provided that the equation of the surface does not contain the time explicitly). The parametric equations of an arbitrary surface may be expressed as

$$x = x(q_1,q_2) \qquad y = y(q_1,q_2) \qquad z = z(q_1,q_2) \tag{13-10}$$

provided that the form of the surface is not changing with time. The differentials dx, dy, and dz may be written

$$dx = \frac{\partial x}{\partial q_1} dq_1 + \frac{\partial x}{\partial q_2} dq_2 \qquad dy = \frac{\partial y}{\partial q_1} dq_1 + \frac{\partial y}{\partial q_2} dq_2, \text{ etc.} \tag{13-11}$$

Squaring these and adding the results, we find ds^2 to be

$$ds^2 = E_1\, dq_1{}^2 + E_2\, dq_2{}^2 + 2E_{12}\, dq_1\, dq_2 \tag{13-12}$$

where

$$E_1 = \left(\frac{\partial x}{\partial q_1}\right)^2 + \left(\frac{\partial y}{\partial q_1}\right)^2 + \left(\frac{\partial z}{\partial q_1}\right)^2 \tag{13-13}$$

$$E_2 = \left(\frac{\partial x}{\partial q_2}\right)^2 + \left(\frac{\partial y}{\partial q_2}\right)^2 + \left(\frac{\partial z}{\partial q_2}\right)^2 \tag{13-14}$$

$$E_{12} = \frac{\partial x}{\partial q_1}\frac{\partial x}{\partial q_2} + \frac{\partial y}{\partial q_1}\frac{\partial y}{\partial q_2} + \frac{\partial z}{\partial q_1}\frac{\partial z}{\partial q_2} \tag{13-15}$$

Thus the kinetic energy of a particle of mass m has the form

$$T = \frac{m}{2}\dot{s}^2 = \frac{m}{2}(E_1\dot{q}_1{}^2 + E_2\dot{q}_2{}^2 + 2E_{12}\dot{q}_1\dot{q}_2) \tag{13-16}$$

which is a homogeneous quadratic function of the velocities $\dot{q}_1$ and $\dot{q}_2$. The quantities E_1, E_2, E_{12} are functions of the coordinates q_1 and q_2 and are independent of the velocities.

At this point it is of interest to inquire as to the form of T if the surface depends explicitly upon the time. Equations (13-10) would then have the form

$$x = x(q_1,q_2,t) \qquad y = y(q_1,q_2,t) \qquad z = z(q_1,q_2,t) \tag{13-17}$$

These give rise to terms, added to Eq. (13-16), such as

$$\frac{m}{2}(2K\dot{q}_1 + 2L\dot{q}_2 + M) \tag{13-18}$$

in which

$$K = \frac{\partial x}{\partial q_1}\frac{\partial x}{\partial t} + \frac{\partial y}{\partial q_1}\frac{\partial y}{\partial t} + \frac{\partial z}{\partial q_1}\frac{\partial z}{\partial t} \tag{13-19}$$

$$L = \frac{\partial x}{\partial q_2}\frac{\partial x}{\partial t} + \frac{\partial y}{\partial q_2}\frac{\partial y}{\partial t} + \frac{\partial z}{\partial q_2}\frac{\partial z}{\partial t} \tag{13-20}$$

$$M = \left(\frac{\partial x}{\partial t}\right)^2 + \left(\frac{\partial y}{\partial t}\right)^2 + \left(\frac{\partial z}{\partial t}\right)^2 \tag{13-21}$$

The quantities K, L, and M, as well as E_1, E_2, and E_{12}, now may be functions of the coordinates q_1, q_2 and also of the time t. Thus we see that, if the form of the surface depends upon the time, the kinetic energy, expressed in terms of the coordinates q_1 and q_2, is no longer a *homogeneous quadratic function* of the velocities but contains terms linear in the velocities and also terms independent of them.

It is interesting to examine, geometrically, the element of arc ds in a given surface. Consider the coordinates q_1 and q_2 of a point moving in the surface. If q_1 is held constant, say to the value c_1, and q_2 allowed to vary, the point will describe the curve $q_1 = c_1$. Similarly if q_2 is held

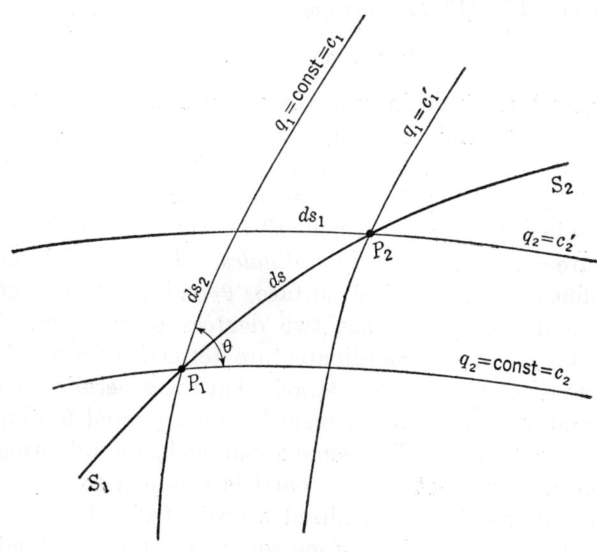

FIG. 13-2

to the constant value c_2 and q_1 permitted to vary, the point will follow the curve $q_2 = c_2$. The two lines $q_1 = c_1$ and $q_2 = c_2$ may be taken to be coordinate lines (see Fig. 13-2), and any point may be considered as the intersection of two such lines. In Fig. 13-2, a point is moving in the surface along the path S_1S_2. During a time dt it moves a distance ds from P_1 to P_2. Clearly, from the infinitesimal parallelogram, we have

$$ds^2 = ds_1^2 + ds_2^2 + 2\,ds_1\,ds_2 \cos\theta \tag{13-22}$$

in which

$$ds_1^2 = E_1\,dq_1^2 \qquad ds_2^2 = E_2\,dq_2^2 \qquad ds_1\,ds_2 \cos\theta = E_{12}\,dq_1\,dq_2 \tag{13-23}$$

The factors E_1, E_2, and E_{12} are as defined above in Eqs. (13-13) to (13-15) and are such that the quantities ds_1^2, and so on, possess the dimensions of length squared. It is evident, from Eq. (13-22), that the square of the velocity will have the form

$$\dot{s}^2 = \dot{s}_1^2 + \dot{s}_2^2 + 2\dot{s}_1\dot{s}_2 \cos\theta \tag{13-24}$$

which is homogeneous quadratic in the velocities $\dot{s}_1$ and $\dot{s}_2$. However, the last term is a cross term in $\dot{s}_1$ and $\dot{s}_2$, which we perceive to be present because the lines $q_1 = c_1$ and $q_2 = c_2$ do not intersect orthogonally. If q_1 and q_2 are orthogonal coordinates, $\cos\theta = 0$ and $\dot{s}^2$ reduces to

$$\dot{s}^2 = \dot{s}_1^2 + \dot{s}_2^2 \tag{13-25}$$

in which the cross term has disappeared. In terms of q_1 and q_2 and their time derivatives, Eq. (13-25) becomes

$$\dot{s}^2 = E_1\dot{q}_1^2 + E_2\dot{q}_2^2 \tag{13-26}$$

13-3. Generalized Coordinates. Lagrange's Equations for a Single Particle. In a holonomic system, if the equations of constraint have been employed to reduce to a minimum the number of coordinates required to describe the system (for example, the elimination of r in the case of the spherical surface in Sec. 13-1), the remaining independent coordinates are called *generalized coordinates*. Thus, in the example of a particle confined to a spherical surface, θ and φ are the generalized coordinates, and the system has two degrees of freedom. The time derivatives of generalized coordinates are termed *generalized velocities*. It is a feature of the Lagrangian methods that the generalized coordinates and generalized velocities are all regarded on an equal footing as independent variables. This will become apparent in the following pages.

Let us consider the motion of a particle confined to a given surface for which the appropriate generalized coordinates q_1 and q_2 have been recognized. The functional relations connecting the coordinates x, y, z of the particle with q_1 and q_2 [for example, Eqs. (6), Appendix 4] are simply the parametric equations of the surface. These are

$$x = x(q_1,q_2,t) \qquad y = y(q_1,q_2,t) \qquad z = z(q_1,q_2,t) \tag{13-27}$$

in which the explicit dependence upon the time here indicates that the form of the surface may be varying with time. It is desired, for example, to obtain from Newton's equations involving x, y, and z, the equation of motion with respect to the generalized coordinate q_1. The Newtonian equations of motion are

$$m\ddot{x} = X + R_x \qquad m\ddot{y} = Y + R_y \qquad m\ddot{z} = Z + R_z \tag{13-28}$$

in which m is the mass of the particle, assumed constant, X, Y, and Z are the rectangular components of the applied force, and R_x, R_y, and R_z are the components of the force of constraint. If we are given the relationships (13-27), we are able to calculate the partial derivatives such as $\partial x/\partial q_1$, $\partial y/\partial q_1$, $\partial z/\partial q_1$. Let us multiply the first of Eqs. (13-28) by $\partial x/\partial q_1$, the second by $\partial y/\partial q_1$, and the third by $\partial z/\partial q_1$. Adding the results, we obtain

$$m\left(\ddot{x}\frac{\partial x}{\partial q_1} + \ddot{y}\frac{\partial y}{\partial q_1} + \ddot{z}\frac{\partial z}{\partial q_1}\right) = \left(X\frac{\partial x}{\partial q_1} + Y\frac{\partial y}{\partial q_1} + Z\frac{\partial z}{\partial q_1}\right) + \left(R_x\frac{\partial x}{\partial q_1} + R_y\frac{\partial y}{\partial q_1} + R_z\frac{\partial z}{\partial q_1}\right) \qquad (13\text{-}29)$$

Forming the total derivative, with respect to time, of the first of Eqs. (13-27), we obtain

$$\dot{x} = \frac{\partial x}{\partial q_1}\dot{q}_1 + \frac{\partial x}{\partial q_2}\dot{q}_2 + \frac{\partial x}{\partial t} \qquad (13\text{-}30)$$

Now $\dot{q}_1$ occurs explicitly only in the first term on the right. Therefore differentiating (13-30) partially with respect to $\dot{q}_1$ gives us

$$\frac{\partial \dot{x}}{\partial \dot{q}_1} = \frac{\partial x}{\partial q_1} \qquad (13\text{-}31)$$

Thus the first term within the parentheses on the left side of Eq. (13-29) may be written $\ddot{x}\,\partial\dot{x}/\partial\dot{q}_1$. Similar expressions for the second and third terms may be obtained. For the sake of brevity in writing let us consider only the first term. We may write

$$\ddot{x}\frac{\partial x}{\partial q_1} = \ddot{x}\frac{\partial \dot{x}}{\partial \dot{q}_1} = \frac{d}{dt}\left(\dot{x}\frac{\partial \dot{x}}{\partial \dot{q}_1}\right) - \dot{x}\frac{d}{dt}\left(\frac{\partial \dot{x}}{\partial \dot{q}_1}\right) \qquad (13\text{-}32)$$

which may be checked merely by carrying out the indicated operations. The right side of (13-32) may be further rewritten

$$\frac{d}{dt}\left(\dot{x}\frac{\partial \dot{x}}{\partial \dot{q}_1}\right) - \dot{x}\frac{d}{dt}\left(\frac{\partial x}{\partial q_1}\right) \qquad (13\text{-}33)$$

Now since $\partial x/\partial q_1$ is, in general, an explicit function of q_1, q_2, t, the last term on the right side of Eq. (13-33) becomes

$$-\dot{x}\left[\frac{\partial}{\partial q_1}\left(\frac{\partial x}{\partial q_1}\right)\dot{q}_1 + \frac{\partial}{\partial q_2}\left(\frac{\partial x}{\partial q_1}\right)\dot{q}_2 + \frac{\partial}{\partial t}\left(\frac{\partial x}{\partial q_1}\right)\right]$$

Accordingly (13-33) assumes the form

$$\frac{d}{dt}\left(\dot{x}\frac{\partial \dot{x}}{\partial \dot{q}_1}\right) - \dot{x}\left(\frac{\partial^2 x}{\partial q_1{}^2}\dot{q}_1 + \frac{\partial^2 x}{\partial q_2\,\partial q_1}\dot{q}_2 + \frac{\partial^2 x}{\partial t\,\partial q_1}\right) \qquad (13\text{-}34)$$

The student may easily verify, employing Eq. (13-30), that the quantity within the second parentheses in Eq. (13-34) is just the partial of $\dot{x}$

with respect to q_1. Hence Eq. (13-32) may now be written

$$\ddot{x}\,\frac{\partial x}{\partial q_1} = \frac{d}{dt}\left(\dot{x}\,\frac{\partial \dot{x}}{\partial \dot{q}_1}\right) - \dot{x}\,\frac{\partial \dot{x}}{\partial q_1} = \frac{d}{dt}\left[\frac{\partial}{\partial \dot{q}_1}\left(\frac{1}{2}\,\dot{x}^2\right)\right] - \frac{\partial}{\partial q_1}\left(\frac{1}{2}\,\dot{x}^2\right) \tag{13-35}$$

The terms in y and z on the left of Eq. (13-29) may be handled in a similar manner. If this is done, the left side of Eq. (13-29), following Eq. (13-35), becomes

$$m\left(\ddot{x}\,\frac{\partial x}{\partial q_1} + \ddot{y}\,\frac{\partial y}{\partial q_1} + \ddot{z}\,\frac{\partial z}{\partial q_1}\right) = \frac{1}{2}\,m\,\frac{d}{dt}\left[\frac{\partial}{\partial \dot{q}_1}\,(\dot{x}^2 + \dot{y}^2 + \dot{z}^2)\right] - \frac{1}{2}\,m\,\frac{\partial}{\partial q_1}\,(\dot{x}^2+\dot{y}^2+\dot{z}^2) \tag{13-36}$$

But the kinetic energy T is $(m/2)(\dot{x}^2 + \dot{y}^2 + \dot{z}^2)$, and thus, if m is constant, the right side of Eq. (13-36) is just

$$\frac{d}{dt}\frac{\partial T}{\partial \dot{q}_1} - \frac{\partial T}{\partial q_1} \tag{13-37}$$

Now T can be expressed as an explicit function of all the q's and $\dot{q}$'s and, in the present case, also of the time t. Thus the left sides of Eqs. (13-28) may be expressed in terms of the q's, their time derivatives, and the time.

Let us now examine the right side of Eq. (13-29). The significance of the first set of terms can be perceived by imagining a small *virtual displacement* δq_1, in q_1, and meanwhile holding q_2 and t constant. This is possible since q_1 and q_2 are completely independent. In such a displacement x, y, and z are changed to

$$x + \frac{\partial x}{\partial q_1}\,\delta q_1 \qquad y + \frac{\partial y}{\partial q_1}\,\delta q_1 \qquad z + \frac{\partial z}{\partial q_1}\,\delta q_1 \tag{13-38}$$

Accordingly, the quantity

$$X\,\frac{\partial x}{\partial q_1}\,\delta q_1 + Y\,\frac{\partial y}{\partial q_1}\,\delta q_1 + Z\,\frac{\partial z}{\partial q_1}\,\delta q_1 = \left(X\,\frac{\partial x}{\partial q_1} + Y\,\frac{\partial y}{\partial q_1} + Z\,\frac{\partial z}{\partial q_1}\right)\delta q_1 \tag{13-39}$$

is the *virtual work* done by the applied force, of components X, Y, and Z, during the virtual displacement δq_1. The quantity within the parentheses is denoted by the symbol Q_1 and is called the *generalized force* corresponding to the coordinate q_1. Clearly it need not have the dimensions of a force. Only if q_1 possesses the dimension of length will Q_1 have the dimensions of force. However, the requirement is that the product $Q_1\,\delta q_1$ must have the dimensions of work (thus Q_1 is a torque if q_1 is an angle).

If R is perpendicular to the surface (smooth surface), the second set of terms on the right side of (13-29) will vanish, since any displacement δq_1 or

δq_2 must lie within the surface and consequently the work done by this quantity must vanish. Alternatively this can be seen since, if q_1 is a length, $\partial x/\partial q_1$, $\partial y/\partial q_1$, and $\partial z/\partial q_1$ are the direction cosines of the element dq_1 with the x, y, and z axes (or a common factor times the direction cosines if q_1 does not have the dimensions of length). Also R_x/R, and so on, are the direction cosines of R. Now, if a, b, c and α, β, γ are the direction cosines of two intersecting vectors, a well-known expression from analytic geometry (see Prob. 1-20) states that the angle between the directions of the two vectors is related to these quantities by

$$\cos\theta = a\alpha + b\beta + c\gamma \tag{13-40}$$

Accordingly the expression

$$R_x\frac{\partial x}{\partial q_1} + R_y\frac{\partial y}{\partial q_1} + R_z\frac{\partial z}{\partial q_1} = R\left(\frac{R_x}{R}\frac{\partial x}{\partial q_1} + \frac{R_y}{R}\frac{\partial y}{\partial q_1} + \frac{R_z}{R}\frac{\partial z}{\partial q_1}\right) \tag{13-41}$$

is some factor times the cosine of the angle included between δq_1 and R. Since R is normal to δq_1, this must be zero. Hence the effect of the reaction of a smooth surface (whether or not the form of the surface varies with time) vanishes from the equations of motion for generalized coordinates. (However, if the surface is not smooth, R is not in general normal to the surface. There may be frictional components present which are not zero. These are obtained by multiplying the reaction normal to the surface by the coefficient of friction.) Thus, by employing (13-37), and since in Eq. (13-29) the first set of terms on the right is Q_1 and the second is zero, Eq. (13-29) may be written

$$\frac{d}{dt}\left(\frac{\partial T}{\partial \dot{q}_1}\right) - \frac{\partial T}{\partial q_1} = Q_1 \tag{13-42}$$

Equation (13-42) is called *Lagrange's equation*, in the *fundamental form*, for the coordinate q_1. Equation (13-42) can be written somewhat differently. If Q_1 is derivable from a potential energy V, that is, if

$$Q_1 = -\frac{\partial V}{\partial q_1} \tag{13-43}$$

Eq. (13-42) becomes

$$\frac{d}{dt}\left(\frac{\partial T}{\partial \dot{q}_1}\right) - \frac{\partial T}{\partial q_1} = -\frac{\partial V}{\partial q_1} \tag{13-44}$$

or

$$\frac{d}{dt}\left(\frac{\partial T}{\partial \dot{q}_1}\right) - \frac{\partial}{\partial q_1}(T - V) = 0 \tag{13-45}$$

Denoting $T - V$ by the symbol L, we have, restricting ourselves, for the moment, to cases in which V is not a function of the velocities (that is, $\partial V/\partial \dot{q} = 0$),

$$\frac{d}{dt}\left(\frac{\partial L}{\partial \dot{q}_1}\right) - \frac{\partial L}{\partial q_1} = 0 \tag{13-46}$$

The quantity L is called the *Lagrangian function* for the system under consideration. The problem is thereby reduced to determining the Lagrangian function L and thence to writing the equations of motion, from Eq. (13-46), for each coordinate q. This usually means finding the kinetic and potential energies. Equation (13-46) is the form usually referred to as *Lagrange's equation.*

Although, in writing Eq. (13-46), we have taken V to be a function of the coordinates only, this restriction is not strictly necessary. For example, in the step from Eq. (13-45) to (13-46), V may be such a function of the velocities that

$$\frac{d}{dt}\left(\frac{\partial V}{\partial \dot{q}_1}\right) - \frac{\partial V}{\partial q_1} = Q_1 \tag{13-47}$$

and Eq. (13-46) will still be true, with L defined as $T - V$. Examples of a potential energy being such a function of the velocity are to be found in electrodynamics.

Although in most cases it will suffice, as above, to define L to be the quantity $T - V$, it seems that more generality is to be achieved by regarding Eq. (13-46) itself to be the definition of L. Hence the Lagrangian function is that expression which, when substituted in Eq. (13-46), provides the correct equation of motion.

It is well to give a word of caution about the use of Lagrange's equations. A careful inspection of the development in this section will reveal that the derivation of Lagrange's equations from Newton's equations is a geometrical one. Some insight can be realized by noticing that, if q has the dimensions of length, the procedure consists in taking the components, along the direction of q increasing, of both sides of all three equations of motion in rectangular coordinates and then adding them together. Thus, the left sides of the equations furnish the total q component of the inertial reaction, and the right sides furnish the *q component* of the applied force. Accordingly we must recognize that the limitations as to the applicability of Newton's equations (for example, T and V must be defined with respect to a nonaccelerated coordinate system) will also obtain for Lagrange's equations.

Finally it should be apparent that the use of Lagrange's equations does not require the presence of constraints. One very useful application is to employ (13-42) and (13-46) in order to obtain the form of the inertial reaction for arbitrary sets of coordinates (spherical polar coordinates, parabolic coordinates, and so on) for the free motion of a particle.

Example 13-1. A particle is moving in a plane, subject to an inverse-square attractive force. It is desired to obtain the equations of motion by the Lagrangian method.

Let us choose plane polar coordinates r, θ for the problem. In rectangular coordinates we have

$$T = \frac{m}{2}(\dot{x}^2 + \dot{y}^2)$$
$$V = -\frac{k}{(x^2 + y^2)^{\frac{1}{2}}} = -\frac{k}{r} \tag{13-48}$$

Now the relations among x, y, r, θ are

$$x = r\cos\theta \qquad y = r\sin\theta$$

from which

$$\dot{x} = \dot{r}\cos\theta - r\dot{\theta}\sin\theta \qquad \dot{y} = \dot{r}\sin\theta + r\dot{\theta}\cos\theta$$

Thus we have

$$\dot{x}^2 + \dot{y}^2 = \dot{r}^2 + r^2\dot{\theta}^2 \tag{13-49}$$

a result with which we are already very familiar. From Eqs. (13-48) and (13-49) we may write the Lagrangian function as

$$L = T - V = \frac{m}{2}(\dot{r}^2 + \dot{r}^2\dot{\theta}^2) + \frac{k}{r}$$

We are now in a position to use Eq. (13-46). Taking $q_1 = r$ and $q_2 = \theta$, we have

$$\frac{d}{dt}\left(\frac{\partial L}{\partial \dot{r}}\right) = m\ddot{r} \qquad \text{and} \qquad \frac{\partial L}{\partial r} = mr\dot{\theta}^2 - \frac{k}{r^2}$$

from which Eq. (13-46) gives us the r equation of motion. We obtain

$$m\ddot{r} - mr\dot{\theta}^2 + \frac{k}{r^2} = 0$$

or

$$m(\ddot{r} - r\dot{\theta}^2) = -\frac{k}{r^2} \tag{13-50}$$

a familiar result indeed. In a similar fashion we may find the θ equation of motion. We have

$$\frac{\partial L}{\partial \dot{\theta}} = mr^2\dot{\theta} \qquad \frac{d}{dt}\left(\frac{\partial L}{\partial \dot{\theta}}\right) = 2mr\dot{r}\dot{\theta} + mr^2\ddot{\theta} \qquad \frac{\partial L}{\partial \theta} = 0$$

from which

$$m(r\ddot{\theta} + 2\dot{r}\dot{\theta}) = \frac{d}{dt}(mr^2\dot{\theta}) = 0$$

and we have an immediate integral

$$mr^2\dot{\theta} = \text{const} = J$$

expressing the constancy of angular momentum.

Example 13-2. It is desired to investigate the motion of a bead of mass m sliding freely on a smooth circular wire of radius b which rotates in a horizontal plane about one of its points, O, with a constant angular velocity ω.

In Fig. 13-3, the plane of rotation is that of the paper, with the rotation in the counterclockwise sense. The angle that the radius, drawn from C to O, makes with the fixed x axis is $\varphi = \omega t$. Produce OC through C to A. The position of m is given by θ, the angle which CB makes with CA. The problem is one involving but a single degree of freedom. We put $\theta = q$, the generalized coordinate of the problem. Now the coordinates x and y of m are easily seen to be

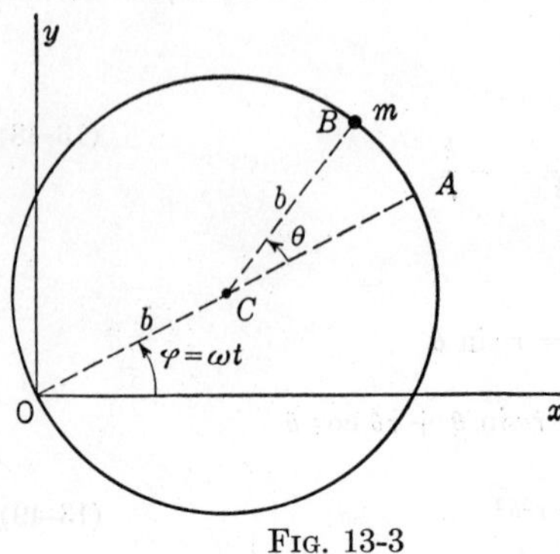

FIG. 13-3

$$x = b \cos \omega t + b \cos (\theta + \omega t) \qquad (13\text{-}51)$$
$$y = b \sin \omega t + b \sin (\theta + \omega t) \qquad (13\text{-}52)$$

Forming $\dot{x}$ and $\dot{y}$, we have

$$\dot{x} = -b\omega \sin \omega t - b(\dot{\theta} + \omega) \sin (\theta + \omega t)$$
$$\dot{y} = b\omega \cos \omega t + b(\dot{\theta} + \omega) \cos (\theta + \omega t)$$

After performing the possible cancellations, we have

$$\dot{x}^2 + \dot{y}^2 = b^2\omega^2 + b^2(\dot{\theta} + \omega)^2 + 2b^2\omega(\dot{\theta} + \omega) \cos \theta$$

from which the kinetic energy T becomes

$$T = \frac{mb^2}{2} [\omega^2 + (\dot{\theta} + \omega)^2 + 2\omega(\dot{\theta} + \omega) \cos \theta] \qquad (13\text{-}53)$$

whence

$$\frac{\partial T}{\partial \dot{\theta}} = mb^2(\dot{\theta} + \omega + \omega \cos \theta)$$
$$\frac{d}{dt}\left(\frac{\partial T}{\partial \dot{\theta}}\right) = mb^2(\ddot{\theta} - \omega\dot{\theta} \sin \theta)$$
$$\frac{\partial T}{\partial \theta} = -mb^2[\omega(\dot{\theta} + \omega) \sin \theta]$$

Therefore Eq. (13-42) becomes, since $Q = 0$ here,

$$mb^2(\ddot{\theta} - \omega\dot{\theta} \sin \theta) + mb^2\omega(\dot{\theta} + \omega) \sin \theta = 0$$

or

$$\ddot{\theta} + \omega^2 \sin \theta = 0 \qquad (13\text{-}54)$$

Hence we see that the bead oscillates about the rotating line OA as a pendulum of length

$$a = \frac{g}{\omega^2}$$

In Example 13-2 it is to be noted that the problem was reduced to one having but one degree of freedom. This was entirely sufficient for determining the position and velocity of the bead as a function of time. If it were desired to find the reaction of the wire by Lagrangian methods, we should have to proceed differently. Instead of reducing the number of degrees of freedom to one, we must now treat the problem as possessing two degrees of freedom, the additional one rendering possible a virtual displacement in a direction perpendicular to the wire. Equations (13-51) and (13-52) would be changed to read

$$x = b \cos \omega t + r \cos (\theta + \omega t) \tag{13-55}$$
$$y = b \sin \omega t + r \sin (\theta + \omega t) \tag{13-56}$$

where, in the final analysis, r will be considered to have the constant

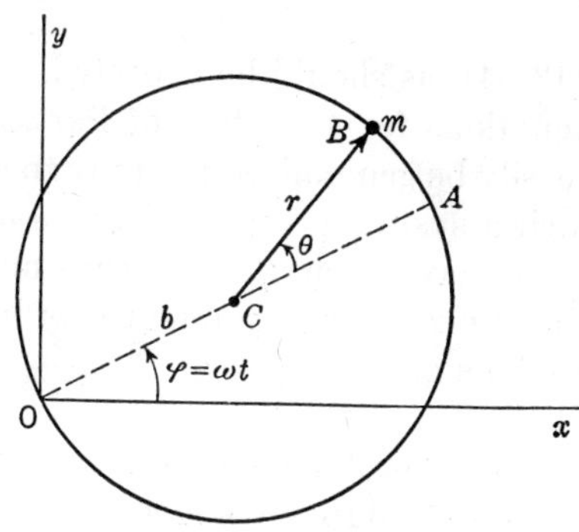

FIG. 13-4

value b (see Fig. 13-4). Therefore

$$\dot{x} = -b\omega \sin \omega t + \dot{r} \cos (\theta + \omega t) - r(\dot{\theta} + \omega) \sin (\theta + \omega t)$$
$$\dot{y} = b\omega \cos \omega t + \dot{r} \sin (\theta + \omega t) + r(\dot{\theta} + \omega) \cos (\theta + \omega t)$$

whence, squaring $\dot{x}$ and $\dot{y}$ and adding, we have finally

$$T = \frac{m}{2} (\dot{x}^2 + \dot{y}^2) = \frac{m}{2} [b^2\omega^2 + \dot{r}^2 + r^2(\dot{\theta} + \omega)^2 + 2b\omega\dot{r} \sin \theta + 2b\omega r(\dot{\theta} + \omega) \cos \theta] \tag{13-57}$$

Now

$$\frac{d}{dt}\left(\frac{\partial T}{\partial \dot{r}}\right) - \frac{\partial T}{\partial r} = R_r \tag{13-58}$$

Therefore, since

$$\frac{\partial T}{\partial \dot{r}} = m\dot{r} + mb\omega \sin \theta$$
$$\frac{d}{dt}\left(\frac{\partial T}{\partial \dot{r}}\right) = m(\ddot{r} + b\omega\dot{\theta} \cos \theta)$$
$$\frac{\partial T}{\partial r} = mr(\dot{\theta} + \omega)^2 + mb\omega(\dot{\theta} + \omega) \cos \theta$$

we have

$$\frac{d}{dt}\left(\frac{\partial T}{\partial \dot{r}}\right) - \frac{\partial T}{\partial r} = m[\ddot{r} + b\omega\dot{\theta} \cos \theta - r(\dot{\theta} + \omega)^2 - b\omega(\dot{\theta} + \omega) \cos \theta] = R_r$$

Inserting the condition that r = const = b, we have

$$R_r = -mb[\omega^2 \cos \theta + (\dot{\theta} + \omega)^2] \tag{13-59}$$

We note that R_r is in the direction of r decreasing. The quantity $\dot{\theta}$ can be found by integrating (13-54) and thus may be eliminated from (13-59).

It also is of interest to observe that if we form

$$\frac{d}{dt}\frac{\partial T}{\partial \dot{\theta}} - \frac{\partial T}{\partial \theta} = 0$$

we again obtain Eq. (13-54), as should be expected.

13-4. Lagrange's Equations for a System of Particles. The material of the last section can easily be generalized to apply to a holonomic system of n particles. Let such a system possess r degrees of freedom after all the equations of condition have been utilized, thus making it possible to represent the system in terms of r independent coordinates $q_1, \ldots, q_r$. There will be $3n$ expressions of the type

$$\begin{aligned} x_i &= x_i(q_1, \ldots, q_r, t) \\ y_i &= y_i(q_1, \ldots, q_r, t) \\ z_i &= z_i(q_1, \ldots, q_r, t) \end{aligned} \tag{13-60}$$

where Eqs. (13-60) provide the relationships between the coordinates x_i, y_i, z_i of the ith particle and the generalized coordinates $q_1, \ldots, q_r$. The equations of motion of the ith particle (for simplicity we omit the forces of constraint, since for smooth constraints these will not appear in Lagrange's equations)

$$m_i\ddot{x}_i = X_i \qquad m_i\ddot{y}_i = Y_i \qquad m_i\ddot{z}_i = Z_i \tag{13-61}$$

We are interested in finding Lagrange's equation for the coordinate q_s. Multiplying the first of these by $\partial x_i/\partial q_s$, the second by $\partial y_i/\partial q_s$, the third by $\partial z_i/\partial q_s$, adding, and summing over all n particles of the system, we have

$$\sum_{i=1}^{n} m_i\left(\ddot{x}_i\frac{\partial x_i}{\partial q_s} + \ddot{y}_i\frac{\partial y_i}{\partial q_s} + \ddot{z}\frac{\partial z_i}{\partial q_s}\right) = \sum_{i=1}^{n}\left(X_i\frac{\partial x_i}{\partial q_s} + Y_i\frac{\partial y_i}{\partial q_s} + Z_i\frac{\partial z_i}{\partial q_s}\right) \tag{13-62}$$

In the same manner as before, the left side can be shown here to be

$$\frac{1}{2}\sum_{i=1}^{n}\left\{m_i\frac{d}{dt}\left[\frac{\partial}{\partial \dot{q}_s}(\dot{x}_i^2 + \dot{y}_i^2 + \dot{z}_i^2)\right]\right\} - \frac{1}{2}\sum_{i=1}^{n}\left[m_i\frac{\partial}{\partial q_s}(\dot{x}_i^2 + \dot{y}_i^2 + \dot{z}_i^2)\right] \tag{13-63}$$

which, for constant masses, becomes

$$\frac{d}{dt}\left(\frac{\partial T}{\partial \dot{q}_s}\right) - \frac{\partial T}{\partial q_s} \tag{13-64}$$

where

$$T = \sum_{i=1}^{n} \frac{m_i}{2}(\dot{x}_i^2 + \dot{y}_i^2 + \dot{z}_i^2) \tag{13-65}$$

Furthermore, the right side of Eq. (13-62) is simply the generalized force Q_s associated with q_s. Accordingly, Lagrange's equation for the coordinate q_s is

$$\frac{d}{dt}\left(\frac{\partial T}{\partial \dot{q}_s}\right) - \frac{\partial T}{\partial q_s} = Q_s \tag{13-66}$$

of which there are r such equations. If Q_s is derivable from a function V, such that

$$\frac{d}{dt}\left(\frac{\partial V}{\partial \dot{q}_s}\right) - \frac{\partial V}{\partial q_s} = Q_s \tag{13-67}$$

we have also the alternate form. Putting $L = T - V$, we obtain

$$\frac{d}{dt}\left(\frac{\partial L}{\partial \dot{q}_s}\right) - \frac{\partial L}{\partial q_s} = 0 \tag{13-68}$$

If Q_s consists of a part Q_s' which is derivable from a potential-energy function V and a part Q_s'' not derivable from V, that is, if

$$Q_s = Q_s' + Q_s'' \tag{13-69}$$

the form (13-68) can still be employed since it is possible to regard L as containing only the part of Q_s derivable from V. Thus, for such a situation, Eq. (13-68) may be written

$$\frac{d}{dt}\left(\frac{\partial L}{\partial \dot{q}_s}\right) - \frac{\partial L}{\partial q_s} = Q_s'' \tag{13-70}$$

13-5. Generalized Momentum. In the preceding sections, the motion of a system was described by a set of Lagrange's equations. One of these was associated with each degree of freedom. Thus a system of r degrees of freedom was described by r equations of the second order. Another feature of the Lagrangian method was that the r generalized coordinates $q_1, \ldots, q_r$ and the r generalized velocities $\dot{q}_1, \ldots, \dot{q}_r$ were regarded together as a duality of independent coordinates. The Lagrangian for such a system may be an explicit function $L(q_1, \ldots, q_r, \dot{q}_1, \ldots, \dot{q}_r, t)$, of the q's, $\dot{q}$'s, and t.

In the present section, a procedure is introduced which is useful if the system possesses a Lagrangian of the type defined by Eq. (13-68), that is, if a Lagrangian can be found such that, when it is substituted in the r equations of the type (13-68), the correct r equations of motion will be completely realized. The use of this method provides a description of a system of r degrees of freedom by $2r$ first-order equations. This is to be contrasted with the Lagrangian procedure, in which r equations of the second order are employed. The process is further characterized by the replacement of the *generalized velocities*, in the duality of independent coordinates, by quantities called *generalized momenta*. The generalized momentum p_s corresponding to (also spoken of as being *conjugate* to) q_s is defined as

$$\frac{\partial L}{\partial \dot{q}_s} = p_s \tag{13-71}$$

In terms of the generalized momentum, it is apparent that the first term on the left side of Eq. (13-68) is simply

$$\frac{d}{dt}\left(\frac{\partial L}{\partial \dot{q}_s}\right) = \dot{p}_s \tag{13-72}$$

An important property of a dynamic system, the Lagrangian of which does not contain a certain coordinate explicitly, follows from Eqs. (13-68) and (13-72). If a coordinate q_s does not appear explicitly in the Lagrangian function, it is evident that the second term in Eq. (13-68), the partial of L with respect to q_s, vanishes. Consequently, in that event the conjugate momentum p_s is a constant of the motion. It must not be inferred, though, that the corresponding generalized velocity, $\dot{q}_s$, may not vary. Note, for example, the case of plane polar coordinates given in Example 13-1. The Lagrangian function does not contain θ explicitly. Hence $p_\theta \equiv mr^2\dot{\theta}$ is a constant of the motion. However, this does not mean that $\dot{\theta}$ is constant. In an elliptic orbit for an inverse-square attractive force, $\dot{\theta}$ may vary quite widely, depending upon the eccentricity of the orbit, whereas p_θ remains constant.

13-6. Motion of a Symmetrical Top from Lagrange's Equations. It is useful to deduce the salient features of the motion of a heavy symmetrical top, with the peg fixed, from Lagrange's equations. To do this, we employ the coordinate system provided for this problem in Sec. 12-14. The kinetic energy may be written

$$\begin{aligned} T &= \tfrac{1}{2}[I_1(\omega_x^2 + \omega_y^2) + I_z(\omega_z + s)^2] \\ &= \tfrac{1}{2}[I_1(\dot{\theta}^2 + \dot{\varphi}^2 \sin^2 \theta) + I_z(\dot{\varphi} \cos \theta + \dot{\psi})^2] \end{aligned} \tag{13-73}$$

the symbols being defined as in Sec. 12-14. The potential energy is

$$V = mgh \cos \theta \tag{13-74}$$

Hence the Lagrangian function becomes

$$L = T - V = \tfrac{1}{2}[I_1(\dot{\theta}^2 + \dot{\varphi}^2 \sin^2 \theta) + I_z(\dot{\varphi} \cos \theta + \dot{\psi})^2] - mgh \cos \theta \quad (13\text{-}75)$$

Neither φ nor ψ appears explicitly in L; hence the momenta, conjugate to φ and ψ, are constants of the motion. We have

$$p_\varphi = \frac{\partial L}{\partial \dot{\varphi}} = I_1 \dot{\varphi} \sin^2 \theta + I_z(\dot{\varphi} \cos \theta + \dot{\psi}) \cos \theta = \text{const} \quad (13\text{-}76)$$

$$p_\psi = \frac{\partial L}{\partial \dot{\psi}} = I_z(\dot{\varphi} \cos \theta + \dot{\psi}) = \text{const} \quad (13\text{-}77)$$

From (13-77), since I_z is constant, we have

$$S_1 \equiv (\dot{\varphi} \cos \theta + \dot{\psi}) = \frac{p_\psi}{I_z} = \text{const} \quad (13\text{-}78)$$

This result is consistent with Eqs. (12-108) and (12-118). From (13-76) we obtain an expression for $\dot{\varphi}$. We have

$$\dot{\varphi} = \frac{p_\varphi - I_z S_1 \cos \theta}{I_1 \sin^2 \theta} = \frac{C - D \cos \theta}{\sin^2 \theta} \quad (13\text{-}79)$$

a result which is in agreement with Eq. (12-126).

In addition to p_φ and p_ψ, given by (13-76) and (13-77), there is a third constant of the motion. This is the total energy W. It is

$$W = T + V = \tfrac{1}{2}[I_1(\dot{\theta}^2 + \dot{\varphi}^2 \sin^2 \theta) + I_z(\dot{\varphi} \cos \theta + \dot{\psi})^2] + mgh \cos \theta \quad (13\text{-}80)$$

From (13-80) we obtain a relationship between $\dot{\theta}$ and $\dot{\varphi}$. This is

$$\dot{\theta}^2 + \dot{\varphi}^2 \sin^2 \theta = \frac{(2W - I_z S_1^2) - 2mgh \cos \theta}{I_1} = A - B \cos \theta \quad (13\text{-}81)$$

where A and B are constants. This is easily seen to be identical to Eq. (12-121). The constants A and B may be shown to have the same significance that is given to them in Chap. 12.

If desired, the angular velocity $\dot{\psi}$ can be obtained from Eq. (13-78).

The above procedure is somewhat more direct and less involved than that employed in the last chapter. There the angular velocities $\dot{\theta}$ and $\dot{\varphi}$ were arrived at via Euler's equations, and considerably more insight into the physics of the problem was required to carry out this procedure to the end results, given by Eqs. (12-121) and (12-126), than has been necessary in the present development.

13-7. The Hamiltonian Function. Hamilton's Equations. Let us now take the differential of $L(q_1, \ldots ,q_r,\dot{q}_1, \ldots ,\dot{q}_r,t)$. We have

$$dL = \sum_{s=1}^{r} \left(\frac{\partial L}{\partial q_s} dq_s + \frac{\partial L}{\partial \dot{q}_s} d\dot{q}_s \right) + \frac{\partial L}{\partial t} dt \tag{13-82}$$

$$= \sum_{s=1}^{r} (\dot{p}_s \, dq_s + p_s \, d\dot{q}_s) + \frac{\partial L}{\partial t} dt \tag{13-83}$$

where the first term within the parentheses in (13-83) follows from Eqs. (13-68) and (13-72) and the second follows from the use of (13-71). Equation (13-83) can be rearranged to take the form

$$d\left(\sum_{s=1}^{r} p_s \dot{q}_s - L \right) = \sum_{s=1}^{r} (\dot{q}_s \, dp_s - \dot{p}_s \, dq_s) - \frac{\partial L}{\partial t} dt \tag{13-84}$$

It is now convenient to define a quantity H, called the *Hamiltonian function*, for the system. This is

$$H = \sum_{s=1}^{r} p_s \dot{q}_s - L(q_1, \ldots ,q_r,\dot{q}_1, \ldots ,\dot{q}_r,t) \tag{13-85}$$

Equation (13-84) therefore may be written

$$dH = \sum_{s=1}^{r} (\dot{q}_s \, dp_s - \dot{p}_s \, dq_s) - \frac{\partial L}{\partial t} dt \tag{13-86}$$

Although the Lagrangian function L is an explicit function of $q_1, \ldots, q_r, \dot{q}_1, \ldots, \dot{q}_r$, and t, it usually is possible to express H as an explicit function only of $q_1, \ldots, q_r, p_1, \ldots, p_r, t$, that is, to eliminate the r generalized velocities from Eq. (13-85). The r equations of the type (13-71) are employed for this. Each gives one of the p's in terms of the $\dot{q}$'s. If these r relationships are such that they can be combined so as to express each $\dot{q}$ in terms of the p's, the elimination of the generalized velocities from (13-85) is possible. This possibility usually exists and is certainly true for the large number of cases in which L is a quadratic function of the $\dot{q}$'s, for then Eq. (13-71) is linear in the $\dot{q}$'s. Accordingly, restricting ourselves to cases in which the elimination of the generalized velocities is possible, we may write

$$H = H(q_1, \ldots ,q_r,p_1, \ldots ,p_r,t) \tag{13-87}$$

The procedure is to consider the generalized coordinates and the generalized momenta as a new duality of independent coordinates upon which together with the time, H may depend explicitly. Consequently, taking the differential dH, we obtain

$$dH = \sum_{s=1}^{r} \left(\frac{\partial H}{\partial q_s} dq_s + \frac{\partial H}{\partial p_s} dp_s \right) + \frac{\partial H}{\partial t} dt \tag{13-88}$$

Comparing Eqs. (13-86) and (13-88), we have the relations

$$\frac{\partial H}{\partial p_s} = \dot{q}_s \qquad \frac{\partial H}{\partial q_s} = -\dot{p}_s \tag{13-89}$$

$$\frac{\partial H}{\partial t} = -\frac{\partial L}{\partial t} \tag{13-90}$$

Equations (13-89) are called *Hamilton's canonical equations.* These are $2r$ in number and provide a complete description of the motion of a system. Hence, for a system having r degrees of freedom, we are able to replace the r Lagrangian equations (13-68) of the second order by the $2r$ Hamiltonian equations of the first order. From the second of Eqs. (13-89) we note that, if any coordinate q_t is not contained explicitly in the Hamiltonian function H, the conjugate momentum p_t is a constant of the motion. A similar circumstance was pointed out above, in Sec. 13-5, in connection with coordinates not explicitly contained in L.

It is interesting to investigate further the nature of the Hamiltonian function H. Consider a system in which L does not contain the time explicitly. From Eq. (13-90) we note that, if L does not explicitly contain t, neither then does H. Consequently $\partial H/\partial t = 0$ in that event, and, dividing Eq. (13-88) through by dt, we obtain

$$\frac{dH}{dt} = \sum_{s=1}^{r} \left(\frac{\partial H}{\partial q_s} \dot{q}_s + \frac{\partial H}{\partial p_s} \dot{p}_s \right) \tag{13-91}$$

$$= \sum_{s=1}^{r} \left(\frac{\partial H}{\partial q_s} \frac{\partial H}{\partial p_s} - \frac{\partial H}{\partial p_s} \frac{\partial H}{\partial q_s} \right) = 0 \tag{13-92}$$

where the step from (13-91) to (13-92) follows from the use of Eqs. (13-89). Accordingly, if H does not contain the time explicitly, it is a constant.

Let us now consider a system in which the explicit dependence upon the generalized velocities enters only through the kinetic energy. For such a case the definition (13-71) reduces to

$$\frac{\partial T}{\partial \dot{q}_s} = p_s \tag{13-93}$$

and H becomes, from Eq. (13-85),

$$H = \sum_{s=1}^{r} \frac{\partial T}{\partial \dot{q}_s} \dot{q}_s - L(q_1, \ldots, q_r, \dot{q}_1, \ldots, \dot{q}_r, t) \tag{13-94}$$

Also, if Eqs. (13-60), for the transformation of coordinates, do not contain the time explicitly (for example, if no time-dependent constraints are present, or if the q's are not rotating with respect to the initial rectangular coordinates, and so on), then T will be a quadratic function of the generalized velocities which is homogeneous in all its terms (see Sec. 13-2). Now Euler's theorem on homogeneous functions states that, if $f(x_1, \ldots, x_r)$ is homogeneous of degree n in the variables $(x_1, \ldots, x_r)$,

$$nf(x_1, \ldots, x_r) = \sum_{s=1}^{r} x_s \frac{\partial f}{\partial x_s} \tag{13-95}$$

Accordingly, the summed term in Eq. (13-94) is just $2T$, since n is 2 here. If the system is such that $L = T - V$, we have

$$H = 2T - T + V = T + V \tag{13-96}$$

Thus, for a system in which $L = T - V$ and in which the transformation equations (13-60) do not explicitly contain the time, H is equal to the total energy of the system. Note that this permits V to contain the time explicitly, so that H is not necessarily constant. Although the conditions just mentioned regarding T and the form of L are sufficient to ensure the validity of (13-96), it is not strictly necessary that L be $T - V$ for (13-96) to be valid. An example of this exception is to be found in *special relativity;* there the form of L which, when substituted in Eq. (13-68), yields the correct equation of motion, is not $T - V$ (see Prob. 13-16).

Example 13-3. A particle of mass m is attracted to a given point by a force of magnitude k/r^2, where k is constant. Express the Hamiltonian function in terms of the coordinates and momenta.

We employ plane polar coordinates. We have

$$T = \frac{m}{2}(\dot{r}^2 + r^2\dot{\theta}^2) \qquad V = -\frac{k}{r} \tag{13-97}$$

Now

$$p_r = \frac{\partial L}{\partial \dot{r}} = m\dot{r} \qquad p_\theta = \frac{\partial L}{\partial \dot{\theta}} = mr^2\dot{\theta} \tag{13-98}$$

Making use of the definition (13-85) of H, we obtain

$$H = p_r\dot{r} + p_\theta\dot{\theta} - \frac{m}{2}(\dot{r}^2 + r^2\dot{\theta}^2) - \frac{k}{r} \tag{13-99}$$

But from Eq. (13-98)

$$\dot{r} = \frac{p_r}{m} \qquad \dot{\theta} = \frac{p_\theta}{mr^2} \tag{13-100}$$

Therefore H becomes

$$H = \frac{p_r^2}{m} + \frac{p_\theta^2}{mr^2} - \frac{m}{2}\left(\frac{p_r^2}{m^2} + \frac{p_\theta^2}{m^2r^2}\right) - \frac{k}{r} = \frac{1}{2m}\left(p_r^2 + \frac{p_\theta^2}{r^2}\right) - \frac{k}{r} \tag{13-101}$$

The second of Eqs. (13-89) gives, therefore,

$$\dot{p}_r = -\frac{\partial H}{\partial r} = \frac{p_\theta^2}{mr^3} - \frac{k}{r^2} \qquad \dot{p}_\theta = -\frac{\partial H}{\partial \theta} = 0 \tag{13-102}$$

Thus $p_\theta \equiv mr^2\dot{\theta}$ is a constant of the motion, a result seen by inspection, since H does not contain θ explicitly. If Eqs. (13-98) are employed to eliminate p_r and p_θ in the first of Eqs. (13-102), the result is seen to be identical to Eq. (13-50).

Example 13-4. Employing Hamilton's equations, deduce an expression for the angle θ_s through which a particle of mass m is scattered by a massive center of repulsive force which is at rest. Let the initial conditions be that the particle is at infinity with a velocity v, so directed that the angular momentum with reference to the scattering center is J.

The situation is represented in Fig. 13-5, with the scattering center at O. Plane polar coordinates r, θ are sufficient, since the problem is cylindrically symmetrical

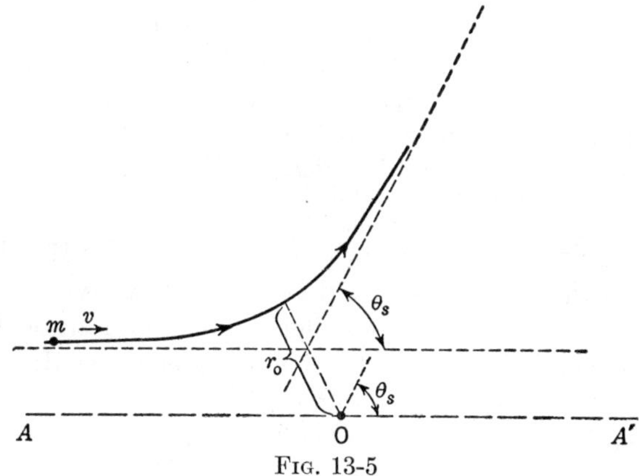

FIG. 13-5

about the line AA' passing through O and parallel to the initial direction of motion of m.

Hamilton's equations for the problem may be written

$$\frac{\partial H}{\partial p_\theta} = \dot{\theta} \qquad \frac{\partial H}{\partial r} = -\dot{p}_r \tag{13-103}$$

whence

$$\frac{\partial}{\partial r}\dot{\theta} = \frac{\partial}{\partial r}\frac{\partial H}{\partial p_\theta} = \frac{\partial}{\partial p_\theta}\frac{\partial H}{\partial r} = -\frac{\partial \dot{p}_r}{\partial p_\theta} = -\frac{\partial \dot{p}_r}{\partial J} \tag{13-104}$$

where the last step follows since $p_\theta = \text{const} = J$. Accordingly

$$\frac{\partial \dot{\theta}}{\partial r} = -\frac{\partial \dot{p}_r}{\partial J}$$

from which

$$\dot{\theta} = -\int_{r_1}^{r_2} \frac{\partial \dot{p}_r}{\partial J}\, dr = -\frac{d}{dt}\int_{r_1}^{r_2} \frac{\partial p_r}{\partial J}\, dr \tag{13-105}$$

in which the limits of integration, for the moment, are r_1 and r_2. Integrating Eq. (13-105) with respect to time, we have

$$\theta = -\int_{r_1}^{r_2} \frac{\partial p_r}{\partial J}\, dr + c \tag{13-106}$$

where c is a constant. Forming the Hamiltonian function, which here is equal to the total energy E $(= \frac{1}{2}mv^2)$, we have

$$H = \frac{1}{2m}\left(p_r^2 + \frac{p_\theta^2}{r^2}\right) + V(r) = E \tag{13-107}$$

or

$$\frac{1}{2m}\left(p_r^2 + \frac{J^2}{r^2}\right) + V(r) = E \tag{13-108}$$

in which $V(r)$ is the potential energy. From Eq. (13-108)

$$p_r = \left\{2m[E - V(r)] - \frac{J^2}{r^2}\right\}^{\frac{1}{2}} \tag{13-109}$$

and Eq. (13-106) yields (since $\partial p_r/\partial J = 0$ at $r = \infty$, $\theta = \pi$)

$$\theta_s = -2\int_{\infty}^{r_0} \frac{\partial}{\partial J}\left\{2m[E - V(r)] - \frac{J^2}{r^2}\right\}^{\frac{1}{2}} dr + \pi \tag{13-110}$$

and the limits $r_1 = \infty$ and $r_2 = r_0$, the closest distance of approach (at which $p_r = 0$), have been inserted. Since the problem is symmetrical about r_0, the total change in θ is obtained by inserting the factor 2 as shown. The scattering angle, or angle of deviation of the particle from the initial direction, is θ_s.

Before concluding the present brief treatment of the Hamiltonian procedure, it is worthwhile to caution that H is not always constant, nor is it always the total energy. We may have any of all four of the following possibilities, depending upon the system being considered:

1. *H not constant and not equal to the total energy.*
2. *H not constant but equal to the total energy.*
3. *H constant but not equal to the total energy.*
4. *H constant and equal to the total energy.*

Also, in passing, it should be pointed out that the most important uses of Hamilton's equations are in the so-called *transformation theory* of dynamics. The method, there, may involve making a transformation from an initial set of coordinates all of which may be explicitly contained in the Hamiltonian function for a system to another set in which one or more of the coordinates do not appear explicitly in the Hamiltonian. For such absent coordinates (termed *cyclic coordinates*) it has been pointed out that the conjugate momenta are constants of the motion. Further consideration of these details is above the level of this text and will be omitted.

Problems

Employing Lagrange's equations, set up the equations of motion for the following:

13-1. Prob. 11-14. **13-2.** Prob. 11-18. **13-3.** Prob. 11-19.
13-4. Prob. 11-20. **13-5.** Prob. 11–21.

13-6. Work Prob. 11-11 by means of Lagrange's equations.

13-7. A smooth straight wire is constrained to rotate in a horizontal plane with a constant angular velocity ω about a vertical axis passing through one end. The point on the wire at a distance b from the axis is the center of an attractive force proportional to the distance (proportionality constant k). A particle of mass m is free to slide on the wire. Find the position of the bead as a function of time. Take the case $m\omega^2 > k$. Is this oscillatory?

13-8. From the kinetic energy, expressed in terms of the Eulerian angles and their time derivatives, deduce with the aid of Eq. (13-42) Euler's dynamic equations for an arbitrary rigid body subjected to arbitrary torques. HINT: Deduce the third of Euler's equations, that giving the rate of change of ω_z; the other two follow from symmetry, that is, from cyclic permutation of the subscripts.

13-9. A smooth wire is bent into the form of a helix the equations of which, in cylindrical coordinates, are $z = a\theta$ and $r = b$, in which a and b are constants. The origin is a center of attractive force which varies directly as the distance. By means of Lagrange's equations find the motion of a bead which is free to slide on the wire.

13-10. In Prob. 13-9 find the components of the reaction of the wire in the r, θ, and z directions.

13-11. A particle moves on the inside surface of a smooth cone of half angle α. The axis of the cone is vertical, with vertex downward. Determine the condition on the angular velocity ω such that the particle can describe a horizontal circle at a height h above the vertex. Gravity g is acting vertically downward. Show also that the period of small oscillation about this circular path is $\dfrac{2\pi}{\cos\alpha}\sqrt{\dfrac{h}{3g}}$.

13-12. Two masses m_1 and m_2 are connected by a massless helical spring of force constant k. The system is set rotating about the center of mass with an angular velocity ω and then released. If the masses are slightly disturbed along the line joining them, show that the angular frequency of oscillation is

$$\left[\frac{3\omega^2 m_1 m_2 + k(m_1 + m_2)}{m_1 m_2}\right]^{\frac{1}{2}}$$

13-13. Work Prob. 10-17, employing the result of Example 13-4.

13-14. A bead of mass m moves on a smooth wire bent into the form of a circle. Initially the radius of circle is a, and it increases as the square of the time (constant of proportionality b). One point of the wire remains fixed. Find by the Lagrangian method the equations of motion of the bead normal to the wire (in its plane) and tangent to the wire.

13-15. A mass m is suspended by a spring of negligible mass, force constant k, and unstretched length b from a point which has a constant upward acceleration a_0. Gravity g is acting vertically downward. Find the Lagrangian function and from it the equation of motion in the vertical direction. Determine also the Hamiltonian function, and write Hamilton's equations. What is the period of the motion?

13-16. For high velocities (relativity theory) a particle has a mass

$$m = \frac{m_0}{[1 - (v^2/c^2)]^{\frac{1}{2}}}$$

where m_0 is the so-called *rest mass*, v the velocity of the particle, and c the velocity of light. Considering the motion to be in one dimension, show that a Lagrangian defined by

$$L = -m_0c^2 \sqrt{1 - \frac{v^2}{c^2}} - V$$

where V, the potential energy, is not velocity-dependent, provides the correct equation of motion. Find the generalized momentum and the Hamiltonian function H. If the relativistic kinetic energy is given to be $m_0c^2/\sqrt{1 - v^2/c^2}$, show that $H = T + V$. (Notice that $L \neq T - V$ here.) Show also that for low velocities the kinetic energy approaches the value $m_0v^2/2$ save for a constant m_0c^2.

13-17. A particle of mass m is attached to a fixed point on a rough horizontal table by a massless spring of force constant k and normal length b. If the coefficient of friction is μ and the acceleration of gravity g, form a Lagrangian function which will give the correct equation of motion when substituted in Eq. (13-46). Restrict the motion to lie along one straight line. Is it equal to $T - V$? Form the Hamiltonian function. Is it equal to the total energy of the particle at any given time? Is it constant?

13-18. A particle of mass m moves in a central field of attractive force of magnitude $(k/r^2)e^{-\alpha t}$, where k and α are constants, t is the time, and r is the distance of m from the force center. Find the Lagrangian and Hamiltonian functions. Is H the total energy? Is it constant?

13-19. A particle of mass m and charge e is moving freely in an electromagnetic field with a velocity $\mathbf{v}$. It is given that the Lagrangian function is

$$L = T - V + \frac{e}{c}(\mathbf{v} \cdot \mathbf{A})$$

where T and V are the kinetic and potential energies, c the velocity of light, and $\mathbf{A}$ a vector (the *vector potential*) with rectangular components A_x, A_y, A_z. Form the Lagrangian equation of motion with respect to the coordinate x. Find the generalized momenta conjugate to x, y, z and also the Hamiltonian function.

13-20. A particle slides down a smooth stationary sphere, under the action of gravity, from a position of rest at the top. Use Lagrange's equation to find the reaction of the sphere on the particle at any value of θ, where θ is the angle between the vertical diameter of the sphere and the normal to the sphere which passes through the particle. Find the value of θ at which the particle falls off.

13-21. Take the case of Prob. 13-17, save that the force constant of the spring depends upon the time in the manner $ke^{-\alpha t}$, where k and α are constants. Form the Hamiltonian function. Is it equal to the total energy of the particle at any time? Is it constant?

13-22. A mass m is suspended by means of a string which passes through a small smooth hole in a table. Initially the length of the string is b. The string is drawn up through the hole at a constant rate α (centimeters per second). Form the Lagrangian function for pendulumlike motion in a vertical plane. What is the Hamiltonian function? Is it equal to the energy? Is it constant?

13-23. A ring of radius a is made to rotate with a constant angular velocity ω about a fixed vertical diameter. A mass m is constrained to move along the smooth periphery of the ring, with gravity g acting vertically downward.

a. Find all equilibrium positions of m.

b. Discuss the stability and instability of these positions for all values of a, ω, and g.

c. Find the frequencies of small oscillations about the stable equilibrium positions.

13-24. In the system of Prob. 13-14 write the Hamiltonian for the motion of the bead subject to the constraint given. Is it constant? Is it equal to the energy? Is the energy constant?

CHAPTER 14

VIBRATING SYSTEMS AND NORMAL COORDINATES

In Chap. 7 was treated the problem of a particle oscillating in a straight line to and fro past an equilibrium position. Such a system has but a single degree of freedom. In the present chapter we shall consider vibrating systems having several degrees of freedom. However, we shall restrict ourselves, for the present, to cases in which the number of these degrees is not large. In the next chapter a system will be treated for which this number is very great.

When the motion of a vibrating system is examined in detail, it is found that the maximum number of distinct frequencies which may be attributed to the system is just equal to its number of degrees of freedom. Those frequencies which are real and different from zero are associated with elastic motions. The time dependence of these motions may be expressed either in terms of sine and cosine functions or, alternatively, in terms of exponentials with imaginary (complex if damping is present) exponents.

It is a feature of vibrating systems that it is possible so to disturb the system initially that in the subsequent free motion but a single frequency is excited. In such a *mode*, or state of motion, all moving parts of the system are oscillating in phase with this one frequency. In many cases an experienced eye is able to perceive, after a brief glance, whether one or a mixture of several modes is being executed by the system. One feature which makes this recognition possible is that, since the particles are moving in phase, they will reach maximum displacements simultaneously and will pass through their equilibrium points simultaneously. Other properties of these motions will be brought out below. It is just this behavior of joint oscillation of the entire system with one frequency which provides a convenient means of analysis of such a vibrating system.

It is to be noticed that, in mentioning the number of characteristic frequencies possessed by a vibrating system, it was stated that the maximum number was equal to the number of degrees of freedom. This precaution was necessary since it may be that not all the frequencies are distinct. It is possible that the frequencies associated with two or more modes of a system all have a common magnitude. When this is the case, the system is said to be *degenerate*. We shall limit ourselves, in the ensuing discussion, to *nondegenerate* systems.

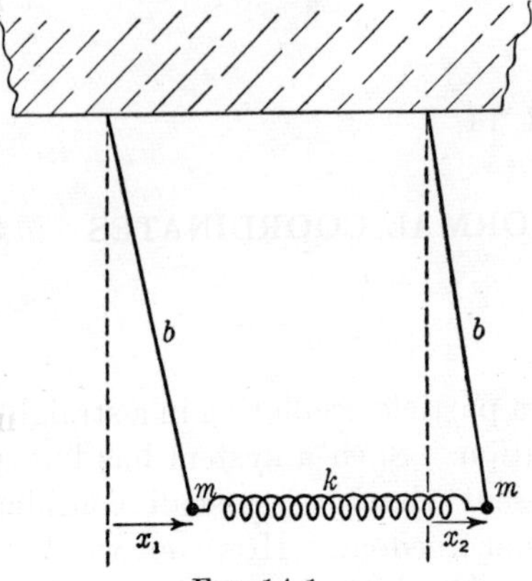

FIG. 14-1

14-1 Coupled Pendulums. An appropriate system with which to introduce the subject of small vibrations is a pair of identical simple pendulums (see Fig. 14-1) coupled by means of a massless spring, of force constant k. We assume that, when both strings are vertical, the spring is just unstretched. The length of the string in each case is b, and the mass of the bob in each case is m. We select as coordinates x_1 and x_2, measured positively to the right from the equilibrium positions of the two masses. Accordingly, for small displacements the equations of motion are, if there is no damping,

$$m\ddot{x}_1 = -\frac{mg}{b}x_1 - k(x_1 - x_2) \tag{14-1}$$

$$m\ddot{x}_2 = -\frac{mg}{b}x_2 + k(x_1 - x_2) \tag{14-2}$$

The procedure is to try solutions of the type

$$x_1 = Ae^{i\omega t} \qquad x_2 = Be^{i\omega t} \tag{14-3}$$

in which the two coordinates are both contributing to an oscillatory motion with a common angular frequency ω; A and B are constants. Moreover, since there is no damping present, ω will be a real quantity if the corresponding motion is oscillatory. We substitute Eqs. (14-3) into Eqs. (14-1) and (14-2), obtaining a pair of simultaneous linear algebraic equations in the coefficients A and B. They are

$$\left(m\omega^2 - \frac{mg}{b} - k\right)A + kB = 0 \tag{14-4}$$

$$kA + \left(m\omega^2 - \frac{mg}{b} - k\right)B = 0 \tag{14-5}$$

Both these equations must be satisfied if the initial assumption [Eqs. (14-3)] as to the form of the solution is correct. Now Eqs. (14-4) and (14-5) may be satisfied by values of A and B different from zero only if the equations are *consistent*. It is well known from the theory of equations that for this last to be true the determinant of the coefficients must be zero. This determinant will contain the so far undetermined quantity ω. It will be possible to make this determinant equal to zero by assigning appropriate values to ω, and consequently Eqs. (14-4) and (14-5) then will be consistent. Accordingly we put

$$\begin{vmatrix} \left(m\omega^2 - \frac{mg}{b} - k\right) & k \\ k & \left(m\omega^2 - \frac{mg}{b} - k\right) \end{vmatrix} = 0 \tag{14-6}$$

Upon expanding the determinant we have

$$\left(m\omega^2 - \frac{mg}{b} - k\right)\left(m\omega^2 - \frac{mg}{b} - k\right) - k^2 = 0 \tag{14-7}$$

an equation for the determination of ω. This may be rewritten

$$\left(m\omega^2 - \frac{mg}{b}\right)\left[\left(m\omega^2 - \frac{mg}{b}\right) - 2k\right] = 0 \tag{14-8}$$

The frequencies allowed by Eq. (14-8) are two in number and are such that

$$m\omega^2 - \frac{mg}{b} - 2k = 0 \qquad m\omega^2 - \frac{mg}{b} = 0 \tag{14-9}$$

from which

$$\pm\omega_1 = \pm\left(\frac{g}{b}\right)^{\frac{1}{2}} \qquad \pm\omega_2 = \pm\left(\frac{g}{b} + \frac{2k}{m}\right)^{\frac{1}{2}} \tag{14-10}$$

In terms of these, the general solutions of Eqs. (14-1) and (14-2) may be written

$$x_1 = A_1 e^{i\omega_1 t} + A_{-1}e^{-i\omega_1 t} + A_2 e^{i\omega_2 t} + A_{-2}e^{-i\omega_2 t} \tag{14-11}$$
$$x_2 = B_1 e^{i\omega_1 t} + B_{-1}e^{-i\omega_1 t} + B_2 e^{i\omega_2 t} + B_{-2}e^{-i\omega_2 t} \tag{14-12}$$

Not all the arbitrary constants present in Eqs. (14-11) and (14-12) are independent. This follows since the differential equations are of the second order and are but two in number. Since the compatibility condition (14-6) determines ω, we may employ the original pair of algebraic equations (14-4) and (14-5) in order to determine A and B. For each allowed value of ω it is possible to determine a ratio between A and B. These will be, from either (14-4) or (14-5),

$$\begin{aligned} &\text{At } \omega = \omega_1 \qquad A = +B \\ &\text{At } \omega = \omega_2 \qquad A = -B \end{aligned} \tag{14-13}$$

Thus Eqs. (14-11) and (14-12) may be written

$$x_1 = A_1 e^{i\omega_1 t} + A_{-1} e^{-i\omega_1 t} + A_2 e^{i\omega_2 t} + A_{-2} e^{-i\omega_2 t} \tag{14-14}$$
$$x_2 = A_1 e^{i\omega_1 t} + A_{-1} e^{-i\omega_1 t} - A_2 e^{i\omega_2 t} - A_{-2} e^{-i\omega_2 t} \tag{14-15}$$

and we have the four arbitrary constants to be expected in the general solutions of two second-order differential equations.

14-2. Normal Coordinates. It is evident, by inspection of Eqs. (14-14) and (14-15), that it is possible to make linear combinations of x_1 and x_2 such that a combination involves but a single frequency. These combinations are especially simple in the present case, being merely the sum and difference of x_1 and x_2. We have

$$x_1 + x_2 = X_1 = 2(A_1 e^{i\omega_1 t} + A_{-1} e^{-i\omega_1 t}) \tag{14-16}$$
$$x_1 - x_2 = X_2 = 2(A_2 e^{i\omega_2 t} + A_{-2} e^{-i\omega_2 t}) \tag{14-17}$$

Thus we see that the linear combinations X_1 and X_2, which we may designate as new coordinates, each varies harmonically with but a single frequency. The quantities X_1 and X_2 are called *normal coordinates.* Except when special boundary conditions exist, both are excited simultaneously. However, if the boundary conditions are such that but one normal coordinate is excited and the other is zero initially, the latter will remain zero for all time. There is no tendency for energy to pass from one normal coordinate to another. They are completely independent.

We may determine the appearance of the *mode of vibration* associated with a given normal coordinate by letting the others (if more than one) be zero. For example let us ascertain the appearance of X_1. This is accomplished by putting $X_2 = 0$. Thus

$$X_2 = 0 = x_1 - x_2$$

from which we have $x_1 = x_2$. The mode associated with X_1 is shown in Fig. 14-2. We see that x_1 and x_2 are in phase, and the frequency of the motion, from Eq. (14-16), is ω_1. From Eq. (14-10) we see that $\omega_1 = \sqrt{g/b}$, which is that of a simple pendulum of length b. This is understandable since the spring is being neither stretched nor compressed during the motion associated with X_1.

The mode of vibration associated with X_2 may be found similarly by suppressing X_1. We have

$$X_1 = 0 = x_1 + x_2$$

from which we have $x_1 = -x_2$. The mode of vibration associated with X_2 is shown in Fig. 14-3. Clearly x_1 and x_2 are in opposite phase. The frequency of the motion, from Eq. (14-17), is ω_2. Consulting Eq. (14-10), we note that ω_2 has the value $(g/b + 2k/m)^{\frac{1}{2}}$. This is greater than that for the simple pendulumlike motion of X_1. This can easily be understood since, in the mode associated with X_2 (see Fig. 14-3), there will be restoring forces contributed by the spring.

A characteristic general feature of normal modes of vibration, anticipated in the introduction, may be noted in Figs. 14-2 and 14-3. This is that, when but one normal coordinate is excited, the original coordinates of position, such as x_1 and x_2, maintain constant ratios throughout the motion, all passing through their equilibrium points simultaneously.

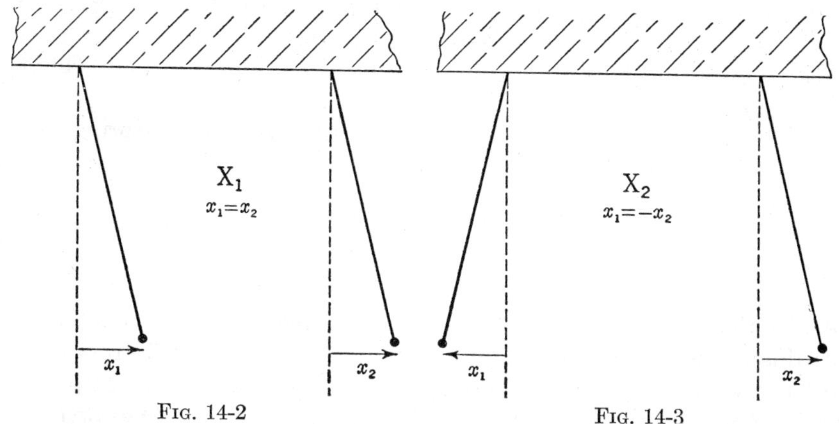

Fig. 14-2

Fig. 14-3

14-3. Equations of Motion and the Energy in Terms of Normal Coordinates. From Eqs. (14-16) and (14-17) we find

$$x_1 = \frac{X_1 + X_2}{2} \qquad x_2 = \frac{X_1 - X_2}{2} \tag{14-18}$$

Substituting these quantities in Eqs. (14-1) and (14-2), we obtain

$$\frac{m}{2}(\ddot{X}_1 + \ddot{X}_2) = -\frac{mg}{2b}(X_1 + X_2) - kX_2 \tag{14-19}$$

$$\frac{m}{2}(\ddot{X}_1 - \ddot{X}_2) = -\frac{mg}{2b}(X_1 - X_2) + kX_2 \tag{14-20}$$

Adding Eqs. (14-19) and (14-20), we find

$$m\ddot{X}_1 = -\frac{mg}{b}X_1 \tag{14-21}$$

a linear equation, in X_1 alone, with constant coefficients. Similarly, subtracting (14-20) from (14-19), we obtain

$$m\ddot{X}_2 = -\left(\frac{mg}{b} + 2k\right)X_2 \tag{14-22}$$

an equation with X_2 as the single dependent variable. Since the coefficients on the right sides are positive quantities, we see that both Eqs.

(14-21) and (14-22) are differential equations of simple harmonic motion, having the frequencies given in Eqs. (14-10). This is likewise a characteristic of normal coordinates. The equations of motion, when expressed in terms of normal coordinates, are linear equations with constant coefficients, and each contains but one dependent variable.

It is worthwhile to express the kinetic and potential energies in terms of the normal coordinates and their derivatives. In terms of the coordinates x_1 and x_2, the potential energy V may be written

$$V = \frac{k}{2}(x_1 - x_2)^2 + \frac{mg}{2b}x_1^2 + \frac{mg}{2b}x_2^2 \tag{14-23}$$

in which the first term is the potential energy stored in the spring when the displacements are x_1 and x_2. (Note that V is a *homogeneous quadratic* function of x_1 and x_2.) Similarly the kinetic energy is

$$T = \frac{m}{2}(\dot{x}_1^2 + \dot{x}_2^2) \tag{14-24}$$

We see that there are no *cross terms* in T but that there is a cross term in V; this arises in the potential energy for the spring. Such cross terms are always present in the energy of a vibrating system, the component parts of which influence each other. However, the *coupling*, or cross, terms are not always in the potential energy. Occasionally they occur in the kinetic energy (see Example 14-1). If we substitute Eqs. (14-18) in Eqs. (14-23) and (14-24) we obtain, respectively,

$$V = \frac{mg}{4b}X_1^2 + \left(\frac{mg}{4b} + \frac{k}{2}\right)X_2^2 \tag{14-25}$$

$$T = \frac{m}{4}(\dot{X}_1^2 + \dot{X}_2^2) \tag{14-26}$$

We note that V has now become a special kind of homogeneous quadratic function. There are no longer any cross terms present. The potential energy is now expressed as a sum of squares of the normal coordinates, multiplied by constant coefficients. The kinetic energy T, in Eq. (14-26), has the form of a sum of squares of the time derivatives of the normal coordinates [however, for this simple case Eq. (14-24) is already a sum of squares of the velocities].

14-4. Transfer of Energy from One Pendulum to the Other. Accepting Eqs. (14-14) and (14-15) as the general solution, let us impose the boundary conditions that at $t = 0, x_1 = 0, \dot{x}_1 = 0, x_2 = a, \dot{x}_2 = 0$. Thus initially we have that the pendulum on the left of Fig. 14-1 is at rest at the origin and the other is at the point of maximum displacement. In short, x_2 is vibrating and x_1 is not. We may evaluate the arbitrary constants in Eqs. (14-14) and (14-15) by means of these boundary conditions. At $t = 0$ we have

$$x_1 = 0 = A_1 + A_{-1} + A_2 + A_{-2} \tag{14-27}$$
$$x_2 = a = A_1 + A_{-1} - A_2 - A_{-2} \tag{14-28}$$
$$\dot{x}_1 = 0 = i\omega_1(A_1 - A_{-1}) + i\omega_2(A_2 - A_{-2}) \tag{14-29}$$
$$\dot{x}_2 = 0 = i\omega_1(A_1 - A_{-1}) - i\omega_2(A_2 - A_{-2}) \tag{14-30}$$

Solving these equations simultaneously, we find

$$A_1 = A_{-1} = \frac{a}{4} \qquad A_2 = A_{-2} = -\frac{a}{4} \tag{14-31}$$

Accordingly Eqs. (14-14) and (14-15) become

$$x_1 = \frac{a}{4}[(e^{i\omega_1 t} + e^{-i\omega_1 t}) - (e^{i\omega_2 t} + e^{-i\omega_2 t})]$$
$$= \frac{a}{2}(\cos\omega_1 t - \cos\omega_2 t) \tag{14-32}$$

and

$$x_2 = \frac{a}{2}(\cos\omega_1 t + \cos\omega_2 t) \tag{14-33}$$

Equations (14-32) and (14-33) may be put in the form

$$x_1 = a\sin\left(\frac{\omega_2 + \omega_1}{2}t\right)\sin\left(\frac{\omega_2 - \omega_1}{2}t\right) \tag{14-34}$$
$$x_2 = a\cos\left(\frac{\omega_2 + \omega_1}{2}t\right)\cos\left(\frac{\omega_2 - \omega_1}{2}t\right) \tag{14-35}$$

a transformation which may be verified readily by expanding Eqs. (14-34) and (14-35). Let us consider the case when ω_2 is almost equal to ω_1. Equations (14-34) and (14-35) may be written

$$x_1 = a\sin\left(\frac{\omega_2 - \omega_1}{2}t\right)\sin\omega_0 t \tag{14-36}$$
$$x_2 = a\cos\left(\frac{\omega_2 - \omega_1}{2}t\right)\cos\omega_0 t \tag{14-37}$$

where the symbol ω_0 replaces $(\omega_1 + \omega_2)/2$. At $t = 0$, x_1 is zero, and $x_2 = a$, as is to be expected. We may regard x_1 and x_2 as executing oscillatory motions, $\sin\omega_0 t$ and $\cos\omega_0 t$, respectively, with the slowly varying amplitudes

$$a\sin\left(\frac{\omega_2 - \omega_1}{2}t\right) \qquad \text{and} \qquad a\cos\left(\frac{\omega_2 - \omega_1}{2}t\right) \tag{14-38}$$

respectively. As the amplitude of x_1 becomes larger, that of x_2 becomes smaller, and vice versa. Thus there is a transfer of energy back and forth, with a period t_0 of transfer such that

$$t_0 = \frac{4\pi}{\omega_2 - \omega_1} \tag{14-39}$$

14-5. Possibility of Expressing an Arbitrary System in Terms of Normal Coordinates. It will be instructive to examine the properties which a mechanical system must possess in order for it to be possible to describe the system in terms of normal coordinates. Consider a vibrating system of r degrees of freedom. It will have some configuration of equilibrium, and we shall describe it in terms of positional coordinates $q_1', \ldots, q_r'$, which are all zero in the equilibrium configuration. We may express the potential energy V in a Taylor's series in these coordinates, expanding about the equilibrium point.

$$V(q_1', \ldots, q_r') = V(q_1' = 0, \ldots, q_r' = 0) + \sum_{s=1}^{r} q_s' \left(\frac{\partial V}{\partial q_s'}\right)_{q_s'=0}$$
$$+ \frac{1}{2} \sum_{s=1}^{r} \sum_{t=1}^{r} q_s' q_t' \left(\frac{\partial^2 V}{\partial q_s' \, \partial q_t'}\right)_{q_s', q_t'=0} + \cdots \tag{14-40}$$

The constant term, that is, the first term on the right, may be set equal to zero by defining V such that it is zero in the equilibrium configuration. Now, since this latter configuration is one of equilibrium, all the forces must be zero there. In other words, all the differential coefficients in the second term must necessarily be zero. Hence the first coefficients different from zero are those of the $q_s'q_t'$ terms in Eq. (14-40). All higher terms may be neglected in comparison, since the displacements q' will be assumed small. Now if the equilibrium is a stable one, the second-order differential coefficients are all positive. Therefore the potential energy is a *homogeneous quadratic* expression in the coordinates which, if the equilibrium is stable, is a *positive definite* expression, that is, it is never negative and is zero only if all the coordinates are zero. The potential energy V for such a system may be written

$$V = a_{11}q_1'^2 + \cdots + a_{rr}q_r'^2 + 2a_{12}q_1'q_2' + \cdots \tag{14-41}$$

every term of which is quadratic in the coordinates. The factors a_{11}, and so on, are all constants. Similarly we have seen that if the kinetic energy T does not contain the time explicitly it will have the form

$$T = b_{11}\dot{q}_1'^2 + \cdots + b_{rr}\dot{q}_r'^2 + 2b_{12}\dot{q}_1'\dot{q}_2' + \cdots \tag{14-42}$$

every term of which is quadratic in the velocities (*homogeneous quadratic*). The motions are assumed to be small, such that the quantities b_{11}, and so on, are approximately constant. Consulting Sec. 13-2, we see that every term in T will be positive (that is, T is *positive definite*). It can be shown that, when T and V have the form of Eqs. (14-41) and (14-42), it is always possible to write r linear combinations, $q_1, \ldots, q_r$, of the original coordinates q_1', of the form

$$\begin{aligned} q_1 &= e_{11}q_1' + \cdots + e_{1r}q_r' \\ &\cdots\cdots\cdots\cdots \\ q_r &= e_{r1}q_1' + \cdots + e_{rr}q_r' \end{aligned} \tag{14-43}$$

such that tbe kinetic and potential energies are reduced to the forms

$$T = \tfrac{1}{2}(m_1\dot{q}_1^2 + m_2\dot{q}_2^2 + \cdots + m_r\dot{q}_r^2) \tag{14-44}$$

$$V = \tfrac{1}{2}(\lambda_1^2 q_1^2 + \lambda_2^2 q_2^2 + \cdots + \lambda_r^2 q_r^2) \tag{14-45}$$

in which the m's and λ's are constants. The *linear combinations* $q_1, \ldots,$ q_r are the *normal coordinates* of the system. Now since the Lagrangian equation for the normal coordinate q_s is

$$\frac{d}{dt}\frac{\partial T}{\partial \dot{q}_s} - \frac{\partial T}{\partial q_s} = -\frac{\partial V}{\partial q_s} \tag{14-46}$$

the equation of motion for q_s, if Eqs. (14-44) and (14-45) are true, is

$$\ddot{q}_s + \omega_s^2 q_s = 0 \tag{14-47}$$

where $\omega_s^2 = \lambda_s^2/m_s$. The solution of Eq. (14-47) is

$$\text{If } \omega_s^2 > 0 \qquad q_s = A_s e^{i\omega_s t} + B_s e^{-i\omega_s t} \tag{14-48}$$

$$\text{If } \omega_s^2 = 0 \qquad q_s = C_s t + D_s \tag{14-49}$$

$$\text{If } \omega_s^2 < 0 \qquad q_s = E_s e^{\omega_s t} + F_s e^{-\omega_s t} \tag{14-50}$$

where the quantities A, B, C, D, E, F are constants.

Equations (14-48) to (14-50) represent the types of solution (for no damping) in terms of the normal coordinates $q_1, \ldots, q_r$. It is apparent that a normal coordinate q_s, for which the value of the associated ω_s^2 is not greater than zero, does not correspond to motion in which the particles oscillate about positions of stable equilibrium. For example, if $\omega_s^2 = 0$, the mode of motion associated with q_s is such that if, in that mode, the particles are slightly displaced from a position of rest there will arise no restoring forces and the subsequent displacements will increase uniformly with time. Such a situation may be apparent as a uniform translation of the center of mass of the entire system (see Example 14-1).

It is evident that the condition for complete stability of the system is that all the quantities $\omega_1^2, \omega_2^2, \ldots, \omega_r^2$ shall be positive, or, what is equivalent, the potential energy shall be positive in all its terms.

In many problems it is convenient to write the potential and kinetic energies, and then the equations of motion, first for an arbitrary displacement which is not necessarily small. Following this the appropriate approximations are made so as to render the kinetic and potential energies homogeneous quadratic, respectively, in the velocities and coordinates and to render the differential equations of motion linear. Such a case is provided by the following example.

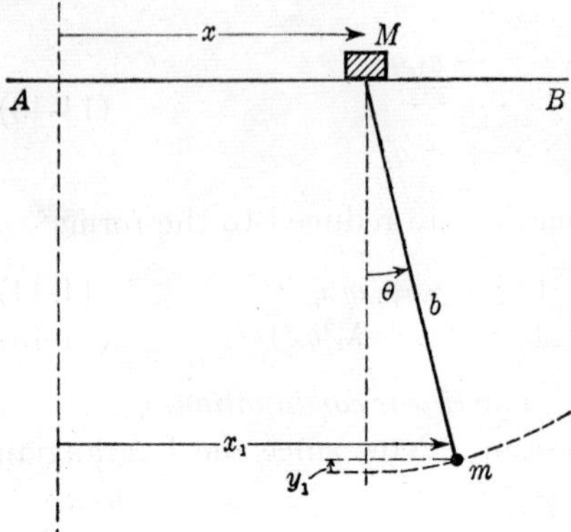

FIG. 14-4

Example 14-1. In Fig. 14-4, the mass M is constrained to slide on the smooth track AB. The mass m is connected to M by a massless inextensible string. It is desired to find the frequencies of the small oscillations of the system, and the modes of vibration associated with them.

We select coordinates as shown, with x and x_1 positive to the right and y_1 positive upward. The angle θ is positive to the right. θ and y_1 are both zero when the string is vertical.

$$T = \tfrac{1}{2}M\dot{x}^2 + \tfrac{1}{2}m(\dot{x}_1^2 + \dot{y}_1^2)$$

Now the system is one having two degrees of freedom, and x_1 and y_1 both may be eliminated in favor of θ. We have

$$x_1 = x + b\sin\theta \qquad y_1 = b(1 - \cos\theta)$$

from which

$$\dot{x}_1 = \dot{x} + b\dot{\theta}\cos\theta \qquad \dot{y}_1 = b\dot{\theta}\sin\theta$$

whence

$$T = \tfrac{1}{2}M\dot{x}^2 + \tfrac{1}{2}m[(\dot{x} + b\dot{\theta}\cos\theta)^2 + b^2\dot{\theta}^2\sin^2\theta] \tag{14-51}$$

The potential energy V is

$$V = mgy_1 = mgb(1 - \cos\theta) \tag{14-52}$$

Now if the normal coordinates of the problem are to be representable as linear combinations of x and θ, that is, if linear transformations to the normal coordinates from x and θ are to be possible, the kinetic and potential energies must be homogeneous quadratic functions, respectively, of the velocities alone and of the coordinates alone. For the present system this will be the case only for small angular displacements θ from an equilibrium configuration, that is, the string can execute but small deviations from a vertical position. Accordingly, if we limit ourselves to small values of θ and of $\dot{\theta}$, Eqs. (14-51) and (14-52) become

$$T \simeq \tfrac{1}{2}M\dot{x}^2 + \tfrac{1}{2}m(\dot{x} + b\dot{\theta})^2 \tag{14-53}$$

$$V \simeq \tfrac{1}{2}mgb\theta^2 \tag{14-54}$$

from which the Lagrangian function is

$$L = \tfrac{1}{2}M\dot{x}^2 + \tfrac{1}{2}m(\dot{x} + b\dot{\theta})^2 - \tfrac{1}{2}mgb\theta^2 \tag{14-55}$$

The equations of motion take the form

$$\ddot{x} + b\ddot{\theta} + g\theta = 0 \tag{14-56}$$

$$(M + m)\ddot{x} + mb\ddot{\theta} = 0 \tag{14-57}$$

[The same results could have been obtained, somewhat more laboriously, by leaving T and V in the original forms (14-51) and (14-52) and then making corresponding approximations in the equations of motion. This would have involved retaining only linear terms in the equations of motion.] Putting

$$x = Ae^{i\omega t} \qquad \theta = Be^{i\omega t}$$

where ω is a quantity to be determined, we have

$$-A\omega^2 - B\omega^2 b + gB = 0 \tag{14-58}$$
$$-(M + m)A\omega^2 - mbB\omega^2 = 0 \tag{14-59}$$

If the equations are to be consistent, we must have

$$\begin{vmatrix} \omega^2 & \omega^2 b - g \\ (M + m)\omega^2 & mb\omega^2 \end{vmatrix} = 0 \tag{14-60}$$

from which we obtain one root at once. This is

$$\pm\omega_1 = 0 \tag{14-61}$$

and the remaining root, from the equation

$$mb\omega^2 - (M + m)(\omega^2 b - g) = 0$$

is

$$\pm\omega_2 = \pm\sqrt{\frac{g(M + m)}{bM}} \tag{14-62}$$

Now since the root ω_1 is zero, the particular solution corresponding to ω_1 will have the form of Eq. (14-49) and is associated with a uniform translation of the center of mass of the system. The complete solutions may be written

$$x = A_1 t + A_1' + A_2 e^{i\omega_2 t} + A_{-2}e^{-i\omega_2 t} \tag{14-63}$$
$$\theta = B_1 t + B_1' + B_2 e^{i\omega_2 t} + B_{-2}e^{-i\omega_2 t} \tag{14-64}$$

Consulting Eqs. (14-58) and (14-59), we see that, at $\omega = \omega_1$, $B = 0$, and, at $\omega = \omega_2$,

$$B = -\frac{M + m}{mb}A$$

Accordingly, Eq. (14-64) takes the form

$$\theta = -\frac{M + m}{mb}(A_2 e^{i\omega_2 t} + A_{-2}e^{-i\omega_2 t}) \tag{14-65}$$

The normal coordinates may be obtained by making linear combinations of (14-63) and (14-65). Evidently these are

$$X_1 = \frac{M + m}{mb}x + \theta = \frac{M + m}{mb}(A_1 t + A_1') \tag{14-66}$$

$$X_2 = -\frac{mb}{M + m}\theta = A_2 e^{i\omega_2 t} + A_{-2}e^{-i\omega_2 t} \tag{14-67}$$

The modes of motion associated with these may be found readily. Putting $X_1 = 0$, we have

$$(M + m)x = -mb\theta \tag{14-68}$$

as the mode of X_2, the frequency of the motion being ω_2. This motion is an oscillation with respect to the center of mass. The mass m moves in one direction and M in the other. The mode associated with X_1 may be found by putting $X_2 = 0$. We obtain

$$\theta = 0 \tag{14-69}$$

a motion in which there is no oscillation relative to the center of mass but simply a uniform translation of the center of mass, following the equation

$$x = A_1 t + A_1'$$

It is to be noticed that, of the pair of space coordinates x and θ employed in the above example, only θ is zero in the equilibrium state. However, the choice is a satisfactory one in spite of the circumstance that x is not zero in the equilibrium configuration. This follows since V does not explicitly contain x, and consequently both T and V are appropriate homogeneous quadratic functions (of $\dot{x}$ and $\dot{\theta}$ in the case of T, and of θ in the case of V).

Example 14-2. Two masses m_1 and m_2 are joined by a massless spring of force constant k and unstretched length a. Obtain the normal coordinates of the system for motion along the line joining the particles.

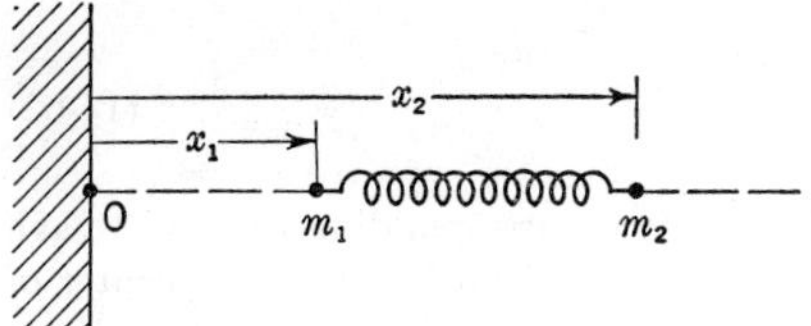

FIG. 14-5

Coordinates x_1 and x_2, as shown in Fig. 14-5, are measured from a fixed point O. The kinetic and potential energies are

$$T = \tfrac{1}{2}(m_1\dot{x}_1^2 + m_2\dot{x}_2^2) \tag{14-70}$$
$$V = \tfrac{1}{2}k(x_2 - x_1 - a)^2 \tag{14-71}$$

Clearly the coordinates are not suitable to permit of a linear transformation to normal coordinates, since the potential energy V is not a homogeneous quadratic function of x_1 and x_2. This difficulty can be overcome by the substitution

$$s = x_2 - a$$

whence

$$T = \tfrac{1}{2}(m_1\dot{x}_1^2 + m_2\dot{s}^2) \tag{14-72}$$
$$V = \tfrac{1}{2}k(s - x_1)^2 \tag{14-73}$$
$$L = T - V = \tfrac{1}{2}(m_1\dot{x}_1^2 + m_2\dot{s}^2) - \tfrac{1}{2}k(s - x_1)^2 \tag{14-74}$$

The equations of motion become

$$m_1\ddot{x}_1 + kx_1 - ks = 0 \tag{14-75}$$
$$m_2\ddot{s} + ks - kx_1 = 0 \tag{14-76}$$

Let

$$x_1 = Ae^{i\omega t} \qquad s = Be^{i\omega t} \tag{14-77}$$

Substituting (14-77) into (14-75) and (14-76), we obtain

$$\begin{aligned} (-m_1\omega^2 + k)A - kB &= 0 \\ -kA + (-m_2\omega^2 + k)B &= 0 \end{aligned} \tag{14-78}$$

The quantity ω is to be determined so that Eqs. (14-78) are mutually consistent. Hence

$$\begin{vmatrix} m_1\omega^2 - k & k \\ k & m_2\omega^2 - k \end{vmatrix} = 0 \tag{14-79}$$

This yields

$$(m_1\omega^2 - k)(m_2\omega^2 - k) - k^2 = 0 \tag{14-80}$$

or

$$\omega^2[m_1m_2\omega^2 - k(m_1 + m_2)] = 0 \tag{14-81}$$

Hence the permitted frequencies are

$$\pm\omega_1 = 0 \qquad \pm\omega_2 = \pm\left[\frac{k(m_1 + m_2)}{m_1m_2}\right]^{\frac{1}{2}} \tag{14-82}$$

Substituting these in either of Eqs. (14-78), we find

$$\begin{aligned} \text{At } \omega &= \omega_1 \qquad A = B \\ \text{At } \omega &= \omega_2 \qquad A = -\frac{m_2}{m_1}B \end{aligned} \tag{14-83}$$

Therefore the general solution may be written

$$x_1 = A_1t + A_1' + A_2e^{i\omega_2t} + A_{-2}e^{-i\omega_2t} \tag{14-84}$$

$$s = A_1t + A_1' - \frac{m_1}{m_2}(A_2e^{i\omega_2t} + A_{-2}e^{-i\omega_2t}) \tag{14-85}$$

Hence the normal coordinates are

$$X_1 = \frac{m_1}{m_2}x_1 + s = \frac{m_1 + m_2}{m_2}(A_1t + A_1') \tag{14-86}$$

$$X_2 = x_1 - s = \frac{m_1 + m_2}{m_2}(A_2e^{i\omega_2t} + A_{-2}e^{-i\omega_2t}) \tag{14-87}$$

As before, the appearance of the mode associated with each coordinate is obtained by suppressing the other coordinate. We have, for the form of X_1,

$$x_1 = s = x_2 - a \tag{14-88}$$

corresponding to a uniform translation of the center of mass. For X_2,

$$x_1 = -\frac{m_2}{m_1}s = -\frac{m_2}{m_1}(x_2 - a) \tag{14-89}$$

corresponding to an oscillation relative to the center of mass. These are depicted in Fig. 14-6. [In part (*b*), the case represented is for $m_1 > m_2$.]

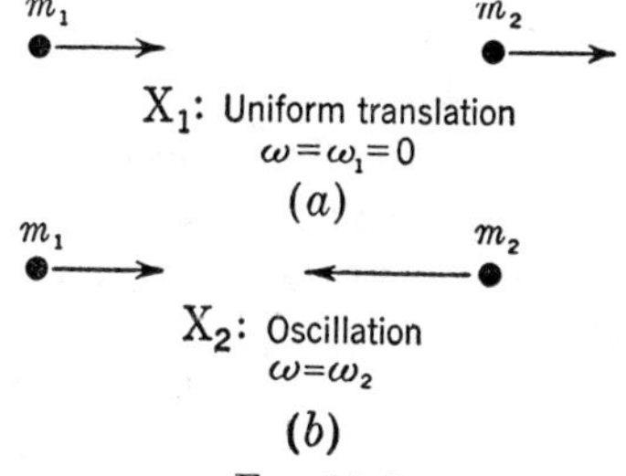

FIG. 14-6

Alternatively the problem could have been solved much more simply by recognizing, from inspection, the existence of the translational coordinate X_1. For the relative motion, a single equation involving the reduced mass μ of the system then can be employed. This at once yields the angular frequency $(k/\mu)^{\frac{1}{2}}$, which is identical to ω_2.

14-6. Dissipative Systems. It is possible, in certain cases, to find normal coordinates for systems in which some of the forces are dissipative. A frequently encountered dissipative system is that in which the motions of the particles are opposed by viscous forces proportional to the first powers of the velocities of the particles. The Newtonian equations of motion for the ith particle, of mass m_i, in such a system are

$$m_i\ddot{x}_i = X_i - k_i\dot{x}_i \qquad m_i\ddot{y}_i = Y_i - k_i\dot{y}_i \qquad m_i\ddot{z}_i = Z_i - k_i\dot{z}_i \tag{14-90}$$

in which k_i is a constant and X_i, Y_i, and Z_i are the components of the resultant of the remaining forces being applied to m_i. We restrict ourselves to a holonomic system containing n particles and in which the forces X_i, Y_i, Z_i, and so on, are derivable from a potential energy which is a homogeneous quadratic function of the coordinates.

Let the system possess r degrees of freedom after all the equations of condition have been utilized, thus making it possible to describe the system in terms of r independent coordinates $q'_1, \ldots, q'_r$. There will be $3n$ expressions of the type

$$\begin{aligned} x_i &= x_i(q'_i, \ldots, q'_r) \\ y_i &= y_i(q'_1, \ldots, q'_r) \\ z_i &= z_i(q'_1, \ldots, q'_r) \end{aligned} \tag{14-91}$$

in which no explicit dependence upon time is permitted, since in that event, following Sec. 13-2, the kinetic energy T would not be a homogeneous quadratic function of the velocities. In the manner of Sec. 13-4, we multiply each of Eqs. (14-90), respectively, by the quantities $\partial x_i/\partial q'_s$, $\partial y_i/\partial q'_s$, $\partial z_i/\partial q'_s$. Adding and summing over all the n particles, we have

$$\begin{aligned} \sum_{i=1}^{n} m_i \left(\ddot{x}_i \frac{\partial x_i}{\partial q'_s} + \ddot{y}_i \frac{\partial y_i}{\partial q'_s} + \ddot{z}_i \frac{\partial z_i}{\partial q'_s}\right) &= \sum_{i=1}^{n} \left(X_i \frac{\partial x_i}{\partial q'_s} + Y_i \frac{\partial y_i}{\partial q'_s} + Z_i \frac{\partial z_i}{\partial q'_s}\right) \\ &\quad - \sum_{i=1}^{n} k \left(\dot{x}_i \frac{\partial x_i}{\partial q'_s} + \dot{y}_i \frac{\partial y_i}{\partial q'_s} + \dot{z}_i \frac{\partial z_i}{\partial q'_s}\right) \end{aligned} \tag{14-92}$$

As before, the left side becomes

$$\frac{d}{dt}\left(\frac{\partial T}{\partial \dot{q}'_s}\right) - \frac{\partial T}{\partial q'_s}$$

Similarly the first sum on the right is the generalized force Q_s (excluding the dissipative forces), which, here, is $-\partial V/\partial q'_s$, where V is the potential energy. The last sum on the right side of Eq. (14-92), by virtue of Eq. (13-31), may be written

$$- \sum_{i=1}^{n} k_i \left(\dot{x}_i \frac{\partial \dot{x}_i}{\partial \dot{q}'_s} + \dot{y}_i \frac{\partial \dot{y}_i}{\partial \dot{q}'_s} + \dot{z}_i \frac{\partial \dot{z}_i}{\partial \dot{q}'_s}\right)$$

This is identical to

$$- \frac{\partial}{\partial \dot{q}'_s}\left[\frac{1}{2} \sum_{i=1}^{n} k_i(\dot{x}_i^2 + \dot{y}_i^2 + \dot{z}_i^2)\right] = - \frac{\partial F}{\partial \dot{q}'_s} \tag{14-93}$$

The quantity F, within the brackets, was termed the *dissipation function* by Rayleigh[1] and is equal to one-half the rate at which energy is being dissipated through the action of the viscous forces (see Sec. 7-9).

In terms of the coordinates $q'_1, \ldots, q'_r$ and the velocities $\dot{q}'_1, \ldots, \dot{q}'_r$ the three quantities V, T, and F have the form

$$V = a_{11}q_1'^2 + \cdots + a_{rr}q_r'^2 + 2a_{12}q'_1q'_2 + \cdots \tag{14-94}$$
$$T = b_{11}\dot{q}_1'^2 + \cdots + b_{rr}\dot{q}_r'^2 + 2b_{12}\dot{q}'_1\dot{q}'_2 + \cdots \tag{14-95}$$
$$F = c_{11}\dot{q}_1'^2 + \cdots + c_{rr}\dot{q}_r'^2 + 2c_{12}\dot{q}'_1\dot{q}'_2 + \cdots \tag{14-96}$$

in which the motions are sufficiently small so that a, b, and c are all approximately constant. As was pointed out in the previous section, the problem of normal modes is to find new coordinates which are linear combinations of $q'_1, \ldots, q'_r$ such that V and T, and here also F, are reduced to sums of squares of the new coordinates in the case of V and to sums of squares of the time derivatives of the new coordinates in the cases of T and F. Although, if F is zero, the possibility of such a transformation exists for V and T of the forms of (14-94) and (14-95), it does not always exist if F is not zero. Under certain circumstances, however, the transformation can be made. Such is the case if the dissipative force on each particle is proportional both to the mass of the particle and to its velocity. Let us consider that k_i, above, is proportional to m_i, and so on. It is to be noted that such a situation is a very special one in which F has the same form as T.

For the circumstance in which k_i is proportional to m_i, the same procedure as before may be employed to find the normal coordinates and frequencies. The solutions (14-48) to (14-50) will be slightly altered by the presence of the damping forces in a manner easily predicted by the simple one-dimensional examples in Chap. 7. Thus one expects to find solutions, which in the undamped problem are oscillations, to become underdamped, critically damped, or overdamped, as the case may be. The normal coordinates are the same, and the phases of the positional coordinates are the same, as those in the corresponding undamped problem. However, there are certain modifications. These are that the amplitudes decrease exponentially with the time and also that the frequencies will differ in the damped case from the corresponding ones in the undamped case (see Chap. 7).

14-7. Forced Oscillations. The preceding sections all have been concerned with free oscillations, that is, those which are caused by an arbitrary set of initial conditions but in which no driving forces are subsequently present. The procedure of normal coordinates occasionally is useful also for a vibrating system which is being subjected to certain

[1] Lord Rayleigh, "The Theory of Sound," p. 102, 1894 ed., reprinted in 1945 by Dover Publications.

types of driving forces. It is necessary, however, that the forces be not so great as to render appreciable the squares of the displacements or velocities, for in that event the equations of motion might no longer be linear. If the forces are constant, as would be the case if the system were in a uniform gravitational field for example, the only change arising would be to alter the equilibrium positions about which the oscillations are being executed. In all other respects the behavior would be that of free oscillations. Another case of driving forces in which the analysis in terms of normal coordinates may be convenient is that in which the force is periodic. As is shown in Appendix 5, a periodic function is equivalent to a superposition of harmonic terms. Thus it is sufficient, for purposes of discussion, to limit ourselves to a single harmonic force of the type $F \cos \omega t$, and where F and ω are constants. In general the effect of such a force applied to a particle of a system is to excite all the normal coordinates except those in which the particle does not execute a motion. Thus those normal coordinates are excited in which the particle normally moves, and those in which the particle is at a *node* are not excited (see, in this regard, Prob. 14-13).

It is possible to apply forces in such a manner that but one normal coordinate is excited. Take, for example, the system of coupled pendulums which has already been considered. Let a force $F \cos \omega t$ be applied to each particle in Fig. 14-1, having the same direction in each case. For the sake of generality suppose damping forces to be acting which are proportional to the velocities of the particles, with constant of proportionality R in each case. The equations of motion (14-1) and (14-2) now become

$$m\ddot{x}_1 + \frac{mg}{b} x_1 + k(x_1 - x_2) = -R\dot{x}_1 + F \cos \omega t \tag{14-97}$$

$$m\ddot{x}_2 + \frac{mg}{b} x_2 - k(x_1 - x_2) = -R\dot{x}_2 + F \cos \omega t \tag{14-98}$$

and Eqs. (14-21) and (14-22), involving the normal coordinates, assume the form

$$\ddot{X}_1 + \frac{R}{m}\dot{X}_1 + \frac{g}{b} X_1 = \frac{2F}{m} \cos \omega t \tag{14-99}$$

$$\ddot{X}_2 + \frac{R}{m}\dot{X}_2 + \left(\frac{g}{b} + \frac{2k}{m}\right) X_2 = 0 \tag{14-100}$$

with solutions (employing the methods of Chap. 7)

$$X_1 = e^{-(R/2m)t}\,(A_1 e^{i\omega'_1 t} + A_{-1} e^{-i\omega'_1 t}) + \frac{2F}{[m^2(\omega_0{}^2 - \omega^2)^2 + \omega^2 R^2]^{\frac{1}{2}}} \cos (\omega t - \varphi) \tag{14-101}$$

$$X_2 = e^{-(R/2m)t}(A_2 e^{i\omega'_2 t} + A_{-2} e^{-i\omega'_2 t}) \tag{14-102}$$

where

$$\omega_0 = \left(\frac{g}{b}\right)^{\frac{1}{2}} \qquad \omega_1' = \left(\frac{g}{b} - \frac{R^2}{4m^2}\right)^{\frac{1}{2}} \qquad \omega_2' = \left[\left(\frac{g}{b} + \frac{2k}{m}\right) - \frac{R^2}{4m^2}\right]^{\frac{1}{2}} \tag{14-103}$$

and $\tan \varphi = \omega R/[m(\omega_0^2 - \omega^2)]$. (We restrict ourselves to the case in which $g/b > R^2/4m^2$.)

It is interesting to note that, although both X_1 and X_2 contain transient terms, only X_1 possesses a steady-state term. This follows as a result of the two driving forces being chosen in the same phase.

Owing to the presence of damping, the transient parts will decay with time, so that ultimately only X_1 will remain excited, regardless of the initial conditions. Furthermore, the frequency associated with the excitation of X_1 will then be that of the driving force. This behavior is quite similar to that of the system with one degree of freedom considered in Chap. 7.

14-8. Vibrations of Molecules. The method of normal coordinates may be applied, with rich rewards, to the study of the vibrations of molecules. It is exceedingly helpful in the physical interpretation of their vibration spectra. The interested student is advised to consult the review article of D. M. Dennison[1] on the subject.

A simple example is provided by the CO_2 molecule, which can be represented as a linear structure (see Fig. 14-7). The determination of the frequencies, and so on, for motion along the line joining the atoms is left to the student in Prob. 14-5. Since there are three degrees of freedom for motion constrained in such a manner, there will be three normal coordinates. It is evident, by inspection, that one will have a zero frequency, $\omega_1 = 0$, and will correspond to simple translation of the center of mass. The modes associated with the remaining two are shown in Fig. 14-8. In Fig. 14-8a, the mode is that in which the carbon atom is stationary, the two oxygen atoms oscillating back and forth in opposite phase, as shown, and with equal amplitudes. In the mode of ω_3, shown in Fig. 14-8b, the carbon atom experiences motion with respect to the mass center and is in opposite phase from that of the two oxygen atoms. It is interesting to point out that, of these two frequencies, only ω_3 will be observed optically. The frequency ω_2 is not observed since, in this mode, the electrical center of the

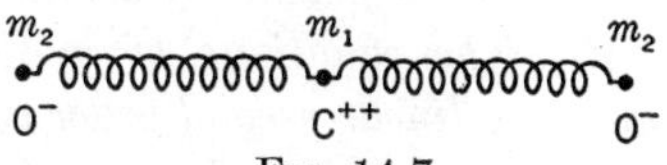

m_2 m_1 m_2
O^- C^{++} O^-

FIG. 14-7

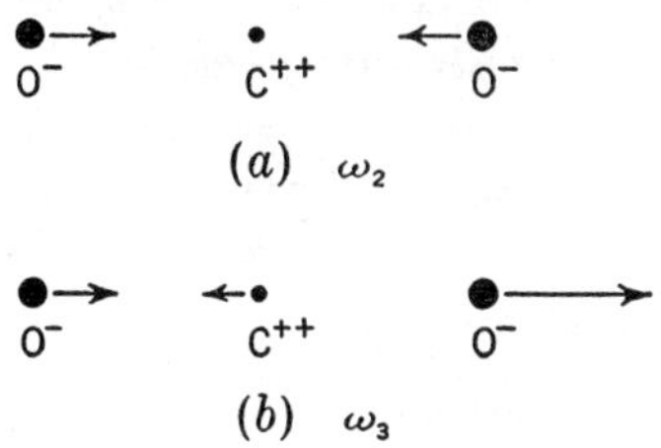

O^- C^{++} O^-
(a) ω_2

O^- C^{++} O^-
(b) ω_3

FIG. 14-8

[1] D. M. Dennison, Infrared Spectra of Polyatomic Molecules, *Revs. Mod. Phys.*, **3**, 280 (1931). Unfortunately for the beginning student, the methods of the quantum mechanics are extensively employed.

system remains always coincident with the center of mass. Thus there is no oscillating electric moment $\Sigma e\mathbf{r}$, and consequently no radiation is emitted corresponding to this mode. In the mode illustrated in Fig. 14-8*b*, there is such a moment, and radiation is emitted.

14-9. Summary of Properties of Normal Coordinates. It is desirable, in conclusion, to summarize some of the main features of normal coordinates. These are as follows:

I. *Any undamped vibrating system having r degrees of freedom, described by the coordinates* $q'_1, \ldots, q'_r$, *and such that the potential energy is a homogeneous quadratic function of these coordinates and the kinetic energy a homogeneous quadratic function of the time derivatives of these coordinates, may be described in terms of r normal coordinates,* $q_1, \ldots, q_r$. *The latter are linear combinations of the primed set. (Under certain conditions of damping the transformation is possible also.) Each normal coordinate* q_s *has a characteristic angular frequency* ω_s *associated with it.*

II. *The potential energy of a vibrating system, in terms of the normal coordinates, is a sum of squares of the normal coordinates, the coefficients of all the terms being constants.*

III. *The kinetic energy of a vibrating system, in terms of the normal coordinates, is a sum of squares of the time derivatives of the normal coordinates, the coefficients of all the terms being constants.*

IV. *The equations of motion, in terms of the normal coordinates, are ordinary linear second-order differential equations with constant coefficients. There is one for each normal coordinate, which then is the single dependent variable occurring in the equation; the single independent variable is the time.*

V. *The normal coordinates are completely independent; no energy is spontaneously transferred to and fro among them.*

VI. *The kinematical picture presented by a vibrating system of particles having one normal coordinate excited is that certain of the particles may be at rest in their equilibrium positions, while the remainder are moving in phase and with a common frequency. All moving particles pass through their equilibrium points simultaneously.*

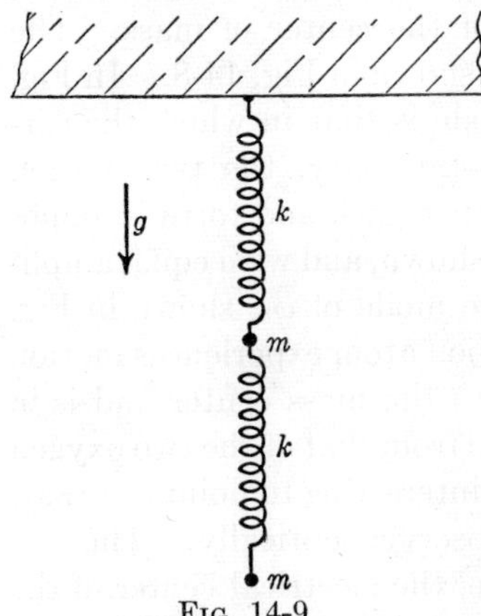

FIG. 14-9

Problems

14-1. A pair of identical pendulums coupled by a spring, as in the text, are moving in a viscous medium which produces a retarding force proportional to the velocity. Find the normal coordinates by transforming the equations of motion directly to ones involving but one dependent variable each.

14-2. Two equal masses are connected as shown in Fig. 14-9, with two identical massless springs of force constant k. Gravity is acting vertically downward. Find the frequencies of the system, considering the motion to be only in the vertical direction. Find the normal coordinates.

14-3. Three equal masses m are connected, as shown in Fig. 14-10, by four equal springs. The tension is S, and the points A and B are fixed. Find the normal modes for transverse motion in one plane. In the equilibrium configuration the length of each spring is a.

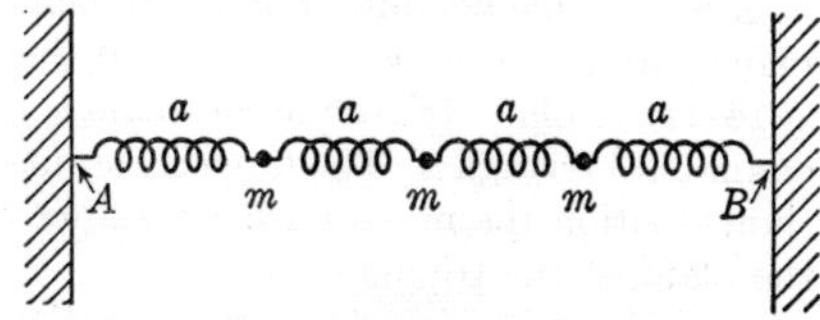

FIG. 14-10

14-4. In the previous problem take the system to be just without tension in the equilibrium position, and find the normal modes for motion along the line joining the masses. Take the force constant of the springs to be k.

14-5. Consider the linear triatomic molecule. The central particle has a mass m_1, and the two end particles each has a mass m_2. Find the normal coordinates and the modes of vibration associated with each. Consider motion only along the line joining the masses. (Assume an elastic bond, of force constant k, to be acting between each m_2 and m_1. Assume no interaction between the end particles.)

14-6. Three masses m, m, and M are interconnected by identical springs of force constant k and placed on a smooth circular loop of wire as shown in Fig. 14-11. The loop is fixed in space. Determine the normal coordinates and frequencies of the system for the general case.

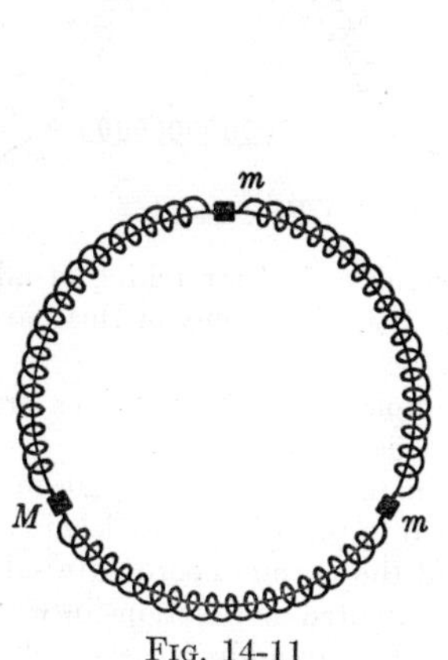

FIG. 14-11

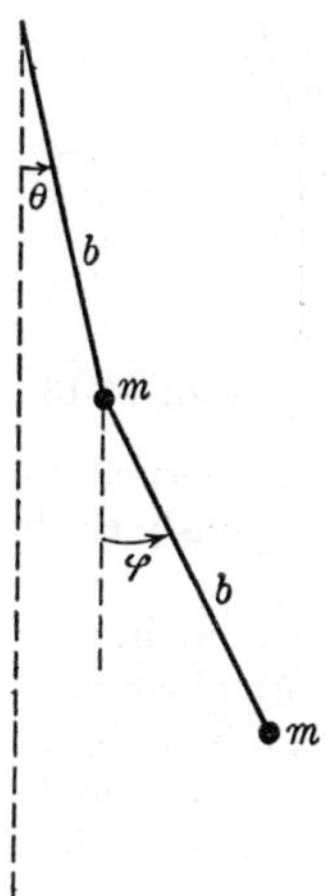

FIG. 14-12

14-7. The balance wheel of a watch oscillates under the action of the spring, with a period p when the case is held fixed. The watch is now laid on a smooth horizontal table. If the moment of inertia of the balance wheel is I and that of the rest of the watch is I_1, both about normal axes passing through the center, find the period of the watch in terms of p, I, I_1. The watch case and the balance wheel are taken coaxial. Will the watch gain or lose?

14-8. In Fig. 14-12, two masses m are connected by weightless inelastic strings of length b. Find the frequencies and the normal coordinates for motion in one plane.

14-9. A smooth wire of mass M bent into the form of a circle is suspended from a point on its circumference. A bead of mass m can slide on the wire. Find the normal modes if the wire is free to swing in its own plane.

14-10. The light rod OA, of length b (Fig. 14-13), is free to move in the xy plane about a pin at O. There is a restoring torque $k\theta$ acting in which k is a constant and θ is measured from the x axis. A mass m is attached to the end A of the rod. Suspended at A is a simple pendulum of length b and mass m which executes vibrations only in a plane parallel to the yz plane. Find the normal modes of oscillation.

14-11. A plane triatomic molecule consists of equal masses m at the vertices of an equilateral triangle. The force constants are identical and equal to k. In the equilibrium position the masses are a distance a apart. Consider the motion to be only in the plane of the triangle.

a. Without writing down the equations of motion, state how many normal modes you expect to find and how many of them are associated with zero frequencies. Give reasons.

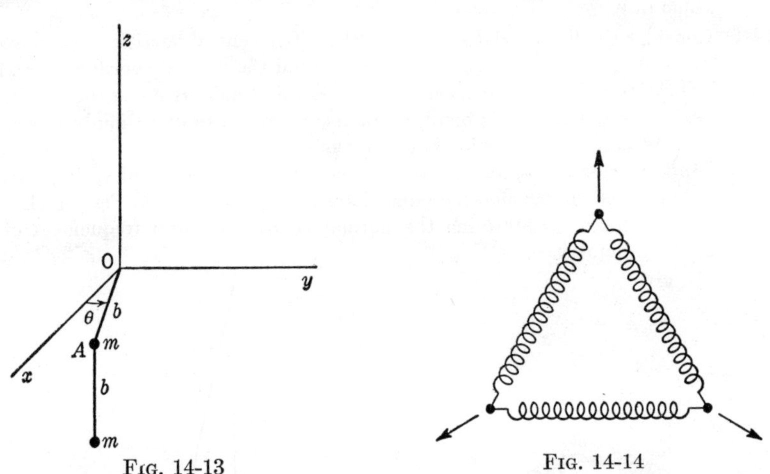

FIG. 14-13 FIG. 14-14

b. One of the normal modes corresponds to a symmetrical stretching of all three bands simultaneously as shown in Fig. 14-14. Find the frequency of this particular mode.

14-12. Work out in detail the plane triatomic molecule of the previous problem, finding the normal coordinates and frequencies associated with each.

14-13. In Prob. 14-3, all particles are initially in the equilibrium position, and a force $F \cos \omega t$, where F and ω are constants, is applied to the middle particle in a direction normal to the line joining the particles. Find the normal coordinate solutions, that is, express those normal coordinates which are excited as functions of time.

14-14. The same as the preceding problem save that, in addition, a small viscous force, proportional to the velocity (small proportionality constant R), is acting on each particle so as to oppose the motion. Express the normal coordinates as functions of time. What are the steady-state solutions? What frequencies are present in the steady state?

14-15. Three equal particles of mass m are joined together by two equal massless springs of force constant k. The whole is free to move along the line joining the three particles and is immersed in a medium which exerts a retarding force proportional to the velocity (proportionality constant $2\sqrt{km}$). Find the normal coordinates and frequencies of the system for motion along the line joining the particles.

14-16. Two equal particles of mass m are connected, as shown in Fig. 14-15, by equal massless springs to two fixed points A and B a distance $3a$ apart. The tension in the system is S, and the whole is immersed in a viscous medium which exerts a retarding force proportional to the velocity (proportionality constant $2\sqrt{mS/a}$). Initially one of the masses is held fixed, while the other is displaced a small distance b

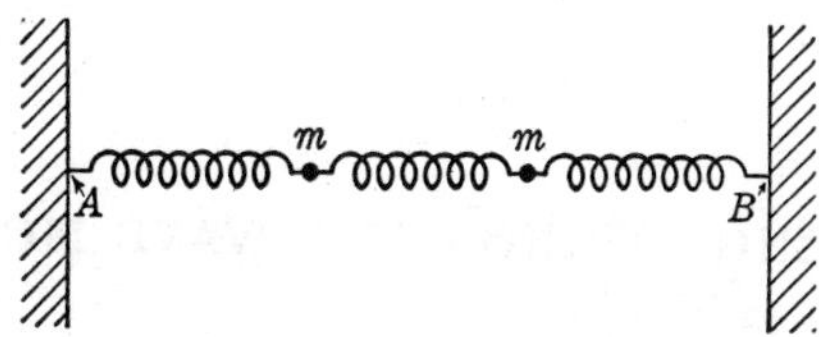

FIG. 14-15

normal to AB. If the two masses are then released, find their subsequent positions as functions of time. Express those normal coordinates which are excited in terms of the initial conditions.

14-17. In Prob. 14-2, let the entire system be situated in a viscous medium which opposes the motions of the masses with a force proportional to the velocity. The constant of proportionality is equal to $[2mk(3 - \sqrt{5})]^{\frac{1}{2}}$.

a. Determine the frequencies and the general solution for an arbitrary disturbance. In a diagram show the appearance of the modes associated with the normal coordinates.

b. If now the initial conditions are that the lower mass is displaced downward a distance b from its equilibrium position, while the upper mass is displaced a distance $2b/(\sqrt{5} - 1)$ upward from its equilibrium position, the whole being then released, find the displacement of each as a function of time, evaluating the arbitrary constants.

CHAPTER 15

VIBRATING STRINGS AND WAVE MOTION

In this chapter we shall consider a vibrating system having a very large number of degrees of freedom, the small lateral motions of a heavy string which is under tension. We shall show that it is possible to represent a string, so disturbed, either in terms of a normal coordinate representation or in terms of transverse waves traveling along the string.

15-1. Equation of Motion. We confine ourselves to a homogeneous string of length l which is fixed at both ends and is under tension. We assume, furthermore, that the lateral displacements are small compared with the length, the angle θ between any small segment of the string and the straight line joining the points of support being taken sufficiently small so that $\sin\theta$ is closely approximated by $\tan\theta$. In the same way the tension T_0 in the string is assumed not to be altered by the small lateral displacements. The coordinate system is such that in its equilibrium position the string is coincident with the x axis, with one point of support at $x = 0$ and another at $x = l$ (see Fig. 15-1). The motion will be restricted to the xy plane.

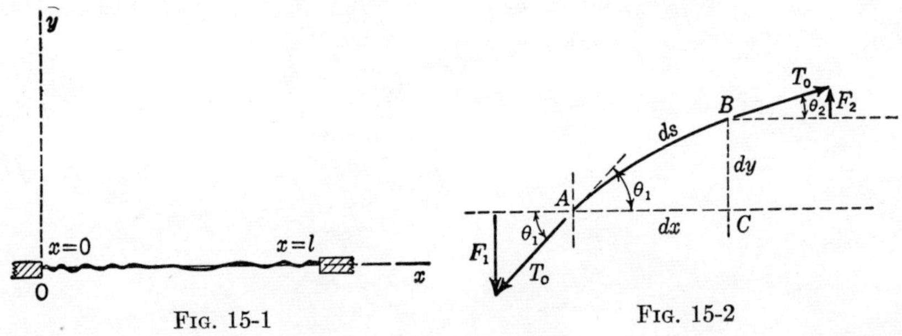

FIG. 15-1 FIG. 15-2

We may obtain the differential equation of motion by considering a small element ds of the string, shown in exaggerated fashion as the segment AB in Fig. 15-2. The y components of the forces acting on ds are the components F_1 and F_2 of the tension T_0 exerted upon ds by the adjacent portions of the string. Clearly if, in the figure, θ_1 and θ_2 are small,

$$F_1 = T_0 \sin \theta_1 \simeq T_0 \tan \theta_1 = T_0 \left(\frac{\partial y}{\partial x}\right)_A \tag{15-1}$$

$$F_2 = T_0 \sin \theta_2 \simeq T_0 \tan \theta_2 = T_0 \left(\frac{\partial y}{\partial x}\right)_B \tag{15-2}$$

where the derivatives are partials, since y depends upon t as well as x. The subscripts A and B signify that the derivatives are to be evaluated, respectively, at points A and B. Now

$$\left(\frac{\partial y}{\partial x}\right)_A = \frac{\partial y}{\partial x} - \left(\frac{\partial}{\partial x}\frac{\partial y}{\partial x}\right)\frac{dx}{2} = \frac{\partial y}{\partial x} - \frac{\partial^2 y}{\partial x^2}\frac{dx}{2} \tag{15-3}$$

$$\left(\frac{\partial y}{\partial x}\right)_B = \frac{\partial y}{\partial x} + \left(\frac{\partial}{\partial x}\frac{\partial y}{\partial x}\right)\frac{dx}{2} = \frac{\partial y}{\partial x} + \frac{\partial^2 y}{\partial x^2}\frac{dx}{2} \tag{15-4}$$

in which the derivatives without subscripts are evaluated at the midpoint of ds. Then the resultant force in the y direction is

$$F_2 - F_1 = T_0 \frac{\partial^2 y}{\partial x^2} dx \tag{15-5}$$

If ρ is the mass per unit length of the string, the inertial reaction of the element ds is $\rho\, ds(\partial^2 y/\partial t^2)$. Also, for small displacements, $ds \simeq dx$. The equation of motion is obtained by equating the inertial reaction to the applied force given by Eq. (15-5). Accordingly, canceling dx on each side, we obtain finally

$$\frac{\partial^2 y}{\partial t^2} = \frac{T_0}{\rho}\frac{\partial^2 y}{\partial x^2} \tag{15-6}$$

15-2. General Solution. For a partial differential equation of the type of Eq. (15-6) it is possible to find a solution of the form

$$y = \Phi(t)\Psi(x) \tag{15-7}$$

in which Φ is a function of t alone and Ψ is a function of x alone. Substituting (15-7) in (15-6), we have

$$\Psi \frac{d^2\Phi}{dt^2} = \frac{T_0}{\rho}\Phi\frac{d^2\Psi}{dx^2} \tag{15-8}$$

Dividing through by $\Phi\Psi$, Eq. (15-8) becomes

$$\frac{1}{\Phi}\frac{d^2\Phi}{dt^2} = \frac{T_0}{\rho}\frac{1}{\Psi}\frac{d^2\Psi}{dx^2} \tag{15-9}$$

Since the left side of Eq. (15-9) depends only on t and the right side depends only on x, the only possible way in which the equality will hold for all values of x and t is for both sides to be equal to a constant. Let this constant be $-\omega^2$. The minus sign signifies that the acceleration of each element of the string is at all times directed toward the equilibrium position, in which the displacement is zero. Accordingly the left side of (15-9) yields

$$\frac{d^2\Phi}{dt^2} + \omega^2\Phi = 0 \tag{15-10}$$

This possesses the general solution

$$\Phi = C \sin \omega t + D \cos \omega t \tag{15-11}$$

in which ω is to be interpreted as an angular frequency. Similarly the right side of Eq. (15-9) furnishes

$$\frac{d^2\Psi}{dx^2} + \frac{\rho\omega^2}{T_0}\Psi = 0 \tag{15-12}$$

having as a general solution

$$\Psi = E \sin \omega \sqrt{\frac{\rho}{T_0}}\, x + F \cos \omega \sqrt{\frac{\rho}{T_0}}\, x \tag{15-13}$$

In Eqs. (15-11) and (15-13) the quantities C, D, E, and F are constants of integration. Thus, from Eq. (15-7),

$$y = (C \sin \omega t + D \cos \omega t)\left(E \sin \omega \sqrt{\frac{\rho}{T_0}}\, x + F \cos \omega \sqrt{\frac{\rho}{T_0}}\, x\right) \tag{15-14}$$

Now, at $x = 0$, $y = 0$. Therefore $F = 0$, and we have left

$$y = (A \sin \omega t + B \cos \omega t) \sin \omega \sqrt{\frac{\rho}{T_0}}\, x \tag{15-15}$$

in which A and B are new constants (CE and DE, respectively). But at $x = l$ we must again have $y = 0$ for any value of the time. This condition will be satisfied provided that

$$\omega \sqrt{\frac{\rho}{T_0}}\, l = n\pi \qquad \text{where} \qquad n = 1, 2, 3, 4, \ldots \tag{15-16}$$

and there is a different solution y_n, for each distinct value of the integer n. The nth particular solution may be written

$$y_n = (A_n \sin \omega_n t + B_n \cos \omega_n t) \sin \omega_n \sqrt{\frac{\rho}{T_0}}\, x \tag{15-17}$$

Since the differential equation (15-6) is linear, the general solution is obtained by adding together all the n particular solutions. Also

$$\omega_n = \sqrt{\frac{T_0}{\rho}}\frac{n\pi}{l} \tag{15-18}$$

Hence

$$y = \sum_{n=1}^{\infty} (A_n \sin \omega_n t + B_n \cos \omega_n t) \sin \frac{n\pi}{l} x \tag{15-19}$$

a solution containing an infinite number of arbitrary constants.

An expression consisting of sums of sines and/or cosines of this type is called a *Fourier series*, and each of the quantities y_n is called a *Fourier component*. The corresponding quantities A_n and B_n are called *Fourier coefficients*. In Fig. 15-3 are shown graphs of y_n for the cases $n = 1, 2, 3, 4$. The integer n is equal to the number of *loops* exhibited by the vibrating string when only y_n is excited. The mode of vibration in which $n = 1$ is called the *fundamental*, or *first harmonic*. The modes in which $n = 2$, 3, 4, and so on, are called, respectively, the *second*, *third*, and *fourth harmonics*. Clearly, from (15-18), the frequency of the nth harmonic is n times the fundamental frequency. It is to be emphasized, in the above, that the allowable frequencies constitute a discrete set, even though there is an infinite number of them. This discreteness is introduced by the presence of the boundary conditions $y = 0$ at $x = 0, l$.

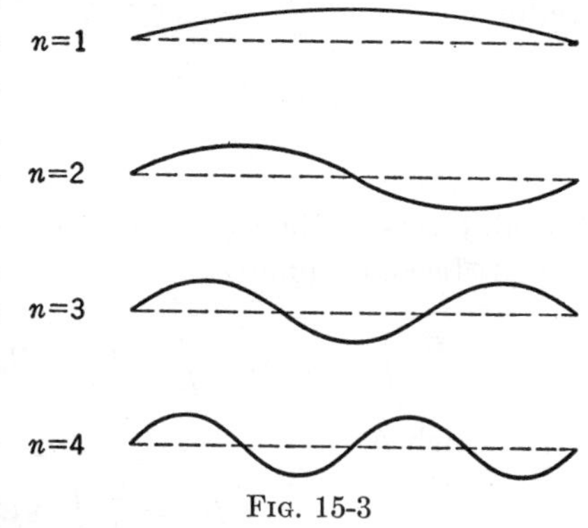

FIG. 15-3

It is sometimes useful to alter slightly the form of Eq. (15-19). By use of obvious trigonometric identities it may be expressed as

$$y = \sum_{n=1}^{\infty} P_n \sin (\omega_n t - \alpha_n) \sin \frac{n\pi}{l} x \tag{15-20}$$

in which

$$\tan \alpha_n = -\frac{B_n}{A_n}$$

When the solution is expressed in the form (15-20), the Fourier coefficients A_n and B_n of the nth harmonic are replaced by the *amplitude* P_n and the *phase constant* α_n of the nth harmonic.

15-3. Determination of the Coefficients A_n and B_n. The general solution (15-19) is completely known when each of the coefficients A_n and B_n is known. These may be determined in terms of the initial conditions, that is, the initial values of y and $\partial y/\partial t$. Each of these may be expressed as functions of x, as

$$(y)_0 = u(x) \qquad \left(\frac{\partial y}{\partial t}\right)_0 = v(x) \tag{15-21}$$

in which the zero subscripts indicate that the values at $t = 0$ are employed. Accordingly, at $t = 0$, from Eq. (15-19),

$$u(x) = \sum_{n=1}^{\infty} B_n \sin \frac{n\pi}{l} x \tag{15-22}$$

Multiplying through by $\sin (m\pi/l)x$ and integrating from $x = 0$ to $x = l$, we have

$$\int_0^1 u(x) \sin \frac{m\pi}{l} x \, dx = \int_0^l \left(\sum_{n=1}^{\infty} B_n \sin \frac{n\pi}{l} x \sin \frac{m\pi}{l} x \right) dx \tag{15-23}$$

But all terms on the right side vanish except that in which $n = m$. Thus (15-23) becomes finally

$$\int_0^l u(x) \sin \frac{m\pi}{l} x \, dx = B_m \int_0^l \sin^2 \frac{m\pi}{l} x \, dx = B_m \frac{l}{2}$$

from which

$$B_m = \frac{2}{l} \int_0^l u(x) \sin \frac{m\pi}{l} x \, dx \tag{15-24}$$

The values of the coefficients A_n can be found in a similar manner. From Eq. (15-19)

$$v(x) = \sum_{n=1}^{\infty} A_n \omega_n \sin \frac{n\pi}{l} x \tag{15-25}$$

Multiplying both sides of Eq. (15-25) by $\sin (m\pi/l)x$ and integrating from $x = 0$ to $x = l$, we obtain finally

$$A_m = \frac{2}{\omega_m l} \int_0^l v(x) \sin \frac{m\pi}{l} x \, dx \tag{15-26}$$

Equations (15-24) and (15-26) show that if the displacements and velocities of all points of the string are given at one time, for example initially, the motion is determined for all later times.

Example 15-1. A string of length l, mass per unit length ρ, and tension T_0 is initially displaced a distance b (where $b \ll l$) at the middle of the string and is then released. It is desired to determine the Fourier coefficients for the subsequent motion.

The initial configuration is given in Fig. 15-4. It is evident that

$$\text{At } 0 < x < \frac{l}{2} \qquad u = \frac{2b}{l}\,x \tag{15-27}$$

$$\text{At } \frac{l}{2} < x < l \qquad u = \frac{2b}{l}\,(l - x) \tag{15-28}$$

Now, since the initial configuration is one of rest, $v(x) = 0$. From this it follows at

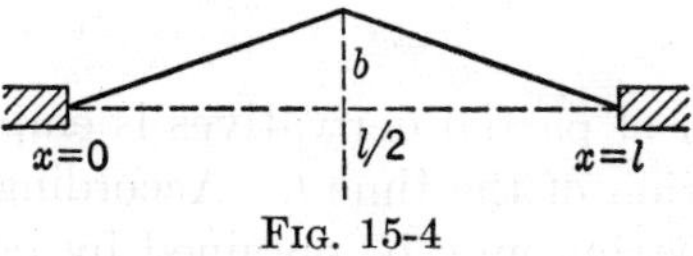

FIG. 15-4

once that $A_n = 0$ for all n. The values of the B's are determined by use of Eqs. (15-27) and (15-28). We have

$$B_m = \frac{2}{l}\left[\frac{2b}{l}\int_0^{l/2} x \sin\frac{m\pi}{l}\,x\,dx + \frac{2b}{l}\int_{l/2}^{l} (l - x)\sin\frac{m\pi}{l}\,x\,dx\right] \tag{15-29}$$

This may be integrated with the aid of tables. The result is

$$B_m = \frac{8b}{m^2\pi^2}\sin m\,\frac{\pi}{2} \qquad m \text{ odd} \tag{15-30}$$

$$B_m = 0 \qquad m \text{ even}$$

Accordingly the general solution has the form

$$y = \frac{8b}{\pi^2}\left(\cos\frac{\pi}{l}\sqrt{\frac{T_0}{\rho}}\,t\,\sin\frac{\pi}{l}\,x - \frac{1}{9}\cos\frac{3\pi}{l}\sqrt{\frac{T_0}{\rho}}\,t\,\sin\frac{3\pi}{l}\,x + \frac{1}{25}\cos\frac{5\pi}{l}\sqrt{\frac{T_0}{\rho}}\,t\,\sin\frac{5\pi}{l}\,x - \cdots\right) \tag{15-31}$$

with only odd harmonics having been excited. Thus none of the harmonics have been stimulated which possess a node at the mid-point.

15-4. Energy and Normal Coordinates of a Vibrating String. It is of interest to find expressions for the kinetic and potential energies of a vibrating string of length l which is fixed at both ends. Consider the element ds of the string shown in Fig. 15-2. In the equilibrium configuration the length of the element is dx. When it is displaced as shown in the figure, the amount of potential energy stored in this element of the string is

$$dV = T_0(ds - dx) = T_0\left(\frac{ds}{dx} - 1\right)dx \tag{15-32}$$

in which the potential energy is zero when the string is unstretched. Now

$$\frac{ds}{dx} = \left[1 + \left(\frac{dy}{dx}\right)^2\right]^{\frac{1}{2}}$$

Substituting this in Eq. (15-32) and expanding by means of the binomial theorem, we have, when $dy/dx \ll 1$,

$$dV = T_0\left[\sqrt{1+\left(\frac{dy}{dx}\right)^2}-1\right]dx = T_0\left[1+\frac{1}{2}\left(\frac{dy}{dx}\right)^2 + \cdots -1\right]dx$$

$$\simeq \frac{T_0}{2}\left(\frac{\partial y}{\partial x}\right)^2 dx \tag{15-33}$$

in which the notation of partial derivatives is employed in the last step since y is also a function of the time t. Accordingly the total potential energy stored in the string may be obtained by integrating Eq. (15-33) over the length of the string. We have

$$V = \frac{T_0}{2}\int_0^l \left(\frac{\partial y}{\partial x}\right)^2 dx \tag{15-34}$$

The expression for the kinetic energy is readily found. The kinetic energy of the element of mass $\rho\, dx$ is $\frac{1}{2}\rho(\partial y/\partial t)^2\, dx$, the difference between ds and dx being neglected. Thus the total kinetic energy of the string may be written

$$T = \frac{\rho}{2}\int_0^1 \left(\frac{\partial y}{\partial t}\right)^2 dx \tag{15-35}$$

Equations (15-34) and (15-35) are useful in identifying the normal coordinates of the vibrating string. For this we rewrite Eq. (15-17) as

$$y_n = \phi_n \sin \omega_n \sqrt{\frac{\rho}{T_0}}\, x \tag{15-36}$$

where ϕ_n is defined as

$$\phi_n = A_n \sin \omega_n t + B_n \cos \omega_n t \tag{15-37}$$

Similarly Eq. (15-19) becomes

$$y = \sum_{n=1}^{\infty} \phi_n \sin \frac{n\pi}{l}\, x \tag{15-38}$$

Let us examine the properties of the quantities ϕ_n. Clearly

$$\frac{\partial y}{\partial x} = \frac{\pi}{l}\sum_{n=1}^{\infty} n\phi_n \cos \frac{n\pi}{l}\, x$$

In view of this, Eq. (15-34) becomes

$$V = \frac{\pi^2 T_0}{2l^2}\sum_{n=1}^{\infty}\sum_{m=1}^{\infty}\left(nm\phi_n\phi_m \int_0^l \cos\frac{n\pi}{l}x \cos\frac{m\pi}{l}x\, dx\right) \tag{15-39}$$

But all terms in the sum vanish upon integrating, save those in which $n = m$. For each of these the value of the integral is simply $l/2$. Hence Eq. (15-39) reduces to

$$V = \frac{\pi^2 T_0}{4l} \sum_{n=1}^{\infty} n^2 \phi_n^2 \tag{15-40}$$

expressed in terms of the functions ϕ_n, where $n = 1, 2, 3, \ldots$. The situation for the kinetic energy is entirely similar. Since

$$\frac{\partial y}{\partial t} = \sum_{n=1}^{\infty} \dot{\phi}_n \sin \frac{n\pi}{l} x$$

Eq. (15-35) takes the form

$$T = \frac{\rho}{2} \sum_{n=1}^{\infty} \sum_{m=1}^{\infty} \left(\dot{\phi}_n \dot{\phi}_m \int_0^l \sin \frac{n\pi}{l} x \sin \frac{m\pi}{l} x \, dx \right) \tag{15-41}$$

Here, as in the case of Eq. (15-39), all terms vanish save those in which $n = m$, and, since in the latter instance the integral again has the value $l/2$, Eq. (15-41) reduces to

$$T = \frac{\rho l}{4} \sum_{n=1}^{\infty} \dot{\phi}_n^2 \tag{15-42}$$

From the above discussion we see that the potential energy is a sum of quantities which are of the form $a_n \phi_n^2$, where a_n is a constant. Similarly we note that the kinetic energy is a sum of quantities of the form $b_n \dot{\phi}_n^2$, in which b_n is a constant. Furthermore, the Lagrangian function can be written, from Eqs. (15-40) and (15-42),

$$L = T - V = \frac{1}{4} \sum_{n=1}^{\infty} \left(\rho l \dot{\phi}_n^2 - \frac{\pi^2 T_0}{l} n^2 \phi_n^2 \right) \tag{15-43}$$

and the Lagrangian equation of motion for ϕ_n becomes

$$\ddot{\phi}_n + \frac{\pi^2 T_0}{\rho l^2} n^2 \phi_n = 0 \tag{15-44}$$

a linear second-order equation involving but one dependent variable ϕ_n and one independent variable t. Clearly the function ϕ_n has the properties of a normal coordinate and may be so regarded. Since, for the general case, n may take all values from 1 to ∞, we note that such a vibrating string is described by an infinite number of normal coordinates. It is evident, also, that the four curves shown in Fig. 15-3 are just the modes of motion associated with the first four of the normal coordinates, ϕ_1, ϕ_2, ϕ_3, and ϕ_4.

The total energy W of the string, expressed in terms of the ϕ's, can be obtained by adding together Eqs. (15-40) and (15-42). We obtain

$$W = T + V = \frac{1}{4}\sum_{n=1}^{\infty}\left(\rho l\dot{\phi}_n^2 + \frac{\pi^2 T_0 n^2}{l}\phi_n^2\right) \tag{15-45}$$

This can be expressed also in terms of the coefficients A_n and B_n. We have, from Eqs. (15-37) and (15-45), and finally from Eq. (15-18),

$$W = \frac{M}{4}\sum_{n=1}^{\infty}[\omega_n^2(A_n^2 + B_n^2)] = \pi^2 M \sum_{n=1}^{\infty}\nu_n^2(A_n^2 + B_n^2) \tag{15-46}$$

in which $M = \rho l$ is the total mass of the string. In the last expression the angular frequency ω_n of the nth mode has been replaced by the frequency $\nu_n = \omega_n/2\pi$.

15-5. Damped and Forced Motion of a Vibrating String. The inclusion of a dissipative force in the problem of a uniform vibrating string is simple for the case in which all points of the string are retarded by a force proportional to the velocity. The force of this type on an element dx of the string may be written as

$$-R\frac{\partial y}{\partial t}dx$$

in which R is the damping coefficient (per unit length). The negative sign signifies that the force is directed oppositely to $\partial y/\partial t$. If, in addition, we assume an applied force $F(x,t)$ per unit length to be acting in the y direction, the equation of motion may be written

$$\rho\frac{\partial^2 y}{\partial t^2} + R\frac{\partial y}{\partial t} - T_0\frac{\partial^2 y}{\partial x^2} = F(x,t) \tag{15-47}$$

The solution of (15-47) may be found readily by the normal coordinate method. Thus, substituting

$$y = \sum_{n=1}^{\infty}\phi_n \sin\frac{n\pi}{l}x \tag{15-48}$$

in Eq. (15-47), we obtain

$$\sum_{n=1}^{\infty}\left[\left(\ddot{\phi}_n + \frac{R}{\rho}\dot{\phi}_n + \frac{n^2\pi^2}{l^2}\frac{T_0}{\rho}\phi_n\right)\sin\frac{n\pi}{l}x\right] = \frac{1}{\rho}F(x,t) \tag{15-49}$$

Multiplying through by sin $(m\pi/l)x$ and integrating over the range of x

in the manner of Sec. 15-3, we have

$$\sum_{n=1}^{\infty}\left[\left(\ddot{\phi}_n + \frac{R}{\rho}\dot{\phi}_n + \frac{n^2\pi^2}{l^2}\frac{T_0}{\rho}\phi_n\right)\int_0^l \sin\frac{n\pi}{l}x \sin\frac{m\pi}{l}x\,dx\right]$$
$$= \frac{1}{\rho}\int_0^l F(x,t)\sin\frac{m\pi}{l}x\,dx \qquad (15\text{-}50)$$

The integral on the left side has a value different from zero only if $n = m$ and in that event has the value $l/2$. Accordingly we have left

$$\ddot{\phi}_m + \frac{R}{\rho}\dot{\phi}_m + \frac{m^2\pi^2}{l^2}\frac{T_0}{\rho}\phi_m = \frac{2}{l\rho}f_m(t) \qquad (15\text{-}51)$$

in which we have written $f_m(t)$ for the quantity

$$\int_0^l F(x,t)\sin\frac{m\pi}{l}x\,dx \qquad (15\text{-}52)$$

the dependence on x disappearing after integrating and evaluating at the limits. It is interesting to note, following Sec. 15-3, that $f_m(t)$ is just the Fourier coefficient, of order m, of the Fourier expansion of $F(x,t)$. Equation (15-51) is the equation of motion of the normal coordinate ϕ_m, and we see that the part of the driving force which is effective in forcing the mode of ϕ_m is the corresponding Fourier component of the force.

Equation (15-51) is quite similar to Eq. (7-37), save that, there, the driving force was less general in nature. The general solution of (15-51) is composed of a complementary function, or transient, solution, plus a particular integral (PI), or steady-state, solution. By inspection of Eqs. (7-7), (7-25), and (7-42), and considering only the underdamped case, we may write the solution of (15-51) as

$$\phi_m = e^{-(R/2\rho)t}(A_m \sin\omega'_m t + B_m \cos\omega'_m t) + \text{PI} \qquad (15\text{-}53)$$

where A_m and B_m are constants of integration, and in which

$$\omega'_m = \sqrt{\omega_m{}^2 - \frac{R^2}{4\rho^2}} = \sqrt{\frac{m^2\pi^2}{l^2}\frac{T_0}{\rho} - \frac{R^2}{4\rho^2}} \qquad (15\text{-}54)$$

Here the last step follows since, from (15-18), $\omega_m = (m^2\pi^2T_0/l^2\rho)^{\frac{1}{2}}$ is the angular frequency of ϕ_m if damping is not present.

The particular integral (PI) can be found by the methods given in Appendix 2. However, if $F(x,t)$ is of the form $g(x)\cos\omega t$, then, after evaluating the integral (15-52), the PI can usually be written immediately by comparing with Eq. (7-42).

Consider for simplicity that $F(x,t)$ has the form $F\cos\omega t$, in which F is a constant, and that it is being applied at a single point $x = b$ of the string. In this event the integrand of Eq. (15-52) possesses a nonvanishing value only at $x = b$. Accordingly $f_m(t)$ becomes

$$f_m(t) = F \sin \frac{m\pi b}{l} \cos \omega t \tag{15-55}$$

Thus the particular integral is

$$\mathrm{PI} = \frac{2}{l} \frac{F}{[\rho^2(\omega_m{}^2 - \omega^2)^2 + \omega^2 R^2]^{\frac{1}{2}}} \sin \frac{m\pi b}{l} \cos (\omega t - \alpha) \tag{15-56}$$

where

$$\tan \alpha = \frac{\omega R}{\rho(\omega_m{}^2 - \omega^2)} \tag{15-57}$$

It is interesting to observe that, if point b is a node for some particular normal coordinate ϕ_s, then ϕ_s is not excited by this force since, in that event, the term $\sin (s\pi b/l)$ in the PI vanishes because sb/l is an integer. Hence the steady-state solutions under the action of such a force are different from zero only for those normal coordinates which do not have a node at the point of application of the force. Moreover, the frequency of each of the forced vibrations (in the steady state) is that of the driving force.

It should be mentioned that, although in the above remarks we have considered only the case of underdamped motion, it is apparent that we may have critically damped and overdamped motions for the string as well. The situation is analogous to that of the oscillator of Chap. 7.

TRANSVERSE WAVE MOTION IN A STRING

15-6. Traveling-wave Solution. The appearance of the Fourier-series solutions of the vibrating string, as shown in Fig. 15-3, with the succession of crests and troughs, suggests the possibility of representing the motion of such a system in terms of traveling waves. Indeed, it will become evident that the normal-mode solution discussed above, much of which was the work of Daniel Bernoulli, has an entirely equivalent representation in terms of superimposed traveling waves. The traveling-wave point of view was supported by Euler and D'Alembert. The equivalence of the two representations was a point of controversy during the eighteenth century. This was finally cleared up in 1807, by Fourier, who showed that an arbitrary function could be represented as a trigonometric series and besides demonstrated how to determine the coefficients (see Appendix 5).

A wave of arbitrary shape which is traveling in the $+x$ direction with a velocity V may be written, as will be shown presently, as an arbitrary function of the coordinate x along the direction of propagation and of the time t, of the type

$$f_1(x - Vt) \tag{15-58}$$

Similarly, a wave traveling with velocity V in the negative x direction may be written as

$$f_2(x + Vt) \tag{15-59}$$

Again, as for the previous discussion, we take the case of a uniform string of mass ρ per unit length which is subjected to a constant tension T_0 along its entire length. Further, we again restrict ourselves to small lateral displacements y from the undistorted configuration along the x axis. As was demonstrated in Sec. 15-1, the equation of motion for small displacements is

$$\frac{\partial^2 y}{\partial t^2} = \frac{T_0}{\rho}\frac{\partial^2 y}{\partial x^2} \tag{15-60}$$

The procedure is to investigate whether or not an arbitrary function of the type of Eq. (15-58) or (15-59) will satisfy Eq. (15-60). More generally, since Eq. (15-60) is a linear partial differential equation of the second order, and since the general solution of such an equation may be written as the sum of two arbitrary functions, we desire to investigate whether or not a superposition

$$y = f_1(x - Vt) + f_2(x + Vt) \tag{15-61}$$

of the two functions (15-58) and (15-59) will satisfy Eq. (15-60). In order to do this, let

$$(x - Vt) = r \qquad (x + Vt) = s \tag{15-62}$$

By employing the variables r and s, Eq. (15-61) may be rewritten

$$y = f_1(r) + f_2(s) \tag{15-63}$$

Now

$$\frac{\partial y}{\partial t} = \frac{df_1}{dr}\frac{\partial r}{\partial t} + \frac{df_2}{ds}\frac{\partial s}{\partial t} = V\left(-\frac{df_1}{dr} + \frac{df_2}{ds}\right)$$

and

$$\frac{\partial^2 y}{\partial t^2} = V^2\left(\frac{d^2 f_1}{dr^2} + \frac{d^2 f_2}{ds^2}\right) \tag{15-64}$$

Similarly

$$\frac{\partial^2 y}{\partial x^2} = \left(\frac{d^2 f_1}{dr^2} + \frac{d^2 f_2}{ds^2}\right) \tag{15-65}$$

Accordingly, substituting (15-64) and (15-65) in Eq. (15-60), we have

$$V^2\left(\frac{d^2 f_1}{dr^2} + \frac{d^2 f_2}{ds^2}\right) = \frac{T_0}{\rho}\left(\frac{d^2 f_1}{dr^2} + \frac{d^2 f_2}{ds^2}\right) \tag{15-66}$$

Thus Eq. (15-61) is a possible solution to Eq. (15-60) for the velocity of propagation V given by

$$V^2 = \frac{T_0}{\rho} \tag{15-67}$$

Moreover it is the general solution, since the functions f_1 and f_2 are arbitrary, no requirements having been placed upon the forms of f_1 and f_2; the only stipulation was that they are functions, respectively, of $x - Vt$ and $x + Vt$. Also, since the forms of f_1 and f_2 are not influenced by the string (except in so far as they are restricted by boundary conditions, as will be brought out later), it is to be inferred that waves of arbitrary shape may be transmitted along a stretched string, the shape remaining unchanged during the process, with a wave velocity given by Eq. (15-67). Such a wave, traveling in the $+x$ direction, is represented in Fig. 15-5.

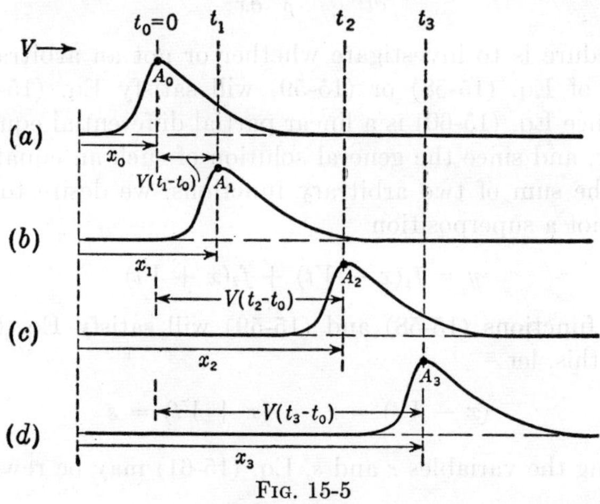

FIG. 15-5

Consider a point of the string such as A_0 in part (a) of the figure. At $t = t_0$ it has the same values of lateral displacement and lateral velocity (that is, the same phase) as do points A_1, A_2, and A_3, respectively, at times $t_1 - t_0$, $t_2 - t_0$, and $t_3 - t_0$ later. Also, during these times the wave moves distances $x_1 - x_0$, $x_2 - x_0$, and $x_3 - x_0$ to the right. For convenience let x_0 and t_0 both be zero. If we represent the form of the wave in part (a) at $t = 0$ by the expression $f_1(x)$, it is evident that in part (b) at a later time t_1 this same form is represented by

$$f_1(x - x_1) = f_1(x - Vt_1) \tag{15-68}$$

Similarly

$$f_1(x - x_2) = f_1(x - Vt_2) = f_1(x - x_3) = f_1(x - Vt_3), \text{ etc.} \tag{15-69}$$

It is apparent from Fig. 15-5 and the above discussion that the expression

$$f_1(x - Vt) \tag{15-70}$$

does represent a wave traveling in the $+x$ direction with a velocity V, thus confirming the initial hypothesis. In the same manner it can be verified that

$$f_2(x + Vt) \tag{15-71}$$

is indeed a wave traveling in the $-x$ direction with a velocity V. The general solution, Eq. (15-61), consists of a wave traveling in the $+x$ direction, plus a wave traveling in the $-x$ direction.

15-7. Terminal Conditions and Initial Conditions. We now consider the case of a taut string of length l which is fixed at both ends. Let the terminals be at $x = 0$ and $x = l$. At $x = 0$, Eq. (15-61) takes the form

$$f_1(-Vt) + f_2(Vt) = 0 \tag{15-72}$$

Accordingly, representing Vt by the symbol z, we note that

$$f_2(z) = -f_1(-z) \tag{15-73}$$

But the forms of f_1 and f_2 are unchanged during the propagation, and, if the relation (15-73) holds at $x = 0$, it holds as well at all other values of x. Hence Eq. (15-61) becomes

$$y = f_1(x - Vt) - f_1[-(x + Vt)] \tag{15-74}$$

At $x = l$, y is again zero, and we have

$$f_1(l - Vt) - f_1(-l - Vt) = 0 \tag{15-75}$$

By employing the symbol z_1 for the quantity $(-l - Vt)$, $l - Vt$ becomes $z_1 + 2l$, and Eq. (15-75) may be written as

$$f_1(z_1) = f_1(z_1 + 2l) \tag{15-76}$$

Equation (15-76) demonstrates that f_1 is a function which is periodic in z_1 with a period $2l$. In a similar fashion f_2 may be shown to be periodic with the same period. Alternatively, in terms of time as a variable, Eqs. (15-75) and (15-76) yield

$$f_1(l - Vt) = f_1(l - Vt - 2l) = f_1\left[l - V\left(t + \frac{2l}{V}\right)\right] \tag{15-77}$$

from which it is evident that f_1 is periodic in t with a period

$$\tau = \frac{2l}{V} \tag{15-78}$$

Thus during one period the wave travels a distance $2l$. The distance the wave travels during one period is defined to be one *wavelength*, **λ.**

λ is related to the velocity of propagation V, the frequency ν, the angular frequency ω, and the period τ by the relation

$$\frac{\lambda}{V} = \tau = \frac{1}{\nu} = \frac{2\pi}{\omega} \tag{15-79}$$

The quantity τ is the fundamental period for the string, being such that the wave motion experiences a node at each end. In the material below we shall designate such quantities as ω, λ, ν, and τ, which are associated with the fundamental wave, by means of the subscript 1. V, of course, is independent of λ and depends only on T_0 and ρ.

It is shown in Appendix 5 that any periodic function can be represented as a sum of sines and cosines. Accordingly it is of interest to investigate sinusoidal waves, such as

$$y = \sin\,[a(x - Vt)] \tag{15-80}$$

which represents a simple harmonic wave traveling in the $+x$ direction. In Eq. (15-80), a is a constant which is determined by the relation

$$\sin\,[a(x - Vt)] = \sin\,[a(x + \lambda - Vt)] \tag{15-81}$$

from which

$$a = \frac{2\pi}{\lambda} = \frac{\omega}{V} \tag{15-82}$$

Consequently Eq. (15-80) becomes

$$y = \sin\omega\left(\frac{x}{V} - t\right) \tag{15-83}$$

We have seen that the general solution consists of the sum of waves traveling in both the $+x$ and $-x$ directions, and in addition it is pointed out in Appendix 5 that, to represent a periodic function in general by means of a trigonometric series, both sines and cosines must be used. This suggests that we try a wave solution of the form

$$y = C\sin\omega\left(\frac{x}{V} - t\right) + D\sin\omega\left(\frac{x}{V} + t\right) + E\cos\omega\left(\frac{x}{V} - t\right) + F\cos\omega\left(\frac{x}{V} + t\right) \tag{15-84}$$

and see whether or not it will satisfy the terminal conditions. Putting $y = 0$ at $x = 0$, and equating the coefficients of the sine terms and those of the cosine terms separately to zero (since the equality must hold for any arbitrary value of t), we obtain

$$C = D \qquad E = -F \tag{15-85}$$

Equation (15-84) therefore becomes

$$y = C\left[\sin \omega\left(\frac{x}{V} - t\right) + \sin \omega\left(\frac{x}{V} + t\right)\right] + E\left[\cos \omega\left(\frac{x}{V} - t\right) - \cos \omega\left(\frac{x}{V} + t\right)\right] \quad (15\text{-}86)$$

Equation (15-86) satisfies the condition at $x = 0$ for any value of ω. On the other hand the condition $y = 0$ at $x = l$ may be satisfied by Eq. (15-86) for arbitrary values of E and C only if ω is suitably restricted. Such a restriction is that contained in Eq. (15-78), which requires that

$$\omega_1 = 2\frac{\pi V}{\lambda_1} = \frac{\pi V}{l} \quad (15\text{-}87)$$

or in terms of the tension T_0 and density ρ of the string, following Eq. (15-67),

$$\omega_1 = \frac{\pi}{l}\sqrt{\frac{T_0}{\rho}} \quad (15\text{-}88)$$

In (15-87) and (15-88) we now employ the subscript 1 since we are speaking of the fundamental. It is to be noted that the fundamental wave concerned here has the same frequency as that of the fundamental mode in the normal-modes treatment of the problem [see Eq. (15-18)].

It is not necessary to restrict ω quite so severely as in Eq. (15-88) since the terminal condition at $x = l$ may be satisfied by an angular frequency which is any integer times ω_1. Thus we may have

$$\omega_n = \frac{n\pi}{l}\sqrt{\frac{T_0}{\rho}} \qquad n = 1, 2, 3, \ldots \quad (15\text{-}89)$$

which is identical to Eq. (15-18). The value $n = 2$ corresponds to a wavelength $\lambda_2 = l$, $n = 3$ to $\lambda_3 = 2l/3$, and so on. Equation (15-86) becomes, for the nth angular frequency ω_n,

$$y_n = C_n\left[\sin \omega_n\left(\frac{x}{V} - t\right) + \sin \omega_n\left(\frac{x}{V} + t\right)\right] + E_n\left[\cos \omega_n\left(\frac{x}{V} - t\right) - \cos \omega_n\left(\frac{x}{V} + t\right)\right] \quad (15\text{-}90)$$

Equation (15-90) may be put into the form of Eq. (15-17) by expanding the trigonometric terms in (15-90), employing the identities for the sum and difference of two angles. We obtain

$$y_n = (2E_n \sin \omega_n t + 2C_n \cos \omega_n t) \sin \omega_n \frac{x}{V} \quad (15\text{-}91)$$

Since $V^2 = T_0/\rho$, Eq. (15-91) is identical to Eq. (15-17) if we put

$$2E_n = A_n \qquad 2C_n = B_n \tag{15-92}$$

where A_n and B_n are the coefficients in Eq. (15-17). Likewise, Eq. (15-19) corresponds to a superposition of waves of the type of Eq. (15-90), with n having all possible (integral) values.

It is shown in Sec. 15-3 that A_n and B_n may be determined from the initial conditions. As was shown there, if $u(x)$ gives the initial displacement of each point of the string and if $v(x)$ gives the initial velocity of each point of the string, we have

$$A_n = \frac{2}{\omega_n l}\int_0^l v(x)\sin\frac{n\pi}{l}x\,dx \qquad B_n = \frac{2}{l}\int_0^l u(x)\sin\frac{n\pi}{l}x\,dx \tag{15-93}$$

with which to evaluate the coefficients A_n and B_n (or A_m and B_m in the notation of Sec. 15-3).

Example 15-2. Find the solution to Example 15-1 in terms of traveling waves. Following Example 15-1, $A_n = 0$ for all n, $B_n = 0$ if n is even; if n is odd,

$$B_n = \frac{8b}{n^2\pi^2}\sin\frac{n\pi}{2} = 2C_n \tag{15-94}$$

Accordingly, in terms of traveling waves the solution becomes

$$y = \frac{4b}{\pi^2}\left\{\left[\sin\omega_1\left(\frac{x}{V}-t\right) + \sin\omega_1\left(\frac{x}{V}+t\right)\right] - \frac{1}{9}\left[\sin\omega_3\left(\frac{x}{V}-t\right) + \sin\omega_3\left(\frac{x}{V}+t\right)\right] + \cdots\right\} \tag{15-95}$$

15-8. More General Discussion of Effect of Initial Conditions. It is to be noted in Example 15-2 that, since initially the string was displaced but not moving (*plucked string*), the cosine terms are not present. The form of the solution is a superposition of pairs of sine waves having equal amplitudes and traveling in opposite directions. [This, in general, is not the case if in addition the string is initially also moving, that is, if neither $u(x)$ nor $v(x)$ is zero.] In Fig. 15-6*a* is shown the initial configura-

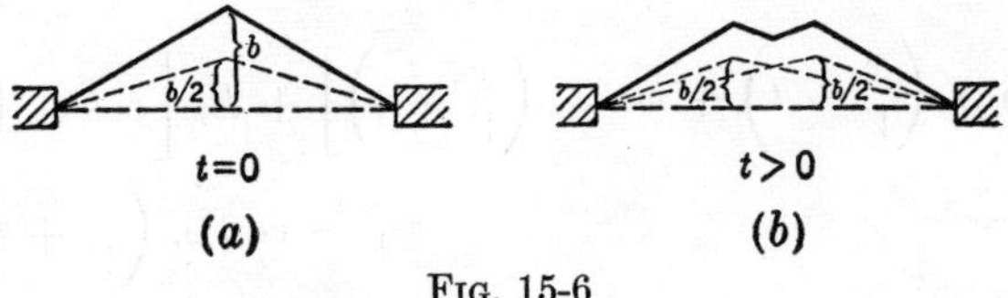

FIG. 15-6

tion of the string (solid line), with the dotted line representing either of the traveling waves. In part (*b*) of the figure, the dotted lines represent the two partial waves a short time later as they are separating, each traveling with the velocity $\sqrt{T_0/\rho}$. The full line again represents the actual configuration of the string.

These features can be perceived also by a consideration of the equation of motion (15-60), where we now put $V^2 = T_0/\rho$, and of the general solution (15-61). These are

$$\frac{\partial^2 y}{\partial t^2} = V^2 \frac{\partial^2 y}{\partial x^2} \tag{15-96}$$

$$y = f_1(x - Vt) + f_2(x + Vt) \tag{15-97}$$

Again we denote

$$x - Vt = r \qquad \text{and} \qquad x + Vt = s \tag{15-98}$$

Now

$$\frac{\partial y}{\partial x} = \frac{df_1}{dr}\frac{\partial r}{\partial x} + \frac{df_2}{ds}\frac{\partial s}{\partial x} = \frac{df_1}{dr} + \frac{df_2}{ds} \tag{15-99}$$

since $\partial r/\partial x = \partial s/\partial x = 1$. Similarly

$$\frac{\partial y}{\partial t} = \frac{df_1}{dr}\frac{\partial r}{\partial t} + \frac{df_2}{ds}\frac{\partial s}{\partial t} = V\left(-\frac{df_1}{dr} + \frac{df_2}{ds}\right) \tag{15-100}$$

We are interested in the initial values of these derivatives, that is, at $t = 0$. Now initially r and s both have the value x, and

$$\left(\frac{df_1}{dr}\right)_0 = \frac{df_1}{dx} \qquad \left(\frac{df_2}{ds}\right)_0 = \frac{df_2}{dx} \tag{15-101}$$

Consequently Eqs. (15-99) and (15-100), evaluated at $t = 0$, have the form

$$\left(\frac{\partial y}{\partial x}\right)_0 = \frac{df_1}{dx} + \frac{df_2}{dx} \qquad \left(\frac{\partial y}{\partial t}\right)_0 = V\left(-\frac{df_1}{dx} + \frac{df_2}{dx}\right) \tag{15-102}$$

Also initially

$$y_0 = u(x) \qquad \text{and} \qquad \left(\frac{\partial y}{\partial t}\right)_0 = v(x) \tag{15-103}$$

By making use of (15-103), Eqs. (15-102) become

$$\frac{df_1}{dx} + \frac{df_2}{dx} = \frac{du}{dx} \qquad -\frac{df_1}{dx} + \frac{df_2}{dx} = \frac{1}{V}v(x) \tag{15-104}$$

from which

$$\frac{df_1}{dx} = \frac{1}{2}\left[\frac{du}{dx} - \frac{1}{V}v(x)\right] \tag{15-105}$$

and

$$\frac{df_2}{dx} = \frac{1}{2}\left[\frac{du}{dx} + \frac{1}{V}v(x)\right] \tag{15-106}$$

A study of Eqs. (15-105) and (15-106) reveals that, since $v(x) = 0$, that is, since there is no initial velocity, the result again is waves of equal amplitudes traveling in opposite directions. It is interesting to note that if the initial conditions are such, for example, as to cause

$$\frac{du}{dx} - \frac{1}{V} v(x) = 0 \tag{15-107}$$

there is only a wave traveling in the negative direction of x. Thus it is possible, by suitably choosing the initial conditions, to produce a wave which has components traveling in but one direction.

15-9. Standing Waves. A solution such as Eq. (15-91), in which all parts of the string vibrate with but one frequency and in which certain points of the string always remain at rest, is said to represent a *standing wave.* Thus, from the wave point of view, the configuration assumed by the string when vibrating in a normal mode is spoken of as a standing wave.

Standing waves also may be produced on a string which is of unlimited length; the only requirement is that continuous sinusoidal waves of common amplitude and frequency be traveling in both positive and negative directions. In the case of a string that is rigidly terminated at one end, for example at $x = 0$, a continuous sine wave which is traveling toward the termination is reflected as a wave of the same amplitude and frequency. The result is a standing wave, no limitations being imposed upon the frequency. If, however, the string is rigidly terminated also at $x = l$, the angular frequency is at once restricted to the range of discrete values given by Eq. (15-89).

Let us consider two sinusoidal waves of arbitrary amplitudes and phases traveling in opposite directions. They shall be restricted, however, to have a common frequency. Such a state of affairs may be represented by

$$y = A \sin \left[\omega \left(\frac{x}{V} - t \right) - \varphi \right] + A' \sin \left[\omega \left(\frac{x}{V} + t \right) - \varphi' \right] \tag{15-108}$$

This may be expanded, employing obvious trigonometric identities, to become

$$\begin{aligned} y = A &\left[\cos \varphi \sin \omega \left(\frac{x}{V} - t \right) - \sin \varphi \cos \omega \left(\frac{x}{V} - t \right) \right] \\ &+ A' \left[\cos \varphi' \sin \omega \left(\frac{x}{V} + t \right) - \sin \varphi' \cos \omega \left(\frac{x}{V} + t \right) \right] \end{aligned} \tag{15-109}$$

If Eq. (15-109) is examined at $x = 0$, at which point we must have $y = 0$, we obtain

$$\begin{aligned} -A \cos \varphi \sin \omega t - A \sin \varphi \cos \omega t + A' \cos \varphi' \sin \omega t \\ - A' \sin \varphi' \cos \omega t = 0 \end{aligned} \tag{15-110}$$

Equating the coefficients of $\sin \omega t$ and those of $\cos \omega t$ separately to zero, we obtain

$$\begin{aligned} -A \cos \varphi + A' \cos \varphi' &= 0 \\ A \sin \varphi + A' \sin \varphi' &= 0 \end{aligned} \tag{15-111}$$

giving

$$\tan \varphi = -\tan \varphi' \qquad \text{or} \qquad \varphi = -\varphi' \tag{15-112}$$

and

$$A = A' \tag{15-113}$$

Thus, in order to satisfy the boundary condition, the two amplitudes must be equal. Equation (15-109) may be rewritten, in view of (15-112) and (15-113), as

$$\begin{aligned} y = A \cos \varphi \left[\sin \omega \left(\frac{x}{V} - t \right) + \sin \omega \left(\frac{x}{V} + t \right) \right] \\ - A \sin \varphi \left[\cos \omega \left(\frac{x}{V} - t \right) - \cos \omega \left(\frac{x}{V} + t \right) \right] \end{aligned} \tag{15-114}$$

which is similar to Eq. (15-90) save for the limitation on the angular frequency ω. If, in addition, the boundary condition $y = 0$ at $x = l$ is imposed, ω is limited as before to the values given by (15-89).

15-10. Behavior at a Junction. Energy Flow. We next investigate the effect of a sudden change in the linear density ρ of the string on a continuous sinusoidal wave. We consider an infinitely long string which has a mass per unit length ρ_1 for all negative values of x and a smaller mass per unit length ρ_2 for positive x. Such a situation, in exaggerated form, is shown in Fig. 15-7. Let the tension be T_0. A wave

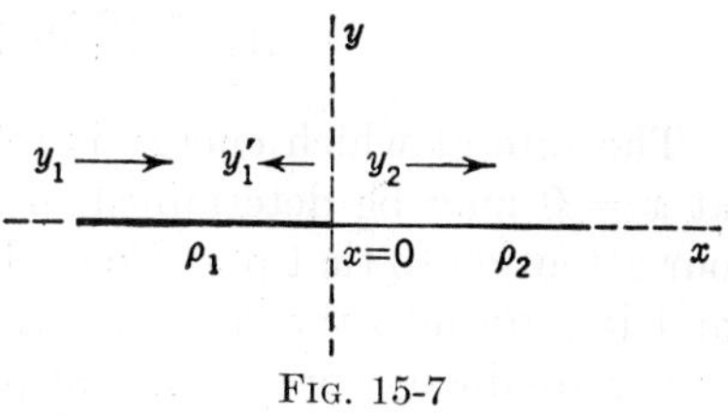

FIG. 15-7

$$y_1 = A_1 \sin \omega \left(\frac{x}{V_1} - t \right) \tag{15-115}$$

is incident on the junction, coming from the $-x$ direction. We postulate the existence of a reflected wave y_1' and a transmitted wave y_2 of the forms

$$y_1' = A_1' \sin \omega \left(\frac{x}{V_1} + t \right) \qquad y_2 = A_2 \sin \omega \left(\frac{x}{V_2} - t \right) \tag{15-116}$$

We desire to determine the values of the reflected and transmitted amplitudes A_1' and A_2 relative to A_1. This is accomplished by imposing the boundary conditions that the lateral displacement y and the slope $\partial y/\partial x$ are both continuous across the junction. The first of these is readily apparent, and the second is equivalent to saying that the restoring force in the string, arising as the result of a displacement y, is the same on each side of the junction. Explicitly the conditions are that, at $x = 0$,

$$(y_1 + y'_1)_{x=0} = (y_2)_{x=0} \tag{15-117}$$

$$\left(\frac{\partial y_1}{\partial x} + \frac{\partial y'_1}{\partial x}\right)_{x=0} = \left(\frac{\partial y_2}{\partial x}\right)_{x=0} \tag{15-118}$$

Substituting (15-115) and (15-116) in these, we have

$$A_1 - A'_1 = A_2 \tag{15-119}$$

$$\frac{1}{V_1}(A'_1 + A_1) = \frac{1}{V_2} A_2 \tag{15-120}$$

Combining these and making use of the fact that $V^2 = T_0/\rho$, we get

$$\frac{A'_1}{A_1} = \frac{V_1 - V_2}{V_1 + V_2} = \frac{\sqrt{\rho_2} - \sqrt{\rho_1}}{\sqrt{\rho_1} + \sqrt{\rho_2}} \tag{15-121}$$

as the relative amplitude of the reflected wave. In a similar fashion the relative amplitude of the transmitted wave is found to be

$$\frac{A_2}{A_1} = \frac{2V_2}{V_1 + V_2} = \frac{2\sqrt{\rho_1}}{\sqrt{\rho_1} + \sqrt{\rho_2}} \tag{15-122}$$

The rate at which energy is reflected from or flows past the junction at $x = 0$ may be determined in a straightforward manner. We direct our attention to that particle of the string which is just situated at $x = 0$ and inquire into the rate dW/dt at which the adjacent portions of the string are doing work upon that particle. This is just the product of the restoring force $-T_0\,\partial y/\partial x$ and the velocity $\partial y/\partial t$ of the particle. But since the particle is located at $x = 0$, these quantities must be evaluated there. Accordingly

$$\frac{dW}{dt} = \left(-T_0\frac{\partial y}{\partial x}\right)_{x=0}\left(\frac{\partial y}{\partial t}\right)_{x=0} \tag{15-123}$$

Considering first the incident and reflected waves, we may write

$$y = y_1 + y'_1 \tag{15-124}$$

Upon computing the partial derivatives and evaluating them at $x = 0$, Eq. (15-123) becomes

$$\begin{aligned}\left(\frac{dW}{dt}\right)_{i+r} &= T_0\frac{\omega^2}{V_1}(A_1 + A'_1)(A_1 - A'_1)\cos^2\omega t \\ &= T_0\frac{\omega^2}{V_1}(A_1{}^2 - A'_1{}^2)\cos^2\omega t\end{aligned} \tag{15-125}$$

We are interested in the average value of this quantity over a period.

This will be, where τ is the period,

$$\frac{1}{\tau}\int_0^\tau \left(\frac{dW}{dt}\right)_{i+r} dt = \frac{T_0\omega^2}{V_1}(A_1^2 - A_1'^2)\frac{1}{\tau}\int_0^\tau \cos^2 \omega t\, dt$$
$$= \frac{T_0\omega^2}{2V_1}(A_1^2 - A_1'^2) \qquad (15\text{-}126)$$

Equation (15-126) represents the mean rate at which energy is transmitted to the particle from the region to the left of the junction. The meanings of the two terms are clear. The quantity $T_0\omega^2 A_1^2/2V_1$ represents the mean rate at which energy is supplied to the junction by the incident wave, and $-T_0\omega^2 A_1'^2/2V_1$ represents the mean rate at which energy is reflected in the wave traveling back in the negative direction.

In a similar manner the average rate at which energy is transmitted past the junction is given by

$$\left[\left(\frac{dW}{dt}\right)_t\right]_{av} = \left[\left(-T_0\frac{\partial y_2}{\partial x}\right)_{x=0}\left(\frac{\partial y_2}{\partial t}\right)_{x=0}\right]_{av} = \frac{T_0\omega^2}{2V_2}A_2^2 \qquad (15\text{-}127)$$

It can easily be shown, by employing Eqs. (15-121) and (15-122), that expression (15-127), representing the transmitted energy, is just equal to the net rate at which energy is supplied to the junction from the left [given by (15-126)].

Problems

15-1. In Example 15-1, add numerically the first three terms of the series at $t = 0$; the first five terms. Plot graphs of the two results.

15-2. A uniform string of length l, mass ρ per unit length, subjected to a tension T_0 is initially displaced in the manner shown in Fig. 15-8 and then released. Find the general solution, evaluating the coefficients of the harmonic terms by means of the initial conditions. Consider that $b \ll l$, such that T_0 is constant and that the string is fixed at the ends.

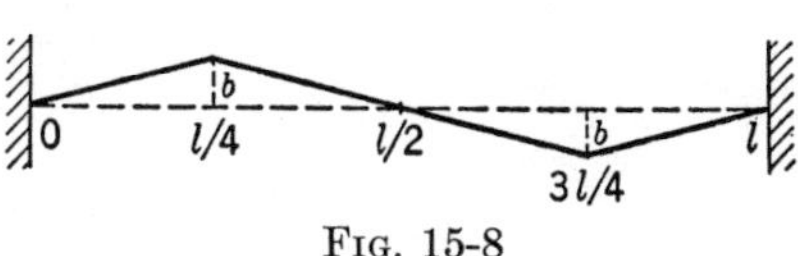

FIG. 15-8

15-3. The string of the preceding problem is initially in the equilibrium position but is initially given a velocity at all points x of the string determined by

$$v = ax \qquad 0 < x < \frac{l}{2}$$
$$v = a(x - l) \qquad \frac{l}{2} < x < l$$

where a is a constant. Find the general solution, evaluating all coefficients.

15-4. A uniform string of length l and mass ρ per unit length is stretched, under a constant tension T_0, along the great circle of a smooth sphere. The end points are fixed. Set up the equations of motion, and find the general solution for arbitrary initial conditions.

15-5. If p_n is the generalized momentum conjugate to ϕ_n of Eq. (15-37), write the Hamiltonian function for a vibrating string which is fixed at both ends.

15-6. A uniform string of length l and mass ρ per unit length is fixed at both ends along the x axis and is under a tension T_0. At the points $x = l/4$ and $l/2$ are situated small particles of iron of negligible mass. Electromagnets situated near these points exert a force $a \cos \omega t$ in the y direction on each particle. The quantities a and ω are constants. Consider that each point of the string is retarded by a viscous force proportional to the velocity (proportionality constant R per unit length) and that initially the string is at rest with the center displaced a small distance b in the y direction. Find the general solution of the problem for the underdamped case, evaluating all constants in terms of the initial conditions.

15-7. In the previous problem find the rate at which the driving force is doing work on the mth normal coordinate. What is the average rate at which this work is being done?

15-8. Verify that Eqs. (15-126) and (15-127) are consistent with energy conservation.

15-9. Consider the reverse case from that discussed in Sec. 15-10, that is, consider the incident wave to be coming from the right. Determine the relative amplitudes of the transmitted and reflected waves. Investigate also the rate at which energy passes to and fro past the junction.

15-10. Work Prob. 15-2 by means of the wave method.

15-11. Work Prob. 15-3 by means of the wave method.

15-12. Verify that the wave equation for motion of a string in a viscous medium, in which the damping force on any particle of the string is proportional to the first power of the velocity of the particle, is satisfied by a solution of the form

$$y = e^{-\alpha x} \sin \omega \left(\frac{x}{V} - t\right)$$

where α is a certain constant.

APPENDIX 1

AREA AND VOLUME ELEMENTS IN COMMON COORDINATE SYSTEMS

1. Plane Polar Coordinates (see Fig. 1). The element of area dA has the plane polar coordinates r, θ, as shown, where

$$x = r\cos\theta \qquad y = r\sin\theta \tag{1}$$

and

$$dA = r\,d\theta\,dr \tag{2}$$

2. Spherical Polar Coordinates (see Fig. 2). The element of volume dV is situated at the point having the spherical coordinates r, θ, φ, where

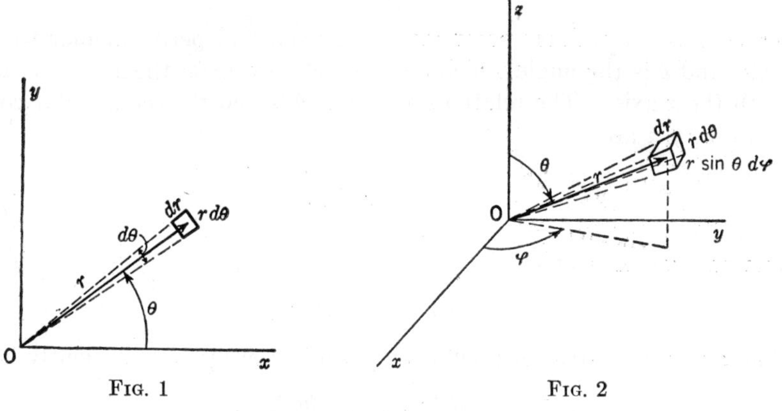

Fig. 1 Fig. 2

r is the radius vector from the origin, θ is the angle which it makes with the z axis, and φ is the angle which the projection of r on the xy plane makes with the x axis. We have

$$x = r\sin\theta\cos\varphi \qquad y = r\sin\theta\sin\varphi \qquad z = r\cos\theta \tag{3}$$

and, by inspection of Fig. 2,

$$dV = r^2\sin\theta\,dr\,d\theta\,d\varphi \tag{4}$$

The element of area for a spherical surface of radius r is seen to be

$$dA_{r\ \text{const}} = r^2 \sin\theta \, d\theta \, d\varphi \tag{5}$$

Other elements of area follow similarly by inspection of Fig. 2.

3. Cylindrical Polar Coordinates (see Fig. 3). The coordinates of the volume element are r, θ, z, where z is the same as in rectangular coordinates, r is the radius vector measured from and perpendicular to the z axis, and θ is the angle which the projection of r on the xy plane makes with the x axis. The relations among r, θ, z and the rectangular coordinates x, y, z are

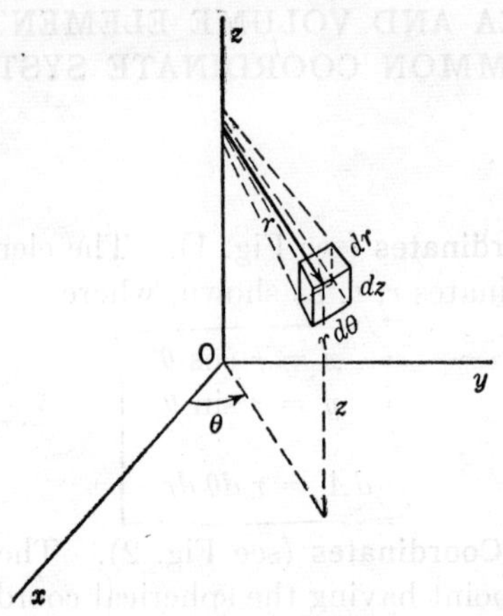

FIG. 3

$$\begin{aligned} x &= r\cos\theta \\ y &= r\sin\theta \\ z &= z \end{aligned} \tag{6}$$

and the volume element is

$$dV = r \, dr \, d\theta \, dz \tag{7}$$

The element of area for a cylindrical surface of radius r evidently is

$$dA_{r\ \text{const}} = r \, d\theta \, dz \tag{8}$$

APPENDIX 2

ELEMENTS OF ORDINARY DIFFERENTIAL EQUATIONS

1. The Nature of Differential Equations. Consider the equations

$$\frac{d^2y}{dx^2} + \left[x^4 \left(\frac{dy}{dx}\right)^3 + ay\right]^{\frac{1}{2}} = 0 \tag{1}$$

$$\frac{\partial^2 y}{\partial t^2} - c^2 \frac{\partial^2 y}{\partial x^2} = 0 \tag{2}$$

where a and c are constants. These are called differential equations since they are equations containing differential coefficients dy/dx, $\partial^2 y/\partial x^2$, and so on. Equation (1) is an *ordinary differential equation*, that is, it contains but one independent variable x. Equation (2) is a *partial differential equation* since it contains two independent variables x and t and consequently contains partial derivatives. We shall be concerned only with ordinary differential equations.

Equation (1) is an equation which is of the *second order* and *second degree*. It is of the second order because the highest-order derivative is d^2y/dx^2, which is of the second order. It is of the second degree since, after the equation has been made *rational* and *integral* so far as the differential coefficients are concerned (in other words, after fractional powers of derivatives have been removed), we have

$$\left(\frac{d^2y}{dx^2}\right)^2 - x^4 \left(\frac{dy}{dx}\right)^3 - ay = 0 \tag{3}$$

The degree of such an equation is the power to which the highest-order differential coefficient is raised, all fractional powers of derivatives having been removed. Accordingly Eq. (3) is of the second degree since this is the power to which the highest-order differential coefficient has been raised. It should be pointed out that this does not require x and y to occur rationally or integrally.

Equation (3) is also a *nonlinear* equation. It is nonlinear because it is of degree higher than the first. However, this must not be interpreted to mean that if it were of the first degree it would necessarily be linear.

For let us alter Eq. (3) to read

$$\frac{d^2y}{dx^2} - x^4 \frac{dy}{dx} - ay^2 = 0 \tag{4}$$

Equation (4) is of the first degree but is nonlinear. It is nonlinear because the dependent variable y is raised to the second power in the last term.

2. The Nature of the Solution of a Differential Equation. A solution of a differential equation is any function $y = f(x)$ which, when substituted in the differential equation, will satisfy it. In fact many times it is possible to write a particular solution to an equation by merely making a clever guess. Consider the equation

$$\frac{d^2y}{dx^2} + a^2y = 0 \tag{5}$$

where a is a positive constant. Let us try the function

$$y = \sin ax \tag{6}$$

and see whether or not it is a solution to Eq. (5). Substituting this in Eq. (5), we have

$$-a^2 \sin ax + a^2 \sin ax = 0 \tag{7}$$

and therefore (6) is indeed a solution of Eq. (5). It is easy to see that if we multiply Eq. (6) by an arbitrary constant A the result

$$y = A \sin ax \tag{8}$$

is also a solution. This may be verified by substituting (8) in (5). To carry this process still further, we may verify that

$$y = A \sin (ax + B) \tag{9}$$

is a solution of Eq. (5) and in fact is the general solution in that it contains the same number of *arbitrary constants* A and B as the order of the differential equation. It will be stated without proof that the most general solution of an ordinary differential equation of order n contains n arbitrary constants.

All particular solutions of Eq. (5) can be obtained from the general solution (9) by giving specific values to A and B. In considering a physical problem for which Eq. (5) is the appropriate differential equation, the arbitrary constants are given those values which enable the solution (9) to fit the boundary conditions of the problem at hand. (There is a certain type of solution called a *singular solution* which cannot be obtained in this manner. These are considered in any text on differential equations.)

3. Formation of Differential Equations by Elimination of Constants. Conversely we may consider a function such as Eq. (9) and determine the differential equation to which it corresponds by eliminating the arbitrary constants A and B. We have

$$y = A \sin (ax + B) \tag{10}$$

from which

$$\frac{dy}{dx} = Aa \cos (ax + B) \tag{11}$$

and

$$\frac{d^2y}{dx^2} = -Aa^2 \sin (ax + B) \tag{12}$$

But, in view of Eq. (10), Eq. (12) becomes

$$\frac{d^2y}{dx^2} = -a^2y \tag{13}$$

and, transposing, we have

$$\frac{d^2y}{dx^2} + a^2y = 0 \tag{14}$$

which is identical to Eq. (5). It should be pointed out that the quantity a in Eq. (1) is not an arbitrary constant. It cannot be eliminated by the process of taking derivatives but occurs as a *parameter* in the differential equation (14). It is usually true that in order to eliminate n arbitrary constants it is necessary to differentiate the solution n times, resulting in a differential equation of order n. Certain exceptions to this rule are mentioned in texts on differential equations and will not be considered here.

EQUATIONS OF THE FIRST ORDER

4. Exact Equations. A differential equation which can be found by simply differentiating a function is called an *exact* equation. Suppose we have

$$F(x,y) = C \tag{15}$$

where, as indicated, the quantity on the left side is a function of both x and y. Differentiating Eq. (15) with respect to x, we have

$$\frac{dF}{dx} = \frac{\partial F}{\partial x} + \frac{\partial F}{\partial y}\frac{dy}{dx} \tag{16}$$

This suggests that if we have an equation of the form

$$f_1(x,y) + f_2(x,y)\frac{dy}{dx} = 0 \tag{17}$$

such that $f_1(x,y)$ and $f_2(x,y)$ are derivable from a function $F(x,y)$, as

$$f_1(x,y) = \frac{\partial F}{\partial x} \qquad f_2(x,y) = \frac{\partial F}{\partial y} \tag{18}$$

then Eq. (17) is an exact equation and can be directly integrated.

A convenient test of this can easily be made. Since the order of differentiation is immaterial, we have

$$\frac{\partial}{\partial y}\left(\frac{\partial F}{\partial x}\right) = \frac{\partial}{\partial x}\left(\frac{\partial F}{\partial y}\right) \qquad \text{or} \qquad \frac{\partial^2 F}{\partial y\,\partial x} = \frac{\partial^2 F}{\partial x\,\partial y} \tag{19}$$

Hence, if in Eq. (17) we find that $f_1(x,y)$ and $f_2(x,y)$ fulfill the condition that

$$\boxed{\frac{\partial f_1}{\partial y} = \frac{\partial f_2}{\partial x}} \tag{20}$$

then f_1 and f_2 are the derivatives of a function $F(x,y)$, as in Eq. (18), and consequently Eq. (17) is an exact equation.

5. Solution by Separation of Variables. If an equation is of the form

$$f(y)\,\frac{dy}{dx} + g(x) = 0 \tag{21}$$

it may be written as

$$f(y)\,dy + g(x)\,dx = 0 \tag{22}$$

in which one term is a function of x alone and the other term is a function of y alone. Equation (21) is a particular case of an exact equation and in principle can be immediately integrated. We obtain

$$\int f(y)\,dy + \int g(x)\,dx = C \tag{23}$$

6. Integrating Factors. Any equation of the type

$$f_1(x,y) + f_2(x,y)\,\frac{dy}{dx} = 0 \tag{24}$$

if not already exact, can be made so by multiplying by a suitable factor, or *integrating factor*, as it is called. Although an integrating factor always exists for each equation of the type (24), it may be troublesome to find it.

A particularly simple case, and one that illustrates the method, is found in the equation

$$x\,\frac{dy}{dx} + 2y + x^2 = 0 \tag{25}$$

Applying the test of Eq. (20), we notice that

$$\frac{\partial}{\partial y}\,(2y + x^2) = 2 \qquad \frac{\partial}{\partial x}\,(x) = 1$$

and therefore Eq. (25), as it stands, is not exact. A possible integrating factor (usually suggested by experience) is x. Multiplying Eq. (25) by x, we have

$$x^2 \frac{dy}{dx} + 2xy + x^3 = 0 \tag{26}$$

Equation (26) integrates directly to

$$x^2y + \frac{x^4}{4} = C \tag{27}$$

Incidentally, Eq. (26) can be seen to satisfy the condition (20) for an exact equation.

An important case in which it is convenient to employ an integrating factor is that of the linear equation of the first order. This type we shall now consider.

7. The Linear Equation of the First Order. If the differential equation is of the first order and is linear in dy/dx and y, that is, if it is of the form

$$\frac{dy}{dx} + f(x)y = g(x) \tag{28}$$

it can easily be integrated after multiplication by the factor

$$e^{\int f(x)\,dx} \tag{29}$$

It is not difficult to verify that (29) is the factor which will always render exact an equation of the form (28), for suppose for the moment that the desired factor is a quantity $R(x)$. Multiplying Eq. (28) by R, we have

$$R\frac{dy}{dx} + Rf(x)y = Rg(x) \tag{30}$$

If this equation is now indeed exact, the term on the extreme left suggests that the entire left side must be

$$\frac{d}{dx}(Ry) = \frac{dR}{dx}y + R\frac{dy}{dx} \tag{31}$$

Comparing Eq. (31) with the left side of Eq. (30), we have

$$\frac{dR}{dx} = Rf(x) \tag{32}$$

Separating the variables and integrating, we have

$$\frac{dR}{R} = f(x)\,dx \qquad \text{or} \qquad \log R = \int f(x)\,dx \tag{33}$$

from which

$$R = e^{\int f(x)\,dx} \tag{34}$$

the predicted result.

It will be instructive to apply this procedure to the simple case of Eq. (25). Rewritten in the form of Eq. (28), it becomes

$$\frac{dy}{dx} + \frac{y}{x} = -x \tag{35}$$

Here $f(x) = 1/x$, and the factor is therefore

$$e^{\int (1/x)\, dx} = e^{\log x} = x \tag{36}$$

Applying this to Eq. (35), we arrive at the same result, Eq. (27), as before.

It is possible to write the general form of the solution to Eq. (28). Applying the integrating factor (29) to Eq. (28), we have

$$e^{\int f(x)\, dx} \frac{dy}{dx} + yf(x)e^{\int f(x)\, dx} = g(x)e^{\int f(x)\, dx} \tag{37}$$

which goes to

$$\frac{d}{dx}\left(ye^{\int f(x)\, dx}\right) = g(x)e^{\int f(x)\, dx} \tag{38}$$

This may be integrated, yielding

$$ye^{\int f(x)\, dx} = \int [g(x)e^{\int f(x)\, dx}]\, dx + C \tag{39}$$

from which we may write the solution y in the form

$$y = e^{-\int f(x)\, dx}\{\int [g(x)e^{\int f(x)\, dx}]\, dx + C\} \tag{40}$$

It is usually convenient to carry out the details of the procedure, as we did for Eq. (35), rather than to employ the type form of Eq. (40).

8. Nonlinear First-order Equations. Only one of this type, an equation which occurs occasionally in physical problems, will be considered here. It is known as Bernoulli's equation and is of the form

$$\frac{dy}{dx} + f(x)y = g(x)y^n \tag{41}$$

where n is a number not necessarily an integer. This equation was studied by Jakob Bernoulli in 1695. It can always be made linear, that is, put into the form (28), for which the integrating factor is well known, by the substitution

$$y = u^{1/(1-n)} \tag{42}$$

When Eq. (42) is substituted in Eq. (41), we obtain

$$\frac{du}{dx} + (1 - n)f(x)u = (1 - n)g(x) \tag{43}$$

which is seen to be of the form of Eq. (28). The integrating factor is

$$e^{\int (1-n)f(x)\, dx} \tag{44}$$

The solution is left to the student.

EQUATIONS OF THE SECOND AND HIGHER ORDERS

9. Linear with Constant Coefficients. Right Side Equal to Zero. Equations of the type

$$a\frac{d^2y}{dx^2} + b\frac{dy}{dx} + cy = g(x) \tag{45}$$

in which a, b, c are constants, have a very wide application in physics. The motions of many kinds of vibrating system obey such an equation. The motion of charges in an electric circuit also follows Eq. (45).

The class to be considered first will be that in which the quantity $g(x)$ on the right side of the equation is zero. This is exemplified in the free (unforced) motion of a vibrating system (see Chap. 7). If $g(x) = 0$, Eq. (45) becomes

$$a\frac{d^2y}{dx^2} + b\frac{dy}{dx} + cy = 0 \tag{46}$$

A linear equation of the type of Eq. (46), with $g(x) = 0$, is called a *homogeneous equation*. In order to solve this we consider first the linear first-order equation with constant coefficients

$$b\frac{dy}{dx} + cy = 0 \tag{47}$$

Separating the variables, we have

$$\frac{dy}{y} = -\frac{c}{b}\,dx$$

which integrates to

$$\log y = -\frac{c}{b}x + \log A$$

where A is an arbitrary constant of integration. This may be written in the form

$$y = Ae^{-(c/b)x} \tag{48}$$

and is the solution of Eq. (47). The appearance of Eq. (48) suggests that Eq. (46) might be satisfied by an expression of the type

$$y = e^{px} \tag{49}$$

where p is a constant. Substituting this in Eq. (46), we obtain

$$e^{px}(ap^2 + bp + c) = 0 \tag{50}$$

and, if p is determined so that it will satisfy the *auxiliary equation*

$$ap^2 + bp + c = 0 \tag{51}$$

it is clear that the form (49) will satisfy Eq. (46). The two roots of Eq. (51) are given by

$$p_1 = \frac{-b + \sqrt{b^2 - 4ac}}{2a} \qquad p_2 = \frac{-b - \sqrt{b^2 - 4ac}}{2a} \tag{52}$$

which in general are unequal. We thus have two solutions to Eq. (46). These are

$$y_1 = Ae^{p_1x} \qquad y_2 = Be^{p_2x} \tag{53}$$

in which A and B are arbitrary constants. It is evident, by trial, that the sum of these two solutions is also a solution of Eq. (46), namely, that

$$y = Ae^{p_1x} + Be^{p_2x} \tag{54}$$

will satisfy Eq. (46). This may be verified by direct substitution. It is important to point out to the student that the situation in which the sum of the two particular solutions is also a solution is a consequence of the fact that Eq. (46) is a linear equation. Also, since Eq. (54) contains two arbitrary constants, it is the general solution to Eq. (46).

10. The Case of Equal Roots. The two roots in Eq. (52) will be identical if

$$b^2 = 4ac \tag{55}$$

Accordingly, Eq. (54) becomes

$$y = (A + B)e^{p_0x} = Ge^{p_0x} \tag{56}$$

where we have put $p_1 = p_2 = p_0$. But $A + B$ is just another constant G, as we have indicated in Eq. (56), and we really now have but a single arbitrary constant of integration. Consequently Eq. (56) is no longer the general solution, and so for the case of *equal roots* we must resort to other tactics in order to obtain the general solution.

Equation (56) provides us with the information that, whatever the general solution is to be, a particular solution will certainly be

$$y = e^{p_0x} \tag{57}$$

Let us try the substitution

$$y = ue^{p_0x} \tag{58}$$

where $u = u(x)$. Substituting (58) in Eq. (46) and collecting terms, we obtain

$$e^{p_0x}\left[a\frac{d^2u}{dx^2} + (2ap_0 + b)\frac{du}{dx} + (ap_0{}^2 + bp_0 + c)u\right] = 0 \tag{59}$$

But, in view of Eq. (51), the coefficient of u in Eq. (59) is identically zero, and, owing to Eqs. (52) and (55), the coefficient of du/dx also vanishes.

We have left, therefore,

$$\frac{d^2u}{dx^2} = 0 \tag{60}$$

from which

$$u = A_0x + B_0 \tag{61}$$

where A_0 and B_0 are constants of integration. On combining this with Eq. (58), the general solution to Eq. (46) for the case $p_1 = p_2 = p_0$ is

$$y = (A_0x + B_0)e^{p_0x} \tag{62}$$

A situation of this type has been seen to follow for the case of a critically damped oscillator (see Chap. 7).

11. The Operator *D*. Consider the linear equation of the *n*th order, with constant coefficients,

$$\frac{d^ny}{dx^n} + a_1\frac{d^{n-1}y}{dx^{n-1}} + \cdots + a_{n-1}\frac{dy}{dx} + a_ny = 0 \tag{63}$$

It is convenient to introduce a new notation, that of the differential operator D, where

$$D \equiv \frac{d}{dx} \qquad D^2 \equiv \frac{d^2}{dx^2}, \text{ etc.} \tag{64}$$

In terms of the operator D, Eq. (63) may be written

$$(D^n + a_1D^{n-1} + \cdots + a_{n-1}D + a_n)y = 0 \tag{65}$$

(Note that D^n signifies d/dx taken n times, and not the nth power of d/dx.)

A better understanding of the use of D can be obtained by considering a simple example. Consider the expression

$$(D^2 + 3D)(x^4 + 3x^2) \tag{66}$$

This states that the quantity $x^4 + 3x^2$ is to be operated upon by the linear *differential operator* of order two, $D^2 + 3D$. We have, arranged in descending powers of x,

$$(D^2 + 3D)(x^4 + 3x^2) = 12x^3 + 12x^2 + 18x + 6 \tag{67}$$

We notice, too, that if we first factor the operator $D^2 + 3D$, we arrive at the same result (67). Thus

$$(D^2 + 3D)(x^4 + 3x^2) = D(D + 3)(x^4 + 3x^2) = 12x^3 + 12x^2 + 18x + 6 \tag{68}$$

and so the expression $D^2 + 3D$ may be factored like an algebraic quantity, and the end result is the same as before. This is a particular case of a general result. A linear differential operator of the nth order [the expression in parentheses in Eq. (65)] may be manipulated according to the laws of algebra provided that the coefficients $(a_1, a_2, \ldots, a_n)$ in Eqs. (63) and (64) are constants.

Evidently the differential operator in Eq. (65) may be regarded as a polynomial of the nth degree. Accordingly, it will have n roots, $p_1, \ldots,$ p_n, in terms of which we may write Eq. (65) as

$$(D - p_1)(D - p_2) \cdots (D - p_n)y = 0 \tag{69}$$

where the order of the factors is immaterial [provided that the a's in Eq. (63) are constants], and $D - p_1$ is understood to be an operator which when applied to y yields zero. Thus

$$(D - p_1)y = Dy - p_1 y = \frac{dy}{dx} - p_1 y = 0 \tag{70}$$

or

$$\frac{dy}{dx} = p_1 y \tag{71}$$

Hence, separating the variables and integrating,

$$y = A_1 e^{p_1 x} \tag{72}$$

A_1 is a constant of integration in Eq. (72). Comparing this result with Eq. (49) and noting Eqs. (50) and (51), we substitute (72) in Eq. (63). We obtain the algebraic equation

$$p_1{}^n + a_1 p_1{}^{n-1} + \cdots + a_{n-1} p_1 + a_n = 0 \tag{73}$$

which is satisfied by p_1. An equation of the same form would be obtained for each of the n roots $p_1, \ldots, p_n$. Accordingly, the n roots of the equation

$$p^n + a_1 p^{n-1} + \cdots + a_{n-1} p + a_n = 0 \tag{74}$$

which is of the same form as

$$D^n + a_1 D^{n-1} + \cdots + a_{n-1} D + a_n = 0 \tag{75}$$

[see Eq. (65)] provide us with n particular solutions to the differential equation (63). Each of these solutions will contain an arbitrary constant and will be of the form of Eq. (72). The general solution may be obtained by adding together all the particular solutions, as

$$y = A_1 e^{p_1 x} + A_2 e^{p_2 x} + \cdots + A_n e^{p_n x} \tag{76}$$

In order to solve Eq. (46) by means of the D operator, we should merely write it as

$$(aD^2 + bD + c)y = 0 \tag{77}$$

and find the roots of the equation

$$aD^2 + bD + c = 0 \tag{78}$$

These are

$$D = \frac{-b \pm \sqrt{b^2 - 4ac}}{2a} \tag{79}$$

Employing the notation

$$p_1 = \frac{-b + \sqrt{b^2 - 4ac}}{2a} \qquad p_2 = \frac{-b - \sqrt{b^2 - 4ac}}{2a} \tag{80}$$

we may write the general solution of Eq. (77) to be

$$y = Ae^{p_1x} + Be^{p_2x} \tag{81}$$

as before.

12. Linear Equations with Constant Coefficients. Right Side Not Equal to Zero. These equations will be of the form

$$\frac{d^ny}{dx^n} + a_1 \frac{d^{n-1}y}{dx^{n-1}} + \cdots + a_{n-1} \frac{dy}{dx} + a_ny = g(x) \tag{82}$$

In the present discussion we shall restrict ourselves to second-order equations. Let us take, as a very simple example of this type, the equation

$$\frac{d^2y}{dx^2} + 3 \frac{dy}{dx} + 2y = 3 + 2x \tag{83}$$

in which $g(x) = 3 + 2x$. Now we see that $y = x$ is one solution of Eq. (83), which fact may be verified by direct substitution. Such a solution without arbitrary constants is called a *particular integral*. Let us now substitute the expression

$$y = x + u \tag{84}$$

in Eq. (83). We obtain

$$\frac{d^2u}{dx^2} + 3 \frac{du}{dx} + 3 + 2(x + u) = 3 + 2x \tag{85}$$

Performing the obvious cancellations, we have left

$$\frac{d^2u}{dx^2} + 3 \frac{du}{dx} + 2u = 0 \tag{86}$$

This is identical to the homogeneous equation

$$\frac{d^2y}{dx^2} + 3 \frac{dy}{dx} + 2y = 0 \tag{87}$$

which is the left side of Eq. (83). We have already seen how to solve this type of equation. The solution may be found, by the substitution e^{px} or by the D-operator method, to be

$$u = Ae^{-x} + Be^{-2x} \tag{88}$$

This part of the solution, the part containing the arbitrary constants, is called the *complementary function.* It is the general solution to the part of Eq. (83) which is homogeneous in y, that is, that part remaining when $g(x) = 0$. The complete solution to Eq. (83) is, therefore,

$$y = Ae^{-x} + Be^{-2x} + x \tag{89}$$

and is the sum of the complementary function and the particular integral.

Let us consider a more general case, however, restricting ourselves to a second-order equation. We write, in terms of the D operator,

$$(D^2 + a_1D + a_2)y = g(x) \tag{90}$$

where a_1 and a_2 are constants. If the roots are designated again as p_1 and p_2, Eq. (90) may be written in the factored form

$$(D - p_1)(D - p_2)y = g(x) \tag{91}$$

Upon making the substitution

$$(D - p_2)y = U \tag{92}$$

Eq. (91) becomes

$$(D - p_1)U = g(x) \tag{93}$$

which is a linear equation of the first order. Employing Eq. (40), we are able to write the solution of (93) at once. It is

$$U = e^{p_1x}\{\textstyle\int[g(x)e^{-p_1x}]\,dx + C_1\} \tag{94}$$

where C_1 is a constant of integration. The entire integral in Eq. (94) is a function of x, say $h(x)$, and so, for convenience, we may write Eq. (94) as

$$U = e^{p_1x}[h(x) + C_1] \tag{95}$$

Substituting this for U in Eq. (92), we are in a position to carry out the remainder of the solution, as

$$(D - p_2)y = e^{p_1x}[h(x) + C_1] \tag{96}$$

for which we again employ Eq. (40), the solution of the linear first-order equation. We obtain

$$y = e^{p_2x}\left\{\int e^{(p_1-p_2)x}[h(x) + C_1]\,dx + C_2\right\} \tag{97}$$

$$= e^{p_2x}\int e^{(p_1-p_2)x}h(x)\,dx + \frac{C_1}{p_1 - p_2}\,e^{p_1x} + C_2e^{p_2x} \tag{98}$$

$$= \underbrace{e^{p_2x}\int e^{(p_1-p_2)x}h(x)\,dx}_{\text{Particular integral}} + \underbrace{Ae^{p_1x} + Be^{p_2x}}_{\text{Complementary function}} \tag{99}$$

where the last step follows by putting $B = C_2$ and $A = C_1/(p_1 - p_2)$. Accordingly, the complete solution to the general second-order equation (90) with constant coefficients consists of the sum of a complementary function and a particular integral.

The solution (99) of Eq. (90) has a very wide application in vibrating physical systems. It is shown in Chap. 7 that the complementary function contains the *characteristic frequencies* of the vibrating system (which die out if damping is present, that is, they are *transient* phenomena), while the particular integral arises only if the vibration is forced. The latter represents the *steady state* which persists after the transient terms have died out.

The above procedure is applicable as well to equations, with constant coefficients, of order higher than the second.

MISCELLANEOUS METHODS FOR EQUATIONS OF ORDER HIGHER THAN THE FIRST

13. The Equation Does Not Contain y Explicitly. An example of this type is the equation

$$x\frac{d^2y}{dx^2} + \frac{dy}{dx} = 4x \tag{100}$$

To solve this, put

$$p = \frac{dy}{dx} \tag{101}$$

(p has a different meaning here, of course, from that which it had before, when it was a symbol denoting the root of an equation.) Differentiating Eq. (101), we obtain

$$\frac{dp}{dx} = \frac{d^2y}{dx^2} \tag{102}$$

whence Eq. (100) becomes

$$x\frac{dp}{dx} + p = 4x \tag{103}$$

a first-order equation (in this case linear in p) with x as the independent variable and p as the dependent variable. This can easily be solved by the methods already given.

14. The Equation Does Not Contain x Explicitly. Such an equation is represented by

$$y\frac{d^2y}{dx^2} = \left(\frac{dy}{dx}\right)^2 \tag{104}$$

Again we put $p = dy/dx$. In this case, it is desirable to eliminate the variable x. We have

$$\frac{d^2y}{dx^2} = \frac{dp}{dx} = \frac{dp}{dy}\frac{dy}{dx} = p\frac{dp}{dy} \tag{105}$$

and Eq. (104) becomes

$$yp\frac{dp}{dy} = p^2 \tag{106}$$

or simply

$$\frac{dp}{p} = \frac{dy}{y} \tag{107}$$

which can easily be solved.

A particular case of this class of equations is frequently found in mechanics. It is

$$\frac{d^2y}{dx^2} = f(y) \tag{108}$$

We are able, of course, to solve this by the substitution $p = dy/dx$. However, instead let us multiply by $2\,dy/dx$. We obtain

$$2\frac{dy}{dx}\frac{d^2y}{dx^2} = 2f(y)\frac{dy}{dx} \tag{109}$$

Integrating, we have

$$\left(\frac{dy}{dx}\right)^2 = 2\int f(y)\,dy + C \tag{110}$$

15. Procedure When One Integral Belonging to the Complementary Function Is Known. If we know one integral, for example, $y = u$, of the homogeneous equation

$$\frac{d^2y}{dx^2} + a\frac{dy}{dx} + by = 0 \tag{111}$$

in which a and b are constants, it is possible to reduce the corresponding inhomogeneous equation

$$\frac{d^2y}{dx^2} + a\frac{dy}{dx} + by = g(x) \tag{112}$$

to an equation of the first order by substituting

$$y = uv \tag{113}$$

Differentiating (113), we obtain

$$\frac{dy}{dx} = \frac{du}{dx}v + u\frac{dv}{dx} \tag{114}$$

$$\frac{d^2y}{dx^2} = \frac{d^2u}{dx^2}v + 2\frac{du}{dx}\frac{dv}{dx} + u\frac{d^2v}{dx^2} \tag{115}$$

Substituting these in Eq. (112), we have, collecting terms appropriately,

$$u\frac{d^2v}{dx^2} + \frac{dv}{dx}\left(2\frac{du}{dx} + au\right) + v\left(\frac{d^2u}{dx^2} + a\frac{du}{dx} + bu\right) = g(x) \tag{116}$$

But the third set of terms on the left side is identically zero since u is a solution of Eq. (111). Furthermore, since u is a known function of x, we may write

$$2\frac{du}{dx} + au = h(x) \tag{117}$$

and if we also put $dv/dx = p$, Eq. (116) reduces to the simple first-order equation

$$u\frac{dp}{dx} + ph(x) = g(x) \tag{118}$$

an equation which is readily solved.

Problems

Solve the following:

1. $\frac{dy}{dx} + \frac{y}{x} = x^2$

2. $xy - \frac{dy}{dx} = y^3e^{-x^2}$

3. $x\frac{d^2y}{dx^2} + \frac{dy}{dx} = 4x$

4. $\frac{dy}{dx} + 2xy = 2e^{-x^2}$

5. $x^2\frac{d^2y}{dx^2} + x\frac{dy}{dx} - y = 8x^3$

6. $\frac{d^2y}{dx^2} + y = 3\cos^2 x + 2\sin^3 x$

APPENDIX 3

NOTE ON HYPERBOLIC FUNCTIONS

It is useful to outline briefly the elementary properties of the class of mathematical functions known as *hyperbolic functions*. The three most commonly employed are defined as

$$\sinh x = \frac{e^x - e^{-x}}{2} \qquad \cosh x = \frac{e^x + e^{-x}}{2} \qquad \tanh x = \frac{e^x - e^{-x}}{e^x + e^{-x}} \tag{1}$$

These are read, respectively, as the *hyperbolic sine, hyperbolic cosine*, and *hyperbolic tangent*, in analogy to the circular functions of similar name. The argument x is a real variable, and the hyperbolic functions merely provide a convenient shorthand way of combining groups of real exponentials. Such real exponentials occur in problems such as the one of the heavy suspended string and in the overdamped motions of elastic systems. The analogy to the corresponding circular functions can be pushed further. We have

$$\begin{aligned} \frac{1}{\sinh x} &= \operatorname{csch} x \qquad & \frac{1}{\cosh x} &= \operatorname{sech} x \\ \tanh x &= \frac{\sinh x}{\cosh x} \qquad & \frac{1}{\tanh x} &= \operatorname{ctnh} x \end{aligned} \tag{2}$$

In a similar manner we have the inverse hyperbolic functions. For example,

$$\text{If } x = \sinh y \qquad \text{then } y = \sinh^{-1} x \tag{3}$$

and so on. Inspection of Eqs. (1) reveals that

$$\sinh 0 = 0 \qquad \cosh 0 = 1 \qquad \tanh 0 = 0 \tag{4}$$

in analogy to the corresponding values for the circular functions. However, the identities involving the hyperbolic functions do not always have precisely the same form as the corresponding ones involving circular functions. A significant example is the relation

$$\cosh^2 x - \sinh^2 x = 1 \tag{5}$$

an equation which may be verified easily by the use of Eqs. (1). It is

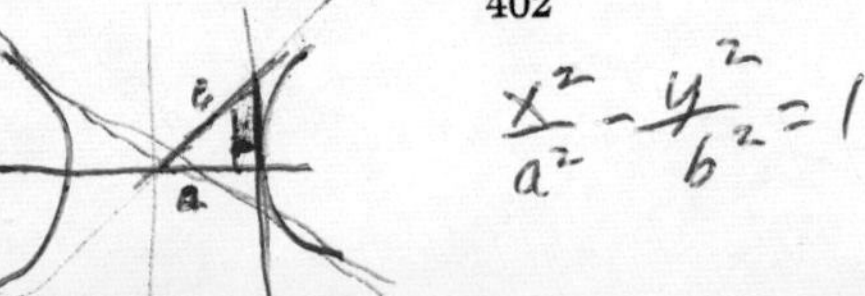

also apparent from Eqs. (1) that

$$\begin{aligned} \frac{d}{dx}(\sinh x) &= \frac{e^x + e^{-x}}{2} = \cosh x \\ \frac{d}{dx}(\cosh x) &= \frac{e^x - e^{-x}}{2} = \sinh x \end{aligned} \tag{6}$$

Other useful identities which may be easily demonstrated are

$$\sinh(x \pm y) = \sinh x \cosh y \pm \cosh x \sinh y \tag{7}$$

$$\cosh(x \pm y) = \cosh x \cosh y \pm \sinh x \sinh y \tag{8}$$

$$\sinh 2x = 2 \sinh x \cosh x \tag{9}$$

$$\cosh 2x = \cosh^2 x + \sinh^2 x \tag{10}$$

$$\sinh^2 x = \tfrac{1}{2}(\cosh 2x - 1) \tag{11}$$

$$\cosh^2 x = \tfrac{1}{2}(\cosh 2x + 1) \tag{12}$$

$$\tanh^2 x + \operatorname{sech}^2 x = 1 \tag{13}$$

$$\operatorname{ctnh}^2 x - \operatorname{csch}^2 x = 1 \tag{14}$$

The above expressions are sufficient for the material of the present text. Further information on the elementary properties of the hyperbolic functions may be obtained from any elementary text on calculus.

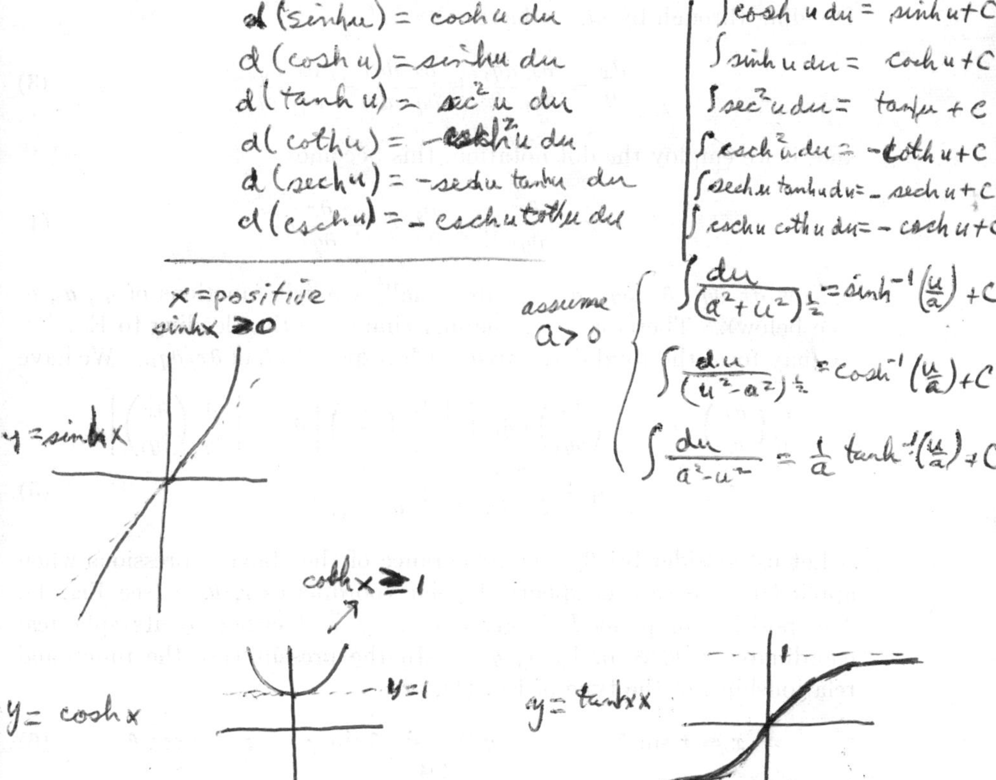

APPENDIX 4

COMMONLY EMPLOYED EXPRESSIONS INVOLVING PARTIAL DERIVATIVES

Suppose the variable x is expressed as an explicit function of the independent variables q_1, q_2, and q_3, each of which may depend upon the time, that is, suppose

$$x = x(q_1,q_2,q_3) \tag{1}$$

We may form the total differential dx as

$$dx = \frac{\partial x}{\partial q_1}\,dq_1 + \frac{\partial x}{\partial q_2}\,dq_2 + \frac{\partial x}{\partial q_3}\,dq_3 \tag{2}$$

Dividing through by dt, we have

$$\frac{dx}{dt} = \frac{\partial x}{\partial q_1}\frac{dq_1}{dt} + \frac{\partial x}{\partial q_2}\frac{dq_2}{dt} + \frac{\partial x}{\partial q_3}\frac{dq_3}{dt} \tag{3}$$

and, if we employ the dot notation, this becomes

$$\dot{x} = \frac{\partial x}{\partial q_1}\,\dot{q}_1 + \frac{\partial x}{\partial q_2}\,\dot{q}_2 + \frac{\partial x}{\partial q_3}\,\dot{q}_3 \tag{4}$$

Now $\partial x/\partial q_1$, $\partial x/\partial q_2$, $\partial x/\partial q_3$ are usually explicit functions of q_1, q_2, q_3 (see below). Therefore, in a manner similar to that leading to Eq. (4), we may form the total derivative, with respect to t, of $\partial x/\partial q_1$. We have

$$\begin{aligned}\frac{d}{dt}\left(\frac{\partial x}{\partial q_1}\right) &= \left[\frac{\partial}{\partial q_1}\left(\frac{\partial x}{\partial q_1}\right)\right]\dot{q}_1 + \left[\frac{\partial}{\partial q_2}\left(\frac{\partial x}{\partial q_1}\right)\right]\dot{q}_2 + \left[\frac{\partial}{\partial q_3}\left(\frac{\partial x}{\partial q_1}\right)\right]\dot{q}_3 \\ &= \frac{\partial^2 x}{\partial q_1{}^2}\,\dot{q}_1 + \frac{\partial^2 x}{\partial q_2\,\partial q_1}\,\dot{q}_2 + \frac{\partial^2 x}{\partial q_3\,\partial q_1}\,\dot{q}_3 \end{aligned} \tag{5}$$

Let us consider briefly the appearance of the above expressions when applied to the case of spherical polar coordinates r, θ, φ (see Fig. 1). The position of point P is given equally well either by its spherical coordinates r, θ, φ, or by x, y, z. In the present case the functional relationships, of the type of Eq. (1), are

$$x = r\sin\theta\cos\varphi \qquad y = r\sin\theta\sin\varphi \qquad z = r\cos\theta \tag{6}$$

For the sake of brevity we take only the case of x. We form $\dot{x}$ in the same way as in Eq. (4). We have

$$\dot{x} = (\sin\theta\cos\varphi)\dot{r} + (r\cos\theta\cos\varphi)\dot{\theta} - (r\sin\theta\sin\varphi)\dot{\varphi} \qquad (7)$$

It is clear, from an examination of Eq. (7), that the derivatives of the type $\partial x/\partial q_1$, and so on, are, in general, functions of the coordinates

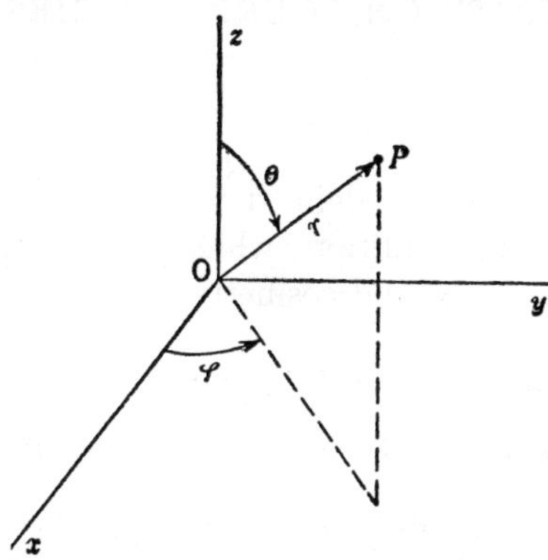

FIG. 1

q_1, q_2, q_3, etc. Similarly in the present instance, Eq. (5) takes the form (with r replacing q_1)

$$\frac{d}{dt}(\sin\theta\cos\varphi) = (\cos\theta\cos\varphi)\dot{\theta} - (\sin\theta\sin\varphi)\dot{\varphi} \qquad (8)$$

APPENDIX 5

NOTE ON FOURIER SERIES

Any function $f(x)$ which is defined within the interval $-\pi < x < \pi$ and is subject to certain conditions which are not too restrictive can be expressed as a series of sine and cosine terms of the form

$$f(x) = \frac{A_0}{2} + A_1 \cos x + A_2 \cos 2x + \cdots + A_n \cos nx + \cdots + B_1 \sin x + B_2 \sin 2x + \cdots + B_n \sin nx + \cdots \quad (1)$$

where A_n and B_n are constants and $n = 1, 2, 3, \ldots$. This also may be written as

$$f(x) = \frac{A_0}{2} + \sum_{n=1}^{\infty} A_n \cos nx + \sum_{n=1}^{\infty} B_n \sin nx \quad (2)$$

The series (2) will represent the function $f(x)$ within the interval $-\pi < x < \pi$ provided that $f(x)$ is single-valued and continuous in the interval and, if discontinuous, provided that there are only a finite number of finite discontinuities within the interval. Under these conditions the series (2) will converge to the function $f(x)$ at all points within the interval $-\pi < x < \pi$. Moreover the series may be integrated term by term, and the series so formed will converge to the integral of $f(x)$ at all points within the interval.

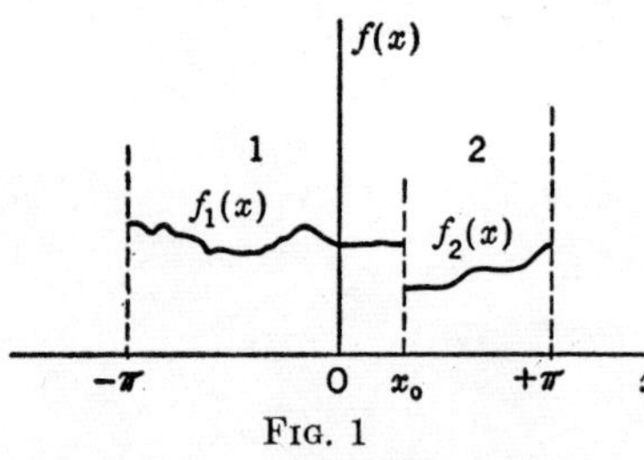

FIG. 1

In the preceding paragraph the term finite discontinuity is mentioned. In Fig. 1 the function $f(x)$ has a finite discontinuity at the point $x = x_0$. In region 1, $f(x)$ has the form $f_1(x)$, in the region 2 the form $f_2(x)$, and, at $x = x_0$, $f_1(x_0)$ and $f_2(x_0)$ are both finite. [It is possible to show that, under the restrictions mentioned above, the series converges at $x = x_0$ to one-half the sum of $f_1(x_0)$ and $f_2(x_0)$.]

At the end points $-\pi$ and $+\pi$, the series converges to the value $\frac{1}{2}[f(-\pi) + f(+\pi)]$, and so the series does not represent the function $f(x)$ at the end points unless $f(-\pi) = f(+\pi)$. However, we can be as close to the end points as we wish, within the interval, and the series will still represent the function.

The question of whether the series (2) may be differentiated term by term and the resulting series still represent the derivative of the function is subject to more restrictions than those stated above. It can be shown that the differentiated series will converge to the derivative of the function $f(x)$ provided that $f(x)$ is single-valued, finite, and continuous, with only a finite number of maxima and minima within the interval, and provided also that, at the end points $-\pi$ and $+\pi$, $f(-\pi) = f(+\pi)$.

It is possible to determine the coefficients in Eq. (2). If both sides of (2) are multiplied by $\cos mx$, where m is an integer, and the result then integrated over the interval, that is, from $-\pi$ to $+\pi$, we have

$$\int_{-\pi}^{\pi} f(x) \cos mx \, dx = \frac{A_0}{2} \int_{-\pi}^{\pi} \cos mx \, dx + \sum_{n=1}^{\infty} A_n \int_{-\pi}^{\pi} \cos nx \cos mx \, dx + \sum_{n=1}^{\infty} B_n \int_{-\pi}^{\pi} \sin nx \cos mx \, dx \tag{3}$$

Now if n and m are integers, for all m and n we obtain

$$\int_{-\pi}^{\pi} \cos nx \cos mx \, dx = 0 \text{ (for } n \neq m); \; = \pi \text{ (for } n = m) \tag{4}$$

$$\int_{-\pi}^{\pi} \sin nx \cos mx \, dx = 0 \qquad \text{all } m \text{ and } n \tag{5}$$

$$\int_{-\pi}^{\pi} \cos mx \, dx = 0 \; (m \neq 0); \; = 2\pi \; (m = 0) \tag{6}$$

Accordingly

$$A_0 = \frac{1}{\pi} \int_{-\pi}^{\pi} f(x) \, dx \tag{7}$$

$$A_m = \frac{1}{\pi} \int_{-\pi}^{\pi} f(x) \cos mx \, dx \qquad m \text{ integral} \tag{8}$$

Similarly if (2) is multiplied through by $\sin mx$ and integrated in the same manner as above, all terms vanish save that involving the square of $\sin mx$. Thus

$$B_m = \frac{1}{\pi} \int_{-\pi}^{\pi} f(x) \sin mx \, dx \tag{9}$$

If there exists a finite discontinuity in $f(x)$ at the point x_0, the coefficients A_0, A_m, B_m are determined by integrating first to $x = x_0$ and then from x_0 to π, as

$$A_0 = \frac{1}{\pi} \left[\int_{-\pi}^{x_0} f(x) \, dx + \int_{x_0}^{\pi} f(x) \, dx \right] \tag{10}$$

$$A_m = \frac{1}{\pi}\left[\int_{-\pi}^{x_0} f(x)\cos mx\,dx + \int_{x_0}^{\pi} f(x)\cos mx\,dx\right] \tag{11}$$

$$B_m = \frac{1}{\pi}\left[\int_{-\pi}^{x_0} f(x)\sin mx\,dx + \int_{x_0}^{\pi} f(x)\sin mx\,dx\right] \tag{12}$$

This procedure may be extended to any finite number of discontinuities.

The series (2), with the coefficients determined in the manner of Eqs. (7), (8), and (9) or (10), (11), and (12), will represent the function $f(x)$ in the interval $-\pi < x < \pi$. If $f(x)$ is periodic, with a period 2π, that is, if

$$f(x + 2\pi) = f(x) \tag{13}$$

then the series (2) represents $f(x)$ at all values of x, save at the discontinuities, as mentioned above.

Example 1. Given $f(x) = -\pi/2$ at $-\pi < x < 0$ and $f(x) = +\pi/2$ at $0 < x < \pi$ Expand $f(x)$ in the interval $-\pi < x < \pi$.

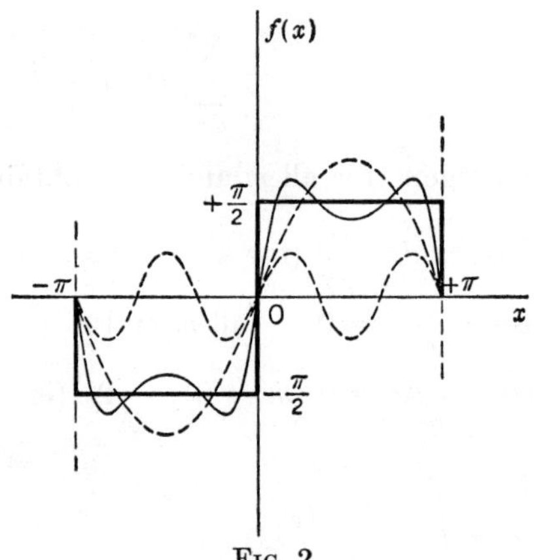

FIG. 2

The function $f(x)$ is shown in Fig. 2 as the heavy line. The coefficients may be found as

$$A_0 = \frac{1}{\pi}\left(-\frac{\pi}{2}\int_{-\pi}^{0} dx + \frac{\pi}{2}\int_0^{\pi} dx\right) = 0$$

$$A_m = \frac{1}{\pi}\left(-\frac{\pi}{2}\int_{-\pi}^{0}\cos mx\,dx + \frac{\pi}{2}\int_0^{\pi}\cos mx\,dx\right)$$

$$= -\frac{1}{2m}\sin mx\Big|_{-\pi}^{0} + \frac{1}{2m}\sin mx\Big|_0^{\pi} = 0$$

$$B_m = \frac{1}{\pi}\left(-\frac{\pi}{2}\int_{-\pi}^{0}\sin mx\,dx + \frac{\pi}{2}\int_0^{\pi}\sin mx\,dx\right)$$

$$= \frac{1}{2m}\cos mx\Big|_{-\pi}^{0} - \frac{1}{2m}\cos mx\Big|_0^{\pi} = \frac{2}{m}\ (m \text{ odd});\ = 0\ (m \text{ even})$$

Accordingly the Fouries-series expansion of $f(x)$ in the interval $-\pi < x < \pi$ may be written

$$f(x) = 2\sin x + \tfrac{2}{3}\sin 3x + \tfrac{2}{5}\sin 5x + \cdots \tag{14}$$

The first two terms are shown dotted in Fig. 2, and their sum is the solid curve. It will be instructive for the student to include additional terms from (14). It is evident that, as more and more terms of the series (14) are included, the sum more and more nearly approaches the shape of $f(x)$.

If we define an odd function $f_0(x)$, that is, if $f_0(-x) = -f_0(x)$, then only sine terms appear in the series expansion of $f_0(x)$ in the interval $\pi < x < \pi$, for

$$_m = \frac{1}{\pi}\int_{-\pi}^{\pi} f_0(x)\cos mx\,dx = \frac{1}{\pi}\left[\int_{-\pi}^{0} f_0(x)\cos mx\,dx + \int_0^{\pi} f_0(x)\cos mx\,dx\right]$$

$$= \frac{1}{\pi}\left[-\int_0^{\pi} f_0(x)\cos mx\,dx + \int_0^{\pi} f_0(x)\cos mx\,dx\right] = 0 \qquad m = 0, 1, 2, \ldots \tag{15}$$

$$B_m = \frac{1}{\pi}\int_{-\pi}^{\pi} f_0(x)\sin mx\,dx = \frac{1}{\pi}\left[\int_{-\pi}^{0} f_0(x)\sin mx\,dx + \int_0^{\pi} f_0(x)\sin mx\,dx\right]$$

$$= \frac{2}{\pi}\int_0^{\pi} f_0(x)\sin mx\,dx \qquad m = 1, 2, 3, \ldots \tag{16}$$

Here we have made use of the fact that $\cos(-mx) = \cos mx$ and

$$\sin(-mx) = -\sin mx$$

Accordingly, the series becomes

$$f_0(x) = B_1 \sin x + B_2 \sin 2x + \cdots \tag{17}$$

Every term in Eq. (17) is an odd function, and the sum represents the odd function $f_0(x)$ in the interval $-\pi < x < \pi$.

On the other hand, if we define an even function $f_e(x)$, that is, if $f_e(-x) = f_e(x)$, then for the interval $-\pi < x < \pi$ only cosine terms appear. In the same manner as for Eqs. (15) and (16), we obtain

$$A_m = \frac{2}{\pi}\int_0^{\pi} f_e(x)\cos mx\,dx \qquad m = 0, 1, 2, \ldots \tag{18}$$

$$B_m = 0 \qquad m = 1, 2, 3, \ldots \tag{19}$$

and the series becomes

$$f_e(x) = \frac{A_0}{2} + A_1\cos x + A_2\cos 2x + \cdots \tag{20}$$

All terms of the series of Eq. (20) are even functions of x, and the series converges to the even function $f_e(x)$ in the interval $-\pi < x < \pi$.

Let us now return to the consideration of an arbitrary function (neither even nor odd). Any such function $f(x)$ can be expressed as a combination of an even function $E(x)$ and an odd function $O(x)$ as

$$\begin{aligned} f(x) &= \tfrac{1}{2}[f(x) + f(-x)] + \tfrac{1}{2}[f(x) - f(-x)] \\ &= E(x) + O(x) \end{aligned} \tag{21}$$

The coefficients for the Fourier expansion in the interval $-\pi < x < \pi$ become, following (18) and (16),

$$A_m = \frac{2}{\pi}\int_0^\pi E(x)\cos mx\,dx \qquad m = 0, 1, 2, \ldots \tag{22}$$

$$B_m = \frac{2}{\pi}\int_0^\pi O(x)\sin mx\,dx \qquad m = 1, 2, 3, \ldots \tag{23}$$

In view of these it is evident that the first two terms on the right side of Eq. (2) represent the even part of $f(x)$ in the interval $-\pi < x < \pi$ and the last term the odd part.

Consider the functions depicted in Figs. 3 and 4. In Fig. 3, $f(x) = x$ in the interval $-\pi < x < \pi$ and the graph is that of an odd function in

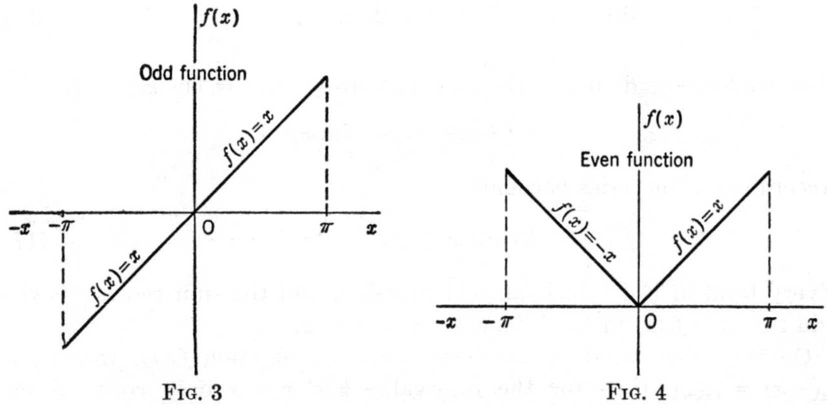

FIG. 3 FIG. 4

$-\pi < x < \pi$. In Fig. 4, $f(x) = -x$ in the interval $-\pi < x < 0$, and $f(x) = x$ in $0 < x < \pi$, an even function in $-\pi < x < \pi$. The Fourier representations of these functions in the interval $-\pi < x < \pi$ are given, respectively, by Eqs. (17) and (20), where

$$B_m = \frac{2}{\pi}\int_0^\pi x\sin mx\,dx \qquad m = 1, 2, 3, \ldots \tag{24}$$

$$A_m = \frac{2}{\pi}\int_0^\pi x\cos mx\,dx \qquad m = 0, 1, 2, \ldots \tag{25}$$

It is apparent from the figures that both functions are identical in the region $0 < x < \pi$. Hence, if it is desired to find a Fourier series representation only in the interval $0 < x < \pi$, either the sine series or the cosine series will do. Either (17) or (20) may be regarded as the expansion of $f(x) = x$ in the interval $0 < x < \pi$, even though the two series do not represent the same function outside of this interval.

It is evident that we need not restrict ourselves to the simple functions depicted in Figs. 3 and 4 and may proceed even further. It is possible to represent an arbitrary function $f(x)$ in the interval $0 < x < \pi$ by either a sine series or a cosine series, for we may define an odd function in $-\pi < x < \pi$, identical to $f(x)$ in $0 < x < \pi$. Its expansion in the interval $-\pi < x < \pi$ will consist of sine terms only and will correctly represent $f(x)$ in the interval $0 < x < \pi$. Alternatively, we are able to define an even function in $-\pi < x < \pi$, identical to $f(x)$ in $0 < x < \pi$; its expansion in the interval $-\pi < x < \pi$ will consist only of cosine terms and will also correctly represent $f(x)$ in the interval $0 < x < \pi$.

A Fourier expansion is not restricted to such intervals as $-\pi < x < \pi$ and $0 < x < \pi$ alone but is possible in any finite interval $-l < x < l$, or $0 < x < l$, and so on. It is necessary only to change the variable, as

$$z = \frac{\pi}{l} x \tag{26}$$

Thus if $f(x)$ is expanded in the interval $-\pi < z < \pi$, the coefficients being determined by expressions of the form of (7) to (9), the coefficients for the expansion of $f(x)$ in the interval $-l < x < l$ may be obtained merely by substituting (26) into these expressions. We have

$$A_m = \frac{1}{l} \int_{-l}^{l} f(x) \cos \frac{m\pi}{l} x \, dx \qquad m = 0, 1, 2, 3, \ldots \tag{27}$$

$$B_m = \frac{1}{l} \int_{-l}^{l} f(x) \sin \frac{m\pi}{l} x \, dx \qquad m = 1, 2, 3, \ldots \tag{28}$$

The function $f(x)$ may be represented by either a cosine series or a sine series in the interval $0 < x < l$. The coefficients for the two cases are, respectively,

$$A_m = \frac{2}{l} \int_{0}^{l} f(x) \cos \frac{m\pi}{l} x \, dx \qquad m = 0, 1, 2, \ldots \tag{29}$$

$$B_m = \frac{2}{l} \int_{0}^{l} f(x) \sin \frac{m\pi}{l} x \, dx \qquad m = 1, 2, 3, \ldots \tag{30}$$

The similarity to Eqs. (16), (18), and (19) should be noted.

This discussion is meant to serve merely as a brief introduction to the subject of Fourier series. Only the points which are deemed useful in the present text have been included. For further study the student is advised to consult other texts.

Problems

Expand the following functions in a Fourier series in the interval indicated, determining all coefficients:

1. $f(x) = 0$ for $-\pi < x < 0$, $f(x) = x$ for $0 < x < \pi$, in the interval $-\pi < x < \pi$. Plot the sum of the first four terms.

2. $f(x) = x$ for $0 < x < \pi/2$, $f(x) = (\pi - x)$ for $\pi/2 < x < \pi$, in the interval $0 < x < \pi$. Find both the sine and cosine series.

3. $f(x) = 1$ for $0 < x < \pi/2$, $f(x) = 0$ for $\pi/2 < x < \pi$, in the interval $0 < x < \pi$. Obtain both the sine and cosine series.

4. $f(x) = x$ for $-\pi < x < \pi$, in the interval $-\pi < x < \pi$. Find another representation valid in the interval $0 < x < \pi$.

SUGGESTED REFERENCES FOR THE STUDENT

The references given below will be found useful by the student. In each case the numerals pertain to the chapter of this text in which they will be particularly pertinent.

W. E. Byerly, "Fourier Series and Spherical Harmonics" (Ginn), Appendix 5.

R. V. Churchill, "Fourier Series and Boundary Value Problems" (McGraw-Hill), Appendix 5.

J. G. Coffin, "Vector Analysis" (Wiley), Chap. 1.

H. C. Corben and P. Stehle, "Classical Mechanics" (Wiley), Chap. 13.

J. P. Den Hartog, "Mechanical Vibrations" (McGraw-Hill), Chap. 7.

L. R. Ford, "Differential Equations" (McGraw-Hill), Appendix 2.

H. Goldstein, "Classical Mechanics" (Addison-Wesley), Chaps. 10, 12, 13, 14.

E. L. Ince, "Ordinary Differential Equations" (Dover), Appendix 2.

J. H. Jeans, "An Elementary Treatise on Theoretical Mechanics" (Ginn), Chaps. 1, 2, 3, 4, 5, 6, 8, 9.

G. Joos, "Theoretical Physics" (Stechert), Chaps. 7, 8, 11, 12, 13.

H. Lamb, "Dynamics" (Cambridge), Chaps. 1, 6, 7, 8, 9, 10.

———, "Introduction to Higher Mechanics" (Cambridge), Chaps. 11, 12, 13, 14.

———, "Statics" (Cambridge), Chaps, 2, 3, 4, 5.

H. Lass, "Vector and Tensor Analysis" (McGraw-Hill), Chap. 1.

R. B. Lindsay, "Physical Mechanics" (Van Nostrand), Chap. 11.

E. Mach, "The Science of Mechanics" (Open Court), Chaps. 1, 2, 3, 5.

W. D. MacMillan, "Statics and Dynamics of a Particle" (McGraw-Hill), Chaps. 10 and 11.

H. Margenau and G. M. Murphy, "The Mathematics of Physics and Chemistry" (Van Nostrand), Chap. 14 and Appendix 2.

P. M. Morse, "Vibration and Sound" (McGraw-Hill), Chap. 15.

F. R. Moulton, "An Introduction to Celestial Mechanics" (Macmillan), Chap. 10.

W. F. Osgood, "Mechanics" (Macmillan), Chap. 8.

L. Page, "Introduction to Theoretical Physics" (Van Nostrand), Chaps. 9, 11, 12, 13.

H. B. Phillips, "Vector Analysis" (Wiley), Chap. 1.

H. T. H. Piaggio, "An Elementary Treatise on Differential Equations and Their Applications" (Open Court), Appendix 2.

A. S. Ramsey, "Dynamics," Vols. 1 and 2 (Cambridge), Chaps. 6, 7, 8, 9, 10.

———, "Statics" (Cambridge), Chaps. 2, 3, 4, 5.

Lord Rayleigh, "The Theory of Sound" (Dover), Chaps. 14 and 15.
E. J. Routh, "Elementary Rigid Dynamics" (Macmillan), Chaps. 9 and 12.
———, "Dynamics of a Particle" (Stechert), Chaps. 6, 7, 8, 10, 11.
———, "A Treatise on Elementary Statics" (Cambridge), Chaps. 2, 3, 4, 5.
J. G. Slater and N. H. Frank, "Mechanics" (McGraw-Hill), Chaps. 7, 10, 12, 13, 15.
I. S. Sokolnikoff and E. S. Sokolnikoff, "Higher Mathematics for Engineers and Physicists" (McGraw-Hill), Appendix 5.
R. J. Stephenson, "Mechanics and Properties of Matter" (Wiley), Chap. 7.
J. L. Synge and B. A. Griffith, "Principles of Mechanics" (McGraw-Hill), Chaps. 1, 2, 3, 4, 5, 6, 7, 9, 10.
E. T. Whittaker, "Analytical Dynamics" (Dover), Chaps. 13 and 14.

INDEX

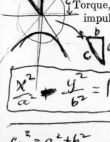

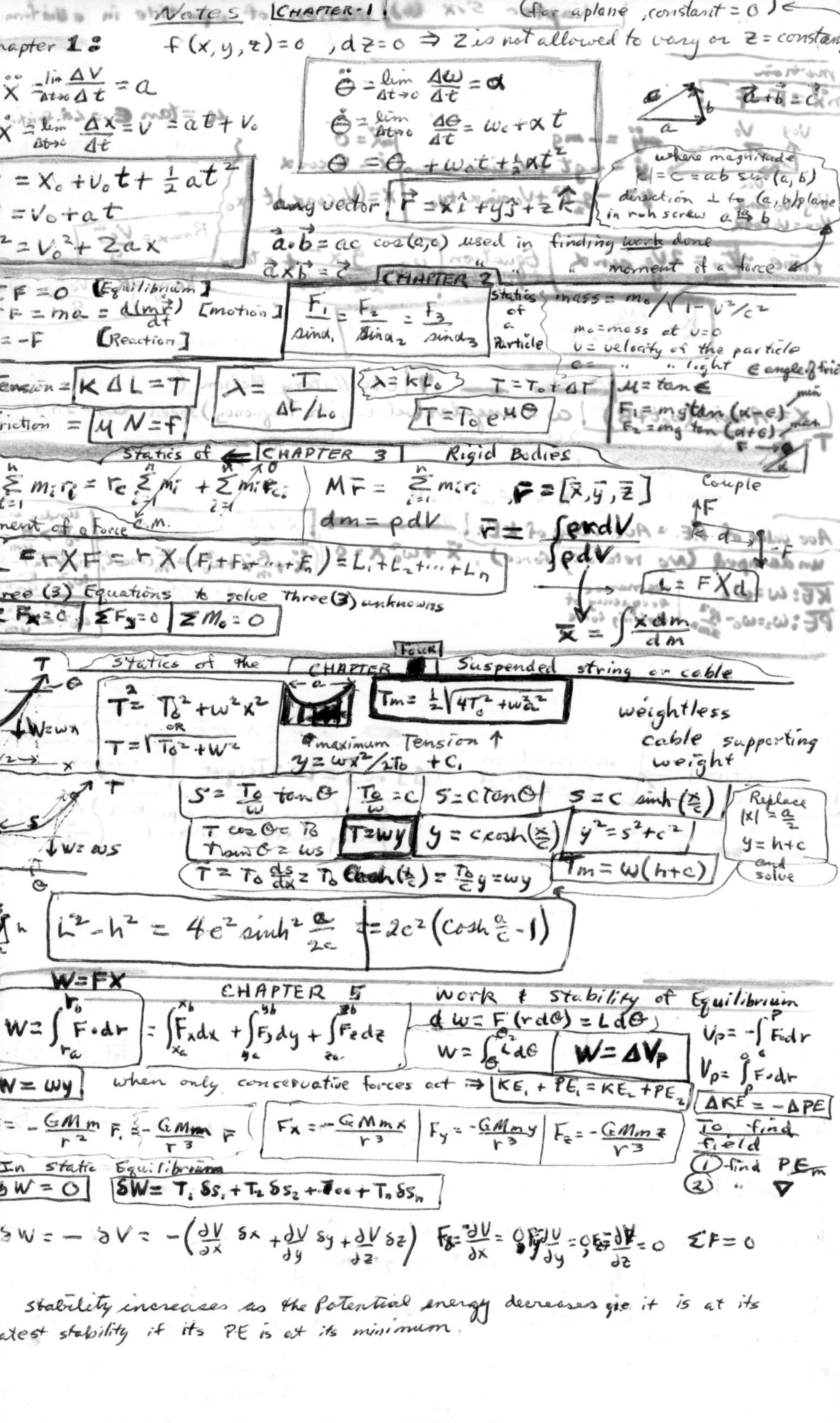

Notes CHAPTER-1 (for a plane, constant = 0)

Chapter 1: $f(x,y,z)=0$, $dz=0 \Rightarrow z$ is not allowed to vary or $z=$ constant

$\ddot{x} = \lim_{\Delta t \to 0} \frac{\Delta v}{\Delta t} = a$

$\dot{x} = \lim_{\Delta t \to 0} \frac{\Delta x}{\Delta t} = v = at + v_0$

$x = x_0 + v_0 t + \frac{1}{2} a t^2$

$v = v_0 + at$

$v^2 = v_0^2 + 2ax$

$\ddot{\theta} = \lim_{\Delta t \to 0} \frac{\Delta \omega}{\Delta t} = \alpha$

$\dot{\theta} = \lim_{\Delta t \to 0} \frac{\Delta \theta}{\Delta t} = \omega_0 + \alpha t$

$\theta = \theta_0 + \omega_0 t + \frac{1}{2}\alpha t^2$

$\vec{a} + \vec{b} = \vec{c}$

any vector $\vec{r} = x\hat{i} + y\hat{j} + z\hat{k}$

where magnitude $|c| = c = ab \sin(a,b)$, direction $\perp$ to (a,b) plane in rh screw a to b

$\vec{a} \cdot \vec{b} = ac \cos(a,c)$ used in finding work done

$\vec{a} \times \vec{b} = \vec{c}$ " " " moment of a force

CHAPTER 2 — Statics of a Particle

$\Sigma F = 0$ [Equilibrium]

$\Sigma F = ma = \frac{d(m\vec{v})}{dt}$ [motion]

$F = -F$ [Reaction]

$\frac{F_1}{\sin \alpha_1} = \frac{F_2}{\sin \alpha_2} = \frac{F_3}{\sin \alpha_3}$

mass $= m_0 / \sqrt{1 - v^2/c^2}$

m_0 = mass at $v=0$

v = velocity of the particle

c = " " light

Tension $= K \Delta L = T$

Friction $= \mu N = f$

$\lambda = \frac{T}{\Delta L / L_0}$

$\lambda = k L_0$

$T = T_0 + \Delta T$

$T = T_0 e^{\mu \theta}$

ϵ angle of friction

$\mu = \tan \epsilon$

$F_1 = mg \tan(\alpha - \epsilon)$ min

$F_2 = mg \tan(\alpha + \epsilon)$ max

Statics of CHAPTER 3 Rigid Bodies

$\sum_{i=1}^{n} m_i r_i = r_c \sum_{i=1}^{n} m_i + \sum_{i=1}^{n} m_i r_{ci}$ (C.M.)

$M\bar{r} = \sum_{i=1}^{n} m_i r_i$, $\bar{r} = [\bar{x}, \bar{y}, \bar{z}]$

$dm = \rho dV$

$\bar{r} = \frac{\int \rho r dV}{\int \rho dV}$

$\bar{x} = \frac{\int x dm}{\int dm}$

Moment of a Force

$L = r \times F = r \times (F_1 + F_2 + \cdots + F_n) = L_1 + L_2 + \cdots + L_n$

Couple: $L = F \times d$

Three (3) Equations to solve three (3) unknowns

$\Sigma F_x = 0$ | $\Sigma F_y = 0$ | $\Sigma M_0 = 0$

Statics of the CHAPTER FOUR Suspended string or cable

$T^2 = T_0^2 + w^2 x^2$ OR $T = \sqrt{T_0^2 + W^2}$

$T_m = \frac{1}{2}\sqrt{4T_0^2 + w^2 a^2}$ maximum Tension

$y = wx^2/2T_0 + C_1$

weightless cable supporting weight

$s = \frac{T_0}{w} \tan \theta$ | $\frac{T_0}{w} = c$ | $s = c \tan \theta$ | $s = c \sinh(\frac{x}{c})$

$T \cos \theta = T_0$, $T \sin \theta = ws$ | $T = wy$ | $y = c \cosh(\frac{x}{c})$ | $y^2 = s^2 + c^2$

$T = T_0 \frac{ds}{dx} = T_0 \cosh(\frac{x}{c}) = \frac{T_0}{c} y = wy$

$T_m = w(h + c)$

Replace $|x| = \frac{a}{2}$, $y = h + c$ and solve

$L^2 - h^2 = 4c^2 \sinh^2 \frac{a}{2c}$ | $= 2c^2(\cosh \frac{a}{c} - 1)$

CHAPTER 5 Work & Stability of Equilibrium

$W = FX$

$W = \int_{r_a}^{r_b} F \cdot dr = \int_{x_a}^{x_b} F_x dx + \int_{y_a}^{y_b} F_y dy + \int_{z_a}^{z_b} F_z dz$

$dW = F(r d\theta) = L d\theta$

$W = \int_{\theta_1}^{\theta_2} L d\theta$ | $W = \Delta V_p$

$V_p = -\int F \cdot dr$, $V_p = \int_p^0 F \cdot dr$

$W = wy$ when only conservative forces act $\Rightarrow KE_1 + PE_1 = KE_2 + PE_2$

$\Delta KE = -\Delta PE$

$F = -\frac{GMm}{r^2}$, $F = -\frac{GMm}{r^3} r$

$F_x = -\frac{GMmx}{r^3}$ | $F_y = -\frac{GMmy}{r^3}$ | $F_z = -\frac{GMmz}{r^3}$

To find field: (1) find PE min, (2) " ∇

In static Equilibrium

$\delta W = 0$ | $\delta W = T_1 \delta s_1 + T_2 \delta s_2 + \cdots + T_n \delta s_n$

$\delta W = -\delta V = -(\frac{\partial V}{\partial x}\delta x + \frac{\partial V}{\partial y}\delta y + \frac{\partial V}{\partial z}\delta z)$ $F_x = \frac{\partial V}{\partial x} = 0$, $\frac{\partial V}{\partial y} = 0$, $\frac{\partial V}{\partial z} = 0$ $\Sigma F = 0$

stability increases as the potential energy decreases, i.e. it is at its greatest stability if its PE is at its minimum.

Chapter Six (6) Motion of a particle in a Uniform

Falling body motion $m\ddot{x} = F - f$

$\ddot{y} = -g$
$\dot{y} = V_0 - gt$
$y = y_0 + V_0 t - \frac{g}{2} t^2$

inclined smooth plane $m\ddot{x} = mg \sin\alpha$

$F = ma$

Atwood Machine $\ddot{x}_1 = \frac{(m_1 - m_2) g}{m_1 + m_2}$

Kinetic Friction $= f = \mu N$

$\mu = \tan\theta = \angle$ of friction

$V_{oy} = V_0 \sin\alpha$
$V_{ox} = V_0 \cos\alpha$

$m\ddot{y} = -mg$
$\dot{y} = -gt + V_0 \sin\alpha$
$y = -\frac{g}{2} t^2 + V_0 \sin\alpha\, t + y_0$

$m\ddot{x} = 0$
$\dot{x} = V_0 \cos\alpha$
$x = (V_0 \cos\alpha) t + x_0$

(1) set $y = 0$, to find t
(2) eliminate t to find equation of path
(3) set $y = 0$ again to find R

$R_{max} = \frac{V_0^2}{g}$

time $= t_0 = \frac{2V_0}{g} \sin\alpha$

Equation of projectile $y = -\frac{g x^2}{2V_0^2 \cos^2\alpha} + x \tan\alpha$

Range $= R = \frac{V_0^2}{g} \sin 2\alpha$

Range in any direction $R = \frac{2V_0^2}{g \cos^2\theta} (\cos\alpha \sin[\alpha - \theta])$

for Air Restant factor see page $\Sigma F = mg - mk\dot{y} = m\ddot{y}$

Chapter Seven (7) Oscillatory Motion (of a partile in one d

$x = A \sin(\omega_0 t + B)$ — ω = angular (velocity or frequency) speed $\omega = 2\pi f$

→ general solution of Undamped Harmonic Oscillator

$T = \frac{1}{f}$

$A = X_{max}$ = Amplitude
B = initial phase
$\omega_0 t$ = phase angle

$x = A \sin B$ = initial value of displacement
$\dot{x} = V_0 = A\omega_0 \cos B$ = " " " velocity $\dot{x}$

Ave value of P.E. = Ave. value of K.E.

undamped (No retarding force) $\ddot{x} + \omega_0^2 x = 0$

Damped → $\ddot{x} + \frac{R}{m}\dot{x} + \omega_0^2 x = 0$

$\ddot{x} + \frac{R}{m}\dot{x} + \omega_0^2 x = \frac{F}{m} \cos\omega t$ — Forced Oscillation

under $= \omega_0^2 > \frac{R^2}{4m^2}$
critical $- \omega_0^2 = \frac{R^2}{4m^2}$
over $- \omega_0^2 < \frac{R^2}{4m^2}$

KE: $\omega = \omega_0$ } resonance frequency of driving force
PE: $\omega_2 = \omega_0 - \frac{R^2}{2m^2}$

$\overline{\dot{W}} = R\overline{(\dot{x})^2}$

Chapter 8 Motion of a System of Particles

$\frac{d(m\dot{r})}{dt} = 0$ → no motion, $F = ma = 0$

⇒ $m\dot{r} = \bar{p}$ = constant, momentum is constant, $\Delta m\dot{r} = 0$

→ $J = \bar{r} \times m\dot{r}$ — moment of momentum = angular momentum

ΔJ or $\dot{J} = L$ = Torque

usually the internal forces are allways central forces

if {no forces act on a particle / acting forces are central} then $\dot{J} = 0$ and J is const

motion of particles can be expressed relative to others or the center of ma

$\bar{F}_2 = m\ddot{r}_2 \rightarrow F_2 = \mu \ddot{r}$ where $\mu = \frac{m_1 m_2}{m_1 + m_2}$

impulse of Compression $= V_1 - V_2 = P\left(\frac{m_1 + m_2}{m_1 m_2}\right)$
" " restitution $= V_1' - V_2' = -P\left(\frac{m_1 + m_2}{m_1 m_2}\right)$

Impulse $= \Delta mV$, the amount imparted to a system

instantaneous ⇒ $P' = eP$

$m(\dot{x}_2 - \dot{x}_1) = P_x = \int_0^\tau F_x dt$
$m(\dot{y}_2 - \dot{y}_1) = P_y = \int_0^\tau F_y dt$
$m(\dot{z}_2 - \dot{z}_1) = P_z = \int_0^\tau F_z dt$

where e = coefficient of restitution

or $V_1' - V_2' = -e(V_1 - V_2)$ | $m_1 V_1' + m_2 V_2' = m_1 V_1 + m_2 V_2$ | $\Delta KE = -KE_0(1 - e^2)$

→ e = coefficient of elasticity — $e = 0$ Perfect in elasticity, $e = 1$ " elasticity

Motion of a Rigid Body in a Plane — Chapter 9

$p(x, y)$

x' and y' are constant
x and y vary with time

$x = x_0 + x' \cos\theta - y' \sin\theta$
$y = y_0 + x' \sin\theta + y' \cos\theta$

$x' = (x - x_0)\cos\theta + (y - y_0)\sin\theta$
$y' = -(x - x_0)\sin\theta + (y - y_0)\cos\theta$

BODY CENTRODE

$\dot{x} = \dot{x}_0 - \dot{\theta}(y - y_0)$
$\dot{y} = \dot{y}_0 + \dot{\theta}(x - x_0)$

SPACE CENTRODE → at $\dot{x} = \dot{y} = 0$: $x = x_0 - \frac{\dot{y}_0}{\dot{\theta}}$, $y = y_0 + \frac{\dot{x}_0}{\dot{\theta}}$

$\dot{r}_P = \dot{s}_Q + \omega \times r_{PQ}$

$\dot{x}_0$ = velocity of moving origin
$\dot{\theta}$ = angular velocity of body rot.

Rotational Kinetic Energy = $T = \frac{I\omega^2}{2}$ $J_z = I\omega$

$L = I\dot{\omega} = I\alpha$

$I = I_{cm} + Mb^2$ cm b

Chapter 10 Motion of a Particle under the action of a Central Force

$T = \frac{1}{2}mv^2$; $v^2 = \dot{r}^2 + r^2\dot{\theta}^2$ $\ddot{\mathbf{r}} = (\ddot{r} - r_1\dot{\theta}^2)\mathbf{r}_1 + (r\ddot{\theta} + 2\dot{r}\dot{\theta})\mathbf{\theta}_1$

$J_z = 2m\dot{A} = mr^2\dot{\theta}$

directrix r θ Polar axis p

$$r = \frac{1}{\left(\frac{km}{J^2}\right) - A\cos\theta}$$

$\ddot{u} + u = \frac{km}{J^2}$ where $k = -r^2 f(r)$

Chapter 11 Accelerated Reference Systems

r R O $[\dot{r}]_0 = \dot{r} + \omega \times r$

$a_0 = [\ddot{R}]_0 + a + \dot{\omega} \times r + 2\omega \times v + \omega \times (\omega \times r)$